Algebraic Surfaces in Positive Characteristics

Purely Inseparable Phenomena
in Curves and Surfaces

Algebraic Surfaces in Positive Characteristics

Purely Inseparable Phenomena
in Curves and Surfaces

Masayoshi Miyanishi
Osaka University, Japan

Hiroyuki Ito
Tokyo University of Science, Japan

World Scientific

NEW JERSEY · LONDON · SINGAPORE · BEIJING · SHANGHAI · HONG KONG · TAIPEI · CHENNAI · TOKYO

Published by

World Scientific Publishing Co. Pte. Ltd.

5 Toh Tuck Link, Singapore 596224

USA office: 27 Warren Street, Suite 401-402, Hackensack, NJ 07601

UK office: 57 Shelton Street, Covent Garden, London WC2H 9HE

Library of Congress Cataloging-in-Publication Data
Names: Miyanishi, Masayoshi, 1940–   author. | Ito, Hiroyuki, 1966–   author.
Title: Algebraic surfaces in positive characteristics : purely inseparable
    phenomena in curves and surfaces / Masayoshi Miyanishi, Hiroyuki Ito.
Description: New Jersey : World Scientific, [2021] | Includes bibliographical references and index.
Identifiers: LCCN 2020020098 | ISBN 9789811215209 (hardcover) | ISBN 9789811215216 (ebook) |
    ISBN 9789811215223 (ebook other)
Subjects: LCSH: Surfaces, Algebraic. | Curves, Algebraic. | Characteristic classes.
Classification: LCC QA571 .M573 2021 | DDC 516.3/52--dc23
LC record available at https://lccn.loc.gov/2020020098

British Library Cataloguing-in-Publication Data
A catalogue record for this book is available from the British Library.

For any available supplementary material, please visit
https://www.worldscientific.com/worldscibooks/10.1142/11690#t=suppl

Printed in Singapore

Preface

Recent progress in algebraic geometry is very fast if it is confined to the case of characteristic zero. Knowledge on curves and surfaces is one of the basic grounds, and the theory of higher-dimensional algebraic varieties is built on these grounds.

Meanwhile, if we consider algebraic varieties defined over a field of positive characteristic, we have a good amount of knowledge on the theory of algebraic curves due to E. Artin and C. Chevalley, Castelnuovo-Enriques-Kodaira classification of algebraic surfaces due to E. Bombieri and D. Mumford and the theory of rational double singularities due to M. Artin and J. Lipman. Notwithstanding all these contributions, one often and easily encounters or steps into an unexplored domain which causes phenomena not parallel to the case of characteristic zero.

Paradoxically, one can say that these theories have been built to check how far the theories established in the case of characteristic zero hold in the case of positive characteristic. Phenomena particular to the case of positive characteristic can occur.

To give a few examples, a fibration on a smooth projective surface can have moving singularities on general fibers like a surface with a quasi-elliptic fibration, and a group scheme is not necessarily reduced like infinitesimal finite group schemes α_p and μ_p. Even a discrete finite group $\mathbb{Z}/p\mathbb{Z}$ behaves like a unipotent group. Hence there appear algebraic surfaces obtained as the quotients of infinitesimal group scheme actions or wild actions of finite p-groups.

In the case of a surface with moving singularities along fibers, the generic fiber of the fibration is a normal algebraic curve defined over the function field K of the base curve and has hidden singular points. Namely, if one changes the base field K to its algebraic closure $\overline{K}$, then the singular points

v

are visualized and move along the fibration. In the case of a quasi-elliptic fibration, the generic fiber with the hidden singular point is a form of the affine line defined over K. If the base curve is a rational curve, the surface with a quasi-elliptic fibration is a unirational surface. One of the most notable facts is that there are plenty of unirational surfaces of various (classification) invariants.

The present book explains the details of these situations by taking, as examples, purely inseparable k-forms of the affine line, unirational quasi-elliptic fibrations or quasi-hyperelliptic fibrations, Zariski surfaces, Artin-Schreier coverings and rational double points.

A plane sandwich is an algebraic surface V which decomposes the Frobenius endomorphism F of the projective plane $\mathbb{P}^2$ by rational mappings f and g:

$$F : \; \mathbb{P}^2 \xrightarrow{\; f \;} V \xrightarrow{\; g \;} \mathbb{P}^2 \;,$$

where $\deg f = \deg g = p$. Replacing $\mathbb{P}^2$ by other surfaces like abelian surfaces, etc., we can consider other kinds of sandwiches. In the case of $\mathbb{P}^2$, since the function field $k(V)$ of V is contained in a purely transcendental field extension $f^* : k(V) \hookrightarrow k(x, y)$, V is a unirational surface, hence $q := h^1(V, \mathcal{O}_V) = 0$ in particular. However, there is a vast world of unirational surfaces which is mostly unexplored. It contains many interesting surfaces of general type.

Extending unirational surfaces, we explain Zariski surfaces V defined by $z^p = f(x, y)$, which are purely inseparable coverings of degree p of $\mathbb{P}^2$. The curve $f(x, y) = 0$ behaves like the branch locus in the case of characteristic zero. If $f(x, y) = 0$ has a singular point, the point of V lying over it has a surface rational singularity as one of hypersurface singularities whose analysis provides a peculiar feature in the case of positive characteristic. We also consider an Artin-Schreier covering of an algebraic surface W which is defined in terms of function fields by $k(V) = k(W)(\xi)$, where $\xi^p - \xi = f \in k(W)$. Even if W is smooth and V is normal, there appear singular points in V, including rational double points, which are mostly hard to analyze. To be short, if one takes a purely inseparable covering or an Artin-Schreier covering of a smooth surface, analysis of rational double singular points will be crucial in understanding the structure of such covering surfaces. Study of these surfaces is not complete, and there are many problems left unsolved.

We often consider actions on a smooth surface V of infinitesimal group schemes α_p, μ_p or a finite simple group $\mathbb{Z}/p\mathbb{Z}$ and consider the quotient surfaces, which will have again singular points to be analyzed. If one looks the given surface V from the quotient surface W as a purely inseparable covering or an Artin-Schreier covering, analysis of singular points on W is different in a subtle manner.

We are motivated by these problems in writing the present book. We intend it to be a comprehensive book but our emphasis tends naturally to be put on an analysis of singularity. We try also to be general in theory but has to be concentrated in specific computation that leads to clarification of singularity. We hope that these concrete computations will help the readers understand the problems we deal with. The book is written not only for researchers with backgrounds in theories of algebraic curves and surfaces but also graduate students who are interested in phenomena in positive characteristic case. For the latter readers, we believe that the book will become a source of many unsolved problems.

In reading chapters in Parts I and II the readers will encounter various surface singularities which are mostly rational double points. Singularity theory in positive characteristic is much more fascinating than in the case of characteristic zero, in view of their complexity and richness.

Part III is devoted to the theory of rational double points. It begins with Artin's basic theory on rational singularities and Lipman's local study of these singularities. We then concentrate ourselves to a detailed study of rational double points, their determinacy and algorithm to determine the types of rational double points.

In Part III, chapter 2, we consider versal deformations and equisingular loci. The results are worked out by Hiroyuki Ito and Natsuo Saito. The authors are very grateful to N. Saito for approving with the inclusion of these results into the present book.

When we refer to a result stated in the same part, we indicate it without mentioning the part number like Lemma 1.2.3. But a result in a different part will be indicated with the part number like Lemma I-1.2.3 or Lemma I.1.2.3.

Our sincere thanks go to Dr. Yuya Matsumoto of Tokyo University of Science and Dr. Natsuo Saito of Hiroshima City University who read the manuscript with great care and found many errors of text as well as typos. Last but not the least, we would like to express our indebtedness to the editor Ms. Kwong Lai Fun of World Scientific Publ. Co. for giving us the chance of writing a book on algebraic surfaces in positive characteristics and constant encouragement during the writing of this book.

M. Miyanishi and H. Ito

Contents

PART I
Forms of the affine line

Chapter 1

Picard scheme and Jacobian variety

1.1 Introduction

Let k be an arbitrary field and let X be a proper k-scheme. Associated with X, there exists a group functor $\mathcal{P}ic_{X/k}$ defined over the category (Sch/k) of all k-schemes, which is called the *Picard functor* of X. We recall the definition after Grothendieck [FGA]. If X is a proper scheme over k, the Picard functor is represented by a k-group scheme $\mathrm{Pic}_{X/k}$ with additional assumption. The connected component $\mathrm{Pic}^0_{X/k}$ of $\mathrm{Pic}_{X/k}$ containing the neutral point e plays an important role in the subsequent theory of forms of the affine line.

Let C be a complete normal curve over k. Then $\mathrm{Pic}^0_{C/k}$ is a smooth k-group scheme. If C is smooth over k, then $\mathrm{Pic}^0_{C/k}$ is an abelian variety of dimension equal to $\dim H^1(C, \mathcal{O}_C)$. It is called the *Jacobian variety* of C. If C is not smooth over k, the curve $\overline{V} := C \otimes_k \overline{k}$ extended to an algebraic closure $\overline{k}$ of k, has finitely many singular points, and $\mathrm{Pic}^0_{\overline{C}/\overline{k}}$ is no longer an abelian variety. It becomes an extension of the abelian variety $\mathrm{Pic}^0_{\widetilde{C}/\overline{k}}$, which is the Jacobian variety of the normalization $\widetilde{C}$ of $\overline{C}$, by an affine algebraic group which reflects the singularities of $\overline{C}$. In this sense, $\mathrm{Pic}^0_{C/k}$ is called the *generalized Jacobian variety*. From the second chapter in this Part, we consider a form of the affine line X, whose normal completion C has a unique one-place singularity at $C \setminus X$. Since $\widetilde{C}$ is isomorphic to the projective line $\mathbb{P}^1_{\overline{k}}$, $\mathrm{Pic}^0_{C/k}$ is an affine algebraic group called a *unipotent group*. It would be an interesting problem to analyze unipotent groups from the viewpoint of the Picard groups of singular points.

1.2   Picard group and Picard group scheme

Let S be a k-scheme. Let $\mathcal{O}_S^*$ be the sheaf of all invertible elements of the structure sheaf $\mathcal{O}_S$. Then the group of all isomorphism classes of invertible sheaves on S is denoted by $\mathrm{Pic}(S)$, which is isomorphic to $H^1(S, \mathcal{O}_S^*)$;

$$\mathrm{Pic}(S) \cong H^1(S, \mathcal{O}_S^*) \, .$$

Let X be a proper scheme over k and let $f : X \to \mathrm{Spec}\,(k)$ be the structure morphism. For the sake of simplicity, we assume that

(i) $H^0(X, \mathcal{O}_X) = k$,
(ii) X has a k-rational point P. Namely, there exists a section $\sigma :$ $\mathrm{Spec}\,(k) \to X$ such that $f \cdot \sigma = 1_k$ and the image of the unique point of $\mathrm{Spec}\,(k)$ by σ is P.

Let S be any k-scheme. Consider the base change

$$f_S : X \times_k S \to S \, ,$$

which has the section $\sigma_S = \sigma \times_k S : S \to X \times_k S$. The S-scheme $X \times_k S$ is denoted by X_S. Then the morphisms f_S and σ_S induce the homomorphisms

$$\mathrm{Pic}(S) \quad \underset{\sigma_S^*}{\overset{f_S^*}{\rightleftarrows}} \quad \mathrm{Pic}(X_S)$$

such that $\sigma_S^* \cdot f_S^* = \mathrm{id}_{\mathrm{Pic}(S)}$. The cokernel of f_S^*, i.e., the kernel of σ_S^*, is denoted by $\mathrm{Pic}(X_S/S)$ and called the *relative Picard group*. By construction, it is clear that every element of $\mathrm{Pic}(X_S/S)$ is represented by an invertible sheaf $\mathcal{L}$ on X_S such that the restriction of $\mathcal{L}$ on $\{P\} \times S$ is trivial. Moreover, we have

$$\mathrm{Pic}(X_S) \cong \mathrm{Pic}(X_S/S) \times \mathrm{Pic}(S).$$

In particular, if k' is a field extension of k, we have $\mathrm{Pic}(X_{k'}) \cong \mathrm{Pic}(X_{k'}/k')$. Furthermore, by assigning to every k-scheme S the group $\mathrm{Pic}(X_S/S)$, we have a contravariant functor on $(\mathcal{S}ch\,/k)$:

$$\mathcal{P}ic_{X/k} : S \in (\mathcal{S}ch\,/k) \longmapsto \mathrm{Pic}(X_S/S).$$

Here we briefly explain the terminology to be used in the subsequent argument. Given a scheme T, a *Grothendieck topology* on T is defined by a set of open coverings $\{U_i \xrightarrow{f_i} T\}_{i \in I}$ of T, where U_i is not necessarily an open set of T in the sense of the Zariski topology. In a broader sense of topology, we consider a set of étale (or flat) morphisms $f_i : U_i \to T$ for $i \in I$ such that the union of the images of f_i covers T. We need some conditions on the set of morphisms f_i so that the axioms of topology are to be satisfied. The category (Sch / k) of schemes over k together with a set of coverings for each k-scheme T of (Sch / k) is called the étale (or flat) *site*. An abelian sheaf on the site is a contravariant functor $\mathcal{F}$ from (Sch / k) to the category of abelian groups such that, for each covering $\{U_i \xrightarrow{f_i} T\}_{i \in I}$, the following sequence is exact

$$0 \longrightarrow \mathcal{F}(T) \longrightarrow \prod_{i \in I} \mathcal{F}(U_i) \xrightarrow{p_1 - p_2} \prod_{(i,j) \in I \times I} \mathcal{F}(U_i \times_T U_j),$$

where p_1 (resp. p_2) is the map induced by the projection from $U_i \times_T U_j$ to the first (resp. second) factor U_i (resp. U_j). The (f.p.q.c)-sheaf is a sheaf on the (f.p.q.c)-site, where we consider as a covering a finite set $\{U_i \xrightarrow{f_i} T\}_{i \in I}$ with flat morphisms f_i. For a concise introduction to the Grothendieck topology, see [51]. Given a contravariant functor $\mathcal{F}$ on the (f.p.q.c)-site, $\mathcal{F}$ being a sheaf is equivalent to the (f.p.q.c)-descent being effective for $\mathcal{F}$. If $\{U_i \xrightarrow{f_i} T\}_{i \in I}$ is a (f.p.q.c)-covering of T, the (f.p.q.c)-descent means the following two conditions.

(S i) Let $\xi_1, \xi_2 \in \mathcal{F}(T)$. Write $(\xi_i)|_{U_i} = \mathcal{F}(f_i)(\xi_i)$ for $i = 1, 2$. If $(\xi_1)|_{U_i} = (\xi_2)|_{U_i}$ for every $i \in I$ then $\xi_1 = \xi_2$.

(S ii) Let $\{\xi_i\}_{i \in I}$ be an element of $\prod_{i \in I} \mathcal{F}(U_i)$. Suppose that $(\xi_i)|_{U_{ij}} = (\xi_j)|_{U_{ij}}$ for every pair (i, j) of $I \times I$, where $U_{ij} = U_i \times_T U_j$. Then there exists an element $\xi \in \mathcal{F}(T)$ such that $\xi_i = \xi|_{U_i}$ for every $i \in I$.

Given a scheme T with a Grothendieck topology and an abelian sheaf $\mathcal{F}$ with respect to the topology, then we can define, as in the usual manner with respect to open coverings, the derived higher cohomology groups $H^i(T, \mathcal{F})$. For a morphism $f : S \to T$ of schemes, which is continuous with respect to the Grothendieck topology on S and T, we can define the relative cohomology groups $R^i f_* \mathcal{G}$ for an abelian sheaf $\mathcal{G}$ on S as well as the spectral sequence $E_2^{p,q} = H^p(T, R^q f_* \mathcal{G}) \Rightarrow H^{p+q}(S, \mathcal{G})$.

Let $G_{m,T}$ be the (f.p.q.c)-sheaf on a scheme T defined by

$$G_{m,T}(T') = \Gamma(T', \mathcal{O}_{T'}^*)$$

for a T-scheme T'. Let $R^q f_{S*}(G_{m,X_S})$ be the q-th derived cohomology considered in the sense of (f.p.q.c)-cohomology. Then we have the spectral sequence

$$E_2^{p,q} = H^p(S, R^q f_{S*}(G_{m,X_S})) \Longrightarrow H^{p+q}(X_S, G_{m,X_S}).$$

In particular, we obtain an exact sequence of lower degree terms

$$0 \to H^1(S, f_{S*}(G_m, X_S)) \xrightarrow{f_S^*} H^1(X_S, G_{m,X_S}) \to H^0(S, R^1 f_{S*}(G_{m,X_S}))$$
$$\to H^2(S, f_{S*}(G_{m,X_S})) \xrightarrow{f_S^*} H^2(X_S, G_{m,X_S}).$$

Since f_S has the section σ_S, we have the following identification:

$$f_{S*}(G_{m,X_S}) = G_{m,S}.$$

In fact, for an S-scheme T, we have $f_{S*}(G_{m,X_S})(T) = \Gamma(X_T, \mathcal{O}^*_{X_T}) = \Gamma(T, \mathcal{O}^*_T) = G_{m,S}(T)$, where the second equality holds by the conditions (i) and (ii). Hence we have the identifications

$$H^1(S, f_{S*}(G_{m,X_S})) = H^1(S, G_{m,S}) = H^1(S, \mathcal{O}^*_S)$$
$$H^1(X_S, G_{m,X_S}) = H^1(X_S, \mathcal{O}^*_{X_S}).$$

Since the existence of σ_S implies that

$$f_S^* : H^2(S, f_{S*}(G_{m,X_S})) \longrightarrow H^2(X_S, G_{m,X_S})$$

is injective, the above sequence gives rise to an exact sequence

$$0 \to H^1(S, \mathcal{O}^*_S) \xrightarrow{f_S^*} H^1(X_s, \mathcal{O}^*_{X_S}) \to H^0(S, R^1 f_{S*}(G_{m,X_S})) \to 0.$$

Therefore we obtain, by the definition of $\mathrm{Pic}(X_S/S)$, the isomorphism

$$\mathrm{Pic}(X_S/S) \cong H^0(S, R^1 f_{S*}(G_{m,X_S})).$$

Namely, under the assumptions (i) and (ii), the functor $\mathcal{P}ic_{X/k}$ is a sheaf $R^1 f_{S*}(G_{m,X_S})$ in the (f.p.q.c)-topology on $(\mathcal{S}ch\,/k)$.

Definition. If the functor $\mathcal{P}ic_{X/k}$ is representable, the representant, i.e., a k-scheme which represents the functor $\mathcal{P}ic_{X/S}$, is called the *Picard scheme* of X and denoted by $\mathrm{Pic}_{X/k}$.

Here $\mathrm{Pic}(X)$ is the Picard group which is the group of isomorphism classes of invertible sheaves on X, while $\mathrm{Pic}_{X/k}$ is a scheme and $\mathrm{Pic}(X) = \mathrm{Pic}_{X/k}(k)$ the set of k-rational points. So, the readers are required to be attentive to the difference.

Since $\mathcal{P}ic_{X/k}$ is identified with the (f.p.q.c)-sheaf $R^1 f_{S*}(G_{m,X_S})$ we have in particular

$$\mathrm{Pic}_{X/k}(k') \cong \mathrm{Pic}(X_{k'}/k') \cong \mathrm{Pic}(X_{k'})$$

for any field extension k' of k. In fact, we have the following result due to Murre [65].

Theorem. *$\mathcal{P}ic_{X/k}$ is representable under the assumptions* (i) *and* (ii). *Moreover, $\mathrm{Pic}_{X/k}$ is a group scheme which is locally of finite type over k.*

1.3 Existence of $\mathrm{Pic}^0_{X/k}$

Now, apply to $\mathrm{Pic}_{X/k}$ a well-known result on a group scheme which is locally of finite type over k, we obtain the following:

Theorem (cf. [12, Chap. II. §5, 1.1]). *Let X be a proper scheme over k satisfying the conditions (i), (ii) of 1.2. Then we have:*

(1) *Let $\mathrm{Pic}^0_{X/k}$ be the open subscheme of $\mathrm{Pic}_{X/k}$ supported by the connected component containing the neutral element e. Then $\mathrm{Pic}^0_{X/k}$ is a group scheme of finite type over k.*

(2) *For any field extension k' of k, $\mathrm{Pic}^0_{X/k} \otimes_k k' \cong \mathrm{Pic}^0_{X_{k'}/k'}$.*

(3) *Every connected component of $\mathrm{Pic}_{X/k}$ is irreducible and has the same dimension as $\mathrm{Pic}^0_{X/k}$.*

1.4 Extendability of a rational map to $\mathrm{Pic}_{X/k}$

Let X be a k-scheme and let P be a closed point of X. Then the residue field $\kappa(P)$ (which we also write $k(P)$ whenever it is clear that X is defined over the field k) is an algebraic extension if X is of finite type over k. We say that X is regular at P if the local ring $\mathcal{O}_{X,P}$ is a regular local ring. Even if X is regular at P, $X \otimes_k k'$ may not be regular at a point of $X \otimes_k k'$ lying over P if $\kappa(P)$ is not separable over k. We say that X is *smooth* at P if $X \otimes_k k'$ is regular at every point of $X \otimes_k k'$ lying over P for any algebraic extension k' of k. We shall prove:

Theorem. *Let X be a k-regular proper integral scheme such that X has a k-rational point P. Let V be a smooth k-scheme. Then any k-rational map $g : V \to \mathrm{Pic}_{X/k}$ is defined everywhere on V.*

Proof. Let U be a dense open set of V on which g is defined, and let $h = g|_U : U \to \mathrm{Pic}_{X/k}$. Then $h \in \mathrm{Pic}_{X/k}(U) = \mathrm{Pic}(X_U/U)$ is represented by an invertible sheaf $\mathcal{L}$ on $X \times U$ such that $\mathcal{L}$ is trivial on $\{P\} \times U$. Note that $X \times U$ is regular (cf. [EGA, Chap. IV]), hence locally factorial. Therefore there is a Cartier divisor D on $X \times U$ such that $\mathcal{L} = \mathcal{O}(D)$. Let $\overline{D}$ be the closure of D on $X \times V$, which is regular too. Then there is an invertible sheaf $\overline{\mathcal{L}}$ on $X \times V$ such that $\overline{\mathcal{L}} = \mathcal{O}(\overline{D})$ and $\mathcal{L} = \overline{\mathcal{L}}|_{X \times U}$. Then $\overline{\mathcal{L}} \otimes f_V^*((\sigma_{V*}\overline{\mathcal{L}})^{-1})$ defines a k-morphism $\overline{g} : V \to \mathrm{Pic}_{X/k}$ such that $\overline{g}|_U = h$. Hence $\overline{g} = g$. Thus g is defined everywhere on V. $\square$

1.5   Lie algebra of $\mathrm{Pic}_{X/k}$

Let X be a proper k-scheme having a k-rational point P. Let $k[\varepsilon]$ be the k-algebra of dual numbers, i.e., $k[\varepsilon] \cong k[t]/(t^2)$. The Lie algebra of $\mathrm{Pic}_{X/k}$ (or equivalently the one of $\mathrm{Pic}^0_{X/k}$) is defined to be the kernel of the canonical homomorphism

$$\mathrm{Pic}_{X/k}(k[\varepsilon]) \longrightarrow \mathrm{Pic}_{X/k}(k)$$

associated with the homomorphism $k[\varepsilon] \to k, \varepsilon \mapsto 0$. Note that

$$\mathrm{Pic}_{X/k}(k[\varepsilon]) \cong \mathrm{Pic}(X \otimes_k k[\varepsilon]), \quad \mathrm{Pic}_{X/k}(k) \cong \mathrm{Pic}(X).$$

Thus, the Lie algebra $\mathrm{Lie}(\mathrm{Pic}_{X/k})$ consists of all invertible sheaves $\mathcal{L}$ on $X \otimes_k k[\varepsilon]$, up tp isomorphisms, such that $\mathcal{L} \otimes_{k[\varepsilon]} k \cong \mathcal{O}_X$.

Then a simple computation using representations of such $\mathcal{L}$'s by transition functions with respect to open coverings of X show that $\mathrm{Lie}(\mathrm{Pic}_{X/k}) \cong H^1(X, \mathcal{O}_X)$. In fact, for an open covering $\mathcal{U} = \{U_i\}_{i \in I}$ of X, let $\widetilde{\mathcal{U}} = \{\widetilde{U}_i\}_{i \in I}$ be the open covering of $X \otimes_k k[\varepsilon]$, where $\widetilde{U}_i$ has the same underlying space as U_i but with the thickened structure sheaf $\mathcal{O}_{\widetilde{U}_i} = \mathcal{O}_{U_i} \oplus \mathcal{O}_{U_i}\varepsilon$. Suppose that $\mathcal{L}|_{\widetilde{U}_i} = \mathcal{O}_{\widetilde{U}_i} e_i$ for $i \in I$. Then $e_j = \widetilde{f}_{ji} e_i$ on the intersection $\widetilde{U}_i \cap \widetilde{U}_j$. Write $\widetilde{f}_{ji} = f_{ji} + g_{ji}\varepsilon$, where $f_{ji} \in \Gamma(U_i \cap U_j, \mathcal{O}_X^*)$ and $g_{ji} \in \Gamma(U_i \cap U_j, \mathcal{O}_X)$. Since $\mathcal{L} \otimes_{k[\varepsilon]} k \cong \mathcal{O}_X$, we may assume that $f_{ji} = 1$ for every pair (i, j). Then it is easy to see that $\{g_{ji}\}$ defines an element of $Z^1(\mathcal{U}, \mathcal{O}_X)$ and the isomorphism class of $\mathcal{L}$ corresponds to the cohomology class of $\{g_{ji}\}$ in $H^1(X, \mathcal{O}_X)$.

1.6   When is $\mathrm{Pic}_{X/k}$ an abelian variety?

We borrow the following result from Grothendieck [FGA, exposé 236, Th. 2.1 and Prop. 2.10]

Theorem. *Let X be a proper scheme over k having a k-rational point. Then the following assertions hold true:*

(1) *If $H^2(X, \mathcal{O}_X) = (0)$, then $\mathrm{Pic}_{X/k}$ is smooth over k.*
(2) *If X is geometrically normal over k, then $\mathrm{Pic}^0_{X/k}$ is proper over k. Hence, if further $H^2(X, \mathcal{O}_X) = (0)$, then $\mathrm{Pic}^0_{X/k}$ is an abelian variety defined over k.*

1.7 $\mathrm{Pic}^0_{C/k}$ for a complete normal curve C

Let C be a complete normal curve defined over k, i.e., C is a proper, normal, geometrically integral k-scheme. Suppose that C has a k-rational point P. Then $\mathrm{Pic}^0_{C/k}$ is k-smooth by Theorem 1.6 above. Hence

$$\dim \mathrm{Pic}^0_{C/k} = \dim \mathrm{Lie}(\mathrm{Pic}^0_{C/k}) = h^1(C, \mathcal{O}_C).$$

This number is the so-called *arithmetic genus* or *k-genus* of C, and denoted by $p_a(C)$.

In general, C is not necessarily k-smooth. However, if the function field $k(C)$ is a regular extension of k[1], there exists a dense open set U of C which is k-smooth[2]. Note that $P \in U$ since P is a k-rational point by assumption. Let D be the divisor $\Delta - \{P\} \times U$ on $U \times U$, where Δ is the diagonal. Since $U \times U$ is regular, D corresponds to an invertible sheaf $\mathcal{L}'$ on $U \times U$. Since $C \times U$ is regular[3], $\mathcal{L}'$ extends to an invertible sheaf $\mathcal{L}$ on $C \times U$. the sheaf $\mathcal{L}'$ defines a k-morphism

$$\iota' : U \longrightarrow \mathrm{Pic}_{C/k}, \quad Q \mapsto \mathcal{L}|_{C \times \{Q\}} \cong \mathcal{O}_C(Q - P)$$

such that $\iota'(P)$ is the neutral point of $\mathrm{Pic}_{C/k}$. Since U is connected, ι' is factored as follows:

$$\iota' : U \xrightarrow{\iota} \mathrm{Pic}^0_{C/k} \hookrightarrow \mathrm{Pic}_{C/k}.$$

The k-morphism ι enjoys the following property:

For any field extension k' of k and for any k'-rational point Q of $U \otimes k'$, $\iota_{k'}(Q)$ is represented by the invertible sheaf $\mathcal{O}_{C \otimes k'}(Q - P)$, where $\iota_{k'} = \iota \otimes k'$. Indeed, $\iota_{k'}(Q)$ corresponds to a Weil divisor on $C \otimes_k k'$,

$$\{\Delta - \{P\} \times U\} \cap (C \times \{Q\}) = (Q - P) \times \{Q\}.$$

---

[1]The existence of a k-rational point P implies that $k(C)$ is a regular extension of k. Indeed, let $\mathcal{O} = \mathcal{O}_{C,P}$, which is a discrete valuation ring. Let $\overline{k}$ be an algebraic closure of k. Then $\overline{\mathcal{O}} := \mathcal{O} \otimes_k \overline{k}$ is a local ring with maximal ideal $\overline{\mathfrak{m}} = \mathfrak{m} \otimes_k \overline{k}$, where $\mathfrak{m}$ is the maximal ideal of $\mathcal{O}$. Since $\mathfrak{m}$ is principal, so is $\overline{\mathfrak{m}}$. This implies that $\overline{\mathcal{O}}$ is a discrete valuation ring. In particular, $\overline{\mathcal{O}}$ is integral and $Q(\overline{\mathcal{O}}) = k(C) \otimes_k \overline{k}$ is an integral domain, i.e., $k(C)$ is a regular extension of k.

[2]This can be seen as follows. Since $k(C)$ is a regular extension of k, it is a separable extension [68]. Namely, there exists a separating transcendence basis $\{t\}$ such that $k(C)/k(t)$ is a separable algebraic extension. Hence, by Part II, Chapter 1, the module of 1-differential forms $\Omega^1_{k(C)/k}$ of $k(C)$ over k is equal to the module $k(C)dt$. Since $\Omega^1_{C/k} \otimes_{\mathcal{O}_C} k(C) \cong \Omega^1_{k(C)/k}$, it follows that $\Omega^1_{C/k}|_U$ is an invertible sheaf for some non-empty open set U of C. Then C is k-smooth everywhere on U.

[3]The first projection $p_1 : C \times U \to C$ is a smooth morphism as the base change of the structure morphism $U \to \mathrm{Spec}\, k$ which is smooth. Since C is regular, i.e., every local ring of C is a regular local ring, $C \times U$ is regular since the regularity lifts up by a smooth morphism.

The open set U is, in fact, contained in the following open set U_0:

Let $\overline{k}$ be an algebraic closure of k. Then the field extension $\overline{k}$ of k gives rise to a faithfully flat morphism $\pi : C_{\overline{k}} \to C$. Then $\pi(C_{\overline{k}} - \mathrm{Sing}\,(C_{\overline{k}}))$ is an open set of C, which we denote by U_0. Here the set $\mathrm{Sing}\,(C_{\overline{k}})$ is the set of singular points of $C_{\overline{k}}$, called the *singular locus* of $C_{\overline{k}}$. The open set U_0 is called the *smooth locus* of C.

Summarizing the above arguments, we have:

Theorem. *Let C be a complete normal curve defined over k. Assume that C has a k-rational point P. Let U_0 be the smooth locus of C. Then there exists a k-morphism*

$$\iota : U_0 \to \mathrm{Pic}^0_{C/k}$$

such that $\iota(P)$ is the neutral point.

In particular, if C is k-smooth and $p_a(C) \geq 1$, then $\mathrm{Pic}^0_{C/k}$ is an abelian variety of dimension $p_a(C)$, and the theory of the Jacobian variety (see Hartshorne [24, p.323]) tells us that $\iota : C \to \mathrm{Pic}^0_{C/k}$ is a closed immersion. $\mathrm{Pic}^0_{C/k}$ is then the *Jacobian variety* of C.

1.8 Generalized Jacobian variety

We have often to consider the case where C is not necessarily k-smooth. Let $\overline{k}$ be an algebraic closure of k and let $\overline{C} := C \otimes_k \overline{k}$. Let $\widetilde{C}$ be the normalization of $\overline{C}$ and let $\varphi : \widetilde{C} \to \overline{C}$ be the normalization morphism. Let $\overline{\Sigma} := \mathrm{Sing}\,(\overline{C})$ and let $\widetilde{\Sigma} := \varphi^{-1}(\overline{\Sigma})$. Let $\overline{\mathfrak{o}} := \bigcap_{\overline{P} \in \overline{\Sigma}} \mathcal{O}_{\overline{C}, \overline{P}}$ and let $\widetilde{\mathfrak{o}} := \bigcap_{\widetilde{P} \in \widetilde{\Sigma}} \mathcal{O}_{\widetilde{C}, \widetilde{P}}$, where both intersections are taken in the field $\overline{k}(\widetilde{C}) = \overline{k}(\overline{C})$.

We recall the definition of the generalized Jacobian variety of $\widetilde{C}$ with equivalence relation defined by $\overline{\mathfrak{o}}$ (see [75]). A divisor D on $\widetilde{C}$ is said to be independent of the places of $\overline{\mathfrak{o}}$ if $\mathrm{Supp}(D) \cap \widetilde{\Sigma} = \emptyset$. For two such divisors D, D' on $\widetilde{C}$, we say that D' is $\overline{\mathfrak{o}}$-linearly equivalent, $D' \sim_{\overline{\mathfrak{o}}} D$ by notation, if there is an element $f \in \overline{\mathfrak{o}}$ such that $D' = D + (f)$. Then the group of all $\overline{\mathfrak{o}}$-linear equivalence classes of divisors on $\widetilde{C}$ of degree 0 which are independent of the places of $\overline{\mathfrak{o}}$ is in one-to-one correspondence with the set of $\overline{k}$-rational points of a group variety $J_{\overline{\mathfrak{o}}}$. The variety $J_{\overline{\mathfrak{o}}}$ is called the *generalized Jacobian variety* of $\overline{C}$ with equivalence relation defined by $\overline{\mathfrak{o}}$.

Note that if D is a divisor on $\widetilde{C}$ independent of the places of $\overline{\mathfrak{o}}$ then the invertible sheaf $\mathcal{O}_{\widetilde{C}}(D)$ descends down to the invertible sheaf $\mathcal{O}_{\overline{C}}(D)$. For divisors D and D' with $\mathrm{Supp}(D) \cap \widetilde{\Sigma} = \mathrm{Supp}(D') \cap \widetilde{\Sigma} = \emptyset$, suppose that

$\mathcal{O}_{\overline{C}}(D) \cong \mathcal{O}_{\overline{C}}(D')$. Then there exists an open covering $\mathfrak{U} = \{U_i\}_{i \in I}$ of $\overline{C}$ such that $U_i \cap \overline{\Sigma}$ consists of only one point and $(\mathrm{Supp}(D) \cup \mathrm{Supp}(D')) \cap U_i = \emptyset$ whenever $U_i \cap \overline{\Sigma} \neq \emptyset$, and that $\mathcal{O}_{\overline{C}}(D' - D)|_{U_i} = \xi_i \mathcal{O}_{U_i}$ with $\xi_i \in \overline{k}(\overline{C})$. Hence if $U_i \cap \overline{\Sigma} \neq \emptyset$, then $\xi_i \in \mathcal{O}_{U_i}^*$. Since $\mathcal{O}_{\overline{C}}(D' - D) \cong \mathcal{O}_{\overline{C}}$, we have $\xi_i \xi_j^{-1} = a_j a_i^{-1}$ for every pair (i, j), where $a_i \in \Gamma(U_i, \mathcal{O}_{U_i}^*)$. Let $g = \xi_j a_j^{-1}$. Then $g \in \overline{k}(\overline{C})$ and $g \in \mathcal{O}_{\overline{C}, \overline{P}}^*$ for every $\overline{P} \in \overline{\Sigma}$. This observation implies the existence of an injective group homomorphism

$$\theta : J_{\overline{\sigma}} \longrightarrow \mathrm{Pic}^0_{\overline{C}/\overline{k}}.$$

On the other hand, by assigning $\varphi^* \mathcal{L}$ to every invertible sheaf $\mathcal{L}$ on $\overline{C}$, we have a homomorphism of k-group schemes,

$$\alpha : J_{\overline{\sigma}} \longrightarrow \mathrm{Pic}^0_{\widetilde{C}/\overline{k}},$$

which is surjective because every divisor D on $\widetilde{C}$ is linearly equivalent to a divisor D' with $\mathrm{Supp}(D') \cap \widetilde{\Sigma} = \emptyset$.

Let D be a divisor on $\widetilde{C}$ of degree 0 which is independent of the places of $\overline{\sigma}$. Suppose that D represents an element of $\mathrm{Ker}\,\alpha$. Then, with D viewed as a divisor on $\widetilde{C}$, there exists an element $g \in k(\widetilde{C})$ such that $D = (g)$ and g is regular and non-zero at every point of $\widetilde{\Sigma}$. Hence g defines an element of $\bigcap_{\widetilde{P} \in \widetilde{\Sigma}} \mathcal{O}_{\widetilde{C}, \widetilde{P}}^*$. The class of D in $J_{\overline{\sigma}}$ corresponds to the class of g up to multiplication of elements of $\bigcap_{\overline{P} \in \overline{\Sigma}} \mathcal{O}_{\overline{C}, \overline{P}}^*$. Hence we have a homomorphism θ_0 from $\mathrm{Ker}\,\alpha$ to $\prod_{\widetilde{P} \in \widetilde{\Sigma}} \mathcal{O}_{\widetilde{C}, \widetilde{P}}^* / \mathcal{O}_{\overline{C}, \overline{P}}^*$ (see Theorem below). Suppose that the image of the class of D by θ_0 is (1), i.e., $g \in \mathcal{O}_{\overline{C}, \overline{P}}^*$ for every $\widetilde{P} \in \widetilde{\Sigma}$ and $\overline{P} = \varphi(\widetilde{P})$. Then $g \in \overline{\sigma}$. Hence $D \sim_{\overline{\sigma}} 0$. This implies that θ_0 is injective.

We recall the definition of unipotent group scheme over k. An algebraic group scheme G defined over the field k is called a *unipotent* group scheme if it is affine and has a central series

$$(1) = G_0 \lhd G_1 \lhd \cdots \lhd G_{n-1} \lhd G_n = G$$

such that the quotient algebraic group scheme G_i/G_{i-1} is isomorphic to a subgroup scheme of the additive group scheme G_a for $1 \leq i \leq n$. A possible non-trivial subgroup scheme is only G_a itself in characteristic zero, but includes α_{p^n} and $\mathbb{Z}/p^n\mathbb{Z}$ if k has positive characteristic p. In characteristic zero, the underlying scheme of a unipotent group is shown to be isomorphic to the affine space $\mathbb{A}^n$ with $n = \dim G$, hence G is connected. This is not the case in positive characteristic. Indeed, there are non-reduced or non-connected, unipotent group schemes. Notwithstanding, subgroup schemes and quotient group schemes of unipotent group schemes are also unipotent. A reference for a general theory of unipotent group schemes is M. Demazure and P. Gabriel [12].

Now we have:

Theorem. *Assume that every point $\overline{P}$ of $\overline{\Sigma}$ is a one-place point. Then we have:*

(1) *The homomorphism θ is an isomorphism.*
(2) *$(\mathrm{Ker}\,\alpha)(\overline{k})$ is isomorphic to $\prod_{\widetilde{P}\in\widetilde{\Sigma}} \mathcal{O}^*_{\widetilde{C},\widetilde{P}}/\mathcal{O}^*_{\overline{C},\overline{P}}$, where $\overline{P} = \varphi(\widetilde{P})$.*
(3) *$\mathrm{Ker}\,\alpha$ is a unipotent group scheme.*

Proof. We have an exact sequence of abelian sheaves on $\widetilde{C}$:

$$1 \to \varphi^*\mathcal{O}^*_{\overline{C}} \to \mathcal{O}^*_{\widetilde{C}} \to S \to 1,$$

where S is a skyscraper sheaf $\prod_{\widetilde{P}\in\widetilde{\Sigma}} \mathcal{O}^*_{\widetilde{C},\widetilde{P}}/\mathcal{O}^*_{\overline{C},\overline{P}}$ with $\overline{P} := \varphi(\widetilde{P})$. Here, by the assumption that every point of $\overline{\Sigma}$ is a one-place point, there is a single point of $\widetilde{C}$ lying over each point $\overline{P}$ of $\overline{\Sigma}$. This gives rise to an exact sequence,

$$1 \longrightarrow H^0(\widetilde{C}, \varphi^*\mathcal{O}^*_{\overline{C}}) \longrightarrow H^0(\widetilde{C}, \mathcal{O}^*_{\widetilde{C}}) \longrightarrow H^0(\widetilde{C}, S)$$
$$\longrightarrow H^1(\widetilde{C}, \varphi^*\mathcal{O}^*_{\overline{C}}) \xrightarrow{\beta'} H^1(\widetilde{C}, \mathcal{O}^*_{\widetilde{C}}) \longrightarrow 1,$$

which is equivalent to an exact sequence

$$1 \longrightarrow \overline{k}^* \xrightarrow{\sim} \overline{k}^* \longrightarrow S \longrightarrow \mathrm{Pic}_{\overline{C}/\overline{k}} \xrightarrow{\beta'} \mathrm{Pic}_{\widetilde{C}/\overline{k}} \longrightarrow 1,$$

where $\beta'(\mathcal{L}) = \varphi^*\mathcal{L}$ for $\mathcal{L} \in \mathrm{Pic}_{\overline{C}/\overline{k}}$. Therefore, we have a commutative exact diagram

$$
\begin{array}{ccccccccc}
0 & \longrightarrow & \mathrm{Ker}\,\alpha & \longrightarrow & J_{\overline{\sigma}} & \xrightarrow{\alpha} & \mathrm{Pic}^0_{\widetilde{C}/\overline{k}} & \longrightarrow & 0 \\
& & \downarrow{\scriptstyle\theta_0} & & \downarrow{\scriptstyle\theta} & & \| & & \\
0 & \longrightarrow & \prod_{\widetilde{P}\in\widetilde{\Sigma}} \mathcal{O}^*_{\widetilde{C},\widetilde{P}}/\mathcal{O}^*_{\overline{C},\overline{P}} & \longrightarrow & \mathrm{Pic}^0_{\overline{C}/\overline{k}} & \xrightarrow{\beta} & \mathrm{Pic}^0_{\widetilde{C}/\overline{k}} & \longrightarrow & 0
\end{array}
$$

where $\beta = \beta'|_{\mathrm{Pic}^0_{\overline{C}/\overline{k}}}$. Now, fix a point $\widetilde{P} \in \widetilde{\Sigma}$ and denote $\mathcal{O}_{\widetilde{C},\widetilde{P}}$ and $\mathcal{O}_{\overline{C},\overline{P}}$ by $\widetilde{\mathcal{O}}$ and $\overline{\mathcal{O}}$, respectively. Denote by $\widehat{\widetilde{\mathcal{O}}}$ and $\widehat{\overline{\mathcal{O}}}$ the completions of $\widetilde{\mathcal{O}}$ and $\overline{\mathcal{O}}$, respectively. Then we have a canonical homomorphism

$$\widetilde{\mathcal{O}}^*/\overline{\mathcal{O}}^* \xrightarrow{\gamma} \widehat{\widetilde{\mathcal{O}}}^*/\widehat{\overline{\mathcal{O}}}^*.$$

We show that γ is an isomorphism. Since every element A of $\widehat{\widetilde{\mathcal{O}}}^*$ is written in the form $A = aB$ with $a \in \widetilde{\mathcal{O}}^*$ and $B \in \widehat{\overline{\mathcal{O}}}^*$, the homomorphism γ is surjective. In order to show that γ is injective, it suffices to show that $\widetilde{\mathcal{O}}^* \cap \widehat{\overline{\mathcal{O}}}^* \subseteq \overline{\mathcal{O}}^*$. This follows from $\widetilde{\mathcal{O}} \cap \widehat{\overline{\mathcal{O}}} \subseteq \overline{\mathcal{O}}$.

In fact, we can identify $\widehat{\widetilde{\mathcal{O}}}$ with $\overline{k}[[t]]$, where we can take t in the maximal ideal $\widetilde{\mathfrak{m}}$ of $\widetilde{\mathcal{O}}$. Then there exists an integer N such that $\widetilde{\mathfrak{m}}^n \subset \overline{\mathfrak{m}}\widehat{\widetilde{\mathcal{O}}}$ for every $n \geq N$, where $\overline{\mathfrak{m}}$ is the maximal ideal of $\overline{\mathcal{O}}$. Write $A = (c_0 + c_1 t + \cdots + c_{N-1}t^{N-1}) + A'$, where $c_0 \neq 0$ and A' is a power series in $k[[t]]$ which is divisible by t^N. Let $a = c_0 + c_1 t + \cdots + c_{N-1}t^{N-1}$ and $B = 1 + a^{-1}A' \in \widehat{\overline{\mathcal{O}}}^*$. Then $A = aB$ and $a \in \widetilde{\mathcal{O}}^*$. We show that $\widetilde{\mathcal{O}} \cap \widehat{\overline{\mathcal{O}}} \subseteq \overline{\mathcal{O}}$. Let $A \in \widetilde{\mathcal{O}} \cap \widehat{\overline{\mathcal{O}}}$. Define an ideal $\mathfrak{c}$ of $\overline{\mathcal{O}}$ by $\mathfrak{c} := \{x \in \overline{\mathcal{O}} \mid xA \in \overline{\mathcal{O}}\}$. Then $\mathfrak{c}$ is a non-zero ideal of $\overline{\mathcal{O}}$ because $Q(\widetilde{\mathcal{O}}) = Q(\overline{\mathcal{O}})$ and $\widetilde{\mathcal{O}}$ is a finite $\overline{\mathcal{O}}$-module. Then $A \in \overline{\mathcal{O}}$ if and only if $1 \in \mathfrak{c}$. Meanwhile, let $\widehat{\mathfrak{c}} = \{X \in \widehat{\overline{\mathcal{O}}} \mid XA \in \widehat{\overline{\mathcal{O}}}\}$. Then $\widehat{\mathfrak{c}} = \mathfrak{c}\widehat{\overline{\mathcal{O}}}$ (exercise !). Since $A \in \widehat{\overline{\mathcal{O}}}$, we have $1 \in \widehat{\mathfrak{c}}$. Hence $1 \in \widehat{\mathfrak{c}} \cap \overline{\mathcal{O}} = \mathfrak{c}$, where the last equality follows from the fact that $\widehat{\overline{\mathcal{O}}}$ is a faithfully flat extension of $\overline{\mathcal{O}}$.

Thus we may assume that $\overline{\mathcal{O}}$ (and hence $\widetilde{\mathcal{O}}$) is a complete local ring. Then $\widetilde{\mathcal{O}} = \overline{k}[[t]]$, a formal power series ring in one variable. The group law on $\widetilde{\mathcal{O}}^*/\overline{\mathcal{O}}^*$ is given by the multiplication on $\widetilde{\mathcal{O}}$. Note that

$$\widetilde{\mathcal{O}}^*/\overline{\mathcal{O}}^* \cong \widetilde{U}/\overline{U},$$

where $\widetilde{U}$ and $\overline{U}$ consist of formal power series in $\widetilde{\mathcal{O}} = \overline{k}[[t]]$ and $\overline{\mathcal{O}}$, respectively, whose constant terms are equal to 1. Let

$$\widetilde{U}^{(n)} = \{A \in U \mid v(A-1) \geq n\}, \quad n = 1, 2, \ldots$$

where v is the t-adic valuation of $\widetilde{\mathcal{O}}$. Then we have a sequence of subgroups

$$\widetilde{U} = \widetilde{U}^{(1)} \supset \widetilde{U}^{(2)} \supset \cdots \supset \widetilde{U}^{(n)} \supset \cdots$$

and, for $A \in \widetilde{U}$, $A \in \overline{U}$ if and only if $A - 1 \in \overline{\mathcal{O}}$. Hence we know that

$$\widetilde{U}^{(n)} \cdot \overline{U}/\widetilde{U}^{(n+1)} \cdot \overline{U} \cong (\widetilde{\mathcal{O}}^{(n)} + \overline{\mathcal{O}})/(\widetilde{\mathcal{O}}^{(n+1)} + \overline{\mathcal{O}})$$
$$\cong \widetilde{\mathcal{O}}^{(n)}/(\widetilde{\mathcal{O}}^{(n+1)} + \widetilde{\mathcal{O}}^{(n)} \cap \overline{\mathcal{O}}),$$

where $\widetilde{\mathcal{O}}^{(n)} = \{A \in \widetilde{\mathcal{O}} \mid v(A) \geq n\}$ and $\widetilde{\mathcal{O}}^{(n)}/\widetilde{\mathcal{O}}^{(n+1)} \cong G_a(k) = k^+$. Furthermore, $\widetilde{U} = \widetilde{U}^{(1)} \cdot \overline{U}$ and $\widetilde{U}^{(n)} \subseteq \overline{U}$ if $n \gg 0$. This implies that

(i) $\widetilde{U}/\overline{U}$ is a unipotent group scheme, and
(ii) $\widetilde{\mathcal{O}}/\overline{\mathcal{O}}$ is the Lie algebra of $\widetilde{U}/\overline{U}$. In particular,

$$\dim \widetilde{\mathcal{O}}^*/\overline{\mathcal{O}}^* = \dim_{\overline{k}} \widetilde{\mathcal{O}}/\overline{\mathcal{O}}.$$

On the other hand, we know by Rosenlicht [75] that

$$\dim \operatorname{Ker} \alpha = \dim_{\overline{k}} \prod_{\widetilde{P} \in \widetilde{\Sigma}} \mathcal{O}_{\widetilde{C}, \widetilde{P}} / \mathcal{O}_{\overline{C}, \overline{P}}.$$

Then θ is an isomorphism by the foregoing exact commutative diagram. $\quad\square$

1.8.1

Corollary. *With the notations and assumptions of Theorems 1.7 and 1.8, the following assertions hold.*

(1) $\operatorname{Pic}^0_{C/k}$ *is a connected algebraic group defined over* k.
(2) *For any field extension* k' *of* k, $\operatorname{Pic}^0_{C \otimes_k k'/k'} \cong \operatorname{Pic}^0_{C/k} \otimes_k k'$.
(3) $\operatorname{Pic}^0_{\overline{C}/\overline{k}}$ *is the generalized Jacobian variety of* $\overline{C}$ *with equivalence relation defined by* $\overline{\mathfrak{o}}$, *and it is an extension of an abelian variety* $\operatorname{Pic}^0_{\widetilde{C}/\overline{k}}$ *by a unipotent algebraic group whose* $\overline{k}$-*rational points are the group* $\prod_{\widetilde{P} \in \widetilde{\Sigma}} \mathcal{O}^*_{\widetilde{C}, \widetilde{P}} / \mathcal{O}^*_{\overline{C}, \overline{P}}$.
(4) *Suppose that the arithmetic genus* $p_a(C)$ *of* C *is positive. Then* $\iota :$ $U_0 \to \operatorname{Pic}^0_{C/k}$ *is a closed immersion.*

Proof. We prove only the assertion (4). If the geometric genus of C, which is equal to $p_a(\widetilde{C})$, is positive, then the composite of $\beta : \operatorname{Pic}^0_{\overline{C}/\overline{k}} \to \operatorname{Pic}^0_{\widetilde{C}/\overline{k}}$ with

$$\overline{\iota} := \iota \otimes_k \overline{k} : \ \overline{U}_0 \to \operatorname{Pic}^0_{C/k} \otimes_k \overline{k} = \operatorname{Pic}^0_{\overline{C}/\overline{k}}$$

is an immersion by the theory of Jacobian variety, where $\overline{U}_0 = U_0 \otimes_k \overline{k} = \widetilde{C} - \widetilde{\Sigma}$. Hence $\overline{\iota}$ is an immersion. Suppose that $p_a(\widetilde{C}) = 0$. Then $\overline{C}$ is a singular rational curve. If $\overline{\iota}$ is not an immersion, there exist distinct points Q_1, Q_2 of $\overline{U}_0$ such that $Q_1 - Q_2 \sim_{\overline{\mathfrak{o}}} 0$. Then $Q_1 \sim Q_2$ as points of $\overline{C}$. Hence $\dim |Q_1| = 1$ and $\overline{C} \cong \mathbb{P}^1_{\overline{k}}$. This is a contradiction to the assumption that $p_a(C) > 0$. Suppose that $\overline{\iota}$ is not a closed immersion. Then there exists a point $Q \in \widetilde{\Sigma}$ such that $\theta([Q - P])$ is a point in the closure of $\overline{\iota}(\overline{U}_0)$ but not in $\overline{\iota}(\overline{U}_0)$, where P is a given k-rational point of C. Then $Q - P \sim_{\overline{\mathfrak{o}}} D$, where D is a divisor of degree 0 on $\widetilde{C}$ which is independent of the places of $\overline{\mathfrak{o}}$. Namely, there exists an element $g \in \overline{\mathfrak{o}}$ such that $Q = P + D + (g)$. But this is a contradiction because the support of the divisor on the right side of the equality contains no points of $\widetilde{\Sigma}$. Hence $\overline{\iota}$ is a closed immersion. $\quad\square$

1.9 Dualizing sheaf

Retaining the notations and assumptions in 1.7, we note that every local ring of C is a regular local ring. Hence C is a Gorenstein k-scheme. Then, for any field extension k' of k, the k'-scheme $C \otimes_k k'$ is Gorenstein. In particular, so is $\overline{C} := C \otimes_k \overline{k}$. Let $\underline{\omega}_{C/k}$ be the dualizing sheaf of C, which is a locally free sheaf of rank 1. We have the following natural isomorphisms

$$\underline{\omega}_{C \otimes_k k'/k'} \cong \mathcal{O}_{C \otimes_k k'} \otimes_{\mathcal{O}_C} \underline{\omega}_{C/k} \cong k' \otimes_k \underline{\omega}_{C/k}.$$

With the notations of 1.8, $\underline{\omega}_{C \otimes_k \overline{k}/\overline{k}}$ is given as follows:

$$\left(\underline{\omega}_{C \otimes_k \overline{k}/\overline{k}}\right)_{\overline{P}} = \begin{cases} \Omega^1_{\mathcal{O}_{\overline{P}}/\overline{k}} & (\overline{P} \notin \overline{\Sigma}) \\ \Omega'_{\overline{P}} & (\overline{P} \in \overline{\Sigma}) \end{cases}$$

where $\mathcal{O}_{\overline{P}} := \mathcal{O}_{\overline{C}, \overline{P}}$ and $\Omega'_{\overline{P}}$ is the $\mathcal{O}_{\overline{P}}$-module formed by all differentials ω of $\overline{k}(\overline{C})$ such that $\operatorname{Res}_{\widetilde{P}}(h\omega) = 0$ for every $h \in \mathcal{O}_{\overline{P}}$ and every point $\widetilde{P}$ of $\widetilde{C}$ lying over $\overline{P}$. We refer to J.-P. Serre [83, p.78] for this result.

1.9.1

Lemma. *Assume that a point $\overline{P}$ of $\overline{\Sigma} := \operatorname{Sing}(\overline{C})$ is a one-place point. Let $\widetilde{P}$ be the point of $\widetilde{C}$ lying over $\overline{P}$ and let s be a uniformisant of $\mathcal{O}_{\widetilde{P}}$. Let d be the degree of the conductor $\mathfrak{c}$ of $\mathcal{O}_{\widetilde{P}}$ into $\mathcal{O}_{\overline{P}}$, i.e., d is the largest integer such that $\mathfrak{c} \subseteq (\mathfrak{m}_{\widetilde{P}})^d$, where $\mathfrak{m}_{\widetilde{P}}$ is the maximal ideal of $\mathcal{O}_{\widetilde{P}}$. Then $\Omega'_{\overline{P}}$ is generated by $\frac{ds}{s^d}$ as $\mathcal{O}_{\overline{P}}$-module.*

Proof. d is the largest positive integer such that $s^{d-1} \notin \widehat{\overline{\mathcal{O}}}$ and $s^n \in \widehat{\overline{\mathcal{O}}}$ for every $n \geq d$, where $\widetilde{\mathcal{O}} = \mathcal{O}_{\widetilde{P}}, \widehat{\widetilde{\mathcal{O}}} = k[[s]]$, etc. Then $\frac{ds}{s^d} \in \Omega'_{\overline{P}}$ and $\operatorname{Res}_{\widetilde{P}}(s^{n-1}\frac{ds}{s^n}) = 1$ if $n > d$. Since $\underline{\omega}_{\overline{C}/\overline{k}}$ is a locally free sheaf of rank 1, we have

$$\widehat{\widetilde{\mathcal{O}}} \otimes_{\widehat{\overline{\mathcal{O}}}} \Omega'_{\overline{P}} = \widehat{\widetilde{\mathcal{O}}} \cdot \left(\frac{ds}{s^m}\right)$$

for some $m > 0$. Hence $\frac{ds}{s^d} = f \cdot \frac{ds}{s^m}$ with $f \in \widehat{\widetilde{\mathcal{O}}}$. The above argument implies that $d = m$. Therefore we have $\Omega'_{\overline{P}} = \widetilde{\mathcal{O}} \cdot \left(\frac{ds}{s^d}\right)$. $\qquad \square$

For any field extension k' of k, let $\omega_{C/k'} := \Gamma(C \otimes_k k', \underline{\omega}_{C \otimes_k k'/k'})$. Then $\omega_{C/k'} = k' \otimes_k \omega_{C/k}$ and $\omega_{C/k}$ is a k-vector space of dimension equal to the arithmetic genus $p_a(C)$ of C, called the *space of Weil differentials of the first kind* of C.

1.9.2

Lemma.   *Assume that every point $\overline{P}$ of $\overline{\Sigma}$ is a one-place point for $\overline{C}$. Then, with the notations of Theorem 1.7, the embedding $\iota : U_0 \to \mathrm{Pic}^0_{C/k}$ induces an isomorphism*

$$\iota^* : \Omega := \mathrm{Lie}(\mathrm{Pic}^0_{C/k}) \xrightarrow{\sim} \omega_{C/k}.$$

Proof.   (I) First of all, let $\Sigma := \mathrm{Sing}\,(C)$ be the set of non-smooth points of C. Let $\pi : \overline{C} \to C$ be the canonical projection. Then $\overline{\Sigma} = \pi^{-1}(\Sigma)$. Indeed, let k_s be the separable closure of k in $\overline{k}$. Then $C_s := C \otimes_k k_s$ is k_s-normal and $\overline{C} = C_s \otimes_{k_s} \overline{k}$. The projection π splits up into a composite

$$\pi : \overline{C} \xrightarrow{\pi_i} C_s \xrightarrow{\pi_s} C.$$

Since the extension $\overline{k}/k_s$ is purely inseparable, we know that, for any $\overline{k}$-rational point $\overline{P}$, $\pi_i^{-1}(\pi_i(\overline{P})) = \{\overline{P}\}$. Moreover, for any $\overline{k}$-rational point Q of C_s, $\pi_s^{-1}(\pi_s(Q))$ is transitive under certain finite group action. Hence, if Q is not smooth, every point of $\pi_s^{-1}(\pi_s(Q))$ is not smooth. This is equivalent to saying that $\pi_s(Q)$ is not smooth. Hence $\overline{\Sigma} = \pi^{-1}(\Sigma)$. Thus, $U_0 = C - \Sigma$.

(II) To prove that the k-linear mapping $\iota^* : \Omega \to \omega_{C/k}$ is an isomorphism, it suffices to show that

$$\overline{k} \otimes \iota^* : \overline{k} \otimes \Omega \to \overline{k} \otimes \omega_{C/k} \cong \omega_{C/\overline{k}}$$

is an isomorphism. As $\overline{k} \otimes \iota^* = (\iota \otimes \overline{k})^*$ and $\mathrm{Pic}^0_{\overline{C}/\overline{k}}$ is identified with the generalized Jacobian variety $J' := J_{\overline{\sigma}}$ as in 1.8, we may, and shall proceed to prove our assertion under the assumption that $k = \overline{k}$ and $\mathrm{Pic}^0_{C/k} = J'$. We follow the proof of Serre [83, Chap. V, Prop. 5, p.97].

Let $X = U_0 := \overline{C} - \overline{\Sigma}$, let $g = p_a(C)$, let $\phi : X^g \to J'$ be the morphism $\phi(x_1, \ldots, x_g) = \sum_{i=1}^g \iota(x_j)$ and let $h_j : X^g \to J'$ be the morphism $h_j(x_1, \ldots, x_g) = \iota(x_j)$. Then $\phi = \sum_{i=1}^g h_j$ and $\phi^*(\omega) = \sum_j h_j^*(\omega)$ for any element ω of Ω. Hence if $\iota^*(\omega) = 0$ then $\phi^*(\omega) = 0$. Meanwhile, the morphism $\phi : X^g \to J'$ is decomposed as a composite of morphisms

$$X^g \xrightarrow{\psi} X^{(g)} \xrightarrow{\rho} J',$$

where $X^{(g)}$ is the symmetric product of g copies of X. Since ψ and ρ are generically separable and dominating, $\phi^*(\omega) = 0$ implies $\omega = 0$. Therefore, the mapping $\omega \mapsto \iota^*(\omega)$ is injective. It is then surjective because

$$\dim_k \Omega = \dim_k \omega_{C/k} = \dim H^1(C, \mathcal{O}_C) = g.$$

It remains to show that $\iota^*(\omega) \in \omega_{C/k}$ if $\omega \in \Omega$. For this, by virtue of [83, IV, no. 9, Prop. 6], we have only to show that $\mathrm{Tr}_a(\iota^*(\omega)) = 0$ for every rational function a such that $a \in \mathcal{O}_P$ for every $P \in \Sigma$ and $a \notin K^p$, where $K = k(C)$. Let $h = \mathrm{Tr}_a(i)$ be the rational mapping from $\mathbb{P}^1_k$ to J' defined in [83, GA; III, no.2]. By [83, III, no.6, Lemma 4 and its remark], $\mathrm{Tr}_a(\iota^*(\omega)) = h^*(\omega)$. On the other hand, it is obvious that $\iota(\mathrm{div}(b)) = 0$ if $b \in \cap_{P \in \Sigma} \mathcal{O}_P$. Therefore, on the basis of [83, III, no.5, Prop.9] slightly modified, we conclude that h is a constant mapping. Hence $h^*(\omega) = 0$. $\qquad\qquad\square$

Chapter 2

Forms of the affine line

2.1 Forms of the rational function field

Let k be an arbitrary field of characteristic $p \geq 0$. Let K be an algebraic function field of one variable containing k as the field of constants. Then there exists a complete normal curve C such that $K = k(C)$ and the correspondence $P \in C \mapsto \mathcal{O}_{C,P}$ is a bijection between C and the set of discrete valuation rings (DVR) over k of K. Thus C is determined uniquely up to k-isomorphisms. We call C a *complete k-normal model* of K. K is said to be a *form of the rational function field* $k(t)$ if a $\overline{k}$-isomorphism

$$\overline{k} \otimes_k K \cong \overline{k} \otimes_k k(t) = \overline{k}(t)$$

exists, where $\overline{k}$ is an algebraic closure of k which we fix throughout this chapter. The definition implies that K is a regular extension of k in the sense of Weil, i.e., $K \otimes_k k'$ is an integral domain for any field extension k' of k. More usable criterion for regularity is that $K \otimes_k k^{p^{-s}}$ is reduced for every $s \geq 0$ and k is algebraically closed in K. Let k' now refer to a finite normal extension of k such that $k' \otimes_k K \cong k'(t)$ (such k' exists by the above definition) and let k'' be the invariant subfield of the field automorphism group $\mathrm{Aut}\,(k'/k)$. Then $k' \otimes_{k''} (k'' \otimes_k K) \cong k'(t)$ with k'/k'' finite separable, so that $k'' \otimes_k K$ has arithmetic genus zero. In fact, let C'' be a regular proper curve defined over k'' such that $k''(C'') = k'' \otimes_k K$. Since k'/k'' is a finite separable extension, $C' := C'' \otimes_{k''} k'$ is étale over C''. Since C' is smooth by definition, so is C''. In particular, $\Omega^1_{C''/k''} \otimes_{k''} k' \cong \Omega^1_{C'/k'}$. Hence $H^0(C', \Omega_{C'/k'}) \cong H^0(C'', \Omega_{C''/k''}) \otimes_{k''} k'$. This show the invariance of arithmetic genera $p_a(C')$ and $p_a(C'')$. It is then well-known that $k'' \otimes_k K$ is k''-isomorphic to either the rational function field or the function field of a smooth conic without k''-rational points. In fact, the anti-canonical linear system $|-K_{C''}|$ with $\mathcal{O}_{C''}(K_{C''}) \cong \Omega^1_{C''/k''}$ embeds C'' isomorphically to

19

a conic in $\mathbb{P}^2_{k''}$. If the conic has a k''-rational point, it is isomorphic to $\mathbb{P}^1_{k''}$.

Here we only consider the former case that $C'' \cong \mathbb{P}^1_{k''}$. Hence $k'' \otimes_k K$ is a *purely inseparable k-form of $k(t)$. Hereafter we assume that k has positive characteristic p and is an infinite field.* Hence there exists a purely inseparable extension k'' of k such that $k'' \otimes_k K \cong k''(t)$. The *height* of K/k is, by definition, the smallest non-negative integer λ' such that there exists an isomorphism

$$k^{p^{-\lambda'}} \otimes K \cong k^{p^{-\lambda'}} \otimes k(t),$$

where $k^{p^{-\lambda'}} = \{u \in \overline{k} \mid u^{p^{\lambda'}} \in k\}$ and the field k is often not indicated in the tensor product whenever it is easily known from the context. We denote such λ' by ht(K) and call it the *height* of K[1]. It is clear that $\lambda' = 0$ if and only if $K \cong k(t)$. The k-form K is then called a *trivial k-form of $k(t)$*.

2.2 Frobenius morphisms

We denote by Θ^n the base change functor induced by the Frobenius homomorphism

$$\varphi^n : k \to k; \quad a \mapsto a^{p^n}, \quad a \in k.$$

For any k-scheme X, there is a canonical morphism of k-schemes

$$F^n_X : X \to \Theta^n X.$$

If $X = \text{Spec}(A)$ is affine, then $\Theta^n X = \text{Spec}((k, \varphi^n) \otimes_k A)$, where (k, φ^n) indicates k viewed as a right k-algebra via φ^n. Furthermore, F^n_X is induced by a k-algebra homomorphism

$$F^n_A : (k, \varphi^n) \otimes_k A \to A, \qquad \alpha \otimes a \mapsto \alpha a^{p^n}.$$

If k' is a purely inseparable extension of exponent $\leq n$ over k, i.e., $k'^{p^n} \subseteq k$, there is a commutative diagram

$$
\begin{array}{ccc}
k & \longrightarrow & k' \\
\varphi^n \downarrow & & \downarrow \overline{\varphi} \\
k & = & k
\end{array}
$$

and we have $\Theta^n X \cong (k, \overline{\varphi}) \otimes_{k'} X_{k'}$. In fact, the ordinary inclusion $k \hookrightarrow k^{p^{-n}}$ is decomposed as $k \hookrightarrow k' \hookrightarrow k^{p^{-n}}$ and $\varphi^n : k \hookrightarrow k^{p^{-n}} \xrightarrow{\varphi^n_k} k$, we obtain $\overline{\varphi}$ as the composite $k' \hookrightarrow k^{p^{-n}} \xrightarrow{\varphi^n_{k'}} k$.

---

[1]In Chapter 6, λ denotes the height of a k-form of $\mathbb{A}^1$ and λ' denotes the height of $k(X)$. For the consistency of the notations, we use λ' from here.

Let K be a purely inseparable k-form of $k(t)$ with height λ'. Since $k^{p^{-\lambda'}} \otimes K \cong k^{p^{-\lambda'}} \otimes k(t)$, we have isomorphisms

$$k(K^{p^{\lambda'}}) \cong (k, \varphi^n) \otimes_k K \cong k(t^{p^{\lambda'}}). \tag{2.1}$$

Conversely, it is clear that if the isomorphisms (2.1) hold we have $k^{p^{-\lambda'}} \otimes K \cong k^{p^{-\lambda'}}(t)$. Therefore the height of the k-form K of $k(t)$ is equal to the smallest positive integer λ' such that

$$(k, \varphi^{\lambda'}) \otimes_k K \cong k(t')$$

for some element t' which is transcendental over k.

2.3 Height

Let X be a k-scheme such that $X \otimes \overline{k}$ is $\overline{k}$-isomorphic to the affine line $\mathbb{A}^1_{\overline{k}}$. Such an X is called a k-form of the affine line $\mathbb{A}^1$. For such an X, one can find a finite normal extension k' of k such that $X \otimes_k k' \cong \mathbb{A}^1_{k'}$. Let k'' be the invariant subfield of the automorphism group $\mathrm{Aut}\,(k'/k)$. Then we have

$$(X \otimes k'') \otimes_{k''} k' \cong X \otimes k' \cong \mathbb{A}^1_{k'}.$$

Since k'/k'' is a finite separable extension, it is well-known that $X \otimes k'' \cong \mathbb{A}^1_{k''}$ (cf. [41]). Thus, each k-form X of $\mathbb{A}^1$ becomes $k^{p^{-\lambda}}$-isomorphic to $\mathbb{A}^1 \otimes k^{p^{-\lambda}}$ for some $\lambda \geq 0$, and the smallest integer $\lambda \geq 0$ for which this isomorphism exists is called the *height* of the k-form X and denoted by $\mathrm{ht}\,(X)$. A k-form X of $\mathbb{A}^1$ of height 0 is called *trivial*, and X is then k-isomorphic to $\mathbb{A}^1$. We note that a k-form X of $\mathbb{A}^1$ is a geometrically integral smooth affine k-scheme of dimension 1 and that the function field $k(X)$ is a k-form of the rational function field $k(t)$.

Furthermore, the height $\lambda = \mathrm{ht}\,(X)$ is the smallest integer $\lambda \geq 0$ such that $\Theta^\lambda X$ is k-isomorphic to $\mathbb{A}^1$, and there exists an inequality

$$\mathrm{ht}\,(X) \geq \mathrm{ht}\,(k(X)).$$

We note that the equality does not necessarily hold as exhibited by the following example-proposition.

2.3.1

Proposition. (1) *Let X be an affine plane curve*

$$y^2 = x + ax^2, \quad a \in k - k^2, \quad p = 2.$$

Then $\mathrm{ht}\,(X) = 1$ *and* $\mathrm{ht}\,(k(X)) = 0$.

(2) Let $k = \mathbb{F}_2(a, b)$, where a and b are algebraically independent over $\mathbb{F}_2$. Let X be an affine plane curve

$$y^2 + b = x + ax^2.$$

Then $\mathrm{ht}\,(X) = \mathrm{ht}\,(k(X)) = 1$.

Proof. (1) Let $a = \alpha^2$ with $\alpha \in k^{1/2}$ and let $\xi = y - \alpha x$. Then $x = \xi^2$ and $y = \xi + \alpha\xi^2$. Hence $X \otimes k^{1/2} = \mathrm{Spec}\,k^{1/2}[\xi]$ and $\mathrm{ht}\,(X) \le 1$. If $X \cong \mathbb{A}_k^1$ then X has two or more k-rational points because $\#(k) \ge 2$. Let (α, β) be a k-rational point. If $\alpha \ne 0$ then $a = (\beta^2 - \alpha)/\alpha^2 \in k$, which contradicts the hypothesis. Hence $\alpha = 0$ and then $\beta = 0$. Namely, X has a unique k-rational point. Thus $X \not\cong \mathbb{A}_k^1$ and $\mathrm{ht}\,(X) = 1$. Set $t = y/x$. Then $x = 1/(t^2 - a)$ and $y = t/(t^2 - a)$. Hence $k(X) = k(t)$ and $\mathrm{ht}\,(k(X)) = 0$.

(2) It is clear that $k^{1/2} \otimes X \cong \mathbb{A}_k^1 \otimes k^{1/2}$ and hence $\mathrm{ht}\,(X) \le 1$. We show that $k(X)$ is not k-rational. Then it follows that $\mathrm{ht}\,(X) = \mathrm{ht}\,(k(X)) = 1$. If $k(X)$ is k-rational, X has a k-rational point. Then we can show that there exist $f, g, h \in \mathbb{F}_2[a, b]$ such that

$$h^2 + bf^2 = gf + ag^2 \quad \text{and} \quad f \ne 0.$$

Suppose $(f, g, h) \ne (0, 0, 0)$. Denote by $\frac{\partial}{\partial a}$ and $\frac{\partial}{\partial b}$ the partial derivations in $\mathbb{F}_2[a, b]$ with respect to a and b, respectively. Then we have

$$f^2 = gf_b + fg_b = (fg)_b$$
$$g^2 = gf_a + fg_a = (fg)_a,$$

where $f_a = \frac{\partial f}{\partial a}$, $g_b = \frac{\partial g}{\partial b}$, etc. Hence $f^2 g^2 = (fg)_a(fg)_b$. Since $(f, g, h) \ne (0, 0, 0)$, we have $(fg)_a \ne 0$ and $(fg)_b \ne 0$. This implies the existence of a nonzero element $q \in \mathbb{F}_2[a, b]$ such that

$$q^2 = q_a \cdot q_b.$$

Write $q = q_0 + q_1 a + q_2 b + q_3 ab$, where $q_i \in \mathbb{F}_2[a^2, b^2]$ for $0 \le i \le 3$. Since

$$q^2 = q_a \cdot q_b = (q_1 + q_3 b)(q_2 + q_3 a) = q_1 q_2 + q_1 q_3 a + q_2 q_3 b + q_3^2 ab,$$

we have $q_3 = 0$ and

$$q_0^2 + q_1^2 a^2 + q_2^2 b^2 = q_1 q_2.$$

Denote by $\deg q_i$ the total degree of q_i with respect to a and b. Then we have $\deg q_1 = \deg q_2$, and the terms of the highest degree of $q_1^2 a^2$ and

$q_2^2 b^2$ cancel each other [2]. However this is impossible because q_1^2 and q_2^2 are polynomials in $\mathbb{F}_2[a^4, b^4]$. Thus we obtain a contradiction. The case $g = 0$ or $h = 0$ can be handled in the same fashion. □

2.3.2

Remark. The example (2) in 2.3.1 gives a smooth non-rational projective plane curve

$$x_2^2 + bx_0^2 = x_0 x_1 + a x_1^2,$$

which has arithmetic genus 0 and becomes a rational curve after the purely inseparable base field extension $k(b^{1/2})/k$.

Notwithstanding this example, we have the following result.

2.3.3

Lemma. *Let C be a smooth complete k-curve such that $C' := C \otimes k'$ is k'-isomorphic to $\mathbb{P}^1$ for a purely inseparable extension k' of k. Assume that C has a k-rational point if $p = 2$. Then C is k-isomorphic to $\mathbb{P}^1$.*

Proof. We may assume that k' is a finite algebraic extension of k. Then there exists a series of extensions

$$k' = k_n \supset k_{n-1} \supset \cdots \supset k_1 \supset k_0 = k$$

such that $[k_i : k_{i-1}] = p$ for $1 \leq i \leq n$. It suffices to show that the assertion steps down for each extension $k_i \supset k_{i-1}$. Thus we may assume that k' is a simple extension of exponent 1, i.e., $[k' : k] = p$. First of all, we show that C has a k-rational point if $p \neq 2$. Indeed, since C' is k'-isomorphic to $\mathbb{P}^1_{k'}$, C' has a k'-rational point P', i.e., $k(P') = k'$. Let $P = \pi(P')$ for the canonical projection $\pi : C' = C \otimes k' \to C$. If $[k(P') : k(P)] = p$ then $k(P) = k$ and P is a k-rational point of C. Suppose that $k(P') = k(P)$. Then the ramification theory for the finite covering $\pi : C' \to C$ implies that t'^p is a uniformisant of $\mathcal{O}_{C,P}$ for a uniformisant t' of $\mathcal{O}_{C',P'}$. Namely the divisor P on C splits as pP' on C'. Let K_C be the canonical divisor

---

[2] In fact, let $m = \deg q_1$ and $n = \deg q_2$. Suppose $m > n$. Then the above equality implies that $\deg q_0 = m + 1$ and the terms of the highest degree of q_0 and $q_1 a$ cancel each other. However, since $q_0, q_1 \in \mathbb{F}_2[a^2, b^2]$, this is impossible. Likewise $n > m$ is impossible. Hence $m = n$. If $\deg q_0 = \deg q_1 + 1 = \deg q_2 + 1$ occurs, the term of the highest degree of q_0 must cancel with terms of $q_1 a + q_2 b$. But this is impossible by the above reason.

on C. Then $\deg(pP') = p$ and $\deg(K_C) = -2$. Here deg is considered on the smooth complete curve $C = C \otimes_k \overline{k}$, where $\overline{k}$ is an algebraic closure of k containing k'. Let n be a positive integer such that $p = 2n + 1$, and consider the divisor $pP' + nK_C$ on C. Then $\deg(pP' + nK_C) = 1$. By the Riemann-Roch theorem on C[3], we have

$$\dim H^0(C, \mathcal{O}(pP' + nK_C)) = 2 \quad \text{and} \quad \dim H^1(C, \mathcal{O}(pP' + nK_C)) = 0.$$

Thus there is a k-rational point Q on C such that Q is linearly equivalent to $pP' + nK_C$.

In both cases ($p = 2$ or $p \neq 2$), C has a k-rational point Q. Then the k-rational map $f = \Phi_{|Q|} : C \to \mathbb{P}^1$ defined by the complete linear system $|Q|$ is a k-isomorphism. $\qquad\square$

2.4 Forms of the affine line and one-place points at infinity

Let X be a k-form of $\mathbb{A}^1$ and let C be a k-normal complete model of X, which is called a *normal completion* or a *regular completion*. The curve C is determined uniquely up to k-isomorphisms. Then C contains X as a dense open set, $C - X$ is a one-place point which might be singular. Namely, for any field extension k' of k, the normalization $\widetilde{C}_{k'}$ of $C_{k'} := C \otimes k'$ contains naturally $X \otimes k'$ and $\widetilde{C}_{k'} - X \otimes k'$ is a single point. Let $P_\infty = C - X$. Such a point P_∞ is called a *geometric one-place point*.

2.4.1

Lemma. *With the above notations, P_∞ is rational over a purely inseparable extension of k.*

Proof. Let k_i be the perfect closure of k. Then $X \otimes k_i$ is isomorphic to $\mathbb{A}^1_{k_i}$ by 2.3. Let C' be a complete k_i-normal model of $X \otimes k_i$. Then C' is k_i-isomorphic to $\mathbb{P}^1$ and hence $C' - X \otimes k_i$ is k_i-rational. Since P_∞ is dominated by the k_i-rational point $C' - X \otimes k_i$, P_∞ is rational over a purely inseparable extension of k. $\qquad\square$

---

[3]Let C be a complete normal curve over a field k. Since every local ring is a regular local ring (hence a Gorenstein local ring), the Riemann-Roch theorem holds for an invertible sheaf $\mathcal{L}$ on C as for a complete smooth curve defined over an algebraically closed field. Namely we have

$$h^0(C, \mathcal{L}) - h^1(C, \mathcal{L}) = \deg \mathcal{L} + 1 - p_a(C), \quad H^1(C, \mathcal{L}) \cong H^0(C, \underline{\omega}_{C/k} \otimes \mathcal{L}^{-1}),$$

where $\underline{\omega}_{C/k}$ is the dualizing invertible sheaf of C and $\deg \mathcal{L}$ is the degree of $\mathcal{L} \otimes_k \overline{k}$ on the normalization $\widetilde{C}$ of $\overline{C} := C \otimes_k \overline{K}$. Here $\overline{k}$ is an algebraic closure of k.

2.4.2

In general, we have the following:

Lemma. *Let C be a complete k-normal curve. Then there exists a finite separable extension k' of k such that every singular point of $C \otimes_k k'$ is a geometric one-place point.*

Proof. Since $C \otimes_k k'$ is k'-normal for a separable algebraic extension k' over k, by replacing C by $C \otimes_k k_s$ with the separable closure k_s of k in $\overline{k}$, we may assume that k is separably closed. Let P be a singular point of C if it exists at all. The local ring $\mathcal{O} := \mathcal{O}_{C,P}$ is a discrete valuation ring, and $k^{p^{-n}} \otimes_k \mathcal{O}$ is $k^{p^{-n}}$-smoothable [4] for some $n > 0$. Let $\widetilde{\mathcal{O}}$ be the normalization of $k^{p^{-n}} \otimes_k \mathcal{O}$. Then $\widetilde{\mathcal{O}}^{p^n} \subseteq \mathcal{O}$. Since $\mathcal{O}$ is a DVR and $\widetilde{\mathcal{O}}$ is the normalization of $\mathcal{O}$ in $Q(k^{p^{-n}} \otimes_k \mathcal{O})$, it is easy to see that $\widetilde{\mathcal{O}}$ is a local ring. Hence P is a geometric one-place point. $\qquad\square$

2.4.3

Theorem. *Let C be a complete k-normal curve and let P_∞ be a point of C. Assume that P_∞ is rational over a purely inseparable algebraic extension of k and $X := C - \{P_\infty\}$ is a k-form of $\mathbb{A}^1$. Further, assume that X has a k-rational point P_0. Then the following conditions are equivalent.*

(1) *The morphism $\iota : X \to \mathrm{Pic}^0_{C/k}$ given in 1.7 is a closed immersion, and $\mathrm{Pic}^0_{C/k}$ is generated as a k-group scheme by the image $\iota(X)$.*
(2) *$\dim \mathrm{Pic}^0_{C/k} > 0$.*
(3) *C is not k-isomorphic to $\mathbb{P}^1$, i.e., C is not k-rational.*
(4) *X cannot be embedded into a smooth complete k-curve.*

Proof. (1) $\Rightarrow$ (2) and (4) $\Rightarrow$ (3) are obvious. (2) $\Rightarrow$ (3): Since $\dim \mathrm{Pic}^0_{C/k} = h^1(C, \mathcal{O}_C)$, we have $p_a(C) > 0$. Hence C is not k-rational. (3) $\Rightarrow$ (2): It is well-known that a complete k-normal curve C with a k-rational point and $p_a(C) = 0$ is k-isomorphic to $\mathbb{P}^1$. In fact, $h^0(C, \mathcal{O}_C(P_0)) = 2$ by the Riemann-Roch theorem and $\Phi_{|P_0|} : C \to \mathbb{P}^1$ is an isomorphism. (3) $\Rightarrow$ (4): If X is embedded into a complete smooth k-curve $\widetilde{C}$, then $\widetilde{C}$ is k-isomorphic to C. Hence $p_a(C) = p_a(C \otimes \overline{k}) = 0$, and C is k-isomorphic to $\mathbb{P}^1$. (2) $\Rightarrow$ (1): Let $\overline{C} := C \otimes \overline{k}$ and let $\widetilde{C}$ be the normalization of $\overline{C}$. Let

---

[4]Namely, the normalization $\widetilde{\mathcal{O}}$ of $k^{p^{-n}} \otimes \mathcal{O}$ is a local ring such that $\widetilde{\mathcal{O}} \otimes k''$ is k''-normal for every field extension k'' of $k^{p^{-n}}$.

$\widetilde{P}_\infty$ be the point of $\widetilde{C}$ lying over P_∞. Let C' be a projective plane curve defined by

$$x_0^3 = x_1^2 x_2$$

in $\mathbb{P}_{\overline{k}}^2$, and let P'_∞ be the unique singular point of C'. We show that there exists a $\overline{k}$-morphism $f : C' \to \overline{C}$. In fact, let $\widetilde{\mathcal{O}}$, $\mathcal{O}'$ and $\overline{\mathcal{O}}$ be the local rings of points $\widetilde{P}_\infty$, P'_∞ and P_∞ on $\widetilde{C}$, C' and $\overline{C}$, respectively. We can identify $\mathcal{O}'$ and $\overline{\mathcal{O}}$ with the subrings of $\widetilde{\mathcal{O}}$. Then it suffices to show that $\mathcal{O}'$ dominates $\overline{\mathcal{O}}$. Identifying $\widetilde{\mathcal{O}}$ with the localization $\overline{k}[t]_{(t)}$ of a polynomial ring $\overline{k}[t]$ at the ideal generated by t, we have $\mathcal{O}' = \overline{k} + t^2\widetilde{\mathcal{O}}$. If $\mathcal{O}'$ does not dominate $\overline{\mathcal{O}}$, then $\overline{\mathcal{O}}$ contains an element of $t\widetilde{\mathcal{O}} - t^2\widetilde{\mathcal{O}}$. Then, if $\widehat{\widetilde{\mathcal{O}}}$ and $\widehat{\overline{\mathcal{O}}}$ denote the completions of $\widetilde{\mathcal{O}}$ and $\overline{\mathcal{O}}$ by the respective maximal ideals, we have $\widehat{\widetilde{\mathcal{O}}} = \widehat{\overline{\mathcal{O}}}$. Hence $\widetilde{\mathcal{O}} = \overline{\mathcal{O}}$ because $\widetilde{\mathcal{O}}$ is a finitely generated $\overline{\mathcal{O}}$-module. This implies that $\dim \mathrm{Pic}_{C/k}^0 = 0$ (cf. the equivalence (2) $\Leftrightarrow$ (3)). therefore, $\mathcal{O}'$ dominates $\overline{\mathcal{O}}$. Now the morphism $f : C' \to \overline{C}$ gives rise to a k-homomorphism

$$\overline{\rho} : \ \mathrm{Pic}_{\overline{C}/\overline{k}}^0 \longrightarrow \mathrm{Pic}_{C'/\overline{k}}^0.$$

Let $\overline{\iota} := \iota \otimes \overline{k} : \ X \otimes \overline{k} \to \mathrm{Pic}_{\overline{C}/\overline{k}}^0$. Then $\overline{\rho} \cdot \overline{\iota}$ is given by $\overline{\rho} \cdot \overline{\iota}(Q) = Q - P_0$ for $Q \in \overline{C}(\overline{k}) - P_\infty$. By 1.8, $\mathrm{Pic}_{C'/\overline{k}}^0$ is isomorphic to $C' - P'_\infty$ [5]. Since $\overline{C} - P_\infty$ and $C' - P_\infty$ are isomorphic to $\mathbb{A}_{\overline{k}}^1$, it is easy to show that $\overline{\rho} \cdot \overline{\iota}$ is a $\overline{k}$-isomorphism. In particular, $\overline{\iota}$ is a closed immersion. Hence it follows that ι is also a closed immersion. Since $\mathrm{Pic}_{\overline{C}/\overline{k}}^0$ is identified with the generalized Jacobian variety of $\overline{C}$ with equivalence relation defined by $\overline{\mathcal{O}}$ (cf. 1.8), the theory of generalized Jacobian varieties implies that $\mathrm{Pic}_{\overline{C}/\overline{k}}^0$ is generated by the image of $\overline{\iota}$. Hence $\mathrm{Pic}_{C/k}^0$ is generated by the image of ι as a k-group scheme. $\qquad\qquad\square$

2.5   Explicit equations

Let K be a purely inseparable k-form of $k(t)$, and let C be a complete k-normal model of K. Then, K is the function field of a k-form of $\mathbb{A}^1$ if and only if C has at most one singular point which is a geometric one-place point. If C has a unique singular point P_∞ which is a geometric one-place

---

[5] Since the normalization of C' is isomorphic to $\mathbb{P}_{\overline{k}}^1$, Theorem 1.8 shows that $\mathrm{Pic}_{C'/\overline{k}}^0$ is isomorphic to $\mathrm{Ker}\,\alpha$ and hence a unipotent group scheme. On the other hand, $\underline{\omega}_{C'/\overline{k}} \cong \mathcal{O}_{C'} dt$, where $C' - P'_\infty = \mathrm{Spec}\,\overline{k}[t]$ by the description of $\Omega'_{P'_\infty}$ in Lemma 1.9.1. Hence $\dim \omega_{C'/\overline{k}} = \dim \mathrm{Lie}(\mathrm{Pic}_{C'/\overline{k}}^0) = 1$. Hence $\theta : C' - P'_\infty \to \mathrm{Pic}_{C'/\overline{k}}^0$ is an isomorphism.

point, then $X := C - P_\infty$ is a nontrivial k-form of $\mathbb{A}^1$. If C is smooth, C is k-isomorphic to $\mathbb{P}^1$ except possibly when $p = 2$ (cf. Lemma 2.3.3), in which case, if P_∞ is any point of C which is purely inseparable over k, $X := C - P_\infty$ gives also a nontrivial k-form of $\mathbb{A}^1$. In this chapter 2, we consider only k-*irrational* k-forms of $k(t)$.

2.5.1

Theorem. *Let K be a purely inseparable k-form of the rational function field $k(t)$ of height λ'. Then K is k-isomorphic to the function field of an affine plane k-curve defined by an equation of the type*

$$y^{p^{\lambda'}} = P(x), \quad P(x) \in k[x]$$

where $P(x) \notin k^p[x]$ and $P(x)$ has only simple factors over $\overline{k}$.

Proof. As explained in 2.2, we have $k(K^{p^{\lambda'}}) = k(x)$ for some $x \in K$, which implies that K is purely inseparable over $k(x)$. Take, on the other hand, a separating transcendence base $\{y\}$ of K/k. Then K is both separable and purely inseparable over $k(x, y)$, whence

$$K = k(x, y), \quad y^{p^{\lambda'}} = \frac{f(x)}{g(x)} \in k(x).$$

Meanwhile, $(yg(x))^{p^{\lambda'}} = f(x)g(x)^{p^{\lambda'}-1}$ and $k(x, y) = k(x, yg(x))$. Hence, replacing y by $yg(x)$, we may set

$$y^{p^{\lambda'}} = P(x) \in k[x], \quad K = k(x, y).$$

Since K/k is a regular extension, clearly $P(x) \notin k[x^p]$. So, the derivative $P'(x)$ is nonzero. Since we can freely substitute $y + a$ for y and $P(x) + a^{p^\lambda}$ for $P(x)$ in the above equation, we may arrange $P(x)$ so that $P(x)$ and $P'(x)$ have no common factors in $\overline{k}[x]$. Therefore, we may assume at our convenience that $P(x)$ has no multiple factors over $\overline{k}$. Finally, the fact that $P(x) \notin k^p[x]$ follows from Lemma 2.5.3 below. $\qquad\square$

2.5.2

Corollary. (a) *Let K be a purely inseparable k-form of $k(t)$. Then, among the fields K' such that $k \subseteq K' \subseteq K$, $K' \cong k(t)$ and K/K' is purely inseparable, there exists a unique maximal one, which is equal to $k(x)$ in the notation of 2.5.1.*

(b) *Let $X = \mathrm{Spec}\,(A)$ be a k-form of $\mathbb{A}^1$. Then, among the k-algebras A' such that $k \subseteq A' \subseteq A$, $A' \cong k[t]$ and A/A' is purely inseparable, there exists a maximal one.*

The result is essentially contained in Russell [78, Lemma 1.3, p.529]. However, we can quickly derive it from Theorem 2.5.1.

Proof. (a) Let K' be an intermediate field satisfying the above conditions. Let $[K : K'] = p^\alpha$. Then $k(K^{p^\alpha}) \subseteq K' \subseteq k(t)$, and hence, by Lüroth's theorem, we have

$$k^{p^{-\alpha}} \otimes K \cong k^{p^{-\alpha}}(t).$$

Hence $\alpha \geq \lambda' = \mathrm{ht}\,(K)$. But $k(K^{p^{\lambda'}}) = k(x^{p^{\lambda'}}, P(x)) = k(x)$, where the last equality follows since the extension $k(x) \supseteq k(x^{p^{\lambda'}}, P(x))$ is purely inseparable and separable (cf. 2.5.1), so $k(K^{p^\alpha}) = k(x^{p^{\alpha-\lambda'}})$. Thus we have

$$[K : k(K^{p^\alpha})] = [K : k(x)][k(x) : k(x^{p^{\alpha-\lambda'}})] = p^{\lambda'} \cdot p^{\alpha-\lambda'} = p^\alpha,$$

and $K' = k(K^{p^\alpha}) \subseteq k(x)$ follows.

(b) Let $K = k(X)$ be the quotient (fraction) field of A and let K' be that of A'. Let $\{\mathcal{O}_i \mid i \in I\}$ be the totality of places of K, i.e., the set of all DVRs of K which contain A. Then A' is determined by K' as

$$A' = \bigcap_{i \in I}(\mathcal{O}_i \cap K').$$

Note that $K' = k(x^{p^\delta})$ for some $\delta \geq 0$ since $A' \cong k[t]$. Take the smallest $\delta \geq 0$ for which

$$\bigcap_{i \in I}(\mathcal{O}_i \cap k(x^{p^\delta})) \cong k[t]$$

holds. Then we have clearly

$$A' \subseteq \bigcap_{i \in I}(\mathcal{O}_i \cap k(x^{p^\delta})) = \bigcap_{i \in I}(\mathcal{O}_i \cap k(K^{p^{\lambda'+\delta}})) = k[A^{p^{\lambda'+\delta}}] \cong k[t],$$

and $k[A^{p^{\lambda'+\delta}}]$ is the maximal subring satisfying the conditions. Hence $\lambda' + \delta = \mathrm{ht}\,(X)$, where $\lambda' = \mathrm{ht}\,(K)$. $\qquad\square$

2.5.3

Lemma. *Let K be the function field of an affine plane k-curve defined by an equation of the type*

$$y^{p^\lambda} = f(x),$$

where $f(x) \notin k[x^p], f(x) \in k^{p^i}[x] \setminus k^{p^{i+1}}[x]$ for $i \geq 0$. Then K is a k-form of the rational function field $k(t)$ of height $\leq \lambda - i$.

Proof. Set $u = x^{p^{-\lambda}}$. Then we can write the given equation as

$$y^{p^\lambda} = \left(f^{p^{-\lambda}}(u) \right)^{p^\lambda}$$

or $y = f^{p^{-\lambda}}(u) \in k^{p^{-\lambda+i}}[u]$. This shows that $k^{p^{-\lambda+i}} \otimes K$ is $k^{p^{-\lambda+i}}$-isomorphic to a subfield of $k^{p^{-\lambda+i}}(u, y) = k^{p^{-\lambda+i}}(u)$. By Lüroth's theorem, $k^{p^{-\lambda+i}} \otimes K \cong k^{p^{-\lambda+i}}(t')$ for some variable t' over $k^{p^{-\lambda+i}}$. $\qquad\square$

2.5.4

Lemma. *Let X be an affine plane curve defined by an equation*

$$y^{p^\lambda} = P(x), \quad P(x) \in k[x] \setminus k[x^p].$$

Then X has only one place at infinity.

Proof. Let $A = k[x, y]/(y^{p^\lambda} - P(x))$, and let $A' := k' \otimes A$ with $k' = k^{p^{-\lambda}}$. Then A' is identified with a k'-subalgebra of $k'[u]$, where $u = x^{p^{-\lambda}}$ and hence $y = P^{p^{-\lambda}}(u)$. Since $k'[u]$ is integral over A', $\operatorname{Spec}(A') = X \otimes k'$ has only one place at infinity. Since k' is purely inseparable over k, X has only one place at infinity. $\qquad\square$

2.6   Smoothness criterion of points

Let K be an algebraic function field of dimension one over k, and let $\mathcal{O}$ be a valuation ring (necessarily a DVR) of K corresponding to a point P on a complete k-normal model of K/k. In this section, we are concerned with the condition for $\mathcal{O}$ (or P) to be smooth over k.

2.6.1

Lemma. *The following conditions are equivalent.*

(1) $\mathcal{O}$ is smooth, i.e., $\mathcal{O}$ is geometrically regular.

(2) $k^{p^{-1}} \otimes \mathcal{O}$ is $k^{p^{-1}}$-normal.
(3) $k[\mathcal{O}^p]$ is k-normal.
(4) $k[\mathcal{O}^p] = \mathcal{O} \cap k[K^p]$.

Proof. (1) $\Leftrightarrow$ (2): We refer to [EGA IV (22.5.8)]. (2) $\Leftrightarrow$ (3): Note that each property of the k-algebra $(k, \varphi) \otimes_k \mathcal{O}$ is equivalent to the corresponding property of the $k^{p^{-1}}$-algebra $k^{p^{-1}} \otimes_k \mathcal{O}$. Moreover, $F_{\mathcal{O}} : (k, \varphi) \otimes_k \mathcal{O} \to \mathcal{O}$ defined in 2.2 sends $(k, \varphi) \otimes_k \mathcal{O}$ isomorphically onto the k-algebra $k[\mathcal{O}^p]$. Thus we have the equivalence. (3) $\Leftrightarrow$ (4): This follows from the fact that $k[\mathcal{O}^p]$ is a local ring of dimension one with the quotient field $k[K^p]$. If $k[\mathcal{O}^p]$ is k-normal, then $k[\mathcal{O}^p]$ is a DVR of $k[K^p]$ contained in $\mathcal{O} \cap k[K^p]$. Hence $k[\mathcal{O}^p] = \mathcal{O} \cap k[K^p]$. Since $\mathcal{O} \cap k[K^p]$ is a DVR of $k[K^p]$, the converse is clear. □

2.6.2

With the notations in 2.6.1, let $\mathcal{O}_0 = \mathcal{O} \cap k[K^p]$. Let $\mathfrak{m}, \mathfrak{m}_0$ be the maximal ideals of $\mathcal{O}, \mathcal{O}_0$, respectively, and let $\mathfrak{k} := \mathcal{O}/\mathfrak{m}, \mathfrak{k}_0 := \mathcal{O}_0/\mathfrak{m}_0$. If $\mathcal{O}_1$ is a DVR dominating a DVR $\mathcal{O}_2$, i.e., $\mathcal{O}_1 \supseteq \mathcal{O}_2$ and $\mathfrak{m}_1 \cap \mathcal{O}_2 = \mathfrak{m}_2$, we denote, as customary, by $e(\mathcal{O}_1 : \mathcal{O}_2)$ and $f(\mathcal{O}_1 : \mathcal{O}_2)$ the ramification index, i.e., $v_1(t_2)$ for a uniformisant t_2 of $\mathcal{O}_2$, where v_1 is the normalized valuation of $\mathcal{O}_1$ and the relative degree, i.e., the degree of the residue field extension.

We note that $\mathcal{O}_0 = k[\mathcal{O}^p]$ (cf. 2.6.1, (4)) if and only if the local ring $k[\mathcal{O}^p]$ has the same residue class field as $\mathcal{O}_0$ and contains a uniformisant of the DVR $\mathcal{O}_0$. Another obvious, but useful fact is that if $\mathcal{O}$ is k-rational, i.e., $\mathfrak{k} = k$, then $\mathcal{O}$ is geometrically regular.

Lemma. (a) *If $\mathfrak{k}_0 = k$ and $f := f(\mathcal{O} : \mathcal{O}_0) = 1$, then $\mathcal{O}$ is smooth.*
(b) *If $\mathfrak{k}_0 \supsetneq k$ and $f = 1$, then $\mathcal{O}$ is not smooth.*

Proof. (a) Since $\mathfrak{k} = k$ by the hypothesis, $\mathcal{O}$ is smooth by the above remark.

(b) One can readily see that the residue field of the local ring $k[\mathcal{O}^p]$ is $k[\mathfrak{k}^p]$ which is equal to $k[\mathfrak{k}_0^p]$ by the hypothesis $f = 1$ in our case. Since $k[\mathfrak{k}_0^p] \subsetneq \mathfrak{k}_0$ as $\mathfrak{k}_0$ is purely inseparable over k and $k \subsetneq \mathfrak{k}$ by the condition, we have $k[\mathcal{O}^p] \subsetneq \mathcal{O}_0$. Hence $\mathcal{O}$ is not smooth by 2.6.1. □

2.7　Case of height ≤ 1

Let $K, \mathcal{O}$ be the same as in 2.6, and let $\mathcal{O}_0 := \mathcal{O} \cap k[K^p]$. We now impose an *additional condition* that $K = k(x, y)$ is the function field of an affine plane k-curve

$$y^p = a_0 + a_1 x + \cdots + a_n x^n = P(x), \tag{2.2}$$

where $P(x) \in k[x] \setminus k[x^p]$. By 2.5.3, K is a k-form of $k(t)$ of height ≤ 1, and $k[K^p] = k(x)$. We continue our study of seeking a smoothness condition for $\mathcal{O}$ under the additional hypothesis.

2.7.1

Lemma. *Assume that $k = k_s$, where k_s is the separable algebraic closure of k in $\overline{k}$, and that $\mathcal{O}_0$ is non-k-rational, i.e., $k \subsetneq \mathfrak{k}_0$. Then $\mathcal{O}$ is smooth if and only if $e := e(\mathcal{O} : \mathcal{O}_0) = 1$.*

Proof. Note that $[K : k[K^p]] = p = ef$. If $\mathcal{O}$ is smooth, then $f \neq 1$ by Lemma 2.6.2, (b) so that $f = p$ and $e = 1$. As for the converse, note that $\mathcal{O}_0$ is realized as a localization of $k[x]$ at a prime ideal of the type $(x^{p^\nu} - d)$ with $d \in k \setminus k^p$ and $\nu > 0$, where the prime element $x^{p^\nu} - d$ of $\mathcal{O}_0$ belongs to $k[\mathcal{O}^p]$. By a remark before Lemma 2.6.2, $\mathcal{O}$ is smooth if and only if $k[\mathcal{O}^p]$ has the same residue field as $\mathcal{O}_0$, i.e., if and only if $k[\mathfrak{k}^p] = k[d^{p^{-\nu}}]$. The last condition holds in the present case because $e = 1$ and $f = [\mathfrak{k} : \mathfrak{k}_0] = p$. In fact, let $\alpha = x \pmod{\mathfrak{m}}$ and $\beta = y \pmod{\mathfrak{m}}$. Then $\mathfrak{k} = k(\alpha, \beta)$ and $\mathfrak{k}_0 = k(\alpha)$. Further, $k[\mathfrak{k}^p] = k(\alpha^p, \beta^p)$, where $\beta^p = P(\alpha)$. Since $f = p$, we have inclusions $k(\alpha^p, \beta^p) \subseteq k(\alpha) \subsetneq k(\alpha, \beta)$ and $[k(\alpha, \beta) : k(\alpha^p, \beta^p)] = p$, whence $k(\alpha^p, \beta^p) = k(\alpha)$. $\qquad\square$

2.7.2

Lemma. *If $P(x)$ is divisible by $(x^{p^h} - c)$ but not by $(x^{p^h} - c)^p$, where $h > 0$ and $c \in k \setminus k^p$, then the point $(x = c^{p^{-h}}, y = 0)$ on the curve $y^p = P(x)$ is not k-smoothable, i.e., the normalization of the local ring is not smooth.*

Proof. Let $\mathcal{O}_0$ be the localization of $k[x]$ by $(x^{p^h} - c)$. Then the unique extension of this valuation ring to $K = k(x, y)$ is the normalization of the local ring of the plane curve $y^p = P(x)$ at $(x = c^{p^{-h}}, y = 0)$. Call this extension $\mathcal{O}$ and its valuation v. Then, since $\mathcal{O}$ is smooth if and only if $\mathcal{O} \otimes k_s$ is smooth, we may assume that $k = k_s$. We have

$$v(y^p) = p \cdot v(y) = (\text{the multiplicity of } x^{p^h} - c \text{ in } P(x)) \cdot v(x^{p^h} - c).$$

Since the first factor of the right side is not divisible by p, the second factor must be divisible by p. Hence $e = v(x^{p^h} - c) = p$ follows. By 2.7.1, $\mathcal{O}$ is then singular, i.e., $\mathcal{O}$ is not smooth. $\qquad\square$

2.7.3

Lemma. *Assume that $\mathcal{O}_0 = k[x]_{(x-c)}$ for some $c \in k$. Then $\mathcal{O}$ is singular if and only if $P'(c) = 0$ and $P(c) \notin k^p$.*

Proof. *Only if* part. If $P'(c) \neq 0$ then $\mathcal{O}$ is smooth by the Jacobian criterion. If $P'(c) = 0$ but $P(c) \in k^p$, then one can write the equation as

$$y^p - d^p = (x - c)^m Q(x)$$

for some $d \in k$ and some $Q(x) \in k[x]$, $Q(c) \neq 0$ and $m > 1$. The place $\mathcal{O}$ is determined by the point $(x = c, y = d)$ on the curve. By the usual change of variables, we may assume without loss of generality that $1 < m < p$. Find then positive integers s, t such that $ms - pt = 1$. Then the element $(y - d)^{ps}/(x - c)^{pt} = (x - c)Q(x)^s$ belongs to $k[\mathcal{O}^p]$ and its order under the valuation of $\mathcal{O}_0$ is 1. Hence $k[\mathcal{O}^p]$ contains a uniformisant of $\mathcal{O}_0$, and $\mathcal{O}$ is therefore smooth by 2.6.2 since $\mathcal{O}_0 = k[\mathcal{O}^p]$ and $\mathfrak{k} = \mathfrak{k}_0 = k$.

If part. Then we can write our equation as

$$y^p - d = (x - c)^m Q(x),$$

where $1 < m$, $Q(c) \neq 0$ and $d \in k \setminus k^p$, and $\mathcal{O}$ is the k-normalization of the local ring of this plane curve at the point $(x = c, y = d^{1/p})$. Since this local ring is $k[x, y]_{(x-c)}$, it is already k-normal because the maximal ideal is principal and hence equal to $\mathcal{O}$. If $\mathcal{O}$ were smooth, $k^{1/p} \otimes \mathcal{O} = k^{1/p} \otimes k[x, y]_{(x-c)}$ would be $k^{1/p}$-normal by 2.6.1, which is not the case by the Jacobian criterion of smoothness. Hence $\mathcal{O}$ is singular. $\qquad\square$

2.7.4

We next consider the place at infinity.

Lemma. *Let K be the function field of the affine plane k-curve defined by the equation (2.2) in 2.7, which is further subject to the condition that $a_n \neq 0$ and $a_n \notin k^p$ if $p \mid n$. Then the unique place $\mathcal{O} \subset K$ at infinity of the plane curve is singular if and only if $p \mid n$ and $a_{n-1} = 0$.*

Proof. *If* part. Suppose that $n = pr$. Let $w = \frac{y}{x^r}$ and $z = \frac{1}{x}$. Then the equation becomes

$$w^p = a_n + a_{n-2}z^2 + \cdots + a_1 z^{n-1} + a_0 z^n.$$

Hence $k(x, y) = k(z, w)$ and $\mathcal{O}$ is the normalization of the local ring of the above plane curve at $(z = 0, w = a_n^{1/p})$. By 2.7.3, we know that such $\mathcal{O}$ is singular.

Only if part. Suppose first that $p \mid n$ and $a_{n-1} \neq 0$. Then, with the same w and z as above, the original plane curve (2.2) is birational to a plane curve

$$w^p = a_n + a_{n-1}z + \cdots + a_1 z^{n-1} + a_0 z^n \quad (a_{n-1} \neq 0),$$

which is, in fact, smooth at $(z = 0, w = a_n^{1/p})$. Hence $\mathcal{O}$ is smooth. Next suppose that $p \nmid n$. Write $n = pr - s$ with $r > 0$ and $0 < s < p$. Then, with the same w and z, the plane curve (2.2) is k-birational to

$$w^p = z^s(a_n + a_{n-1}z + \cdots + a_1 z^{n-1} + a_0 z^n)$$

and $\mathcal{O}$ is the k-normalization of the local ring at $(z = 0, w = 0)$, which is smooth by 2.7.3. $\quad\square$

2.7.5

We apply the foregoing results to describe the function field of a k-form of $\mathbb{A}^1$ of height 1. We may start off with an affine plane k-curve

$$y^p = a_0 + a_1 x + \cdots + a_n x^n = P(x) \in k[x]$$

and assume that

(i) $P(x)$ has only simple factors (cf. 2.5.1), and
(ii) $a_n \neq 0$ and $a_n \notin k^p$ if $p \mid n$ (cf. 2.7.4).

Then the function field $K = k(x, y)$ gives a nontrivial k-form of $\mathbb{A}^1$ if and only if K has a unique singular place unless $p = n = 2$. In the last case, the complete k-normal model of K is smooth.

Case where $p \mid n$ and $a_{n-1} = 0$. Then, by 2.7.4, the place at infinity is singular, so all places at finite distance must be smooth, or equivalently, all points of the given affine plane curve must be k-smoothable. Let $Q = (x = c, y = P(c)^{1/p})$ with $c, P(c) \in \overline{k}$ be a point on the curve. If $P'(c) \neq 0$ then Q is smooth. If $P'(c) = 0$ and $c \in k_s$, then Q is smooth if and only if

$P(c) \in k_s^p$, i.e., Q is k_s-rational (cf. 2.7.3). If $P'(c) = 0$ and $c \notin k_s$, then a criterion of k-smoothability for the point Q is given by 2.6.2 or 2.7.1.

Case $p \nmid n$ or $p \mid n$ and $a_{n-1} \neq 0$. We apply Lemmas 2.6.2, 2.7.1 and 2.7.3 to ensure that one and only one place of K at finite distance of the above curve is singular.

2.8   Forms of $\mathbb{A}^1$ with arithmetic genus 0 or 1

In this section, we give a complete classification of all k-forms of $\mathbb{A}^1$ with arithmetic genus 0 or 1 under the assumption that there exists a k-rational point on each form.

2.8.1

First of all, consider the case of arithmetic genus zero. As for the existence of a k-rational point in this case, we note that if $p \geq 3$ all k-forms have k-rational points (cf. 2.3.3). Moreover, by the same Lemma 2.3.3 or its proof, we know that a nontrivial k-form of $\mathbb{A}^1$ with arithmetic genus zero and without k-rational points exists only in the case where $p = 2$ and the form is embedded as a dense open set into a smooth conic in $\mathbb{P}^2$. Hence, for such a k-form X of $\mathbb{A}^1$, we have $\mathrm{ht}\,(X) = \mathrm{ht}\,(k(X)) = 1$.

In the case under consideration, a k-form of $\mathbb{A}^1$ with arithmetic genus zero has the function field k-isomorphic to $k(t)$. So, we call such a form a *k-rational k-form of $\mathbb{A}^1$.*

Let a be an element of $k \setminus k^p$ and let n be a positive integer. Let t be an inhomogeneous variable of $\mathbb{P}^1$ and let P_∞ be a point defined by $t^{p^n} = a$. Let $\Phi : \mathbb{P}^1 \to \mathbb{P}^{p^n}$ be the embedding given by

$$t \longmapsto (1, t, \ldots, t^{p^n - 1}, t^{p^n} - a).$$

We denote by $X_{a,n}$ the image $\Phi(\mathbb{P}^1 - P_\infty)$. Then we have:

Theorem. (1) *Every k-rational k-form of $\mathbb{A}^1$ is k-isomorphic to either $\mathbb{A}^1$ or $X_{a,n}$ for suitable $a \in k \setminus k^p$ and $n \in \mathbb{Z}^+$.*

(2) *$X_{a,n}$ is a k-rational k-form of $\mathbb{A}^1$ of height n not k-isomorphic to $\mathbb{A}^1$.*

(3) *$X_{a,n}$ is k-isomorphic to $X_{b,m}$ if and only if $m = n$ and there exist $\alpha, \beta, \gamma, \delta$ in k^{p^n} such that $\alpha\delta - \beta\gamma \neq 0$ and $b = (\alpha a + \beta)/(\gamma a + \delta)$.*

Proof. (1) Let X be a k-rational k-form of $\mathbb{A}^1$ and let C be a complete k-normal model of X. By the hypothesis, C is a curve with $p_a(C) = 0$

and a k-rational point. Hence C is k-isomorphic to $\mathbb{P}^1$ and $P_\infty := C - X$ is rational over a purely inseparable algebraic extension of k (cf. 2.4.1). Choose a variable t of C ($\cong \mathbb{P}^1$) so that t is finite at P_∞. Suppose that P_∞ is given by $t^{p^n} = a$ with $a \in k \setminus k^p$. If $n = 0$, P_∞ is a k-rational point, whence X is k-isomorphic to $\mathbb{A}^1$. Assume that $n > 0$. The divisor $p^n P_\infty$ is a prime k-rational cycle on C of degree p^n. Hence $h^0(C, \mathcal{O}_C(p^n P_\infty)) = p^n + 1$. Since

$$\left\{ 1, \frac{1}{t^{p^n} - a}, \frac{t}{t^{p^n} - a}, \dots, \frac{t^{p^n - 1}}{t^{p^n} - a} \right\}$$

is a basis of the complete linear system $|p^n P_\infty|$, the embedding $\Phi : C \to \mathbb{P}^{p^n}$ defined by $|p^n P_\infty|$ is given by

$$t \longmapsto (1, t, \dots, t^{p^n - 1}, t^{p^n} - a).$$

Then X is isomorphic to $\Phi(C - P_\infty)$, which is $X_{a,n}$.

(2) Let $\Phi : \mathbb{P}^1 \to \mathbb{P}^{p^n}$ be the embedding which defines $X_{a,n}$, and let P_∞ be a point of $\mathbb{P}^1$ given by $t^{p^n} = a$. Then the point $\Phi(P_\infty)$ is not k-rational, but rational over $k(a^{p^{-n}})$. Therefore, $X_{a,n}$ is a k-rational k-form of $\mathbb{A}^1$, not isomorphic to $\mathbb{A}^1$. By the choice of a, P_∞ is not rational over $k^{p^{-n+1}}$, whence $X_{a,n}$ has height n.

(3) If $X_{a,n}$ is k-isomorphic to $X_{b,m}$, a k-isomorphism $\Psi : X_{a,n} \to X_{b,m}$ extends to a k-isomorphism $\overline{\Psi}$ between their complete k-normal models $\overline{X}_{a,n}$ and $\overline{X}_{b,m}$, which sends the point ($t^{p^n} = a$) of $\overline{X}_{a,n}$ to the point ($t^{p^m} = b$) of $\overline{X}_{b,m}$. If $\overline{X}_{a,n}$ and $\overline{X}_{b,m}$ are identified with $\mathbb{P}^1_k$, $\overline{\Psi}$ is given by

$$t \longmapsto t' = \frac{\alpha' t + \beta'}{\gamma' t + \delta'}$$

with $\alpha', \beta', \gamma', \delta' \in k$ and $\alpha' \delta' - \beta' \gamma' \neq 0$. Then, setting $\alpha = \alpha'^{p^n}, \dots, \delta = \delta'^{p^n}$, we have $(\alpha a + \beta)/(\gamma a + \delta) = b$. Clearly, we have $m = n$. The converse is obvious. $\qquad\square$

2.8.2

We next consider the case where the arithmetic genus is equal to 1. It is known that a complete k-normal curve C of arithmetic genus 1 having a k-rational point is k-isomorphic to a projective plane curve defined by an equation of type [6]

$$x_1^2 x_2 + \lambda x_0 x_1 x_2 + \mu x_1 x_2^2 = x_0^3 + \alpha x_0^2 x_2 + \beta x_0 x_2^2 + \gamma x_2^3 \qquad (2.3)$$

---

[6]Let P_0 be a k-rational point of C. Then $\Phi_{|3P_0|} : C \to \mathbb{P}^2$ is a closed immersion to a curve of degree 3. Its equation is obtained by choosing bases $\{1, x\}$ and $\{1, x, y\}$ for $|2P_0|$ and $|3P_0|$ respectively and finding a relation in the generating system $\{1, x, x^2, x^3, y, xy, y^2\}$ of $|6P_0|$. Finally find a homogeneous form of the relation for x, y.

with $\lambda, \mu, \alpha, \beta, \gamma \in k$, where we assume that $\lambda = \mu = \alpha = 0$ if $p \neq 2, 3$ and that $\lambda = \mu = 0$ if $p \neq 2$.

Theorem. *Let C be a plane cubic given by the above equation (2.3). Assume that C is k-normal. Assume further that C has a one-place singular point. Then the following assertions hold.*

(1) *The characteristic p of the field k is either 2 or 3.*
(2) *If $p = 3$, the equation (2.3) is reduced to $x_1^2 x_2 = x_0^3 + \gamma x_2^3$, where $\gamma \notin k^3$.*
(3) *If $p = 2$, the equation (2.3) is reduced to $x_1^2 x_2 = x_0^3 + \beta x_0 x_2^2 + \gamma x_2^3$, where $\beta \notin k^2$ or $\gamma \notin k^2$.*

Proof. (1) The curve C has arithmetic genus 1, but the normalization of $C \otimes \overline{k}$ is rational. Hence the genus drops by 1. Meanwhile, there is a famous result of Tate [92] which says that the genus drop occurs by a multiple of $(p-1)/2$. Hence 1 is a multiple of $(p-1)/2$. This occurs only in the case $p = 3$ or $p = 2$.

(2) Note that if $x_2 = 0$ then there is only one point P_0 on C which is defined by $(x_0, x_1, x_2) = (0, 1, 0)$. Let $u = x_0/x_1$ and $z = x_2/x_1$. Then the equation (2.3) is written as

$$z = u^3 + \alpha u^2 z + \beta u z^2 + \gamma z^3,$$

which is written also as

$$z(1 - \alpha u^2 - \beta u z - \gamma z^2) = u^3.$$

Since the point P_0 is defined by $(u, z) = (0, 0)$, the curve C is smooth at P_0. Hence, in order to find a singular point, we may assume that $x_2 \neq 0$. Let $x = x_0/x_2$ and $y = x_1/x_2$. The equation (2.3) is written as

$$y^2 = x^3 + \alpha x^2 + \beta x + \gamma.$$

By the Jacobian criterion, a singular point, say P_∞, must satisfy the condition $2y = -\alpha x + \beta = 0$. By the assumption, the singular point P_∞ exists and is not k-rational. If $\alpha \neq 0$, then P_∞ is given as $(x, y) = (\beta/\alpha, 0)$, which is a k-rational point. So, $\alpha = 0$ and hence $\beta = 0$. Then the equation (2.3) is $y^2 = x^3 + \gamma$, where $\gamma \notin k^3$, for otherwise P_∞ is k-rational.

(3) Suppose $p = 2$. The computation is more or less similar to the case (2). If $x_2 = 0$, there is only one point P_0 on C, which is given by $(x_0, x_1, x_2) = (0, 1, 0)$, and the point P_0 is k-smooth. If $x_2 \neq 0$, set $x = x_0/x_2$ and $y = x_1/x_2$. Then the equation (2.3) is written as

$$y^2 + \lambda xy + \mu y = x^3 + \alpha x^2 + \beta x + \gamma.$$

By the Jacobian criterion, the singular point P_∞, which we assume exists on C, must satisfy the condition $\lambda x + \mu = \lambda y + x^2 + \beta = 0$. If $\lambda \neq 0$, then P_∞ is defined by $(x, y) = (\mu \lambda^{-1}, \mu^2 \lambda^{-3} + \beta \lambda^{-1})$, whence P_∞ is k-rational. So, $\lambda = \mu = 0$, and the equation (2.3) is

$$y^2 = x^3 + \alpha x^2 + \beta x + \gamma.$$

By replacing x by $x + \alpha$, the equation (2.3) is reduced to the one with $\alpha = 0$. So, write it anew as

$$y^2 = x^3 + \beta x + \gamma.$$

Then the singular point P_∞ is given by $(x, y) = (\beta^{1/2}, \gamma^{1/2})$, which is k-rational if and only if $\beta^{1/2} \in k$ and $\gamma^{1/2} \in k$. $\qquad\square$

2.8.3

Let C be anew the plane cubic curve defined by one of the above equations and let P_∞ be the singular point of C. Then $P_\infty = (-\gamma^{1/3}, 0, 1)$ in the first case and $P_\infty = (\beta^{1/2}, \gamma^{1/2}, 1)$ in the second case. Let $X := C - P_\infty$. Then X is a k-form of $\mathbb{A}^1$. Furthermore, by Theorem 2.4.3, X has a k-group scheme structure. The equation of C (or X) can be written as follows to elucidate the k-group structure:

(2)$'$ $p = 3$ and $y^3 = x - \gamma x^3$ with $\gamma \notin k^3$,
(3)$'$ $p = 2$ and $y^4 = x + \beta x^2 + \gamma^2 x^4$ with $\beta \notin k^2$ or $\gamma \notin k^2$.

It is easy to see that (2)$'$ is obtained from (2) by setting $y = x_0/x_1$ and $x = x_2/x_1$. However it is not straightforward to see that (3)$'$ corresponds to (3). To see this correspondence, write the equation (3)$'$ in a homogeneous form

(3)$''$ $$T^4 + UV^3 + \beta U^2 V^2 + \gamma^2 U^4 = 0$$

by setting $y = T/V$ and $x = U/V$. The singular point of the curve (3)$''$ is given by $(T, U, V) = (\gamma^{1/2}, 1, 0)$. Let $t = T/U$ and $v = V/U$. Then (3)$''$ becomes

$$t^4 + v^3 + \beta v^2 + \gamma^2 = 0.$$

Hence we have

$$\left(\frac{(t^2 + \gamma)}{v} \right)^2 = v + \beta.$$

Let $u = (t^2 + \gamma)/v$. Then we have $t^2 = u^3 + \beta u + \gamma$. This is the equation obtained from (3) by setting $t = x_1/x_2$ and $u = x_0/x_2$.

The k-group scheme X is a unipotent k-group scheme called a *k-group of Russell type*, which we are going to study in details in the next chapter. Summarizing the above arguments, we have the following theorem:

Theorem. *Every k-form of $\mathbb{A}^1$ of arithmetic genus one and with a k-rational point is k-isomorphic to one of the following k-groups of Russell type:*

(1) $p = 3$ *and* $y^3 = x - \gamma x^3$ *with* $\gamma \notin k^3$,
(2) $p = 2$ *and* $y^4 = x + \beta x^2 + \gamma^2 x^4$ *with* $\beta \notin k^2$ *or* $\gamma \notin k^2$.

2.8.4

Let X be a k-form of $\mathbb{A}^1$ of arithmetic genus one but having not necessarily a k-rational point. Then, as shown in 3.3.1, X is a principal homogeneous space of one of the k-groups of Russell type listed in Theorem 2.8.2.

Chapter 3

Groups of Russell type

We owe to Russell [78] most results treated in this chapter. For a k-group scheme G, we denote by $\overline{G}$ the underlying k-scheme.

3.1 Forms of the additive group

We denote by G_a the *additive k-group scheme*, whose underlying scheme is $\operatorname{Spec} k[t]$ and group laws are given by the following k-homomorphisms:

$$\Delta(t) = t \otimes 1 + 1 \otimes t \quad \text{comultiplication,}$$
$$\iota(t) = -t \qquad\qquad \text{coinverse,}$$
$$\varepsilon(t) = 0 \qquad\qquad \text{counit.}$$

Equivalently, G_a is a k-group scheme which represents the k-group functor:

$$A \longmapsto A^+,$$

where A is a k-algebra and A^+ denotes A with the group law induced by the addition of A. A k-group scheme G is called a *k-form of G_a* if the k'-group scheme $G \otimes_k k'$ is k'-isomorphic to G_a for some finite algebraic extension k' of k. A k-form of G_a is called a *k-group of Russell type*.

3.1.1

Lemma. (1) $\mathbb{A}^1_k$ *carries a unique k-group scheme structure with a given k-rational point as the origin, which is $G_{a,k}$.*

(2) *Let X be a k-form of $\mathbb{A}^1$. Then any k-group scheme G with the underlying scheme X is a k-form of G_a. The k-group scheme structure on X (if it exists) is unique up to the choice of the origin.*

Proof. (1) Write $\mathbb{A}^1_k = \operatorname{Spec} k[t]$. A k-group scheme structure on $\mathbb{A}^1_k$ is given by k-algebra homomorphisms $\Delta, \iota, \varepsilon$ as above which are determined

39

by polynomials

$$\Delta(t) = f(t_1, t_2) \in k[t_1, t_2], \quad \iota(t) = g(t) \in k[t]$$
$$\varepsilon(t) = \alpha \in k$$

which are subject to the relations

$$f(f(t_1, t_2), t_3) = f(t_1, f(t_2, t_3)), \quad f(t, g(t)) = f(g(t), t) = \varepsilon(t)$$
$$f(t, \varepsilon(t)) = f(\varepsilon(t), t) = t.$$

By a change of variable, we may assume that $\varepsilon(t) = 0$. Since $g(g(t)) = t$, we can write $g(t) = \beta t$ with $\beta \in k$, $\beta^2 = 1$. Write

$$f(t_1, t_2) = \sum_{\lambda, \mu} a_{\lambda\mu} t_1^\lambda t_2^\mu$$

and let (m, n) be the largest pair with respect to the lexicographic order in the set of all pairs (λ, μ) with $a_{\lambda\mu} \neq 0$. Then, comparing the monomials in the equality $f(f(t_1, t_2), t_3) = f(t_1, f(t_2, t_3))$, we have

$$(a_{mn})^{m+1} t_1^{m^2} t_2^{mn} t_3^n = (a_{mn})^{n+1} t_1^m t_2^{mn} t_3^{n^2}.$$

Hence $m = 0$ or 1. However, we can easily exclude the case $m = 0$. Thus $m = 1$. Similarly, we have $n = 1$ for the largest pair (m, n) with respect to the anti-lexicographic order. Therefore, we may write

$$f(t_1, t_2) = a + bt_1 + ct_2 + dt_1 t_2, \quad a, b, c, d \in k.$$

Since $f(t, g(t)) = 0$, we have $d = 0$ and $b + \beta c = 0$. By the equality $f(f(t_1, t_2), t_3) = f(t_1, f(t_2, t_3))$, we obtain $b^2 = b$, $c^2 = c$ and $bc \neq 0$. Hence $b = c = 1$ and $\beta = -1$. Since $f(t, \varepsilon(t)) = t$ and $\varepsilon(t) = 0$, it follows that $a = 0$. Note that once a variable t is chosen the group structure, which is determined by Δ, ι and ε, is uniquely determined. Hence the given k-group scheme structure makes $\mathbb{A}^1$ the additive k-group scheme $G_{a,k}$.

(2) Let G be a k-group scheme such that $\overline{G} = X$, and let k' be a purely inseparable algebraic extension of k such that $X \otimes k'$ is k'-isomorphic to $\mathbb{A}^1_{k'}$. Then $G \otimes k'$ is a k'-group scheme and by the assertion (1), $G \otimes k'$ is isomorphic to $G_{a,k'}$. If we fix a k'-isomorphism $\psi : X \otimes k' \xrightarrow{\sim} \mathbb{A}^1_{k'}$, the group law on $G \otimes k'$ is the one transferred from the group law on $G_{a,k'}$ by ψ^{-1}. This implies that there is a unique k-group scheme structure on X up to the choice of the origin. $\qquad \square$

3.1.2

Lemma. *Let G be a k-form of G_a, and let n be a nonnegative integer. Then the following conditions are equivalent.*

(1) *n is the smallest nonnegative integer such that $G \otimes k^{p^{-n}} \cong G_{a,k^{p^{-n}}}$.*

(2) *n is the smallest nonnegative integer such that $\Theta^n G \cong G_{a,k}$ (see 2.2 for Θ^n).*

(3) *$n = \operatorname{ht}(\overline{G})$.*

Proof. (1) $\Leftrightarrow$ (3): If $G \otimes k^{p^{-n}} \cong G_{a,k} \otimes k^{p^{-n}}$, we have clearly $\operatorname{ht}(\overline{G}) \leq n$. Conversely, if $n \geq \operatorname{ht}(\overline{G})$, Lemma 3.1.1 implies that $G \otimes k^{p^{-n}} \cong G_{a,k} \otimes k^{p^{-n}}$.

(2) $\Leftrightarrow$ (3): It suffices to note that $\Theta^n \overline{G} \cong \mathbb{A}_k^1$ if and only if $\overline{G} \otimes k^{p^{-n}} \cong \mathbb{A}_k^1 \otimes k^{p^{-n}}$. $\square$

The integer n defined in Lemma 3.1.2 is called the *height* of G and denoted by $\operatorname{ht}(G)$.

3.2 k-groups of Russell type as subgroups of G_a^2

Let $\mathfrak{R} := \operatorname{Hom}_{k-gr}(G_a, G_a)$ be the ring of all k-group endomorphisms of G_a. If we identify G_a with $\operatorname{Spec} k[t]$ having group laws as given in 3.1, every k-group scheme endomorphism in $\mathfrak{R}$ is given by a polynomial $f(t) \in k[t]$ such that

$$f(t_1 + t_2) = f(t_1) + f(t_2).$$

Such a polynomial $f(t)$ is written in the form

$$f(t) = a_0 t + a_1 t^p + \cdots + a_m t^{p^m}$$

and called a *p-polynomial*. Conversely, it is clear that a p-polynomial gives rise to an element of $\mathfrak{R}$. On the other hand, since ΘG_a is identified with $G_{a,k}$, $F := F_{G_a}$ is an element of $\mathfrak{R}$. Noting that a p-polynomial $f(t)$ of the above type corresponds to an element

$$a_0 \cdot 1 + a_1 F + \cdots + a_m F^m,$$

where 1 is the identity automorphism, we know that the ring $\mathfrak{R}$ is a ring $k[F]$ of noncommutative polynomials with relations $Fa = a^p F$ for $a \in k$. For $\tau = a_0 + a_1 F + \cdots + a_m F^m$ and an integer n, we set

$$\tau^{(n)} = a_0^{p^n} + a_1^{p^n} F + \cdots + a_m^{p^n} F^m.$$

Then we have $F^n \tau = \tau^{(n)} F^n$. We define the ring $\widehat{\mathfrak{R}} = k[[F]]$ of power series in a similar fashion. We set $\mathfrak{R}^* = \mathfrak{R} \setminus \mathfrak{R}F$ and $\mathfrak{R}^{**} = 1 + \mathfrak{R}F$.

Let $G = \operatorname{Spec} A$ be a k-group scheme with the comultiplication $\Delta_G :$ $A \to A \otimes A$. Then $\operatorname{Hom}_{k-gr}(G, G_a)$ is identified with the set

$$\{f : G \to G_a \mid f(gg') = f(g)f(g'),\ g, g' \in G(R),\ \text{for all } k\text{-algebra } R\}$$
$$= \{f^* : k[t] \to A \mid k\text{-algebra hom. such that } (f^* \otimes f^*) \cdot \Delta = \Delta_G \cdot f^*\}$$
$$= \{a \in A \mid \Delta_G(a) = a \otimes 1 + 1 \otimes a\}.$$

3.2.1

Theorem (Russell [78, Th.2.1]). *Let G be a k-form of G_a. Then G is k-isomorphic to a subgroup of $G_a^2 = \operatorname{Spec} k[x, y]$ defined by an equation of the type*

$$y^{p^n} = a_0 x + a_1 x^p + \cdots + a_m x^{p^m}, \quad a_0 \neq 0, \quad n \geq \operatorname{ht}(G).$$

Equivalently, G is a fiber product

$$
\begin{array}{ccc}
G & \xrightarrow{\ \xi\ } & G_a \\
{\scriptstyle \eta}\big\downarrow & & \big\downarrow{\scriptstyle \tau} \\
G_a & \xrightarrow{\ F^n\ } & G_a
\end{array}
$$

where $\tau = a_0 + a_1 F + \cdots + a_m F^m \in \mathfrak{R}^*$ and ξ, η are the natural projections. Conversely, a k-group scheme defined in the above way is a k-form of G_a.

Proof. Let $G = \operatorname{Spec} A$, let $\Delta : A \to A \otimes A$ be the comultiplication of G and let n be an integer such that $n \geq \operatorname{ht}(G)$. Then,

$$\overline{\Delta} := (k, \varphi^n) \otimes \Delta\ :\ (k, \varphi^n) \otimes_k A \to\ (k, \varphi^n) \otimes_k (A \otimes A)$$
$$\cong ((k, \varphi^n) \otimes_k A) \otimes_k ((k, \varphi^n) \otimes_k A)$$

induces a k-group scheme structure on $\Theta^n G$. Since $n \geq \operatorname{ht}(G)$, $\Theta^n G \cong G_a$ and $(k, \varphi^n) \otimes A \cong k[t]$ with $\overline{\Delta}(t) = t \otimes 1 + 1 \otimes t$. Write $t = \sum_{i=1}^{r} a_i \otimes y_i$ with $a_i \in k$ and $y_i \in A$. We choose the a_i in such a way that the a_i are linearly independent over k^{p^n}. Then the a_i form a free basis of $(k, \varphi^n) \otimes_k (A \otimes A)$ over $1 \otimes_k (A \otimes A)$. Since

$$\overline{\Delta}(t) = t \otimes 1 + 1 \otimes t = \sum_i (a_i \otimes y_i \otimes 1 + a_i \otimes 1 \otimes y_i)$$
$$= \sum_i \overline{\Delta}(a_i \otimes y_i) = \sum_i a_i \otimes \Delta(y_i),$$

we have $\Delta(y_i) = y_i \otimes 1 + 1 \otimes y_i$ for every i. Hence y_i (resp. $1 \otimes y_i$) as an element of $(k, \varphi^n) \otimes_k A$ defines a k-group scheme homomorphism

$\eta_i : G \to G_a$ (resp. $\Theta^n \eta_i : G_a \to G_a$). In particular, we may write $1 \otimes y_i = f_i(t)$, where $f_i(t)$ is a p-polynomial in t. Write $x = F_A^n(t)$, where $F_A^n : (k, \varphi^n) \otimes_k A \to A$ is a k-algebra homomorphism $\alpha \otimes a \mapsto \alpha a^{p^n}$ (cf. 2.2). Then we have $y_i^{p^n} = f_i(x)$.

On the other hand, since the k-subalgebra $A_0 := k[y_1, \ldots, y_r] \subseteq A$ satisfies $(k, \varphi^n) \otimes_k A_0 = (k, \varphi^n) \otimes_k A$, we have $A_0 = A$. Hence, by MacLane's theorem (cf. Zariski-Samuel [95, Ch. II, §13, Theorem 30], we can choose a separating transcendence basis, say y, of $k(G)$ over k among $y_1, \ldots, y_r$. Write
$$y^{p^n} = f(x) = a_0 x + a_1 x^p + \cdots + a_m x^{p^m}.$$
Since x is separable over $k(y)$, we have $a_0 \neq 0$. Since the affine plane curve defined by $y^{p^n} = f(x)$ is then smooth, the ring $k[x, y]$ (with the relation $y^{p^n} = f(x)$) is, in particular, k-normal. Since A is purely inseparable over $k[x]$ because $k[A^{p^n}] = k[x]$, the quotient field $k(G) = Q(A)$ is separable and purely inseparable over $k(x, y)$. Hence $Q(A) = k(x, y)$. This implies that $A = k[x, y]$. Another interpretation of this argument is that the commutative diagram in the statement of the theorem is cartesian, i.e., the diagram makes G a fiber product, where ξ (resp. η) is the k-homomorphism corresponding to the inclusion $k[x] \hookrightarrow A$ (resp. $k[y] \hookrightarrow A$).

Conversely, given a k-algebra $A = k[x, y]$ with the relation $y^{p^n} = f(x)$ as above, the comultiplication $\Delta : A \to A \otimes A$ given by $\Delta(x) = x \otimes 1 + 1 \otimes x$ and $\Delta(y) = y \otimes 1 + 1 \otimes y$ defines a k-group scheme structure on Spec A, which we denote by G. We show that G is a k-form of G_a. Replacing x by $a_0^{-1} x$, we may assume that $f(x)$ has the form
$$f(x) = x + a_1 x^p + \cdots + a_m x^{p^m}.$$
In $(k, \varphi^n) \otimes_k A$, we have
$$1 \otimes x = \left(1 \otimes y^{p^{n-1}} - \left(a_1^{p^{n-1}} \otimes x + \cdots + a_m^{p^{n-1}} \otimes x^{p^{m-1}} \right) \right)^p := t_1^p.$$
Hence we have
$$1 \otimes y^{p^{n-1}} = t_1 + a_1^{p^{n-1}} t_1^p + \cdots + a_m^{p^{n-1}} t_1^{p^m}.$$
Then it follows by a similar substitution that
$$t_1 = \left(1 \otimes y^{p^{n-2}} - \left(a_1^{p^{n-2}} t_1 + \cdots + a_m^{p^{n-2}} t_1^{p^{m-1}} \right) \right)^p := t_2^p$$
and
$$1 \otimes y^{p^{n-2}} = t_2 + a_1^{p^{n-2}} t_2^p + \cdots + a_m^{p^{n-2}} t_2^{p^m}.$$
Construct the elements $t_1, t_2, \ldots, t_n \in (k, \varphi^n) \otimes_k A$ by repeating the above construction. Then, by setting $t := t_n$, we have $1 \otimes x = t^{p^n}$ and $1 \otimes y = f(t)$. Hence $(k, \varphi^n) \otimes_k A \cong k[t]$. $\square$

A k-form G of G_a determined by F^n and τ is denoted by $G = (F^n, \tau)$ as an abbreviation (see the cartesian product in the above theorem).

3.2.2

Lemma. *Let $G = (F^n, \tau), G_1 = (F^{n_1}, \tau_1)$ be k-forms of G_a with $n_1 \leq n$. Then $G \cong G_1$ if and only if there exist $\rho \in \mathfrak{R}^*$, $\sigma \in \mathfrak{R}$ and $c \in k^*$ such that*

$$\tau_1^{(n-n_1)} = \left(\rho^{(n)} \tau + F^n \sigma \right) c^{-1},$$

where ρ can be chosen as a polynomial in F of degree $< n$.

Proof. Write the coordinate ring $k[G]$ of G with the comultiplication in the form:

$$k[G] = k[x, y], \quad y^{p^n} = f(x) = a_0 x + a_1 x^p + \cdots + a_m x^{p^m}, \quad a_0 \neq 0$$
$$\Delta(x) = x \otimes 1 + 1 \otimes x, \quad \Delta(y) = y \otimes 1 + 1 \otimes y.$$

It is then straightforward to ascertain that every element $g = g(x, y) \in k[x, y]$ with $\Delta(g) = g \otimes 1 + 1 \otimes g$ is written as $g = g_1(x) + g_2(y)$, where $g_1(x)$ (resp. $g_2(y)$) is a p-polynomial in x (resp. y). This implies that, setting ξ, η as in Theorem 3.2.1, we have

$$\text{Hom}_{k-gr}(G, G_a) = \mathfrak{R}\xi + \mathfrak{R}\eta, \quad F^n \eta = \tau \xi$$
$$\tau = a_0 + a_1 F + \cdots + a_m F^m.$$

By a straightforward computation, one can show that $\text{Hom}_{k-gr}(G, G_a)$ is a *torsion-free*, left $\mathfrak{R}$-module.

Let $\psi : G \to G_1$ be a k-isomorphism. Consider a commutative diagram

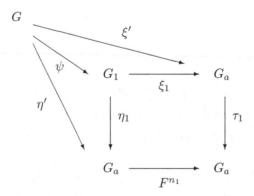

where $\xi' := \xi_1 \cdot \psi$ and $\eta' := \eta_1 \cdot \psi$. We may write $\eta' = \rho\eta + \sigma\xi$ for some $\rho, \sigma \in \mathfrak{R}$, where $\rho \in \mathfrak{R}^*$ because η' gives a separating transcendence basis of $k(G)$ over k. If one writes $\rho = \rho_1 + \rho_2 F^n$ then $\eta' = \rho_1 \eta + (\sigma + \rho_2 \tau)\xi$.

Hence we may assume that ρ has F-degree $< n$. Consider first the case $n = n_1$. Since ψ induces an isomorphism

$$k[x_1] \cong (k, \varphi^n) \otimes_k k[G_1] \cong (k, \varphi^n) \otimes_k k[G] \cong k[x],$$

we have $\xi' = \xi_1 \psi = c\xi$ with $c \in k^*$. Now, we compute

$$\tau_1 \xi_1 \psi = F^n \eta_1 \psi = F^n \eta' = F^n \rho \eta + F^n \sigma \xi = \rho^{(n)} F^n \eta + F^n \sigma \xi$$
$$= (\rho^{(n)} \tau + F^n \sigma)\xi = (\rho^{(n)} \tau + F^n \sigma)c^{-1} \xi_1 \psi.$$

Since $\mathrm{Hom}_{k-gr}(G, G_a)$ is a torsion-free left $\mathfrak{R}$-module, we have $\tau_1 = (\rho^{(n)} \tau + F^n \sigma)c^{-1}$. Conversely, if $\tau_1 = (\rho^{(n)} \tau + F^n \sigma)c^{-1}$, define $\xi', \eta' \in \mathrm{Hom}_{k-gr}(G, G_a)$ by $\xi' = c\xi$ and $\eta' = \rho\eta + \sigma\xi$. Then $F^n \eta' = \tau_1 \xi'$. Hence we obtain a k-group scheme homomorphism $\psi : G \to G_1$ such that $\xi' = \xi_1 \cdot \psi$ and $\eta' = \eta_1 \cdot \psi$. Since $\rho \in \mathfrak{R}^*$, ρ is invertible in $\widehat{\mathfrak{R}}$. Write $\rho^{-1} = \rho_1 + \sigma_2 F^n$ with $\rho_1 \in \mathfrak{R}^*$ of F-degree $< n$ and $\sigma_2 \in \widehat{\mathfrak{R}}$. Then $\tau = (\rho_1^{(n)} \tau_1 + F^n \sigma_1)c$ with $\sigma_1 = (\sigma_2 \rho^{(n)} \tau - \rho_1 \sigma)c^{-1}$. Since $F^n \sigma_1 c \in \mathfrak{R}$, we know $\sigma_1 \in \mathfrak{R}$. Reversing the roles of G and G_1, we obtain a k-group scheme homomorphism $\psi_1 : G_1 \to G$, which is, in fact, the inverse of ψ. Consider next the case where $n_2 = n - n_1 \geq 0$. Consider the commutative diagram

$$
\begin{array}{ccccc}
G_1 & \xrightarrow{\xi_1} & G_a & \xrightarrow{F^{n_2}} & G_a \\
\downarrow{\eta_1} & & \downarrow{\tau_1} & & \downarrow{\tau_1^{(n_2)}} \\
G_a & \xrightarrow{F^{n_1}} & G_a & \xrightarrow{F^{n_2}} & G_a
\end{array}
$$

where both (small) squares are cartesian. Hence the big square is cartesian, and $G_1 \cong (F^n, \tau_1^{(n_2)})$. Now apply the previous arguments. $\square$

3.2.3

Corollary. *Let $G = (F^n, \tau)$ be a k-form of G_a. Then we have:*

(1) $G \cong G_a$ *if and only if* $\tau = \tau_1^{(n)} c^{-1}$ *for some* $c \in k^*$ *and* $\tau_1 \in \mathfrak{R}$.
(2) *For* $0 \leq m \leq n$, $\Theta^m G = (F^{n-m}, \tau)$.

Proof. (1) Since $G_a = (F^n, 1)$, we have $\tau c = \rho^{(n)} + F^n \sigma$ by 3.2.2 with $\rho, \sigma \in \mathfrak{R}$ and $c \in k^*$. Hence $\tau c = \tau_1^{(n)}$ with $\tau_1 = \rho + \sigma F^n$. If $\tau c = \tau_1^{(n)}$, then $\tau_1 \in \mathfrak{R}^*$ and we can write $1 = \rho' \tau_1 + \sigma' F^n$ with $\rho' \in \mathfrak{R}^*$ of F-degree $< n$ and $\sigma' \in \widehat{\mathfrak{R}}$. Since ρ' and τ_1 have finite F-degrees, it follows that $\sigma' \in \mathfrak{R}$. Hence $1 = 1^{(n)} = (\rho'^{(n)} \tau + F^n \sigma' c^{-1})c$. By 3.2.2, $(F^n, 1) \cong (F^n, \tau)$.

(2) Clearly, $\Theta^m G = (F^n, \tau^{(m)})$ and, by the arguments in the proof of Theorem 3.2.1, $(F^n, \tau^{(m)}) \cong (F^{n-m}, \tau)$. $\square$

3.2.4

Lemma. *Let $G = (F^n, \tau), G_1 = (F^{n_1}, \tau_1)$ be k-forms of G_a, and let K be a finitely generated, separable extension of k such that k is algebraically closed in K. Then G is k-isomorphic to G_1 if and only if $G_K := G \otimes_k K$ is K-isomorphic to $G_{1K} := G_1 \otimes_k K$.*

Proof. Since the *only if* part is obvious, it suffices to prove the *if* part. By the proof of Theorem 3.2.1, we may assume that $n = n_1$, $\tau = 1 + \sum_i a_i F^i \in \mathfrak{R}^{**}$ and $\tau_1 = 1 + \sum_i b_i F^i \in \mathfrak{R}^{**}$. Assume that $(\rho^{(n)} \tau + F^n \sigma) x_0^{-p^n} = \tau_1$ for $\rho = \sum_i x_i F^i \in \mathfrak{R}_K$ and $\sigma = \sum_i y_i F^i \in \mathfrak{R}$, where we may assume that $x_i = 0$ whenever $i \geq n$ (see Lemma 3.2.2). Then we obtain the relations

$$\left(\sum_{j=0}^{i-1} x_j^{p^n} a_{i-j}^{p^j} + x_i^{p^n} + y_{i-n}^{p^n} \right) x_0^{-p^{n+i}} = b_i \in k \tag{3.1}$$

for $i \geq 1$, where we set $y_i = 0$ if $i < 0$. We shall show that the same relations hold for some $x_i, y_i \in k$. If $G \cong G_{a,k}$ then $G_1 \otimes K \cong G_a \otimes K$. Since $G \cong G_{a,k}$, we may assume that $a_i = 0$ for all i. Then the equality (3.1) implies that $x_i \cdot x_0^{-p^i} \in K \cap k^{p^{-n}} = k$ for $0 \leq i < n$ since G_1 is a k-form of G_a of height n. Similarly, $y_{i-n} \cdot x_0^{-p^i} \in k$ for $i \geq n$. Replacing ρ, σ by $\rho' := \rho \cdot x_0^{-1}, \sigma' := \sigma \cdot x_0^{-p^n}$, respectively, we have $\tau_1 = \rho'^{(n)} + F^n \sigma'$ with $\rho' \in \mathfrak{R}^*$ and $\sigma' \in \mathfrak{R}$. Hence $G_1 \cong G_{a,k}$. Therefore, we may assume that $G \not\cong G_{a,k}$. Then not all $a_i \in k^{p^n}$ (cf. 3.2.3, (1)), and there is an integer $r \geq 1$ such that $a_1, \ldots, a_{r-1} \in k^{p^n}$ but $a_r \notin k^{p^n}$. If $r > 1$, we may replace τ by $(1 - a_1 F)\tau = (1 - a_1^{p^{-n}} F)^{(n)} \tau$, which has a zero F-linear term. Then, by an induction argument, we can assume that $a_1 = \cdots = a_{r-1} = 0$. The equality (3.1) for $i = r$ gives

$$a_r x_0^{p^n - p^{n+r}} + x_r^{p^n} x_0^{-p^{n+r}} + y_{r-n}^{p^n} x_0^{-p^{n+r}} = b_r.$$

Set $u = x_0^{-1}$, $v = x_r x_0^{-p^r}$ if $r < n$, and $u = x_0^{-1}$, $v = y_{r-n} x_0^{-p^r}$ if $r \geq n$. Then we have

$$a_r u^{(p^r-1)p^n} + v^{p^n} = b_r.$$

Replacing this relation by the p-th roots as long as both a_r and b_r have p-th roots in k, we obtain a relation of the type

$$a u^{(p^r-1)p^{n'}} + v^{p^{n'}} = b$$

$$a, b \in k, \quad n' \geq 1, \quad a \notin k^p \text{ or } b \notin k^p.$$

Suppose $u \notin k$. Then u is transcendental over k by the assumption, and

$$k(u, v) = k(u)[V] / \left(V^{p^{n'}} + a u^{(p^r-1)p^{n'}} - b \right),$$

where $V^{p^{n'}} + au^{(p^r-1)p^{n'}} - b$ is irreducible over $k(u)$. However, if $k' := k(a^{p^{-1}}, b^{p^{-1}})$, then $K' \otimes_k k(u,v)$ contains a nonzero nilpotent element, which contradicts the separability of K over k. Therefore, $x_0 = u^{-1} \in k$. Considering the equality (3.1) for $i = 1, \ldots, n-1$, we see that $x_i \in k$, and then $y_{i-n} \in k$ follows for $i \geq n$. (Use the relation $K \cap k^{p^{-n}} = k$.) $\qquad\square$

3.3 G-torsors for k-groups of Russell type G

We shall consider the following problem: Given a k-form X of $\mathbb{A}^1$, under what conditions does X become a k-group of Russell type? Let G be a k-group scheme and let X be a k-scheme on which G acts from the left by the k-morphism $\sigma : G \times_k X \to X$. We call X a *principal homogeneous space* for G (or G-*torsor*, in short) if $(\sigma, p_2) : G \times_k X \to X \times_k X$, $(g,x) \mapsto (gx, x)$ is a k-isomorphism, where p_2 is the projection onto the second factor. Then, X is k-isomorphic to G if and only if X has a k-rational point. For general and relevant results, we refer to Démazure-Gabriel [12].

3.3.1

Lemma. *Let X be a k-form of $\mathbb{A}^1$ such that $X \otimes k_s$ admits a k_s-group scheme structure, where k_s is the separable algebraic closure of k in $\overline{k}$. Then there exists a k-form G of G_a, unique up to a k-isomorphism, and X is a principal homogeneous space for G. If G is identified with a closed subgroup of G_a^2 defined by $v^{p^n} = f(u)$, $f(u)$ being a p-polynomial, then X is k-isomorphic to a closed subscheme of $\mathbb{A}_k^2 = \operatorname{Spec} k[x,y]$ defined by an equation of the type $y^{p^n} = b + f(x)$ with $b \in k$. Conversely, if X and G are defined as above, then X is a principal homogeneous space for G.*

Proof. Let $B = k[X]$ be the coordinate ring of X, and let $n = \operatorname{ht}(X)$. Then, $(k, \varphi^n) \otimes_k B \cong k[t]$. Write $t = \sum_{i=1}^r a_i \otimes y_i$, where the a_i are linearly independent over k^{p^n}. Since $1 \otimes y_i \in k[t]$, we can write $1 \otimes y_i = g_i(t)$, where $g_i(t) \in k[t]$ which is not necessarily a p-polynomial. As in the proof of 3.2.1, $y_i^{p^n} = g_i(x) \in k[x]$ with $x := F_B^n(t)$ and $B = k[y_1, \ldots, y_r]$. Let $Q \in X \otimes k_s$ be a k_s-rational point and let $c_i := y_i(Q) \in k_s$. Let $y_i' := y_i - c_i$, $t' = t - c$ and $x' = x - c$ with $c = \sum_{i=1}^r a_i c_i^{p^n}$. Then $t' = \sum_i a_i \otimes y_i'$ and Q lies over the point $t' = 0$ of $\Theta^n(X \otimes k_s) \cong \mathbb{A}^1 \otimes k_s$. We can choose Q as the origin of a k_s-group scheme structure on $X \otimes k_s$. Let Δ be the comultiplication. Then $\Delta(t') = t' \otimes 1 + 1 \otimes t'$, and since the a_i are linearly independent over $k_s^{p^n}$ as well, we conclude that $y_i'^{p^n} = f_i(x') \in k_s[x']$, where $f_i(x')$ is a

p-polynomial (cf. the proof of 3.2.1). Hence $g_i(x) = y_i^{p^n} = b_i + f_i(x)$ with $b_i = c_i^{p^n} - f_i(c)$. Since $g_i(x) \in k[x]$, we have $b_i \in k$ (as $b_i = g_i(0)$) and $f_i(x) \in k[x]$ (as $f_i(x) = g_i(x) - b_i$). Choose a separating transcendence basis, say y, of $k(X)$ over k among $y_1, \ldots, y_r$. Then $y^{p^n} = b + f(x)$ as above, where the linear term of $f(x)$ is nonzero. Hence $k[x, y]$ is k-normal and $B = k[x, y]$ (cf. the proof of 3.2.1). Let $G = \operatorname{Spec} A$ with $A = k[u, v]$ having a relation $v^{p^n} = f(u)$. Then G is a k-form of G_a. Define a G-action σ on X by the k-algebra homomorphism $s : B \to B \otimes_k A$ defined by $s(x) = x \otimes 1 + 1 \otimes u$ and $s(y) = y \otimes 1 + 1 \otimes v$. Then $(\sigma, p_2) : G \times_k X \to X \times_k X$ corresponds to the k-algebra homomorphism $\bar{s} : B \otimes_k B \to B \otimes_k A$ given by $\bar{s}(w \otimes z) = (w \otimes 1)s(z)$. Since $k_s \otimes \bar{s}$ is an isomorphism, $\bar{s}$ is a k-isomorphism. Hence (σ, p_2) is a k-isomorphism, i.e., X is a principal homogeneous space for G. Finally, assume that this is also the case for a k-form G_1 of G_a. Then $(\sigma_{G_1}, p_2)^{-1} \cdot (\sigma_G, p_2) : G \times_k X \to G_1 \times_k X$ is an isomorphism of X-group schemes. Let $K = k(X)$. Then we get, in particular, an isomorphism of K-group schemes $G \otimes K \overset{\sim}{\to} G_1 \otimes K$. By 3.2.4, G is then k-isomorphic to G_1. □

3.3.2

Let C be a complete k-normal curve. Then the automorphism functor $\mathcal{A}ut_{C/k}$ is defined by

$$S \in (\mathcal{S}ch/k) \longmapsto \operatorname{Aut}_S(C \times_k S) = \{S\text{-automorphisms of } C \times_k S\}$$

and $\mathcal{A}ut_{C/k}$ is, in fact, represented by a k-group scheme $\operatorname{Aut}_{C/k}$ which is locally of finite type over k; we refer to Grothendieck [FGA, 221–19]. Let $\operatorname{Aut}^0_{C/k}$ be the neutral component which is a connected k-group scheme of finite type. Let X be a k-form of $\mathbb{A}^1$, and let C be a complete k-normal model of X. It is then well-known that every element of $\operatorname{Aut}_k(X)$ extends to an element of $\operatorname{Aut}_k(C) = \operatorname{Aut}_{C/k}(k)$.

3.3.3

Lemma (Rosenlicht [76, Lemma, p.5] and Russell [78, Th.4.2]). *Let X be a k-form of $\mathbb{A}^1$ and let C be its complete k-normal model. Assume that the arithmetic genus $p_a(C)$ is positive and that $\operatorname{Aut}_{C/k}(k_s)$ is an infinite group. Then $X \otimes k_s$ carries a k_s-group scheme structure.*

Proof. Obviously, we may assume that $k = k_s$. Let $P_\infty := C \setminus X$, which is not rational over k but rational over a purely inseparable extension of

k. Let $\lambda = \operatorname{ht}(k(X))$ (cf. 2.1). Since $(k, \varphi^i) \otimes_k k(X) \cong k(k(X)^{p^i})$ for $0 \le i \le \lambda$, every element of $\Gamma := \operatorname{Aut}_k(C)$ induces a k-automorphism of $k(k(X)^{p^i})$. Let $K := k(k(X)^{p^{\lambda-1}})$. Then the k-genus of K, i.e., the arithmetic genus of a complete k-normal model of K, is positive because the assumption $k = k_s$ implies that K has a k-rational place[1]. Let P be the place of K induced by P_∞. Then P is a unique singular place of K. Hence $\sigma(P) = P$ for any k-automorphism σ of K. On the other hand, we have $k(K^p) = k(y)$ with $y \in K$, and

$$\sigma(y) = \frac{ay+b}{cy+d}, \quad a, b, c, d \in k, \quad ad - bc \ne 0$$

for σ as above. In particular, $\sigma(P) = P$ and $\sigma(y)$ is written as above for every element σ of Γ. We may assume that y is finite[2] at P, i.e., $v_P(y) \ge 0$ for the valuation associated with the place P. Consider the following two cases separately: (1) $y(P) \notin k$ and (2) $y(P) \in k$.

CASE (1). Then $\alpha := y(P)$ is purely inseparable over k. In fact, $y \in \mathcal{O}_{C,P}$ and hence $y(P) \in k(P)$ the residue field of P. Since $k(P)/k$ is a purely inseparable extension, α is purely inseparable over k. For any $\sigma \in \Gamma$, we have

$$\sigma(\alpha) = \frac{a\alpha+b}{c\alpha+d} = \alpha, \quad i.e., \quad c\alpha^2 + (d-a)\alpha - b = 0.$$

Hence $p = 2$, $a = d$, $c\alpha^2 = b$ and $\alpha^2 \in k$. Moreover, $\operatorname{ht}(X) = \lambda + 1$. Let $t = (y^2 + \alpha^2)^{-1}$. Then it follows that

$$\sigma(t) = t + \frac{c(\sigma)^2}{a(\sigma)^2 + c(\sigma)^2 \alpha^2},$$

where a, b, c, d are denoted by $a(\sigma), b(\sigma), c(\sigma), d(\sigma)$, respectively, since they depend on σ but α is independent of σ, and where

$$\frac{c(\sigma)}{a(\sigma) + c(\sigma)\alpha} = \frac{c(\sigma')}{a(\sigma') + c(\sigma')\alpha} \quad \text{if and only if} \quad \sigma = \sigma'.$$

---

[1] Let K be a separable extension of a field k of transcendence degree one. Let x be a separating transcendence basis over k. Since $K/k(x)$ is then a separable algebraic extension, hence a simple extension, we can write $K = k(x, y)$ with the relation of minimal degree $F(y) = a_0(x)y^n + a_1(x)y^{n-1} + \cdots + a_{n-1}(x)y + a_n(x) = 0$ with all $a_i(x) \in k[x]$. Write $F(y) = a_0(x) \prod_{i=1}^n (y - \alpha_i)$. Then the discriminant of $F(y)$, which is $D = a_0(x)^N \prod_{i,j}(\alpha_i - \alpha_j)^2$ with a suitable integer $N > 0$, is an element of $k[x]$. Assume that $k = k_s$. Then k is an infinite field. Hence, except for finitely many elements of k, $a_0(c) \ne 0$ and $D(c) \ne 0$ with $c \in k$. Then $F(y, c) = 0$ is a separable equation. Hence $F(y, c) = 0$ has n distinct roots $\gamma_1, \ldots, \gamma_n$. Then, for a general element c of k, the points (c, γ_i) for $1 \le i \le n$ are k-rational points of a complete k-normal model C of K.

[2] Otherwise, consider y^{-1} instead of y.

CASE (2). $\alpha := y(P) \in k$. Replacing y by $y - \alpha$, we may assume that $\alpha = 0$. Note that $[K : k(K^p)] = p$. In fact, if we write $K = k(\xi, \eta)$ with a separating transcendence basis ξ and η separable over $k(\xi)$, then $k(K^p) = k(\xi^p, \eta^p)$ and K is purely inseparable over $k(K^p)$. Then η is separable and purely inseparable over $k(K^p)(\xi)$. Hence $K = k(K^p)(\xi)$ with $\xi \notin k(K^p)$. Hence $[K : k(K^p)] = p$. Let z be anew a separating transcendence basis of K over k. Then $K \supseteq k(y, z) \supset k(y) = k(K^p)$, where $K/k(y, z)$ is separable and purely inseparable. Hence $K = k(y, z)$. Further $z \notin k(y)$ and $z^p = f(y) \in k[y]$, where we may assume that $f(y) \in k[y]$. Let e, f be respectively the ramification index and the residue class field degree of P over $k(y)$. Since P_∞ is the unique place of K lying over the place $y = 0$ of $k(y)$, we have $ef = p$. If $f = 1$, then P is k-rational, which is a contradiction to the assumption that P is a singular place. Hence $f = p$, $e = 1$ and $v_P(y) = 1$, where v_P is the normalized valuation of P. For $\sigma \in \Gamma$, we can write

$$\sigma(y) = \frac{ay}{1 + cy}, \quad a, c \in k, \quad a \neq 0.$$

Since $z^p = f(y)$, we have

$$\sigma(z)^p = f(\sigma(y)) = f\left(\frac{ay}{1 + cy}\right).$$

Let $m = \deg f(y)$, where we choose z, $f(y)$ in such a way that $m := \deg f(y)$ is minimal, and choose the integer i such that $(i - 1)p < m \leq ip$. Then $i > 0$ and

$$\left((1 + cy)^i \sigma(z)\right)^p = (1 + cy)^{pi} f\left(\frac{ay}{1 + cy}\right) \in k[y].$$

Since $\sigma(P) = P$, we have $\left((1 + cy)^i \sigma(z)\right)(P) = \sigma(z)(P) = z(P)$. So, $v_P\left((1 + cy)^i \sigma(z) - z\right) > 0$. Hence the only possible pole of the function $\left((1 + cy)^i \sigma(z) - z\right) \cdot y^{-1}$ is at the point $y^{-1} = 0$. Thus, we have

$$\left(\frac{(1 + cy)^i \sigma(z) - z}{y}\right)^p = \frac{(1 + cy)^{pi} f\left(\frac{ay}{1+cy}\right) - f(y)}{y^p}$$

is a polynomial in y of degree $\leq pi - p < m$. By the minimality of m, we have $((1+cy)^i \sigma(z) - z) \cdot y^{-1} \in k(y)$, Indeed, if $z' := ((1+cy)^i \sigma(z) - z) \cdot y^{-1} \notin k(y)$, then we can take z', y instead of z, y. Since z'^p is a polynomial in y of degree $< m$, we have a contradiction to the minimality of m. So, we can write $(1 + cy)^i \sigma(z) = z + g(y)$ with $g(y) \in k[y]$. Hence we have

$$f(y) + g(y)^p = (1 + cy)^{pi} f\left(\frac{ay}{1 + cy}\right).$$

By differentiating this equality by y, we have

$$f'(y) = a(1 + cy)^{pi-2} f'\left(\frac{ay}{1 + cy}\right).$$

Since y is separable over $k(z)$, we have $f'(y) \neq 0$, so we write $f'(y) = y^r h(y)$ with $r \geq 0$, $h(y) \in k[y]$, $h(0) \neq 0$. Then we have

$$h(y) = a^{r+1}(1 + cy)^{pi-2-r} h\left(\frac{ay}{1 + cy}\right).$$

Setting $y = 0$, we obtain $a^{r+1} = 1$. Since the mapping $\sigma \in \Gamma \mapsto a(\sigma) \in \overline{k}^*$ is a group homomorphism, there exists a normal subgroup H of Γ of finite index such that $a = 1$ for every $\sigma \in H$. Let $t = y^{-1}$. Then $\sigma(t) = t + c$ for $\sigma \in H$ and, for $\sigma, \sigma' \in H$, we have $c(\sigma) = c(\sigma')$ if and only if $\sigma = \sigma'$. Finally note that $\operatorname{ht}(X) = \operatorname{ht}(k(X))$ in the latter case.

By the above arguments, we have shown the following result: Let $n = \operatorname{ht}(X)$ and let $(k, \varphi^n) \otimes_k k[X] \cong k[t]$. Then there exists a normal subgroup H of Γ of finite index such that $\sigma(t) = t + b(\sigma)$ for every $\sigma \in H$ and the representation of H given by $\sigma \mapsto b(\sigma) \in k$ is faithful[3]. Furthermore, we may choose t so that the point Q of X lying over the point $t = 0$ is rational over k. Write $t = \sum_{i=1}^r a_i \otimes v_i$ with $a_i \in k$, $v_i \in k[X]$ and $v_i^{p^n} = f_i(t) \in k[t]$. Let $c_i = v_i(Q)$. Then $f_i(0) = c_i^{p^n} \in k^{p^n}$. Assuming that the a_i are linearly independent over k^{p^n}, the equation $\sum_{i=1}^r a_i f_i(0) = 0$ implies that $f_i(0) = 0$ for every i. On the other hand, we have

$$t + b(\sigma) = \sigma(t) = \sum_{i=1}^r a_i \otimes \sigma(v_i)$$

for every $\sigma \in H$. Let $b_i(\sigma) = \sigma(v_i)(Q)$. Then $b(\sigma) = \sum_{i=1}^r a_i b_i(\sigma)^{p^n}$ and $t + b(\sigma) = \sum_{i=1}^r a_i \otimes (v_i + b_i(\sigma))$. Hence $\sigma(v_i) = v_i + b_i(\sigma)$ for $1 \leq i \leq r$ and every $\sigma \in H$. Thus $f_i(t + b(\sigma)) = f_i(t) + b_i(\sigma)^{p^n}$, where $b_i(\sigma)^{p^n} = f_i(b(\sigma))$ (set $t = 0$). Since H is an infinite group (because so is Γ and $[G : H] < \infty$), f_i is a p-polynomial in t. Therefore, X has a k-group scheme structure such that $\Delta(v_i) = v_i \otimes 1 + 1 \otimes v_i$ and $\Delta(t) = t \otimes 1 + 1 \otimes t$ because $k[X] = k[v_1, \ldots, v_r]$. □

3.3.4

Theorem. *Let X be a k-form of $\mathbb{A}^1$ and let C be a complete k-normal model of X. Assume that the arithmetic genus of C is positive and that X has a k-rational point P_0. Then the following conditions are equivalent:*

---

[3]Note that the notation for $c(\sigma)$ above is changed to $b(\sigma)$

(1) X is the underlying scheme of a k-group of Russell type.

(2) $\operatorname{Aut}_{C/k}(k_s)$ is an infinite group, where k_s is the separable algebraic closure of k in $\overline{k}$.

(3) There exists a surjective homomorphism ρ of k-group schemes from $\operatorname{Pic}^0_{C/k}$ to a one-dimensional unipotent k-group scheme G such that $\rho \cdot \iota : X \to G$ is an isomorphism such that $(\rho \cdot \iota)(P_0)$ is the neutral point of G. (See 1.7 for the definition of ι.)

Proof. (1) $\Rightarrow$ (2). Let G be a k-group of Russell type such that $X = \overline{G}$. Note that $G \otimes k_s$ is a k_s-group of Russell type with $X \otimes k_s = \overline{G \otimes k_s}$ and $C \otimes k_s$ is a complete k_s-normal model of $X \otimes k_s$. Then the translations by elements of $G(k_s)$ give rise to k_s-automorphisms of $C \otimes k_s$, so we have an injective homomorphism $G(k_s) \hookrightarrow \operatorname{Aut}_{C/k}(k_s)$. Hence $\operatorname{Aut}_{C/k}(k_s)$ is an infinite group.

(2) $\Rightarrow$ (1). This follows from Lemmas 3.3.1 and 3.3.3.

(1) $\Rightarrow$ (3). Let G be as above. We may assume that P_0 is the neutral point of G. For any field extension k'/k, define $(\rho \cdot \iota)(k') : X(k') \to G(k')$ by $P_0 \mapsto$ the neutral point of G and $Q \mapsto Q - P_0$ (subtraction according to the group law of G). Then $\rho \cdot \iota$ is a k-isomorphism. Since $\operatorname{Pic}^0_{C/k}$ is generated as a group scheme by the image $\iota(X)$, $\rho \cdot \iota$ extends to a surjective homomorphism $\rho : \operatorname{Pic}^0_{C/k} \to G$.

(3) $\Rightarrow$ (1). Obvious. $\qquad\square$

3.3.5

Lemma. Let G be a k-group of Russell type. Then the following assertions hold.

(1) $\operatorname{ht}(G) = \operatorname{ht}(\overline{G}) = \operatorname{ht}(k(G))$ unless $p = 2$ and G is defined by an equation of the form

$$y^{2^n} = x + a_1 x^2 + \cdots + a_m x^{2^m}, \quad a_1 \in k \setminus k^2, \ a_i \in k^2 \ (2 \le i \le m).$$

(2) Assume that $G \not\cong G_{a,k}$. Then $k(G)$ is k-rational if and only if $p = 2$ and G is defined by an equation of the form

$$y^2 = x + ax^2, \quad a \in k \setminus k^2.$$

Proof. (1) By Lemma 3.1.2, $\operatorname{ht}(G) = \operatorname{ht}(\overline{G})$. Note that $\operatorname{ht}(G) = \operatorname{ht}(G \otimes k_s)$ and $\operatorname{ht}(k(G)) = \operatorname{ht}(k_s \otimes_k k(G))$. If $k = k_s$, G has infinitely many

k-rational points, and the translations by those points give rise to k-automorphisms of a complete k-normal model of $\overline{G}$. Then, by the proof of Lemma 3.3.3, ht $(\overline{G}) = $ ht $(k(G))$ unless $p = 2$ and ht $(\overline{G}) = $ ht $(k(G)) + 1$. Also, in the last exceptional case, $\Theta^{n-1}G$ is a non-trivial k-form of G_a such that $k(\Theta^{n-1}G)$ is k-rational. Then, the assertion (1) follows from the assertion (2), Corollary 3.2.3 and Lemma 3.2.2. Thus, it suffices to prove the assertion (2).

(2) Since the *if* part is obvious [4], we will prove the *only if* part. Since $k(G)$ is k-rational, G has a k-rational point, say P, other than the neutral point. The translation by P induces a k-automorphism σ of the underlying scheme of G, hence the complete k-normal model $\mathbb{P}^1_k$ of G. Choose an inhomogeneous parameter t of $\mathbb{P}^1$ such that t is finite on $P_\infty := \mathbb{P}^1 \setminus G$. Then, we have

$$\sigma(t) = \frac{at+b}{ct+d}, \quad a,b,c,d \in k, \quad ad - bc \neq 0.$$

Let $t(P_\infty) = \alpha$. Then $\alpha \notin k$ because $G \not\cong G_{a,k}$. Further, $p = 2, a = d$ and $b = c\alpha^2$ (cf. the proof of Lemma 3.3.3). Set anew $a = \alpha^2$ with $a \in k \setminus k^2$. Then the rational mapping Φ defined by the linear system $|2P_\infty|$ is an embedding of $\mathbb{P}^1$ into $\mathbb{P}^2$ given by $t \mapsto (1, t, t^2 - a)$ (cf. 2.8.1). Then the image $\Phi(\mathbb{P}^1 \setminus P_\infty)$ is a curve in $\mathbb{A}^2$ defined by $y^2 = x + ax^2$, $a \in k \setminus k^2$. $\quad\square$

---

[4] $k(G) = k(t)$ for $t = y/x$ since $x = 1/(t^2 - a)$ and $y = t/(t^2 - a)$.

Chapter 4

Hyperelliptic forms of the affine line

In this chapter, k is assumed to be an *infinite field*. Let K be an algebraic function field of one variable over k. Assume that K has k-genus $g \geq 2$. If K is a quadratic extension of a subfield K_0 of genus 0, then K is called *hyperelliptic*. It is known that the subfield K_0 is uniquely determined by K[1]. A k-form X of $\mathbb{A}^1$ is called a *hyperelliptic k-form of $\mathbb{A}^1$* if $k(X)$ is hyperelliptic over k. A goal of this chapter is to describe all hyperelliptic k-forms X of $\mathbb{A}^1$ under the assumption that X has a k-rational point.

4.1 Birational forms of hyperelliptic curves

Lemma (E. Artin [1, XVI, §7]). *Every hyperelliptic k-curve of genus $g \geq 2$ carrying a k-rational point is birationally equivalent to an algebraic plane curve of either type:*

(i) $p \geq 2$, $y^2 = f(x) \in k[x]$, *where* $\deg f = 2g + 1$ *or* $2g + 2$.

(ii) $p = 2$, $y^2 + f(x)y + h(x) = 0$, *where* $f(x), h(x) \in k[x]$, $\deg f \leq g + 1$ *and* $\deg h = 2g + 2$.

In both cases (i), (ii), *we may assume that the points at infinity of the plane curve are k-smoothable.*

Proof. Since the unique subfield K_0 of genus 0 is k-rational[2], it is well-known that a given hyperelliptic curve is represented by a plane curve

---

[1]If C is hyperelliptic, the canonical linear system $|K_C|$ is composed of a pencil, i.e., the rational map $\Phi_{|K_C|}$ is a $2:1$ map from C to $\mathbb{P}^1$. Since K_C is uniquely determined, so is the canonical mapping $\Phi_{|K_C|}$.

[2]Let C and C_0 be complete k-normal models of K and K_0. Then the inclusion $K_0 \hookrightarrow K$ gives rise to a k-morphism $q : C \to C_0$. Since C has a k-rational point P, the image $q(P)$ is a k-rational point of C_0. Hence C_0 is k-isomorphic to $\mathbb{P}^1$.

defined by one of the above equations. So, what is to be proved in the lemma is that any point at infinity is k-smoothable.

Let K be the function field over k of the given curve. Then $K = k(x, y)$ and $K_0 = k(x)$. Let $\mathcal{O}$ (identified with a DVR of K) be a place *at infinity* of the curve in question. $\mathcal{O}$ is such a place if and only if $k[x, y] \not\subseteq \mathcal{O}$, which is equivalent to $x \notin \mathcal{O}$ because y is integral over $k[x]$. Now observe that if each place at infinity (there might exist several such places) is smooth (cf. 2.6.1) then the corresponding point at infinity is k-smoothable, and *vice versa*. In either case (i) or (ii), after a suitable change of coordinates $(x, y) \mapsto (x-a, y)$ with $a \in k$, we may assume that the points $(x = 0, y = \pm\sqrt{f(0)})$ in case (i) and $(x = 0, y = \text{roots of } Y^2 + f(0)Y + h(0) = 0)$ in case (ii) are all smooth points since k is an infinite field. Let $\mathcal{O}_1, \mathcal{O}_2$ be the smooth places corresponding to these points, where it could happen that $\mathcal{O}_1 = \mathcal{O}_2$. Consider the following k-birational substitution $x = 1/\xi$ and $y = \eta/\xi^{g+1}$, whence $k(x, y) = k(\xi, \eta)$. Then one can easily check that the equations

(i') $\eta^2 = \xi^{2g+2} f(1/\xi) = \varphi(\xi) \in k[\xi], \quad \deg\varphi = 2g + 1 \text{ or } 2g + 2,$

(ii') $\eta^2 + \xi^{g+1} f(1/\xi)\eta + \xi^{2g+2} h(1/\xi) = \eta^2 + \varphi(\xi)\eta + \psi(\xi) = 0,$

$\varphi(\xi), \ \psi(\xi) \in k[\xi], \ \deg\varphi \leq g + 1, \ \deg\psi = 2g + 2$

correspond to (i), (ii), respectively, where we take the above a to be a sufficiently general element of k. Since x belongs to the maximal ideals of $\mathcal{O}_1, \mathcal{O}_2$ and $x = 1/\xi$, the places $\mathcal{O}_1, \mathcal{O}_2$ are at infinity of the (ξ, η)-plane curves (i'), (ii'), and these are clearly the only places at infinity. Passing from the equations (i), (ii) to (i'), (ii') if necessary, we may assume that our k-curve is k-smoothable at infinity. $\qquad\square$

4.2 k-normality of the birational form

Lemma. *Let C be a hyperelliptic affine plane k-curve of genus $g \geq 2$, defined by an equation of either type (i) or (ii) of 4.1 and having a k-smoothable point at infinity. Then C is k-normal.*

Proof. Consider first the case $p > 2$ and the equation is of type (i). By the standard homogenization followed by a suitable dehomogenization, the point at infinity may be identified with $(t = 0, u = 0)$ of multiplicity $n - 2$ of the plane curve

$$t^{n-2} = a_0 u^n + a_1 u^{n-1} t + \cdots + a_{n-1} u t^{n-1} + a_n t^n, \quad a_0 \neq 0,$$

where we put $n = 2g + 1$ or $n = 2g + 2$ according as $\deg f = 2g + 1$ or $\deg f = 2g + 2$. The blowing-up of this curve at $(t = 0, u = 0)$ is achieved by setting $v = t/u$ and the proper transform has an equation

$$v^{n-2} = u^2(a_0 + a_1 v + \cdots + a_n v^n).$$

The point $(u = 0, v = 0)$ of multiplicity 2 is the only point of this curve lying over the point $(t = 0, u = 0)$. Now perform $g - 1$ blowing-ups of the point $(u = 0, v = 0)$ in succession. This operation amounts to introducing variables $u_1, \ldots, u_{g-1}$ which are subject to the relations

$$u = vu_1, \ u_1 = vu_2, \ldots, u_{g-2} = vu_{g-1}.$$

In each step, we obtain a unique singular point of multiplicity 2 until we end up with equations

$$v = u_{g-1}^2(a_0 + a_1 v + \cdots + a_n v^n) \ \text{if} \ n = 2g + 1$$
$$v^2 = u_{g-1}^2(a_0 + a_1 v + \cdots + a_n v^n) \ \text{if} \ n = 2g + 2.$$

The point $(u_{g-1} = 0, v = 0)$ is in either case the unique point lying over the original point $(t = 0, u = 0)$. It is smooth in the case $n = 2g + 1$, and an ordinary double point in the case $n = 2g + 2$. By one more blowing-up $u_{g-1} = vu_g$, this double point is resolved into two smooth points which are k-rational if $a_0^{1/2} \in k$ and non-k-rational if $a_0^{1/2} \notin k$. After the foregoing series of blowing-ups, the arithmetic k-genus of the proper transform is

$$\frac{(n-1)(n-2)}{2} - \frac{(n-2)(n-3)}{2} - \alpha = g,$$

where $\alpha = g - 1$ if $n = 2g + 1$ and $\alpha = g$ if $n = 2g + 2$. Since C has k-genus g by the assumption, the curve C cannot have any other non-k-normal points.

Consider next the case $p = 2$ and the equation is of type (ii). Since the computations are very similar to the preceding case, we merely outline the necessary steps in what follows. By homogenization and dehomogenization of the equation (ii) we can bring the original point at infinity to the point of origin, which turns out to have multiplicity $2g$ on the transformed curve. By a single blowing-up, we obtain a unique point with multiplicity 2 lying over the origin. Then we continue to blow up the singular points in succession, altogether g times, so that the resulting sequence of multiplicities is $(2g, 2, \ldots, 2)$. At the last step, the (local) equation for the singular point $(u = 0, v = 0)$ is written as

$$u^2 + (a_d + a_{d-1}u + \cdots + a_0 u^d)u^{g+2-d}v$$
$$+ (b_{2g+2} + b_{2g+1}u + \cdots + b_0 u^{2g+2})v^2 = 0,$$

where $f(x) = \sum_{i=0}^{d} a_i x^i$ with $d \leq g+1$, $a_d \neq 0$ and $h(x) = \sum_{j=0}^{2g+2} b_j x^j$ with reference to the equation (ii) of 4.1. Blow up the point $(u = 0, v = 0)$ once more, and we obtain either two smooth (possibly non-k-rational) points or one possibly singular point according as $d = g+1$ or $d < g+1$. All through $g + 1$ blowing-ups thus far, the arithmetic k-genus has dropped down to

$$\frac{(2g + 2 - 1)(2g + 2 - 2)}{2} - \frac{2g(2g - 1)}{2} - g = g.$$

Therefore, the last proper transform of C is k-normal, which implies the k-normality of C. $\square$

4.2.1

Remark. We have actually shown that, in case $p > 2$, the point at infinity of a k-curve defined by the equation (i) is always k-smoothable. We can prove that, when $p = 2$, the curve defined by the equation (ii) is k-smoothable at infinity except in the case where either $d = g$, $b_{2g+1} \neq 0$ and $b_{2g+2} a_d^2 + b_{2g+1}^2 = 0$ or $d < g$, $b_{2g+1} = 0$ and $b_{2g+2} \notin k^2$ with the notations in the previous proof. In the first case, k-smoothability is determined by further conditions including other coefficients of $f(x)$ and $h(x)$. In the second case, if $b_{2g+2} \notin k^2$, the singular point is given by $(u = 0, t = 1/\sqrt{b_{2g+2}})$, where $t = v/u$. Since this point is not k-rational, it can not be further desingularized by the blowing-up over k. But, if $b_{2g+2} \in k^2$, then one more blowing-up is possible over k.

4.3 Hyperelliptic k-forms of $\mathbb{A}^1$

Theorem. *Let k be a non-perfect field of characteristic $p > 2$. Then, a hyperelliptic k-form of $\mathbb{A}^1$ of k-genus $g \geq 2$ and carrying a k-rational point is k-birationally equivalent to an affine plane curve of the type*

$$y^2 = x^{p^m} - a, \quad a \in k \setminus k^p,$$

for which $g = \frac{1}{2}(p^m - 1)$. Conversely, the complete k-normal model C of every such plane curve has a unique singular point P and $C \setminus \{P\}$ is a k-form of $\mathbb{A}^1$ of k-genus $\frac{1}{2}(p^m - 1)$.

Proof. Let C be a hyperelliptic k-curve of k-genus g defined by an equation $y^2 = f(x)$ as in 4.1, (i) for which the point at infinity is k-smoothable. Decompose $f(x)$ into a product of irreducible factors over k:

$$f(x) = \prod_i P_i(x)^{\mu_i} \times \prod_j Q_j(x)^{\nu_j}$$

with distinct non-separable irreducible polynomials $P_i(x)$ and distinct separable irreducible polynomials $Q_j(x)$. Note that one can assume $\mu_i = \nu_j = 1$ for all i, j. In fact, if $f(x) = g(x)^{2r+s}h(x)$ with $g(x)$ irreducible, $g(x) \nmid h(x)$ and $s = 0$ or 1, then by the birational change of variables $x' = x$, $y' = y/(g(x))^r$ the equation $y^2 = f(x)$ is transformed into $y'^2 = g(x')^s h(x')$ and the places at infinity of the new plane curve are exactly the same as the places at infinity of the old curve. Write anew the above equation as

$$y^2 = \prod_{i=1}^{t} P_i(x) \times \prod_{j=1}^{u} Q_j(x). \tag{4.1}$$

Assume that C is birationally equivalent to a k-form of $\mathbb{A}^1$. This assumption is equivalent to assuming the following two conditions:

(a) The function field $K := k(x, y)$ of C has exactly one singular place.
(b) The $\bar{k}$-genus of C is zero.

We shall deduce the consequences of the conditions (a) and (b).

Since the places at infinity of the curve defined by the equation (4.1) are all smooth by Lemma 4.1, by the condition (a) above, there exists exactly one singular place at finite distance of the curve. This implies that $t = 1$ and $P_1(x)$ has the form $x^{p^m} - a$ with $m > 0$ and $a \in k \setminus k^p$. Indeed, if $P_i(x) = b \cdot \prod_{j=1}^{r}(x^{p^m} - a_j)$ is a decomposition of $P_i(x)$ over k_s with $b \in k^*$, $a_j \in k_s$ $(1 \le j \le r)$ [3] , then the point $(x = a_j^{p^{-m}}, y = 0)$ is a singular point if $m > 0$ by the Jacobian criterion, but its local ring has the maximal ideal (y) which is principal and hence is a discrete valuation ring. Thus each $P_i(x)$ yields one singular point $(x = a_i^{p^{-m_i}}, y = 0)$ $(1 \le i \le r)$. So, only one factor $P_i(x)$ is present and it is of the form $b(x^{p^m} - a)$ with $b \in k^*$, $a \in k \setminus k^p$ and $m > 0$. Thus our equation becomes

$$y^2 = b(x^{p^m} - a)Q_1(x) \cdots Q_u(x). \tag{4.2}$$

Let us now extend the scalar field k to the algebraic closure $\bar{k}$, and view (4.2) as the defining equation of a $\bar{k}$-curve. The $\bar{k}$-rational point $(x = a^{p^{-m}}, y = 0)$ is not $\bar{k}$-normal and has multiplicity 2. Perform the $\bar{k}$-normalization by a series of blowing-ups whose first step is the substitution $y = wt$ with $w = x - a^{p^{-m}}$. After each blowing-up, we obtain a non-$\bar{k}$-normal point of multiplicity 2 or a $\bar{k}$-normal point. In the non-$\bar{k}$-normal

---

[3] An irreducible polynomial $f(x) \in k[x]$ is called *separable* if the derivative $f'(x)$ is a nonzero polynomial. Such a polynomial decomposes as $f(x) = a \cdot \prod_{i=1}^{n}(x - \alpha_i)$, where the α_i are mutually distinct elements of k_s. Given an irreducible polynomial $F(x) \in k[x]$ there exist a non-negative integer $m \ge 0$ and a separable polynomial $f(x) \in k[x]$ such that $F(x) = f(x^{p^m})$.

case, blow it up. Each blowing-up drops the genus by 1. After $\frac{1}{2}(p^m - 1)$ times blowing-ups, the equation becomes

$$T^2 = bW(W - \beta_1) \cdots (W - \beta_s)$$

with $b \in k^*$ and nonzero, mutually distinct $\beta_1, \ldots, \beta_s \in \overline{k}$, which represents a $\overline{k}$-normal affine curve. In this process of blowing-ups over $\overline{k}$, the arithmetic genus drops by $\frac{1}{2}(p^m - 1)$ which is equal to the number of blowing-ups conducted thus far. Then the condition (b) implies that $g = \frac{1}{2}(p^m - 1)$. Meanwhile, the degree of the curve defined by the equation (4.2) is

$$p^m + \sum_{j=1}^{u} \deg Q_j = 2g + 1 \quad \text{or} \quad 2g + 2,$$

whence $\sum_{j=1}^{u} \deg Q_j = 0$ or 1. Suppose that $\sum_{j=1}^{u} \deg Q_j = 1$. Then the equation (4.2) has the form

$$y^2 = b(x^{p^m} - a)(x - c)$$

with $c \in k$, and we may assume $c = 0$ by replacing x by $x + c$ and a by $a - c^{p^m}$. Then the equation can be written as

$$\frac{y^2}{x^{p^m+1}} = b(1 - ax^{-p^m}).$$

Therefore, by the birational transformation

$$(x, y) \mapsto \left(z = \frac{1}{x}, w = \frac{y}{x^{(p^m+1)/2}} \right),$$

we have

$$w^2 = b(1 - az^{p^m}).$$

All in all, the function field can be identified with that of an affine plane k-curve of the type

$$y^2 = b(x^{p^m} - a), \quad b \in k^*, \ a \in k \setminus k^p \tag{4.3}$$

whose k-genus is $g = \frac{1}{2}(p^m - 1)$.

Finally, the birational transformation $(x, y) \mapsto (bx, b^{(p^m-1)/2}y)$ changes the equation (4.3) to

$$y^2 = x^{p^m} - ab^{p^m}.$$

This gives the form of the equation we look for.

From the foregoing arguments, the converse is obvious. Namely, the k-normal completion of the affine plane curve $y^2 = x^{p^m} - a$, after its unique singular point is removed, gives an affine k-curve of genus $g = (p^m - 1)/2$ which is a k-form of $\mathbb{A}^1$.   □

4.4 Hyperelliptic k-forms of $\mathbb{A}^1$ in characteristic 2

Theorem. *Let k be a non-perfect field of characteristic 2. Then a hyperelliptic k-form of $\mathbb{A}^1$ of k-genus $g \geq 2$ and carrying a k-rational point is k-birationally equivalent to an affine plane k-curve of one of the following types:*

(A) $y^2 + (x^{2^i} + a)^{2^\ell} y + b + c(x^{2^i} + a)^{2^{\ell+1}} = 0$, *where* $i \geq 0, \ell \geq 0$, $a, c \in k$, $b \in k^*$; $a \notin k^2$ *if* $i > 0$ *and* $b \notin k^2$ *if* $\ell > 0$; $g = 2^{i+\ell} - 1$.

(B) $y^2 = x(x + \alpha)^{2g} + E(x)$, *where* $\alpha \in \overline{k}$, $(x + \alpha)^{2g} \in k[x]$, $E(x) \in k[x]$ *is an even polynomial*[4] *of degree $2g + 2$ and $E(\alpha) \notin k^2$ if $\alpha \in k$.*

The proof will be given in the subsections 4.4.1 $\sim$ 4.4.7.

4.4.1

Let C be a hyperelliptic k-curve of genus g defined by an equation

$$y^2 + f(x)y + h(x) = 0 \qquad (4.4)$$

subject to the conditions in Lemma 4.1. Assume that C is k-birationally equivalent to a k-form of $\mathbb{A}^1$. As in the earlier case, this is equivalent to assuming that

(a) The function field $K := k(x, y)$ of C has one singular place at finite distance.

(b) The $\overline{k}$-genus of C is zero.

Then C is k-normal by Lemma 4.2 and smooth except at precisely one point since the points at infinity being k-smoothable.

4.4.2

Lemma. *With the foregoing notations and assumptions, the coefficient $f(x)$ of y in the equation (4.4) is either 0 or $c(x+\alpha)^{g+1}$, where $c \in k^*$ and $\alpha \in \overline{k}$.*

Proof. Suppose that $f(x) \neq 0$. Let $S = (x = \alpha, y = \beta)$ be the unique k-normal singular point on C which is a one-place point. Note that the same situation holds even if we change the base from k to k_s. The point S

---

[4]A polynomial of the form $f(x^2)$.

is clearly not k_s-rational. By the Jacobian criterion used over $\overline{k}$ it follows that the elements $\alpha, \beta \in \overline{k}$ must satisfy

$$f(\alpha) = 0, \quad f'(\alpha)\beta + h'(\alpha) = 0, \quad \beta^2 + h(\alpha) = 0. \tag{4.5}$$

If $\alpha \in k_s$, then $f'(\alpha) = 0$. For otherwise, $\beta = f'(\alpha)^{-1}h'(\alpha) \in k_s$ and S is k_s-rational. This is a contradiction. If $\alpha \notin k_s$, then $f(x) = (x^{2^i} + a)f_1(x)$ for some $f_1(x) \in k_s[x]$, $\alpha^{2^i} = a \in k_s$ with $i > 0$, and $f'(\alpha) = 0$ again. In either case,

$$f(x) = (x + \alpha)^m f_1(x) \tag{4.6}$$

with $2 \leq m \leq g + 1$, where $f_1(x) \in k_s[x]$ and $f_1(\alpha) \neq 0$. By setting $X = x + \alpha$, $Y = y + \beta$, we can rewrite the equation (4.4) of $\overline{C} := C \otimes_k \overline{k}$ as

$$Y^2 + X^m f_1(X + \alpha)Y + h_1(X) = 0, \tag{4.7}$$

where $h_1(X) = f(X + \alpha)\beta + \beta^2 + h(X + \alpha)$ and $h_1(0) = 0$ as seen by the equation (4.5). Since the $\overline{k}$-curve $\overline{C}$ must have the point $(X = 0, Y = 0)$ as a singular point with multiplicity sequence

$$(\underbrace{2, 2, \ldots, 2}_{m-1}, 1, \ldots, 1)$$

which becomes a $\overline{k}$-smooth point only after g blowing-ups, it follows that $m = g + 1$ by Lemma 4.4.3 below [5].

Hence it follows from the equation (4.6) that $f(x) = c(x + \alpha)^{g+1}$, where $c \in k^*$ clearly.

4.4.3

Lemma. *Let $\overline{B}$ be an affine $\overline{k}$-curve defined by*

$$Y^2 + X^m F(X)Y + (b_0 + b_1 X + \cdots + b_n X^n) = 0, \tag{4.8}$$

where $m > 1$, $F(X) \in \overline{k}[x]$, $F(0) \neq 0$, $H(X) := b_0 + b_1 X + \cdots + b_n X^n \in \overline{k}[X]$. Then, the point $(X = 0, Y = b_0^{1/2})$ on $\overline{B}$ is a singular point of multiplicity 2 if and only if $b_1 = 0$. Furthermore, in that case, one blowing-up centered at $(X = 0, Y = b_0^{1/2})$ transforms $\overline{B}$ into the curve defined by

$$Y^2 + X^{m-1}F(X)Y + b_0^{1/2}X^{m-2}F(X) + b_2 + b_3 X + \cdots + b_n X^{n-2} = 0.$$

---

[5]If m blowing-ups are possible after the blowing-up with center $(X = 0, Y = 0)$, the curve after m blowing-ups is defined by an equation of the form $Y^2 + F(X)Y + H(X) = 0$ with $F(0) \neq 0$. Since this equation is separable over $k[X]$, there are two points lying over the point $(X = 0, Y = 0)$, which contradicts the assumption that the point $(X = 0, Y = 0)$ is geometrically a one-place point.

Proof. The first assertion is an immediate consequence of the Jacobian criterion. Suppose now $b_1 = 0$. Then, setting $Z = Y - b_0^{1/2}$, we rewrite the equation (4.8) as

$$Z^2 + X^m F(X)Z + b_0^{1/2} X^m F(X) + (b_2 X^2 + \cdots + b_n X^n) = 0.$$

Then, the blowing-up centered at $(X = 0, Z = 0)$ is achieved by setting $Z = XY_1$, and we obtain

$$Y_1^2 + X^{m-1} F(X)Y_1 + b_0^{1/2} X^{m-2} F(X) + (b_2 + \cdots + b_n X^{n-2}) = 0.$$

$\square$

4.4.4

We assume that the affine plane k-curve C defined by an equation

$$y^2 + c(x + \alpha)^{g+1} y + h(x) = 0, \quad c \neq 0, \quad \deg h(x) \leq 2g + 2 \qquad (4.9)$$

is k-birationally equivalent to a k-form of $\mathbb{A}^1$ of genus $g \geq 2$ and that the unique singular point is at finite distance. Clearly, we may assume $c = 1$ by replacing y by y/c.

4.4.4.1. Lemma. *With the above notations and assumptions, we have:*

(1) *By a k-automorphism of the (x, y)-plane, we can transform the equation (4.9) for C to the one for which $h(x)$ has the form*

$$h(x) = b(x + \alpha)^{2g+2} + h_1(x), \quad b \in k, \quad h_1(x) \in k[x], \quad \deg h_1(x) \leq g.$$

(2) *There exists a separable extension k' of k with $[k' : k] \leq 2$ such that the equation (4.9) for $C \otimes k'$ is transformed into*

$$y^2 + (x + \alpha)^{g+1} y + h_1(x) = 0, \quad h_1(x) \in k[x], \quad \deg h_1(x) \leq g$$

by a k'-automorphism $(x, y) \mapsto (x, y + a(x + \alpha)^{g+1})$ of the (x, y)-plane, where $a \in k'$.

Proof. (1) Since $(x + \alpha)^{g+1} \in k[x]$ and $\deg h(x) \leq 2g + 2$, we may write $h(x) = b(x + \alpha)^{2g+2} + $ (a polynomial in x of degree $< 2g + 2$). If we write

$$h(x) = b(x + \alpha)^{2g+2} + b'x^d + \text{(terms of degree} < d)$$

with $g + 1 \le d < 2g + 2$, then, by the substitution of $y + b'x^{d-g-1}$ for y in the equation (4.9), we have

$$(y + b'x^{d-g-1})^2 = y^2 + b'^2 x^{2(d-g-1)},$$
$$(x + a)^{g+1}(y + b'x^{d-g-1}) = (x + a)^{g+1}y + b'x^d + \text{(terms of degree} < d),$$
$$d - 2(d - g - 1) = 2g + 2 - d > 0.$$

Hence the term $b'x^d$ has been eliminated. By these substitutions as long as $g + 1 \le d$, we can transform the equation (4.9) to the desired form.

(2) Let k' be the extension of k adjoined a root γ of $\zeta^2 + \zeta + b = 0$, where b is as in the assertion (1). Then the substitution $(x, y) \mapsto (x, y + \gamma(x+a)^{g+1})$ eliminates the term $b(x + a)^{2g+2}$ in $h(x)$. □

4.4.4.2. We assume, for the time being, that $\deg h(x) \le g$ in the equation (4.9) for C. We then claim that $h(x)$ is an even polynomial. To see this, write $Y = y, X = x + a$ in the equation (4.9) to obtain the equation of the curve C,

$$Y^2 + X^{g+1}Y + H(X) = 0,$$

where $H(X) := h(X + a) = \beta_0 + \beta_1 X + \cdots + \beta_g X^g \in \overline{k}[X]$. We now apply Lemma 4.4.3 repeatedly. If $g + 1 = 2r$, blow up the unique singular point $(X = 0, Y = \beta_0^{1/2})$ of C and its $(r - 1)$ infinitely near singular points in succession, which is possible because $(g + 1) - r = r > 1$ as $g \ge 2$. The resulting equation over $\overline{k}$ is

$$Y^2 + X^{g+1-r}Y + \beta_0^{1/2} + \beta_2^{1/2}X + \cdots + \beta_{g-1}^{1/2}X^{g-r} = 0$$

with $\beta_1 = \beta_3 = \cdots = \beta_g = 0$. If $g = 2r$, repeat the same process r times to get

$$Y^2 + X^{g+1-r}Y + \beta_g + \beta_0^{1/2}X + \cdots + \beta_{g-2}^{1/2}X^{g-r} = 0$$

with $\beta_1 = \beta_3 = \cdots = \beta_{g-1} = 0$. In both cases, $H(X)$ is therefore an even polynomial, and hence $h(x)$ is an even polynomial.

4.4.4.3. One can push forward the arguments of the paragraph 4.4.4.2. After $r = \left[\frac{(g+1)}{2}\right]$ blowing-ups, the coefficient of Y in the equation is X^{g+1-r}. If $g + 1 - r > 1$, then perform $\left[\frac{(g+1-r)}{2}\right]$ blowing-ups which will allow us to deduce that $\beta_{2i} = 0$ for all odd i if g is odd, and that $\beta_{2i} = 0$ for all even i if g is even. This process continues until the second term of the equation becomes XY and then altogether g blowing-ups are performed.

Consequently, all but one β_i is nonzero and i is even by 4.4.4.2. Hence we have $h(X + \alpha) = H(X) = \beta_{2m} X^{2m}$ or equivalently $h(x) = \beta_{2m}(x + \alpha)^{2m}$.

4.4.4.4. Lemma. *A plane k-curve defined by*

$$y^2 + f(x)y + h(x) = 0$$

such that $\deg f(x) = g+1$ *and* $\deg h(x) \leq g$ *has a unique one-place point P at infinity whose singularity is of multiplicity g. The point P_1 which is an infinitely near point of P in the first infinitesimal neighborhood is smooth. The genus of the curve is $\leq g$, and the equality holds if and only if the curve is k-normal.*

The proof is an easy exercise of blowing-up. So we omit it.

4.4.5

Assuming that $\deg h(x) \leq g$ in the equation (4.9) for C, we have shown that the equation (4.9) takes the form

$$y^2 + (x + \alpha)^{g+1}y + b(x + \alpha)^{2m} = 0, \ b \in k^*. \tag{4.10}$$

4.4.5.1. We consider first the case where $\alpha \in k$. Then C is clearly k-isomorphic to C_1 defined by an equation $y^2 + x^{g+1}y + bx^{2m} = 0$. If $m > 0$, this curve is k-birationally equivalent to a curve C_1' defined an equation

$$z^2 + x^{g+1-m}y + b = 0$$

by a substitution $z = y/x^m$, so that by Lemma 4.4.4.4 the k-genus of $k(C) = k(C_1')$ is not greater than $g - m$, which is a contradiction. Hence $m = 0$ follows. Thus the equation (4.9) is reduced to

$$y^2 + x^{g+1}y + b = 0. \tag{4.11}$$

Moreover, the resolution of the unique singularity of C shows that $g+1 = 2^\ell$ with $\ell \geq 2$, and we must have $b \in k \setminus k^2$. Conversely, the affine plane k-curve defined by the equation (4.11) is k-birationally equivalent to a k-form of $\mathbb{A}^1$ of genus g, whose unique singular place is k-rational when restricted to the unique rational subfield $k(x)$.

4.4.5.2. We consider next the case where $\alpha \notin k$. Since $(x + \alpha)^{g+1} \in k[x]$, $g + 1 \equiv 0 \pmod{2}$. Then $(x^2 + \alpha^2)^{(g+1)/2} \in k[x]$. If $\alpha^2 \notin k$ then $\frac{(g+1)}{2} \equiv 0 \pmod{2}$ and $(x^{2^2} + \alpha^{2^2})^{(g+1)/2^2} \in k[x]$. This argument implies that α is purely inseparable over k. Write $a := \alpha^{2^i} \in k$ and $\alpha^{2^{i-1}} \notin k$ for some $i > 0$. Then $2^i \mid (g+1)$. Put $u := x^2$ and rewrite the equation (4.10) as

$$y^2 + (u + \alpha^2)^{(g+1)/2}y + b(u + \alpha^2)^m = 0, \qquad (4.12)$$

which defines a plane curve $C^{(1)}$ in the (u, y)-plane. The inclusion $k[u, y] \hookrightarrow k[x, y]$ induces a natural finite k-morphism $C \to C^{(1)}$. The only possible singular point of $C^{(1)}$ is $(u = \alpha^2, y = 0)$ if $m > 0$ and $(u = \alpha^2, y = b^{1/2})$ if $m = 0$, which is dominated by the unique singular point of C. Denote by $\mathcal{O}^{(1)}$ and $\mathcal{O}$ the local rings of these points. Then $\mathcal{O} = \mathcal{O}^{(1)}[x]$ and $\mathcal{O}$ is a faithfully flat extension of $\mathcal{O}^{(1)}$ of rank 2. Since $\mathcal{O}$ is k-normal by assumption, so is $\mathcal{O}^{(1)}$. Thus $C^{(1)}$ is everywhere k-normal and hence $k(C^{(1)})$ has k-genus exactly equal to $\frac{(g+1)}{2} - 1 = \frac{(g-1)}{2}$.

The field $k(C^{(1)})$ is contained in $k(C)$ which is a k-form of the rational function field $k(t)$. Hence $k(C^{(1)})$ is a k-form of $k(t^2)$ by Lüroth's theorem. The possible singularity of $C^{(1)}$ is a one-place singularity since the corresponding singularity of C is such a one-place singularity. Hence $C^{(1)}$ defines a (possibly trivial) k-form of $\mathbb{A}^1$ of genus $\frac{(g-1)}{2}$. Further this implies the following assertions.

(1) If $\alpha^2 \in k$, i.e., $i = 1$, then, by the results in 4.4.5.1, we have $m = 0$, $\frac{(g-1)}{2} = 2^\ell - 1$ with $\ell \geq 1$ and $b \in k \setminus k^2$. Then $g + 1 = 2^{\ell+1}$.

(2) If $\alpha^2 \notin k$, i.e., $i > 1$, then m is even by 4.4.4.3, and $\frac{(g+1)}{2}$ is an even integer, too.

In the above case (2), we can make another substitution $v := u^2$ in the equation (4.12) to obtain $C^{(2)}$ which defines a k-form of $\mathbb{A}^1$ of genus $\frac{(g+1)}{4} - 1$. Continuing this argument, we can reach to a k-normal curve $C^{(i)}$ defined by

$$y^2 + (u + a)^N y + b(u + a)^M = 0,$$

where $N = \frac{(g+1)}{2^i}$, $M = \frac{m}{2^{i-1}}$, $M < N$, and $C^{(i)}$ defines a (possibly trivial) k-form of $\mathbb{A}^1$ of genus $N - 1$. If $N - 1 \geq 2$ then $M = 0$ and $N - 1 = 2^\ell - 1$ with $\ell \geq 2$ by 4.4.5.1. If $N - 1 = 1$ (hence $M \leq 1$), $C^{(i)}$ has a unique singular point determined by $u = a$. Hence $M = 0$. If $N - 1 = 0$ then $M = 0$ because $M < N$. In any case we have $M = 0$ (whence $m = 0$), and $N - 1 = 2^\ell - 1$ with $\ell \geq 0$. Thus $g = 2^{i+\ell} - 1$ with $i > 0$, $\ell \geq 0$ and $i + \ell \geq 2$. Unless $N = 1$, the condition $b \in k \setminus k^2$ is required for $C^{(i)}$ to have genus $N - 1$. Hence the equation (4.9) is brought down to

$$y^2 + (x^{2^i} + a)^{2^\ell} y + b = 0 \qquad (4.13)$$

under the condition that $\deg h(x) \leq g$ and $\alpha \notin k$, where $a \in k \setminus k^2$, $b \in k$, $i > 0$, $\ell \geq 0$, $i + \ell \geq 2$, $g = 2^{i+\ell} - 1$, and $b \in k \setminus k^2$ if $\ell > 0$.

Conversely, the affine plane k-curve defined by the equation (4.13) is k-birationally equivalent to a k-form of $\mathbb{A}^1$ of genus $g = 2^{i+\ell} - 1$, whose unique singular place induces a k-rational place on the unique rational subfield $k(x)$.

4.4.6

By Lemma 4.4.4.1, we may assume that the equation (4.9) has the form

$$y^2 + (x + \alpha)^{g+1} y + b(x + \alpha)^{2g+2} + h_1(x) = 0, \quad \deg h_1(x) \leq g \quad (4.14)$$

and that, over the extension field $k' = k(\gamma)$ of k adjoined a root γ of the equation $\zeta^2 + \zeta + b = 0$, the equation (4.9) is transformed to

$$y^2 + (x + \alpha)^{g+1} y + h_1(x) = 0 \qquad (4.15)$$

by a k'-automorphism $(x, y) \mapsto (x, y + \gamma(x + \alpha)^{g+1})$. The equation (4.15) is the one we started with in 4.4.5, and we have shown that it coincides with either (4.11) or (4.13). We note that both the curves (4.14) and (4.15) are defined over k, though the automorphism sending (4.14) to (4.15) is defined over k'. Hence, if we know the equation (4.15), it is then easy to get back the equation (4.14), which is the one we classified as the type (A) in Theorem 4.4.

4.4.7

We shall now return to the situation in 4.4.1 and treat the remaining case where $f(x) = 0$ in the equation (4.4).

4.4.7.1. Lemma. *Assume that the point at infinity of the curve C defined by*

$$y^2 = h(x) = b_0 + b_1 x + \cdots + b_{2g+2} x^{2g+2}, \quad b_{2g+2} \neq 0 \qquad (4.16)$$

ie either smooth or k-smoothable. Write $h(x) = E_1(x) + E_2(x)x$ where $E_1(x), E_2(x)$ are even polynomials in $k[x]$. Then we have the following assertions.

(1) *If $k(C)$ is the function field of a k-form of $\mathbb{A}^1$ of k-genus g, then*

$$E_2(x) = b(x + \alpha)^{2g} \qquad (4.17)$$

with $b \in k^*$, $\alpha \in \overline{k}$, and $E_1(\alpha) \notin k^2$ if $\alpha \in k$.

(2) *Conversely, if $F_2(x)$ is as in (4.17) above with the accompanying conditions, then $k(C)$ is the function field of a k-form of $\mathbb{A}^1$ of k-genus $\leq g$.*

Proof. (1) Suppose that C is k-birationally equivalent to a k-form of $\mathbb{A}^1$. Let $S = (x = \alpha, y = \beta)$ be the unique k-normal, non-k-rational singular point. Hence all other points are smooth. Then $\beta^2 = E_1(\alpha)$, $E_2(\alpha) = 0$ and (α, β) is the unique pair in $\overline{k} \times \overline{k}$ satisfying these two conditions. Therefore $E_2(x)$ is as in (4.17), where $b \neq 0$ by Lemma 2.7.4 if $b_{2g+2} \notin k^2$ (for otherwise the place at infinity is singular) and $b \neq 0$ if $b_{2g+2} \in k^2$ because C has k-genus g. Furthermore, since S is non-k-rational, either $\alpha \notin k$ or $E_1(\alpha) \notin k^2$.

(2) Conversely, suppose that $h(x) = E_1(x) + E_2(x)x$ satisfies the stated conditions. Then $k(C)$ is clearly a k-form of the rational function field $k(t)$, and its complete k-normal model $\widetilde{C}$ has a single one-place singular point S at finite distance on C. Hence $C \setminus \{S\}$ is a k-form of $\mathbb{A}^1$. The genus drop effected by S brings down the k-genus to no more than g. □

4.4.7.2. In the assertion (2) above, we have no results to decide whether S is k-normal or not. If it is not k-normal, the k-genus of C becomes less than g. So, we can say only that C has k-genus $\leq g$.

The foregoing subsections 4.4.1 $\sim$ 4.4.7 establish the validity of Theorem 4.4.

4.4.8

Remark. In the case (A) of Theorem 4.4, assume that $i = 0$, i.e., the place of $k(x)$ induced by the singular place of the curve is k-rational. Then we may assume that $a = 0$. The substitution $x = x_1^{-1}$ and $y = x_1^{2^\ell} y_1$ brings the equation to

$$y_1^2 + y_1 + c = bx_1^{2^{\ell+1}}.$$

This is a principal homogeneous space of a k-group G of Russell type defined by

$$y_1^2 + y_1 = bx_1^{2^{\ell+1}},$$

which has height $\ell + 1$.

4.5   Existence theorem of hyperelliptic forms of $\mathbb{A}^1$

In Theorems 4.3 and 4.4, we put the assumptions that k is an infinite field and that a hyperelliptic k-form X of $\mathbb{A}^1$ under consideration carries a k-rational point. Nevertheless, the second condition has been used, so far, only to ensure that the unique subfield K_0 of the function field $K = k(X)$ is rational over k. Hence we may weaken the second condition to the effect that K_0 is rational over k. Thus, there remains still open the problem of determining a hyperelliptic k-form X of $\mathbb{A}^1$ whose function field K has the unique non-k-rational subfield of genus zero.

In particular, if a (hyperelliptic) k-form X of $\mathbb{A}^1$ has k-genus 2, its complete k-normal model $\widetilde{C}$ is a double covering of $\mathbb{P}^1_k$, the covering morphism being given by $H^0(\widetilde{C}, \omega_{\widetilde{C}/k})$, where $\omega_{\widetilde{C}/k}$ is the dualizing invertible sheaf of $\widetilde{C}$. Therefore, the unique subfield K_0 of genus zero is automatically rational over k. As an immediate consequence of Theorems 4.3 and 4.4, we have the following result.

Theorem. *The k-forms of $\mathbb{A}^1$ of genus 2 exist only if the characteristic of the infinite field k is either 2 or 5. Such a k-form is k-birationally equivalent to one of the following k-normal affine plane curves:*

(I) *In case $p = 2$, $y^2 = x(x+\alpha)^4 + E(x)$, where $\alpha^4 \in k$, $E(x) \in k[x]$ is an even polynomial of degree 6, and either $\alpha \notin k$ or $E(\alpha) \notin k^2$.*

(II) *In case $p = 5$, $y^2 = x^5 + a$ with $a \in k \setminus k^5$.*

Chapter 5

Automorphisms

Let X be a nontrivial k-form of $\mathbb{A}^1$. By Lemmas 3.3.1 and 3.3.3, X is a principal homogeneous space for a k-group of Russell type if and only if $\operatorname{Aut}_{k_s}(X \otimes k_s)$ is an infinite group. If $k(X)$ has genus 0, use Theorem 2.8 and the proof of Lemma 3.3.5. In this chapter, we shall consider the case where $\operatorname{Aut}_{k_s}(X \otimes k_s)$ is a finite group.

5.1   Finite $\operatorname{Aut}_k(X)$ for a k-form X of $\mathbb{A}^1$

Let K be a purely inseparable k-form of the rational function field $k(t)$ of height $\lambda' \geq 0$, and let C be a complete k-normal model of K. Then, for any finite separable extension k' of k, since $C \otimes k'$ is a complete k'-normal model of $k' \otimes K$, there exists clearly an isomorphism $\operatorname{Aut}_{k'}(k' \otimes K) \cong \operatorname{Aut}_{k'}(C \otimes k')$ which is compatible with the extension. Since $(k, \varphi^{\lambda'}) \otimes_k K \cong k(K^{p^{\lambda'}}) = k(x)$ (with the notations in 2.5.1), the subfield $k(x)$ is invariant under $\operatorname{Aut}_k(K)$, which acts by $1 \otimes \sigma$ on $(k, \varphi^{\lambda'}) \otimes_k K$. Hence, for any $\sigma \in \operatorname{Aut}_k(K)$, $\sigma(x) = (\alpha x + \beta)(\gamma x + \delta)^{-1}$ with $\alpha, \beta, \gamma, \delta \in k$ and $\alpha\delta - \beta\gamma \neq 0$. Suppose that K is the function field of a nontrivial k-form X of $\mathbb{A}^1$. Then $X = C \setminus \{P_\infty\}$ with a non-k-rational point P_∞. Clearly, there exists an injective homomorphism $\operatorname{Aut}_k(X) \to \operatorname{Aut}_k(C)$, compatible with finite separable extension k'/k. An element σ of $\operatorname{Aut}_k(C)$ belongs to $\operatorname{Aut}_k(X)$ if and only if $\sigma(P_\infty) = P_\infty$.

5.1.1

Lemma.   *With the notations and assumptions as above, we have the following assertions.*

(1) *If the k-genus of K is positive, then $\operatorname{Aut}_k(X) = \operatorname{Aut}_k(C)$.*

(2) *If the k-genus of K is zero and X is non-isomorphic to a k-group of Russell type G defined by $y^2 = x + ax^2$ nor a principal homogeneous space for G defined by $y^2 = x + ax^2 + b$, where $p = 2$ and $a, b \in k \setminus k^2$, then $\mathrm{Aut}_k(X) = \{1\}$.*

Proof. (1) If the k-genus of K is positive, then P_∞ is a unique singular point of C. Hence every element σ of $\mathrm{Aut}_k(C)$ fixes the point P_∞. This means that $\mathrm{Aut}_k(X) = \mathrm{Aut}_k(C)$.

(2) Suppose first that K is rational over k, i.e., C is isomorphic to $\mathbb{P}^1_k$. Choose an inhomogeneous coordinate t of $\mathbb{P}^1_k$ such that $t^{p^n} = a$ at P_∞ with $a \in k \setminus k^p$, where $n = \mathrm{ht}(X)$ (cf. 2.8). Take an element σ of $\mathrm{Aut}_k(X)$. Then σ, extended to an element of $\mathrm{Aut}_k(C)$, fixes the point P_∞. Write $\sigma(t) = (bt + c)(dt + e)^{-1}$ with $b, c, d, e \in k$ and $be - cd \neq 0$. Then, for $\alpha \in k^{p^{-n}}$ with $\alpha^{p^n} = a$, we have $d\alpha^2 + (e - b)\alpha - c = 0$. Since X is not isomorphic to G of the said type, either $p \neq 2$ or $p = 2$ and $n \geq 2$ (cf. the proof of Lemma 3.3.5). Then $1, \alpha$ and α^2 are linearly independent over k. Hence $d = c = 0$ and $e = b$, i.e., σ is the identity. Suppose next that K has k-genus zero but K is not rational over k. Then $p = 2$ by Lemma 2.3.3. Clearly, $n = \mathrm{ht}(X) = \mathrm{ht}(X \otimes k_s)$ and $k_s \otimes K$ is rational over k_s. Hence, by the above case, $\mathrm{Aut}_{k_s}(X \otimes k_s) = \{1\}$ unless $X \otimes k_s \cong G \otimes k_s$. Therefore, by Lemma 3.3.1, $\mathrm{Aut}_k(X) = \{1\}$ unless X is a principal homogeneous space for G. $\qquad\square$

5.1.2

Corollary. *Assume that $\mathrm{ht}(X) > \mathrm{ht}(K)$, and that either $p > 2$ or $\mathrm{ht}(X) - \mathrm{ht}(K) \geq 2$. Then $\mathrm{Aut}_k(X) = 1$.*

Proof. Let $\lambda' := \mathrm{ht}(K)$ and let $X' := X \otimes k'$ with $k' := k^{p^{-\lambda'}}$. If $p > 2$ then $\mathrm{ht}(X') > 0$ by the assumption $\mathrm{ht}(X) > \mathrm{ht}(K)$. If $p = 2$ then $\mathrm{ht}(X') \geq 2$ again by the assumption and hence X' is not k'-isomorphic to the k'-group of Russell type or its principal homogeneous space in Lemma 5.1.1, (2). Hence $\mathrm{Aut}_{k'}(X') = \{1\}$ by Lemma 5.1.1, (2). Since $\mathrm{Aut}_k(X)$ is a subgroup of $\mathrm{Aut}_{k'}(X')$, we have $\mathrm{Aut}_k(X) = \{1\}$. $\qquad\square$

5.1.3

Lemma. *If the group $\mathrm{Aut}(X)$ for a k-form X of $\mathbb{A}^1$ is elementwise of finite order not divisible by p, then $\mathrm{Aut}_k(X)$ is a finite cyclic group.*

Proof. Write $X = \operatorname{Spec} A$. Let $\lambda := \operatorname{ht}(X)$. Then $k[A^{p^\lambda}] = k[u]$. An element σ of $\operatorname{Aut}_k(X)$ determines a k-automorphism σ^* of A and hence a k-automorphism of $k[u]$ via $a^{p^\lambda} \mapsto (\sigma^*(a))^{p^\lambda}$. Thus there exists an injective group homomorphism $\operatorname{Aut}_k(X) \to \operatorname{Aut}_k(\mathbb{A}^1)$, where $\operatorname{Aut}_k(\mathbb{A}^1)$ satisfies an exact sequence

$$1 \longrightarrow k^+ \longrightarrow \operatorname{Aut}_k(\mathbb{A}^1) \overset{\rho}{\longrightarrow} k^* \longrightarrow 1.$$

The restriction of ρ onto the subgroup $\operatorname{Aut}_k(X)$ is injective because any element of $\operatorname{Aut}_k(X)$ has order n which is prime to p. Therefore $\operatorname{Aut}_k(X)$ is a finite cyclic group. $\square$

5.2 Quotients of k-forms of $\mathbb{A}^1$ by finite groups

We consider now the quotient scheme of a k-form of $\mathbb{A}^1$ under a finite group action.

5.2.1

Lemma. *Let G be a finite group of k-automorphisms of a polynomial ring $k[t]$. Then the G-invariant subring $k[t]^G$ is k-isomorphic to a polynomial ring $k[u]$.*

Proof. The action of G on $k[t]$ is given by $g(t) = a(g)t + b(g)$ for $g \in G$ with $a(g) \in k^*$ and $b(g) \in k$. Further, $a(hg) = a(h)a(g)$ and $b(hg) = a(g)b(h) + b(g)$ for $g, h \in G$. In particular, $a : g \in G \mapsto a(g) \in k^*$ is a group homomorphism and the image $a(G)$ is a finite cyclic subgroup of k^*. Let $H = \operatorname{Ker} a$. Then H is a normal subgroup of G and embedded into an additive subgroup of k^+ by an injective group homomorphism $b : H \to k^+$. Hence H is a vector subgroup of finite rank over the prime field $\mathbb{F}_p$. Let H_1 be a subgroup of H isomorphic to $\mathbb{Z}/p\mathbb{Z}$. Then $k[t]^H = \left(k[t]^{H_1}\right)^{H/H_1}$ and H_1 acts on $k[t]$ via a translation $h(t) = t + \alpha$, where $H_1 = \langle h \rangle$ and $\alpha \in k^*$. Then $k[t]^{H_1} = k[t_1]$ with $t_1 = t^p - ct$ with $c = \alpha^{p-1}$. By induction on $\mathbb{F}_p$-rank of H, it follows that $k[t]^H = k[u]$ and $k[t]^G = k[u]^{G/H}$, where $G/H \cong \mathbb{Z}/n\mathbb{Z}$ with $\gcd(n, p) = 1$. Then $k[u]^{G/H} = k[v]$ with $v = u^n$. Thus $k[t]^G$ is a polynomial ring $k[v]$. $\square$

5.2.2

Theorem. *Let G be a finite group of k-automorphisms of a k-form of $\mathbb{A}^1$. Then the quotient scheme X/G is a k-form of $\mathbb{A}^1$.*

Proof. Write $X = \operatorname{Spec} A$. Then G acts naturally on A and the action can be extended to $\overline{k} \otimes A$ in a natural fashion. We show that $\overline{k} \otimes A^G \cong (\overline{k} \otimes A)^G$. In fact, consider a k-module homomorphism $(\pi - \delta) : A \to \prod_{g \in G} A$ defined by

$$a \in A \longmapsto (g_1(a) - a, g_2(a) - a, \ldots, g_n(a) - a) \in \underbrace{A \times A \times \cdots \times A}_{|G|} \;,$$

where $G = \{g_1, g_2, \ldots, g_n\}$. Then $A^G = \operatorname{Ker}(\pi - \delta)$. Hence we have an exact sequence of k-modules

$$0 \longrightarrow A^G \longrightarrow A \overset{\pi - \delta}{\longrightarrow} \prod_{g \in G} A \;.$$

Since $\overline{k}/k$ is a flat extension, we have an exact sequence of $\overline{k}$-modules

$$0 \longrightarrow \overline{k} \otimes A^G \longrightarrow \overline{k} \otimes A \overset{\pi - \delta}{\longrightarrow} \prod_{g \in G} (\overline{k} \otimes A).$$

Hence $(\overline{k} \otimes A)^G \cong \overline{k} \otimes A^G$. Since $(\overline{k} \otimes A)^G \cong \overline{k}[t]^G \cong \overline{k}[v]$ by 5.2.1, it follows that $\overline{k} \otimes A^G \cong \overline{k}[v]$ and $X/G = \operatorname{Spec}(A^G)$ is a k-form of $\mathbb{A}^1$. $\qquad\square$

5.3 Case of hyperelliptic forms of the affine line

We shall consider $\operatorname{Aut}_k(X)$ for a hyperelliptic k-form X of $\mathbb{A}^1$ under the hypothesis that the base field k is an infinite field and X has a k-rational point.

5.3.1

Theorem. *Assume $p > 2$ and the above hypothesis. Then $|\operatorname{Aut}_k(X)| = 2$ for every hyperelliptic k-form X of $\mathbb{A}^1$.*

Proof. By Theorem 4.3, X is k-birationally equivalent to an affine plane curve $y^2 = x^{p^m} - a$ with $m \geq 1$ and $a \in k \setminus k^p$. Then the subfield $k(x)$ is invariant under any birational k-automorphism of $k(X)$. Hence each automorphism $\sigma \in \operatorname{Aut}_k(X)$, restricted on $k(x)$, is written in the form

$$\sigma(x) = \frac{bx + c}{dx + e}, \quad b, c, d, e \in k.$$

Let $\alpha^{p^m} = a$ with $\alpha \in \overline{k}$. Since $(x = \alpha, y = 0)$ determines the unique singular place S of $k(X)$, σ fixes the point $(x = \alpha, y = 0)$. Hence we have

$$d\alpha^2 + (e - b)\alpha - c = 0.$$

Since $1, \alpha, \alpha^2$ are linearly independent over k, $d = c = 0$ and $e = b$. Hence $\sigma(x) = x$. Then we must have $\sigma(y)^2 = y^2$. Hence $\sigma(y) = \pm y$. Since $\sigma(x) = x$ and $\sigma(y) = -y$ defines a nontrivial element of $\mathrm{Aut}\,_k(X)$, we have $|\mathrm{Aut}\,_k(X)| = 2$. $\qquad\square$

5.3.2

Theorem. *Assume that $p = 2$ and that, with the notations in Theorem 4.4, $i > 1$ in the case (A) and $\alpha^2 \notin k$ in the case (B). Then $\mathrm{Aut}\,_k(X) \cong \mathbb{Z}/2\mathbb{Z}$ in the case (A) and $\mathrm{Aut}\,_k(X) = \{1\}$ in the case (B).*

Proof. By Theorem 4.4, the assumption that $i > 1$ in the case (A) and $\alpha^2 \notin k$ in the case (B) implies that the k-genus of X is larger than one. Hence the subfield $k(x)$ is invariant under any automorphism of $\mathrm{Aut}\,_k(X)$ by the arguments in 5.1. So, $\sigma(x) = x$. In the case (A), let $z = y / \left(x^{2^i} + a\right)^{2^\ell}$. Then the equation of X is written as

$$z^2 + z = b\left(x^{2^i} + a\right)^{-2^{\ell+1}} + c.$$

Hence we obtain

$$(\sigma(z) + z)(\sigma(z) + z + 1) = 0.$$

This implies that either $\sigma(z) = z$ or $\sigma(z) = z + 1$. In the latter case, σ is determined by $\sigma(x) = x$ and $\sigma(z) = z + 1$, whence $\sigma(y) = y + \left(x^{2^i} + a\right)^{2^\ell}$. Then σ determined this way is clearly an element of $\mathrm{Aut}\,_k(X)$. Thus $\mathrm{Aut}\,_k(X) \cong \mathbb{Z}/2\mathbb{Z}$. In the case (B), $\sigma(x) = x$ and $\sigma(y)^2 = y^2$. Hence $\sigma(y) = y$. So, $\sigma = \mathrm{id}$ and $\mathrm{Aut}\,_k(X) = \{1\}$. $\qquad\square$

5.4 Examples with concrete automorphism groups

We shall construct a k-form X of $\mathbb{A}^1$ with $|\mathrm{Aut}\,_k(X)| = m$ for every pair of characteristic $p > 2$ and positive integer m such that $\gcd(p, m) = \gcd(p, m + 1) = 1$.

5.4.1

Let K be an algebraic function field of one variable over k, and let α be an element of $k^{1/p}$ not in k. Let $k' = k(\alpha)$ and $K' = k' \otimes K$. If $\sigma \in \mathrm{Aut}\,_k(X)$, σ is extended canonically to an element σ of $\mathrm{Aut}\,_{k'}(K')$. Let D be a k-trivial derivation of k' uniquely determined by the condition $D(\alpha) = 1$. Then D can be extended canonically to a K-trivial derivation of K' by setting

$$D(\sum \lambda_i \otimes a_i) = \sum D(\lambda_i) \otimes a_i$$

for $\sum \lambda_i \otimes a_i \in K'$. It is then clear that $\sigma D = D\sigma$.

5.4.2

Theorem. *Assume $p > 2$ and $k = k_s$. Let $m > 1$ be an integer such that $\gcd(m, p) = \gcd(m + 1, p) = 1$, and let B be an affine plane curve defined by*

$$y^m = x^{p^n} + a, \quad n > 0, \quad a \in k \setminus k^p.$$

Then B is k-birationally equivalent to a k-form X of $\mathbb{A}^1$ with $\operatorname{ht}(X) = n$, and $\operatorname{Aut}_k(X) \cong \mathbb{Z}/m\mathbb{Z}$.

Proof. Note that B has a unique k-normal singular point P defined by $(x, y) = (-\alpha, 0)$, where $\alpha \in \overline{k}$ and $\alpha^{p^n} = a$. Since $\gcd(p, m) = 1$, it is straightforward to show that the point at infinity of B is a k-smoothable, one-place point. Hence, if C is a complete k-normal model of $k(B)$, the point P gives a unique singular place of C. Since $k(\alpha) \otimes_k B$ is rational over $k(\alpha)$, the affine curve $X := C \setminus \{P\}$ is a k-form of $\mathbb{A}^1$. It is clear that $\operatorname{ht}(X) = n$. Note that we used so far only the assumption that $\gcd(p, m) = 1$.

Let ξ be a primitive m-th root of unity. Define a birational k-automorphism τ of $k(B)$ by $\tau(x) = x$ and $\tau(y) = \xi y$. Then $F := \{\tau^i \mid 0 \le i < m\}$ is a subgroup of $\operatorname{Aut}_k(X)$. We shall show that $F = \operatorname{Aut}_k(X)$. Let $X' := X \otimes_k (k, \varphi^{n-1})$. Since $\operatorname{Aut}_k(X)$ is a subgroup of $\operatorname{Aut}_k(X')$, it suffices to show that $F = \operatorname{Aut}_k(X')$. Thus we may assume that $n = 1$. Then $\alpha \in k^{1/p}$. Let $k' := k(\alpha)$, $K := k(B)$ and $K' := k' \otimes_k K$. Let r and s be positive integers such that $rm - sp = 1$, and let $t := (x + \alpha)^r/y^s$. Then $y = t^p$ and $x + \alpha = t^m$. Take any element σ of $\operatorname{Aut}_k(K)$ and extend σ to an element of $\operatorname{Aut}_{k'}(K')$. Since $K' = k'(t)$ and σ fixes the point $(x, y) = (-\alpha, 0)$, we may write $\sigma(t) = t/(\gamma t + \delta)$ with $\gamma, \delta \in k'$. Define a derivation D as in 5.4.1. Then $D\sigma = \sigma D$. Since $x + \alpha = t^m$, it follows that $D(t) = rt^{-m+1}$. Now computing both sides of $\sigma D(t) = D\sigma(t)$, we obtain

$$r(\gamma t + \delta)^{m+1} = -\gamma' t^{m+1} - \delta' t^m + r\delta, \tag{5.1}$$

where $\gamma' := D(\gamma)$ and $\delta' := D(\delta)$. Since $m > 1$ and $\gcd(m + 1, p) = 1$, we have $r\gamma \delta^m = 0$ by comparison of the coefficients of t in both sides of the above relation (5.1). Note that $\delta \neq 0$, for otherwise $\sigma(t) = 1/\gamma$ and σ would not be an automorphism of K. Hence it follows that $\gamma = 0$. Furthermore, comparing the constant terms in the relation (5.1), we obtain $\delta^m = 1$. Hence

$$\sigma(t) = \delta^{-1}t, \quad \sigma(x) = \sigma(t^m - \alpha) = \delta^{-m}t^m - \alpha = t^m - \alpha = x,$$
$$\sigma(y) = \sigma(t^p) = \delta^{-p}t^p = \delta^{-p}y.$$

Therefore, σ is an element of F. Thus, we have shown that $\operatorname{Aut}_k(X) = F$.

$\square$

5.4.3

Proposition. *In the settings of 5.4.2, assume that $p \mid (m+1)$. Then $\operatorname{Aut}_k(B) \cong \operatorname{Aut}_k(X)$ and the following assertions hold.*

(1) *If $m + 1 = p^\ell m'$ with $\ell \geq 1, \gcd(p, m') = 1$ and $m' > 1$, then $|\operatorname{Aut}_k(B)| = m$.*

(2) *If $n = 1$ and $m + 1 = p^\ell$ with $\ell \geq 1$, then $\operatorname{Aut}_k k(B)$ is isomorphic to a group*

$$\mathfrak{A} = \left\{ (b, c, \delta) \mid b, c, \delta \in k, \quad c = b^{p^\ell} + a^{p^\ell - 1}c^{p^\ell}, \quad \delta = \delta^{p^\ell} \right\},$$

where the multiplication in $\mathfrak{A}$ is given by

$$(b, c, \delta) \cdot (b', c', \delta') = (b + b'\delta, c + c'\delta, \delta\delta')$$

and the action of $\sigma = (b, c, \delta) \in \mathfrak{A}$ on B is given by

$$\sigma(x) = \frac{(by^{p^{\ell-1}} + \delta x)}{(cy^{p^{\ell-1}} + \delta)}, \quad \sigma(y) = \frac{y}{((b^p + c^p a)y + \delta^p)}.$$

Since $k = k_s$, $\operatorname{Aut}_k k(B)$ is an infinite group. Therefore, B is k-birationally equivalent to the underlying scheme of a k-group G of Russell type defined by an equation

$$y^{p^\ell} = x - a^{p^\ell - 1}x^{p^\ell}.$$

We have an exact sequence of k-group schemes

$$0 \longrightarrow G \longrightarrow \mathfrak{A} \xrightarrow{p_3} \mathbb{Z}/m\mathbb{Z} \longrightarrow (1),$$

where $p_3 : \mathfrak{A} \to \mathbb{Z}/m\mathbb{Z}$ is the projection $(b, c, \delta) \mapsto \delta$.

Proof. We follow the proof of Theorem 5.4.2.

(1) The arguments until the equality (5.1) is the same including the reduction to the case $n = 1$, and the equality (5.1) becomes

$$r\left(\gamma^{p^{\ell}}t^{p^{\ell}} + \delta^{p^{\ell}}\right)^{m'} = -\gamma' t^{p^{\ell}m'} - \delta' t^m + r\delta. \tag{5.2}$$

Comparing the coefficients of $t^{p^{\ell}}$ in the both sides, we obtain $r\gamma^{p^{\ell}}\delta^{p^{\ell}(m'-1)} = 0$, whence $\gamma = 0$ and $\sigma(t) = \delta^{-1}t$. Again, comparing the constant terms in (5.2), $\delta^{m+1} = \delta$. The rest is the same.

(2) Since $m = p^{\ell} - 1$, the equality $rm - sp = 1$ holds for $r = p - 1$ and $s = p^{\ell} - p^{\ell-1} - 1$. Hence the equality (5.1) becomes

$$-\left(\gamma^{p^{\ell}}t^{p^{\ell}} + \delta^{p^{\ell}}\right) = -\gamma' t^{p^{\ell}} - \delta' t^{p^{\ell}-1} - \delta. \tag{5.3}$$

Hence $\delta^{p^{\ell}} = \delta$, $\delta' = 0$, and $\gamma^{p^{\ell}} = \gamma'$. The last relation implies that $\gamma = c\alpha + b$ with $b, c \in k$. The relation $\gamma^{p^{\ell}} = \gamma'$ implies the relation

$$c = b^{p^{\ell}} + a^{p^{\ell-1}}c^{p^{\ell}}.$$

Let $\sigma \in \operatorname{Aut}_k(X)$ and let $\sigma(t) = t/((c\alpha+b)t+\delta)$. Let σ' be another element of $\operatorname{Aut}_k(X)$, and write $\sigma'(t) = t/((c'\alpha + b')t + \delta')$. Then we have

$$\sigma'\sigma(t) = \frac{t}{((c + c'\delta)\alpha + (b + b'\delta))t + \delta\delta'}.$$

Hence $\operatorname{Aut}_k(X)$ is identified with the group $\mathfrak{A}$.

Let $\sigma(t) = t/((c\alpha + b)t + \delta)$. Since $y = t^p$, we have

$$\sigma(y) = \frac{t^p}{(c^p a + b^p)t^p + \delta^p} = \frac{y}{(c^p a + b^p)y + \delta^p}.$$

Since $y^{p^{\ell}-1} = x^p + a$, we have

$$\sigma(x)^p = -a + \sigma(y)^{p^{\ell}-1}$$

$$= -a + \frac{y^{p^{\ell}-1}}{((c^p a + b^p)y + \delta^p)^{p^{\ell}}} \cdot ((c^p a + b^p)y + \delta^p)$$

$$= \frac{y^{p^{\ell}-1}\{(c^p a + b^p)y + \delta^p\} - a((c^{p^{\ell+1}}a^{p^{\ell}} + b^{p^{\ell+1}})y^{p^{\ell}} + \delta^{p^{\ell+1}})}{(c^{p^{\ell+1}}a^{p^{\ell}} + b^{p^{\ell+1}})y^{p^{\ell}} + \delta^{p^{\ell+1}}}$$

$$= \frac{(c^p a + b^p)y^{p^{\ell}} + \delta^p y^{p^{\ell}-1} - a(c^p y^{p^{\ell}} + \delta^p)}{c^p y^{p^{\ell}} + \delta^p}$$

$$= \frac{\delta^p(y^{p^{\ell}-1} - a) + b^p y^{p^{\ell}}}{c^p y^{p^{\ell}} + \delta^p} = \frac{\delta^p x^p + b^p y^{p^{\ell}}}{c^p y^{p^{\ell}} + \delta^p},$$

where we used the relations $c = b^{p^{\ell}} + a^{p^{\ell-1}}c^{p^{\ell}}$ and $\delta^{p^{\ell}} = \delta$. Hence we have

$$\sigma(x) = \frac{by^{p^{\ell-1}} + \delta x}{cy^{p^{\ell-1}} + \delta}.$$

Since $\operatorname{Ker} p_3$ consists of elements $(b, c, 1)$ with $\delta = 1$ and $c = b^{p^\ell} + a^{p^{\ell-1}} c^{p^\ell}$, it follows that $\operatorname{Ker} p_3$ is isomorphic to the k-group G of Russell type. Since $\operatorname{Ker} p_3$ has a k-group scheme structure and X has a k-rational point P_0 as $k = k_s$, the orbit $(\operatorname{Ker} p_3) P_0$ must be X. Hence X is k-isomorphic to the underlying scheme of the group scheme G. $\square$

5.4.4

By an argument similar to the one in 5.4.3, we can prove the following result.

Proposition. (1) *Assume $p > 2$. Let C be a projective plane curve defined by*

$$X^p Z + a Z^{p+1} = Y^p(Y + aZ), \quad a \in k \setminus k^p.$$

Then C is a k-normal curve with only one singular point P defined by $(X, Y, Z) = (-\alpha, 0, 1)$, where $\alpha^p = a$. Furthermore, $C \setminus \{P\}$ is a non-k-rational k-form of $\mathbb{A}^1$, and $\operatorname{Aut}_k(C) = \{1\}$.

(2) *In case $p = 2$, the above curve C is k-birationally equivalent to a k-group of Russell type $y^2 + y^4 = x + ax^2 + a^2 x^4$. Hence $|\operatorname{Aut}_k(C)| = \infty$.*

(3) *Assume $p > 3$. Let B be an affine plane curve defined by*

$$x^p + a = y^{2p}(a + y^2), \quad a \in k \setminus k^p.$$

Then B is k-birationally equivalent to a k-form of $\mathbb{A}^1$, and $|\operatorname{Aut}_k(k(B))| = 2$. The subfield of elements of $k(B)$ invariant under $\operatorname{Aut}_k(k(B))$ is the function field of the above curve C. Hence B is not a hyperelliptic curve.

(4) *Assume $p = 3$. Then, for an affine plane curve B defined by the same equation as in the assertion (3) above, $\operatorname{Aut}_k k(B)$ is isomorphic to the symmetric group S_3.*

Proof. (1) Setting $x = X/Z$ and $y = Y/Z$, we have an inhomogeneous form $x^p + a = y^p(y + a)$ of C. The point P defined by $(x, y) = (-\alpha, 0)$ is the unique singular point of C by the Jacobian criterion, where $\alpha^p = a$. The local ring of C at P has y as a uniformisant because the maximal ideal $\mathfrak{m}_{C,P}$ is generated by $x^p + a$ and y, where $x^p + a \in (\mathfrak{m}_{C,P})^p$. Hence C is a k-normal complete plane curve. Then the k-genus of C is $(p(p-1))/2 > 0$, whence C is non-k-rational. We shall show that $|\operatorname{Aut}_k(C)| = 1$. Let $k' = k(\alpha)$, $K = k(C)$ and $K' = k' \otimes_k K$. Let D be a k-trivial derivation of k' such that $D(\alpha) = 1$. Then D is extended canonically to a K-trivial derivation of K'. Let $\sigma \in \operatorname{Aut}_k K$ and extend it to a birational k'-automorphism of

K'. As in 5.4.3, we have $D\sigma - \sigma D$. Let $t = (x + \alpha - \alpha y)/y$. Then $y = t^p$ and $x = -\alpha + \alpha t^p + t^{p+1}$. Hence $D(t) = (1 - t^p)/t^p$. Since σ fixes the singular point P, $\sigma(t) = t/(\gamma t + \delta)$ with $\gamma, \delta \in k'$, where $\delta \neq 0$. Since $\sigma D(t) = D\sigma(t)$, we obtain

$$\{\gamma^2(\gamma^p - 1) + \gamma'\}t^{p+2} + \{2\gamma\delta(\gamma^p - 1) + \delta'\}t^{p+1}$$
$$+ \delta\{(\gamma^p - 1)\delta + 1\}t^p + \delta^p\gamma^2 t^2 + 2\gamma\delta^{p+1}t$$
$$+ \delta(\delta^{p+1} - 1) = 0.$$

Since $p > 2$, each t-term in the above equation has distinct degree. The coefficient of the term t must be zero. Since $\delta \neq 0$, it follows that $\gamma = 0$. The coefficient of the term t^p being zero then implies $\delta = 1$. Hence σ is the identity automorphism. Thus $|\text{Aut}_k K| = 1$.

(2) Since $p = 2$, the above equation in γ and δ reads

$$\{\gamma^2(\gamma^2 + 1) + \gamma'\}t^4 + \delta't^3 + (\delta^2 + \delta)t^2 + \delta(\delta^3 + 1) = 0.$$

Since all the coefficients are zero, we have $\delta = 1$ and $\gamma' = \gamma^2(\gamma^2 + 1)$. Write $\gamma = b\alpha + c$ with $b, c \in k$. Then $\gamma' = \gamma^2 + \gamma^4$ implies that $c^2 + c^4 = b + ab^2 + a^2b^4$. By the same argument as in the case (2) of Proposition 5.4.3, we conclude that $X := C \setminus \{P\}$ is isomorphic to the k-group of Russell type

$$y^2 + y^4 = x + ax^2 + a^2x^4.$$

(3) The curve B has two singular points P_0 and P_∞ defined respectively by $(X, Y, Z) = (-\alpha, 0, 1)$ and $(X, Y, Z) = (1, 0, 0)$, where $\{X, Y, Z\}$ is a system of homogeneous coordinates of $\mathbb{P}^2$ and $a = \alpha^p$. It is easy to see that the point P_∞ is k-smoothable and the point P_0 is a singular, one-place point. Hence $X := B \setminus \{P_0\}$ is a k-form of $\mathbb{A}^1$. Since $x^p + a = y^{2p}(y^2 + a)$, set $t = (x + \alpha)/y^2$. Then $x + \alpha = t(t^p - a)$ and $y^2 = t^p - a$. Set $u = y/(t - \alpha)^n$ with $p = 2n + 1$. Then $t = u^2 + \alpha$, $x = -\alpha + u^{2p}(u^2 + \alpha)$ and $y = u^p$. Let σ be an element of $\text{Aut}_k X$. Then σ fixes the singular point P_0. Hence $\sigma(u) = u/(\gamma u + \delta)$ with $\gamma, \delta \in k' := k(\alpha)$. We consider the relation $D\sigma(u) = \sigma D(u)$, where D is a k-derivation of k' such that $D(\alpha) = 1$ and extended canonically to $k' \otimes_k k(B)$. Since $D(u) = (1 - u^{2p})/(2u^{2p+1})$, we obtain the following relation:

$$-\delta u^{2p} - \gamma'u^2 - \delta'u + \delta$$
$$= \gamma^3(\gamma^{2p} - 1)u^{2p+3} + 3\gamma^2\delta(\gamma^{2p} - 1)u^{2p+2} + 3\gamma\delta^2(\gamma^{2p} - 1)u^{2p+1}$$
$$+ \delta^3(\gamma^{2p} - 1)u^{2p} + 2\gamma^{p+3}\delta^p u^{p+3} + 6\gamma^{p+2}\delta^{p+1}u^{p+2} + 6\gamma^{p+1}\delta^{p+2}u^{p+1}$$
$$+ 2\gamma^p\delta^{p+3}u^p + \gamma^3\delta^{2p}u^3 + 3\gamma^2\delta^{2p+1}u^2 + 3\gamma\delta^{2p+2}u + \delta^{2p+3}. \tag{5.4}$$

Since $p > 3$, comparison of terms of degree 0 and 1 in u implies that $\gamma\delta^{2p+2} = 0$. Since $\delta \neq 0$, we have $\gamma = 0$. Comparison of terms of degree $2p$ in u implies $\delta^2 = 1$. So, $\sigma(u) = \delta^{-1}u$, $\sigma(x) = x$ and $\sigma(y) = \delta^{-1}y$. Hence $\operatorname{Aut}_k X \cong \operatorname{Aut}_k(B) \cong \mathbb{Z}/2\mathbb{Z}$. The rest of the assertion is clear in view of Theorem 4.3.

(4) The relation (5.4) reads as

$$-\delta u^6 - \gamma'u^2 - \delta'u + \delta = \gamma^3(\gamma^6 - 1)u^9 - \delta^3 u^6 + \delta^9,$$

whence we have $\gamma(\gamma^2 - 1) = 0$ and $\delta^2 = 1$. Conversely, if γ and δ satisfy these relations, the equality $D\sigma(u) = \sigma D(u)$ holds. Then it follows that $\sigma(z) \in \operatorname{Ker} D$ for $z \in \operatorname{Ker} D$. Namely, the automorphism σ of $k' \otimes_k k(B) = k'(u)$ induces an automorphism of $\operatorname{Ker} D = k(B)$, and $|\operatorname{Aut}_k(k(B)| \leq 6$. Define the k'-automorphisms σ and τ of $k'(B)$ by

$$\sigma(u) = \frac{u}{u+1}, \quad \tau(u) = -u.$$

Then $\sigma^3 = 1 = \tau^2$ and $\sigma\tau\sigma = \tau$. Since σ and τ commutes with the derivation D, they induce k-automorphisms of $k(B)$. Hence $\operatorname{Aut}_k k(B) \cong S_3$. $\qquad\square$

5.4.5

The assertions (3) and (4) of Proposition 5.4.4 can be generalized to the case of characteristic $p > 2$ by a similar computation.

Proposition. *Assume that $p > 2$. Let B be an affine plane curve defined by*

$$x^p + a = y^{np}(y^n + a), \quad n > 0, \quad \gcd(p, n) = 1, \quad a \in k \setminus k^p.$$

Then B is k-birationally equivalent to a k-form of $\mathbb{A}^1$. If $p \nmid n + 1$, then $\operatorname{Aut}_k(B) \cong \mathbb{Z}/n\mathbb{Z}$ which is generated by a k-automorphism τ such that $\tau(x) = x$ and $\tau(y) = \zeta y$, where ζ is a primitive nth root of the unity. If $n + 1 = p$, then $\operatorname{Aut}_k k(B)$ is a dihedral group of order $2p$.

Proof. Let $a = \alpha^p$. Let $k' = k(\alpha)$ and $K' = k' \otimes_k k(B)$. Then $K' = k'(u)$ with $x = u^p$ and $x = -\alpha + u^{pn}(u^n + \alpha)$. If $n + 1 = p$, define k'-automorphisms σ and τ of K' by

$$\sigma(u) = \frac{u}{u+1}, \quad \tau(u) = -u.$$

Then $\sigma^p = 1, \tau^2 = 1$ and $\sigma\tau\sigma = \tau$. The computations are almost the same as in the case $p = 3$. $\qquad\square$

5.5 A remark

Let X be a k-form of $\mathbb{A}^1$. Since $\overline{X} := X \otimes \overline{k}$ is isomorphic to $\mathbb{A}^1_{\overline{k}}$, $\mathrm{Aut}_k(X)$ is realized as a subgroup of $\mathrm{Aut}_{\overline{k}}(\mathbb{A}^1_{\overline{k}})$ and also of $\mathrm{PGL}(2, \overline{k})$ which consists of elements σ such that $\sigma^*(A) \subseteq A$, where $X = \mathrm{Spec}(A)$. More precisely, let C be a complete normal model of $k(X)$ and let $P_\infty := C - X$. As the normalization of $\overline{C} := C \otimes \overline{k}$ is $\mathbb{P}^1_{\overline{k}}$, let $\widetilde{P}_\infty$ be the point of $\mathbb{P}^1_{\overline{k}}$ lying over the point P_∞. Choose an inhomogeneous parameter t of $\mathbb{P}^1_{\overline{k}}$ such that $t = 0$ at $\widetilde{P}_\infty$. Then $\mathrm{Aut}_k(X)$ identified with a subgroup of $\mathrm{PGL}(2, \overline{k})$ consists of an automorphism σ such that $\sigma^*(A) \subseteq A$ and $\sigma(\widetilde{P}_\infty) = \widetilde{P}_\infty$. Hence $\mathrm{Aut}_k(X)$ cannot be finite subgroups $\mathrm{PGL}(2, \mathbb{F}_q)$ or $\mathrm{PSL}(2, \mathbb{F}_q)$, where q is a power of p. Propositions 5.4.4 and 5.4.5 shows that a dihedral group of order $2p$ can be the group $\mathrm{Aut}_k(X)$. A classification of finite subgroups of $\mathrm{PGL}(2, \overline{k})$ was made by Faber [15]. It is an interesting problem to classify $\mathrm{Aut}_k(X)$ as a subgroup of $\mathrm{PGL}(2, \overline{k})$ and give a k-form X of $\mathbb{A}^1$ which gives the subgroup.

Chapter 6

Divisor class groups

We shall introduce various invariants of a form of $\mathbb{A}^1$ in order to study the divisor class group. Let $X = \mathrm{Spec}\,(A)$ be a k-form of $\mathbb{A}^1$ and C a complete k-normal model of $K := k(X)$. We assume that X has a k-rational point P. Let $K' = k(x)$ be the unique maximal rational subfield over which K is purely inseparable, and let $A' = k[u]$ be the unique maximal polynomial subring over which A is purely inseparable (cf. Corollary 2.5.2). Let $\gamma : C \to \mathbb{P}^1_K$ be the morphism induced by the inclusion $K' \hookrightarrow K$. Let $P_\infty := C - X$, $P'_\infty := \gamma(P_\infty)$ and $\mathcal{O}_\infty, \mathcal{O}'_\infty$ the places corresponding to P_∞, P'_∞, respectively. Let $\widetilde{\mathcal{O}}_\infty$ be the integral closure of $\mathcal{O}_\infty$ in $\overline{k} \otimes K$.

We denote by $C\ell\,(X)$ (or $C\ell\,(A)$) the divisor class group of X. Clearly it holds that $C\ell\,(X) \cong \mathrm{Pic}(X)$.

6.1   Various invariants and their interrelations

We put now:

$\lambda := \mathrm{ht}\,(X) =$ the height of X (cf. 2.3)

$\lambda' := \mathrm{ht}\,(K) =$ the height of K (cf. 2.1)

$p^\varepsilon := [k(P'_\infty) : k] =$ the degree of P'_∞

$p^\eta := e(\mathcal{O}_\infty : \mathcal{O}'_\infty) =$ the ramification index of $\mathcal{O}_\infty$ over $\mathcal{O}'_\infty$

$p^\nu := e(\widetilde{\mathcal{O}}_\infty : \mathcal{O}_\infty) =$ the ramification index of $\widetilde{\mathcal{O}}_\infty$ over $\mathcal{O}_\infty$.

6.1.1

Lemma. *With the notations as above, we have:*

(1) *There are the following two exact sequences*

(i) $0 \longrightarrow \mathbb{Z} \overset{j}{\longrightarrow} \mathrm{Pic}(C) \overset{\rho}{\longrightarrow} C\ell\,(X) \longrightarrow 0$

(ii) $0 \longrightarrow \mathrm{Pic}^0_{C/k}(k) \longrightarrow C\ell\,(X) \longrightarrow \mathbb{Z}/p^\nu\mathbb{Z} \longrightarrow 0,$

where ρ is the restriction map and j is defined by assigning $1 \in \mathbb{Z}$ to P_∞.

(2) $\mathrm{Pic}^0_{C/k}$ *is a unipotent algebraic group.*

(3) $\mathrm{C}\ell(X)$ *is an abelian p-group with bounded p-exponent.*

Proof. (1) Let $\mathcal{L}$ be an invertible sheaf on C. Then $\rho(\mathcal{L})$ is given as the restriction $\mathcal{L}|_X$. Then the first exact sequence is easily verified to exist. As for the existence of the second one, consider an exact sequence

$$0 \longrightarrow \mathrm{Pic}^0_{C/k}(k) \longrightarrow \mathrm{Pic}(C) \longrightarrow \mathbb{Z} \longrightarrow 0 ,$$

where a generator of $\mathbb{Z} \cong \mathrm{Pic}(C)/\mathrm{Pic}^0_{C/k}(k)$ is given by the invertible sheaf on C corresponding to the k-rational point P, which we assume to exist. Since $\mathrm{Pic}^0_{C/k}(k) \cap j(\mathbb{Z}) = (0)$ and since $P_\infty \sim p^\nu P$ (linear equivalence), we obtain the exact sequence (ii).

(2) $\mathrm{Pic}^0_{C \otimes \overline{k}/\overline{k}}$ is identified with the generalized Jacobian variety of $C \otimes \overline{k}$ with equivalence relation defined by $\overline{k} \otimes \mathcal{O}_\infty$. Since the $\overline{k}$-genus of C is zero, $\mathrm{Pic}^0_{C \otimes \overline{k}/\overline{k}} = \mathrm{Pic}^0_{C/k} \otimes_k \overline{k}$ is a unipotent algebraic group, hence so is $\mathrm{Pic}^0_{C/k}$.

(3) This follows from the exact sequence (ii) and the assertion (2). $\qquad\square$

We define μ to be the smallest integer ≥ 0 such that $p^\mu \mathrm{C}\ell(X) = (0)$ and call μ the *divisorial p-exponent* of X.

6.1.2

Lemma. *We have the following relations among the invariants $\lambda, \lambda', \varepsilon, \eta$ and ν.*

(1) $\lambda = \lambda' + \varepsilon$ *with* $k(u) = k(x^{p^\varepsilon})$.

(2) $\lambda = \eta + \nu$.

(3) $p^\nu = [k(P_\infty) : k]$.

(4) $\lambda \geq \nu \geq \lambda - \lambda'$.

Proof. (1) In the proof of Corollary 2.5.2, we have shown that $K' = k(x)$, $A' = k[A^{p^\lambda}]$, $k(u) = k(x^{p^\delta})$ and $\lambda = \lambda' + \delta$ for $\delta \geq 0$. On the other hand, $\mathbb{P}^1_k - P'_\infty$ is a k-rational k-form of $\mathbb{A}^1$ with height equal to ε (cf. Theorem 2.8.1). Therefore, the equality $\varepsilon = \delta$ follows.

(2) Let $\widetilde{\mathcal{O}}'_\infty$ be the integral closure of $\mathcal{O}'_\infty$ in $\overline{k} \otimes K'$. Then, if we write $K' = k(t)$ for a suitable t, the place $\mathcal{O}'_\infty$ corresponds to a point defined by $t^{p^\varepsilon} - a = 0$ with $a \in k \setminus k^p$. Hence $\widetilde{\mathcal{O}}'_\infty$ is given by $t - a^{p^{-\varepsilon}} = 0$, so

$e(\widetilde{O}'_\infty : O'_\infty) = p^\varepsilon$. Also, we have

$$p^{\lambda'} = [K : K'] = [\overline{k} \otimes K : \overline{k} \otimes K'] = e(\widetilde{O}_\infty : \widetilde{O}'_\infty)$$

because $f(\widetilde{O}_\infty : \widetilde{O}'_\infty) = 1$. Consequently,

$$p^\nu \cdot p^\eta = e(\widetilde{O}_\infty : O'_\infty) = e(\widetilde{O}_\infty : \widetilde{O}'_\infty) \cdot e(\widetilde{O}'_\infty : O'_\infty) = p^{\lambda'} \cdot p^\varepsilon \ ,$$

and the equality $\nu + \eta = \lambda' + \varepsilon$ follows.

(3) We have

$$[k(P_\infty) : k] = [k(P_\infty) : k(P'_\infty)] \cdot [k(P'_\infty) : k]$$
$$= f(O_\infty : O'_\infty) \cdot p^\varepsilon = [K : K'] \cdot p^{-\eta} \cdot p^\varepsilon = p^{\lambda' - \eta + \varepsilon} = p^\nu \ .$$

(4) Clear from (3) and the definition of ε. □

6.1.3

Lemma. *We have $\lambda = \mu \geq \nu$.*

The proof consists of three steps which take the following three paragraphs.

6.1.3.1. Lemma. *Let O be a smooth place of K/k whose residue field is purely inseparable over k, and let $O_1 := k^{1/p} \otimes O$. Then $e(O_1 : O) = 1$ if and only if O is k-rational.*

Proof. Let k' be a finite extension of k such that $k' \subseteq k^{1/p}$ and $e(k' \otimes O : O) = e(O_1 : O)$, where $k' \otimes O$ is k'-normal (cf. Lemma 2.6.1). Write $O' := k' \otimes O$, $\mathfrak{k}' :=$ the residue field of O' and $\mathfrak{k} :=$ the residue field of O. Then,

$$[\mathfrak{k}' : k] = [\mathfrak{k}' : k'][k' : k] = [\mathfrak{k}' : k'][k' \otimes K : K] = [\mathfrak{k}' : k']e(O' : O)f(O' : O)$$
$$= [\mathfrak{k}' : \mathfrak{k}][\mathfrak{k} : k] = f(O' : O)[\mathfrak{k} : k] \ ,$$

whence follows

$$[\mathfrak{k}' : k']e(O' : O) = [\mathfrak{k} : k].$$

This implies that if $e(O' : O) > 1$ then O is non-k-rational and proves the "if" part. Suppose that $e(O_1 : O) = 1$. This implies clearly $\mathfrak{k}' \cong k' \otimes \mathfrak{k}$. If $\mathfrak{k} \supsetneq k$, choose k' so that $k' \cap \mathfrak{k} \supsetneq k$. Then $k' \otimes \mathfrak{k}$ would contain a nilpotent element, which contradicts the last isomorphism. Therefore, $\mathfrak{k} = k$ and O is k-rational. □

6.1.3.2. Claim. $p^\lambda \geq p^\mu$.

Proof. If we replace X by $X \otimes k_s$, the invariant λ is unchanged (cf. Lemma 2.4.1). On the other hand, $\mathrm{Pic}(C)$ is a subgroup of $\mathrm{Pic}(C \otimes k_s)$, where $C \otimes k_s$ is a complete k_s-normal model of $k_s(X)$. By the exact sequence (i) of Lemma 6.1.1, $C\ell\,(X)$ is then a subgroup of $C\ell\,(X \otimes k_s)$. Hence if the claim is proved for $X \otimes k_s$, we have the claim proved for X. So, we may assume that $k = k_s$.

For each $i \geq 1$, let $k_i := k^{p^{-i}}$, let $X_i := X \otimes k_i$, and let C_i be the normalization of C in $K_i := k_i \otimes K$, where $K = k(X)$. Further, let P_i be the point of C_i lying over P_∞ of C. Since X_λ is k_λ-isomorphic to $\mathbb{A}^1$, P_λ is k_λ-rational. Let Q be an arbitrary point on X, and for $i \geq 1$, let Q_i be the point of $X \otimes k_i$ lying over Q. Suppose first that Q is k-rational. We can find an inhomogeneous parameter t on $\mathbb{P}^1_{k_\lambda}$, such that $t = 0$ at Q_λ and $t = \infty$ at P_λ. Since $t^{p^\lambda} \in K$, we have a linear equivalence $p^\lambda Q \sim p^{\lambda-\nu}P_\infty$. Therefore the divisor class in $C\ell\,(X)$ represented by a k-rational point Q is annihilated by p^λ. Suppose next that Q is not a k-rational point. Let $\mathcal{O}$ be the local ring of Q and let $\mathcal{O}_i$ be the local ring of Q_i. We know that Q_i is non-k_i-rational if and only if $e(\mathcal{O}_{i+1} : \mathcal{O}_i) = p$ (see 6.1.3.1). Therefore, if r is the smallest positive integer such that Q_r is a k_r-rational point, and if $\lambda \geq r$, then $p^{\lambda-r}Q \sim p^{\lambda-\nu}P_\infty$ by an argument similar to the preceding case. If $r > \lambda$, $Q_\lambda \sim p^{r-\lambda}Q'_\lambda$ with a k_λ-rational point Q'_λ lying over a k-rational point Q' of X. Hence $Q \sim p^r Q' \sim p^{r-\lambda}P_\infty$. Therefore, in each case the divisor class in $C\ell\,(X)$ represented by a point Q is annihilated by p^λ. □

6.1.3.3. Claim. $p^\mu \geq p^\lambda$.

Proof. We identify $(k, \varphi^i) \otimes_k A$ with a k-subalgebra $A^{[i]} := k[A^{p^i}]$ of A. Then $A^{[i]}$ is a k-subalgebra of $A^{[j]}$ $(0 \leq j < i)$ as well. Let P be a k-rational point of $X = \mathrm{Spec}\,(A)$ and $\mathfrak{p}$ the prime ideal of A corresponding to P. Let $\mathfrak{p}_i := \mathfrak{p} \cap A^{[i]}$. Since the divisor class in $C\ell\,(X)$ represented by P is annihilated by p^μ, it is not hard to show that $\mathfrak{p}_\mu A = \mathfrak{p}^{p^\mu} A$ is a principal ideal. Write $\mathfrak{p}_\mu A = \xi A$.

We shall show that $\xi \in A^{[\mu]}$. In fact, consider $\overline{A} := \overline{k} \otimes A$. Then $\overline{A} = \overline{k}[t]$ for some element $t \in \overline{A}$, and $\overline{A}^{[i]} = \overline{k} \otimes A^{[i]} = \overline{k}[t^{p^i}]$. Moreover, $\mathfrak{p}\overline{A}$ and $\mathfrak{p}_i\overline{A}^{[i]}$ are maximal ideals of $\overline{A}$ and $\overline{A}^{[i]}$, respectively, because P is a k-rational point of X (cf. 6.1.3.1). We may therefore assume that

$\mathfrak{p}\overline{A} = t\overline{A}$. Since $e(A_{\mathfrak{p}} : A^{[i]}_{\mathfrak{p}_i}) = p^i$, we know that $\xi = \alpha t^{p^{\mu}}$ with $\alpha \in \overline{k}$. Hence $\xi \in \overline{A}^{[\mu]} \cap K = A^{[\mu]}$. Since A is a faithfully flat $A^{[\mu]}$-module, this implies that $\mathfrak{p}_\mu = \xi A^{[\mu]}$. Let $C^{(\mu)}$ be a complete k-normal model of $X^{[\mu]} := \operatorname{Spec} A^{[\mu]}$, and let $P^{(\mu)}_\infty$ be its point at infinity. Then, $\mathfrak{p}_\mu = \overset{\circ}{\xi} A^{[\mu]}$ implies that the point P_μ of $C^{(\mu)}$ corresponding to $\mathfrak{p}_\mu$ is linearly equivalent to $P^{(\mu)}_\infty$ over k. Hence, $P^{(\mu)}_\infty$ is a k-rational point. Therefore, $X^{[\mu]} = C^{(\mu)} - P^{(\mu)}_\infty$ is a trivial k-form of $\mathbb{A}^1$, so we have $p^\mu \geq p^\lambda$. □

The preceding two inequalities show that $\lambda = \mu$, while $\lambda \geq \nu$ is proved in Lemma 6.1.2. This completes the proof of Lemma 6.1.3.

6.1.4

Example. (1) Let C be a projective plane curve defined by

$$Y^p Z - a Z^{p+1} = X^p(Y + X)$$

with $a \in k \setminus k^p$. Then C is a k-normal curve with only one singular point $P_\infty = (0, a^{1/p}, 1)$. Let $X = C - P_\infty$. Then X is a nontrivial k-form of $\mathbb{A}^1$ with $\lambda = 2$ and $\lambda' = \nu = 1$.

(2) Let C be a projective plane curve defined by

$$X^p Z - a Z^{p+1} = (Y^p - b Z^p)(X + Y)$$

with $a, b \in k \setminus k^p$ and $a + b + 1 \neq 0$. Then C is a k-normal curve with only one singular point $P_\infty = (a^{1/p}, b^{1/p}, 1)$. Let $X = C - P_\infty$. Then X is a k-form of $\mathbb{A}^1$ with $\lambda = \nu = 2$ and $\lambda' = 1$.

(3) Let X be a k-group of Russell type given by

$$y^{p^2} = x + a x^p \quad (p > 2, \ a \in k \setminus k^p).$$

Then $\lambda = \lambda' = 2$ and $\nu = 1$.

6.2 Divisor class groups

Let X be a nontrivial k-form of $\mathbb{A}^1$, let $X_i := X \otimes k^{p^{-i}}$ and let C_i be a complete $k^{p^{-i}}$-normal model of X_i. Let $P_i := C_i - X_i$ with local ring $\mathcal{O}_i \subset k^{p^{-i}}(C_i)$.

6.2.1

Lemma. *The following conditions are equivalent.*

(1) $\lambda = \nu$.

(2) $e(\mathcal{O}_{i+1} : \mathcal{O}_i) = \begin{cases} p \text{ for } & 0 \le i < \nu \\ 1 \text{ for } & \nu \le i \,. \end{cases}$

Proof. $(1) \Rightarrow (2)$: Since X_λ is $k^{p^{-\lambda}}$-isomorphic to $\mathbb{A}^1$, P_λ is $k^{p^{-\lambda}}$-trivial. Therefore, $e(\mathcal{O}_{i+1} : \mathcal{O}_i) = 1$ if $i \ge \lambda$ by Lemma 6.1.3.1. Then $e(\mathcal{O}_{i+1} : \mathcal{O}_i)$ must be equal to p for $0 \le i < \nu$ provided $p^\lambda = p^\nu$.

$(2) \Rightarrow (1)$: By the assumption and Lemma 6.1.2, (3), we can find a finite extension k' of k such that $k \subseteq k(P_\infty) \subseteq k' \subseteq k^{p^{-\nu}}$ and that the place P'_∞ of $X \otimes k'$ has ramification index p^ν over P_∞. The last assertion follows from $e(\mathcal{O}_\nu : \mathcal{O}_\infty) = p^\nu$. Let $d := [k' : k]$ and let $\mathcal{O}'_\infty$ be the local ring of P'_∞. Since $[k' \otimes K : K] = d$ and $e(\mathcal{O}'_\infty : \mathcal{O}_\infty) = p^\nu$, it follows that $f(\mathcal{O}'_\infty : \mathcal{O}_\infty) = d \cdot p^{-\nu}$. But $[k(P_\infty) : k] = p^\nu$ and $[k' : k] = d$, so $[k' : k(P_\infty)] = d \cdot p^{-\nu}$, too. Therefore, $k'(P'_\infty) = k'$ and P'_∞ is smooth. Then P_ν which dominates P'_∞ is smooth, too, and $X \otimes k^{p^{-\nu}}$ is isomorphic to $\mathbb{A}^1$ over $k^{p^{-\nu}}$. So, $\nu \ge \lambda$, whence $\nu = \lambda$. $\qquad\square$

6.2.2

Let $C\ell^0(X)$ be the subgroup of $C\ell(X)$ consisting of divisor classes of degree zero on X, and let $\rho : \mathrm{Pic}(C) \to C\ell(X)$ be the canonical map (cf. Lemma 6.1.1). Then we have:

Lemma. (1) *The following sequence*

$$0 \longrightarrow C\ell^0(X) \longrightarrow C\ell(X) \xrightarrow{d} \mathbb{Z}/p^\lambda\mathbb{Z} \longrightarrow 0$$

is an exact sequence, where d is the degree map.

(2) $C\ell^0(X) \subseteq \mathrm{Pic}^0_{C/k}(k)$ *and* $\rho(\mathrm{Pic}^0_{C/k}(k))/C\ell^0(X) \cong \mathbb{Z}/p^{\lambda-\nu}\mathbb{Z}$.

Proof. Let P_0 be a k-rational point of X. By the proof of Claim 6.1.3.3, p^λ is the smallest positive integer such that $p^\lambda P_0 \sim 0$ on X. Since every point on X is linearly equivalent to an integral multiple of P_0 modulo a divisor of degree zero, we obtain the exact sequence in (1). In the assertion (2), the inclusion $C\ell^0(X) \subseteq \mathrm{Pic}^0_{C/k}(k)$ is obvious. Since $p^\nu P_0 - P_\infty$ represents an element of $\mathrm{Pic}^0_{C/k}(k)$, $\rho(\mathrm{Pic}^0_{C/k}(k))$ is generated by the divisor class of $\rho(p^\nu P_0 - P_\infty) = p^\nu P_0$ modulo $C\ell^0(X)$. This yields the isomorphism between two groups. $\qquad\square$

6.2.3

Corollary. *The following conditions are equivalent.*

(1) $\lambda = \nu$.
(2) $\rho(\mathrm{Pic}^0_{C/k}(k)) = C\ell^0(X)$.

6.3 Picard varieties as unipotent groups

Now we will observe the behavior of the Picard group schemes under the Frobenius morphisms. For any k-scheme Y, we denote by $Y^{(p)}$ the scheme $Y \otimes (k, \varphi)$, and by $F_y : Y \to Y^{(p)}$ the Frobenius k-morphism of Y. With the notation in 2.2, we have $Y^{(p)} = \Theta Y$. If G is a commutative k-group scheme, the *multiplication by p* endomorphism of G is decomposed into a composite of k-homomorphisms

$$p: G \xrightarrow{F_G} G^{(p)} \xrightarrow{V_G} G \,.$$

The homomorphism V_G is called the *Verschiebung* of G. It is well-known that G is smooth over k if and only if F_G is faithfully flat. If G is smooth over k, V_G is uniquely determined. Namely, if $V' : G^{(p)} \to G$ is a k-homomorphism of k-group schemes such that $V' \cdot F_G = p$, then $V' = V_G$ [1].

Let X be a k-form of $\mathbb{A}^1$ of genus $g > 0$ and carrying a k-rational point, let C be a complete k-normal model of X, and let $P_\infty := C - X$. For $G = \mathrm{Pic}^0_{C/k}$, we consider the Frobenius k-homomorphism $F_G : \mathrm{Pic}^0_{C/k} \to (\mathrm{Pic}^0_{C/k})^{(p)}$. Let $i : X \hookrightarrow \mathrm{Pic}^0_{C/k}$ be the closed immersion defined in Theorem 1.7 by means of a k-rational point P_0 of X. Then we have the following obvious result.

6.3.1

Lemma. *The following diagram is commutative.*

$$
\begin{array}{ccc}
X & \xrightarrow{\;i\;} & \mathrm{Pic}^0_{C/k} \\
{\scriptstyle F_X}\downarrow & & \downarrow{\scriptstyle F_G} \\
X^{(p)} & \xrightarrow{\;i^{(p)}\;} & (\mathrm{Pic}^0_{C/k})^{(p)} \,,
\end{array}
$$

where $i^{(p)} := i \otimes (k, \varphi)$, and F_G is faithfully flat.

---

[1]The existence and the uniqueness of V_G are enough to understand the subsequent arguments. For the proof of these properties, the readers are referred to [43] and Demazure [11].

6.3.2

Let C' be the complete k-normalization of $X^{(p)}$ in the function field of $C^{(p)}$. Then $F_C : C \to C^{(p)}$ decomposes into a composite

$$C \xrightarrow{f} C' \xrightarrow{g} C^{(p)} .$$

The k-morphisms f and g induce k-homomorphisms of the Picard schemes, where the first isomorphism below follows from Theorem 1.3, (2):

$$g_0 : \text{Pic}^0_{C^{(p)}/k} \cong (\text{Pic}^0_{C/k})^{(p)} \to \text{Pic}^0_{C'/k} ,$$

$$f_0 : \text{Pic}^0_{C'/k} \to \text{Pic}^0_{C/k} .$$

Let $P'_\infty := f(P_\infty)$ and $P''_\infty := g(P'_\infty)$. Then $C' - P'_\infty$ is k-isomorphic to $C^{(p)} - P''_\infty$. Let $P'_0 := f(P_0)$ and $P''_0 := g(P'_0)$. The closed immersion $i^{(p)}$ is, in fact, defined by $i^{(p)}(Q'') = Q'' - P''_0$ for a point Q'' of $X^{(p)}$ which is rational over an extension field k' of k. The k-rational point P'_0 defines a k-morphism $i' : X^{(p)} \to \text{Pic}^0_{C'/k}$, which is given by $i'(Q') = Q' - P'_0$ for any point Q' of $X^{(p)}$ which is rational over an extension field k' of k. The morphism i' is a closed immersion if the arithmetic genus g' of C' is positive, but i' is the zero map if $g' = 0$. Clearly we have $i' = g_0 \cdot i^{(p)}$. *We claim that $f_0 \cdot g_0$ is the Verschiebung V_G of $G := \text{Pic}^0_{C/k}$.* We denote $f_0 \cdot g_0$ by $G(F_C)$. The claim follows from the following general fact.

Lemma. *For any complete normal k-curve C with the Picard scheme $G := \text{Pic}^0_{C/k}$, it holds that $G(F_C) = V_G$.*

Proof. By the remark before Lemma 6.3.1, it suffices to show that $G(F_C) \cdot F_G$ is the "multiplication by p" map $p \cdot \text{id}_G$. In order to prove this assertion, we may assume that $k = k_s$. Let U be a dense subset of smooth k-rational points on C such that the canonical rational mapping $i : C \to G$ is defined on U (cf. Theorem 1.7). Thus, $i : U \to G$ and $i^{(p)} : U^{(p)} \to G^{(p)}$ are defined by $i(Q) = Q - P$ and $i^{(p)}(Q^{(p)}) = Q^{(p)} - P^{(p)}$ with the previous notations, where $Q^{(p)} = F_C(Q)$ and $P^{(p)} = F_C(P)$. Since $F_C^*(Q^{(p)} - P^{(p)}) = p(Q - P)$ with $(Q^{(p)} - P^{(p)})$ and $Q - P$ considered as divisors on $C^{(p)}$ and C, respectively. Therefore,

$$p \cdot i(Q) = G(F_C) \cdot i^{(p)} \cdot F_C(Q) = G(F_C) \cdot F_G(i(Q)) .$$

So, $p \cdot \text{id}_G$ and $G(F_C) \cdot F_G$ agree on the subset $i(U)$ which generates a dense subgroup of $G(k)$. Hence $p \cdot \text{id}_G = G(F_C) \cdot F_G$. $\square$

6.3.3

Lemma. *Let C be a complete k-normal model of a nontrivial k-form X of $\mathbb{A}^1$, and let $G := \mathrm{Pic}^0_{C/k}$. Let H be the kernel of the multiplication by p endomorphism $p \cdot \mathrm{id}_G : G \to G$. Then H is geometrically connected.*

Proof. Since $H \otimes \overline{k}$ is the kernel of $p \cdot \mathrm{id}_{G \otimes \overline{k}}$ on $G \otimes \overline{k} \cong \mathrm{Pic}^0_{C \otimes_k \overline{k}/\overline{k}}$, it suffices to show that $H(\overline{k})$ is connected, so we may assume that $k = \overline{k}$. Then C is not normal. The normalization $\widetilde{C}$ of C is isomorphic to $\mathbb{P}^1_k$. Let $\widetilde{P}_\infty$ be the point of $\widetilde{C}$ lying over P_∞, and choose an inhomogeneous parameter s of $\mathbb{P}^1_k$ such that $s = 0$ at $\widetilde{P}_\infty$. Let $\widetilde{\mathcal{O}}$ and $\mathcal{O}$ be the local rings of $\widetilde{P}_\infty$ and P_∞, respectively. Let $\widehat{\widetilde{\mathcal{O}}}$ and $\widehat{\mathcal{O}}$ be respectively the completions of $\widetilde{\mathcal{O}}$ and $\mathcal{O}$ with respect to their linear topologies. Then $\widehat{\widetilde{\mathcal{O}}} = k[[s]]$. Let $\widetilde{U}$ and U be the groups with multiplicative group law defined by

$$\widetilde{U} := \{ f \in \widehat{\widetilde{\mathcal{O}}} \mid f(\widetilde{P}_\infty) = 1 \}, \quad U := \{ f \in \widehat{\mathcal{O}} \mid f(P_\infty) = 1 \} .$$

Then we have the following group isomorphisms (cf. the proof of Theorem 1.8)

$$\mathrm{Pic}^0_{C/k}(k) \cong \widetilde{\mathcal{O}}^* / \mathcal{O}^* \cong \widetilde{U}/U ,$$

and so, the endomorphism $p \cdot \mathrm{id}_G$ on $\mathrm{Pic}^0_{C/k}$ is transformed via the above isomorphisms to the endomorphism φ on $\widetilde{U}/U$ defined by $f(s) \mapsto f(s)^p$.

Let $f(s)$ be an element of $\widetilde{U} \setminus U$ such that $f(s)^p \in U$, and write $f(s) = 1 + g(s)$ with $g(0) = 0$. Then $g(s) \notin \widehat{\mathcal{O}}$ and $g(s)^p \in \widehat{\mathcal{O}}$. Then, for any element $\lambda \in k^*$, $\lambda g(s) \notin \widehat{\mathcal{O}}$ and $(\lambda g(s))^p \in \widehat{\mathcal{O}}$. Thus, $f_\lambda(s) := 1 + \lambda g(s) \in \widetilde{U} \setminus U$ and $f_\lambda(s)^p \in U$. This shows that the element of $\widetilde{U}/U$ represented by $f(s)$ is connected to the origin (represented by $f_0 := 1$ in $\widetilde{U}$) by an affine line contained in $H(k)$. Hence $H(k)$ is connected. $\square$

6.3.4

Lemma. *With the notations of 6.3.2, the kernel of $f_0 : \mathrm{Pic}^0_{C'/k} \to \mathrm{Pic}^0_{C/k}$ is geometrically connected.*

Proof. We may assume that $\mathrm{Pic}^0_{C'/k} \neq (0)$. Then $g_0 : (\mathrm{Pic}^0_{C/k})^{(p)} \to \mathrm{Pic}^0_{C'/k}$ is faithfully flat. In fact, since both Picard group schemes are k-smooth, it suffices to show that g_0 is surjective. For this proof, we may assume that $k = k_s$. Then $i^{(p)}(X^{(p)}(k))$ and $i'(X^{(p)}(k))$ generate dense subgroups in $(\mathrm{Pic}^0_{C/k})^{(p)}$ and $\mathrm{Pic}^0_{C'/k}$, respectively, and $i' = g_0 \cdot i^{(p)}$. Hence

g_0 is surjective. Since F_G and g_0 are faithfully flat and $f_0 \cdot g_0 \cdot F_G = V_G \cdot F_G = p \cdot \mathrm{id}_G$ (cf. 6.3.2), it follows that $\mathrm{Ker}\, f_0 = (g_0 \cdot F_G)(\mathrm{Ker}\,(p \cdot \mathrm{id}_G))$. Since $\mathrm{Ker}\,(p \cdot \mathrm{id}_G)$ is geometrically connected by Lemma 6.3.3, so is $\mathrm{Ker}\, f_0$. $\square$

6.3.5

Lemma. *With the notations of 6.3.2, $f_0 : \mathrm{Pic}^0_{C'/k} \to \mathrm{Pic}^0_{C/k}$ is a closed immersion.*

Proof.   If $\mathrm{Pic}^0_{C'/k} = (0)$ then the assertion is obvious. If $\mathrm{Pic}^0_{C'/k} \neq (0)$, it suffices to show that the Lie algebra homomorphism

$$\mathrm{Lie}(f_0) : \mathrm{Lie}(\mathrm{Pic}^0_{C'/k}) \to \mathrm{Lie}(\mathrm{Pic}^0_{C/k})$$

is injective, for $\mathrm{Ker}\, f_0$ is geometrically connected.   Meanwhile, $\mathrm{Lie}(\mathrm{Pic}^0_{C'/k}) \cong H^1(C', \mathcal{O}_{C'})$, $\mathrm{Lie}(\mathrm{Pic}^0_{C/k}) \cong H^1(C, \mathcal{O}_C)$ and $\mathrm{Lie}(f_0)$ is identified with $f^* : H^1(C', \mathcal{O}_{C'}) \to H^1(C, \mathcal{O}_C)$. Therefore we have only to show the next result.

6.3.5.1. Lemma. $f^* : H^1(C', \mathcal{O}_{C'}) \to H^1(C, \mathcal{O}_C)$ *is injective.*

Proof.   Since H^1 commutes with the base extension $\overline{k}/k$, it suffices to prove that $H^1(\overline{C}', \mathcal{O}_{\overline{C}'}) \to H^1(\overline{C}, \mathcal{O}_{\overline{C}})$ is injective, where $\overline{C} := C \otimes \overline{k}$ and $\overline{C}' := C' \otimes \overline{k}$. Meanwhile, as in 1.8, we have

$$H^1(\overline{C}, \mathcal{O}_{\overline{C}}) \cong \tilde{\mathcal{O}}_\infty / \overline{\mathcal{O}}_\infty, \quad H^1(\overline{C}', \mathcal{O}_{\overline{C}'}) \cong \tilde{\mathcal{O}}'_\infty / \overline{\mathcal{O}}'_\infty \ ,$$

where $\mathcal{O}_\infty$, $\mathcal{O}'_\infty$ are the places at infinity of X, $X^{(p)}$, respectively, $\overline{\mathcal{O}}_\infty = \overline{k} \otimes \mathcal{O}_\infty$, $\overline{\mathcal{O}}'_\infty = \overline{k} \otimes \mathcal{O}'_\infty$, and $\tilde{\mathcal{O}}_\infty$, $\tilde{\mathcal{O}}'_\infty$ are the normalizations of $\overline{\mathcal{O}}_\infty$, $\overline{\mathcal{O}}'_\infty$, respectively. Since the mapping

$$\overline{f}^* : \tilde{\mathcal{O}}'_\infty / \overline{\mathcal{O}}'_\infty \longrightarrow \tilde{\mathcal{O}}_\infty / \overline{\mathcal{O}}_\infty$$

is induced by the canonical mapping

$$F_K : (k, \varphi) \otimes K \to K, \quad \xi \otimes a \mapsto \xi a^p,$$

for $K = k(X)$, it follows that $\overline{f}^*$ decomposes as

$$\tilde{\mathcal{O}}'_\infty / \overline{\mathcal{O}}'_\infty \xrightarrow{\sim} \tilde{\mathcal{O}}_1 / \overline{\mathcal{O}}_1 \longrightarrow \tilde{\mathcal{O}}_\infty / \overline{\mathcal{O}}_\infty \ ,$$

where $\mathcal{O}_1 := \mathcal{O}_\infty \cap k[K^p]$, $\overline{\mathcal{O}}_1 = \overline{k} \otimes \mathcal{O}_1$ and $\tilde{\mathcal{O}}_1$ is the integral closure of $\overline{\mathcal{O}}_1$ and the mapping $\tilde{\mathcal{O}}_1 / \overline{\mathcal{O}}_1 \to \tilde{\mathcal{O}}_\infty / \overline{\mathcal{O}}_\infty$ is induced by the natural inclusion $\mathcal{O}_1 \hookrightarrow \mathcal{O}_\infty$. Thus, it suffices to show that $\tilde{\mathcal{O}}_1 / \overline{\mathcal{O}}_1 \to \tilde{\mathcal{O}}_\infty / \overline{\mathcal{O}}_\infty$ is injective, i.e., $\tilde{\mathcal{O}}_1 \cap \overline{\mathcal{O}}_\infty \subseteq \overline{\mathcal{O}}_1$. Let $\sum_i \xi_i \otimes y_i \in \overline{\mathcal{O}}_\infty = \overline{k} \otimes \mathcal{O}_\infty$ with $\xi_i \in \overline{k}$ and $y_i \in \mathcal{O}_\infty$, where we choose $\xi_i \in \overline{k}$ so that $\{\xi_i\}$ are linearly independent over k. If $\sum_i \xi_i \otimes y_i \in \tilde{\mathcal{O}}_1 \subseteq \overline{k} \otimes k[K^p]$, then $y_i \in \mathcal{O}_\infty \cap k[K^p] = \mathcal{O}_1$ for all i. Here we note that $\{\xi_i \otimes 1\}$ as elements of $\overline{k} \otimes_k K$ are linearly independent over K because K is a regular extension of k. Hence $\sum_i \xi_i \otimes y_i \in \overline{k} \otimes \mathcal{O}_1 = \overline{\mathcal{O}}_1$. $\square$

6.3.6

Summarizing the results in 6.3.1 $\sim$ 6.3.5, we obtain the following result:

Theorem. *In the setting of 6.3.2, we have:*

(1) $f_0 \cdot g_0 = V_G$, *and* f_0 *is a closed immersion.*
(2) $f_0(\mathrm{Pic}^0_{C'/k}) = p(\mathrm{Pic}^0_{C/k})$, *the image of* $G := \mathrm{Pic}^0_{C/k}$ *by* $p \cdot \mathrm{id}_G$.

6.3.7

Corollary. *The exponent of* $\mathrm{Pic}^0_{C/k}$ *is equal to* $p^{\lambda'}$, *where* $\lambda' = \mathrm{ht}\, k(X)$.

Proof. Let $C^{(i)}$ be the normalization of $C \otimes (k, \varphi^i)$ in its function field. Then $C^{(i)}$ is k-isomorphic to $\mathbb{P}^1_k$ if and only if $C \otimes k^{p^{-i}}$ is $k^{p^{-i}}$-birational to $\mathbb{P}^1_{k^{p^{-i}}}$ because X has a k-rational point (cf. 2.2). By Theorem 6.3.6, $\mathrm{Pic}^0_{C^{(i)}/k}$ is isomorphic to $p^i(\mathrm{Pic}^0_{C/k})$, the image of $G := \mathrm{Pic}^0_{C/k}$ under $p^i \cdot \mathrm{id}_G$. Since $\mathrm{Pic}^0_{C^{(i)}/k} = (0)$ if and only if $C^{(i)}$ is k-isomorphic to $\mathbb{P}^1_k$, we infer that the exponent of $\mathrm{Pic}^0_{C/k}$ agrees with $p^{\lambda'}$. $\square$

6.3.8

Corollary. *With the notations of 6.2.2, the exponent of* $C\ell^0(X)$ *is less than or equal to* $p^{\lambda'}$, *where the equality holds provided* X *has two* k-*rational points.*

Proof. By Lemma 6.2.2 and Corollary 6.3.7, $p^{\lambda'}$ annihilates $C\ell^0(X)$. Let r be the p-exponent of $C\ell^0(X)$. Then $r \leq \lambda'$. Let P and Q be distinct k-rational points of X and let $\mathfrak{p}$ and $\mathfrak{q}$ be the prime ideals of A corresponding to P and Q, respectively, where $X = \mathrm{Spec}\,(A)$. Let $\mathfrak{p}_r = \mathfrak{p} \cap A^{[r]}$ and $\mathfrak{q}_r = \mathfrak{q} \cap A^{[r]}$, where $A^{[r]} := k[A^{p^r}]$. Then $\mathfrak{p}_r A = \mathfrak{p}^{p^r} A$ and $\mathfrak{q}_r A = \mathfrak{q}^{p^r} A$. Moreover, there exists an element ξ of $K := k(X)$ such that $(\mathfrak{p}\mathfrak{q}^{-1})^{p^r} A = \xi A$. We shall show that $\xi \in K^{[r]} := k(K^{p^r})$ and $\mathfrak{p}_r \mathfrak{q}_r^{-1} = \xi A^{[r]}$. In fact, put $\overline{A} := \overline{k} \otimes A = \overline{k}[t]$ and $\overline{K} := \overline{k} \otimes K = \overline{k}(t)$. Then P and Q are given by $t = \alpha$ and $t = \beta$ with $\alpha, \beta \in \overline{k}$, respectively. Therefore, $(\mathfrak{p}\mathfrak{q}^{-1})\overline{A} = (t-\alpha)(t-\beta)^{-1}\overline{A}$. Hence $\xi = \gamma((t-\alpha)(t-\beta)^{-1}))^{p^r}$ with $\gamma \in \overline{k}$. This implies that $\xi \in K \cap \overline{K}^{[r]} = K^{[r]}$. Then $\mathfrak{p}_r \mathfrak{q}_r^{-1} A = \xi A$ implies that $\mathfrak{p}_r \mathfrak{q}_r^{-1} = \xi A^{[r]}$. Since $\mathfrak{p}_r$ and $\mathfrak{q}_r$ correspond to k-rational points of $X^{[r]} := \mathrm{Spec}\,(A^{[r]})$, $X^{[r]}$

is k-rational. Hence $r \geq \lambda'$. Since $\lambda' \geq r$ has been proved, we have $r = \lambda'$.

□

6.4   An example of unipotent Picard variety

Let X be a k-form of $\mathbb{A}^1$ of k-genus $g > 0$ and let C be a complete k-normal model of X. By Lemma 6.1.1, $\mathrm{Pic}^0_{C/k}$ is a unipotent algebraic group defined over k, and it is an interesting problem to find out the structure of this unipotent group. If X is a hyperelliptic k-form of $\mathbb{A}^1$ of the type

$$(*) \qquad y^2 = x^p + a, \quad a \in k \setminus k^p, \quad p = 2\ell + 1 > 2 \ ,$$

we can describe the structure of $\mathrm{Pic}^0_{C/k}$ as follows, by using the purely inseparable descent technique of Samuel [79] and logarithmic differentials.

Theorem. *Let C be a complete k-normal model of the affine plane curve defined by the above equation $(*)$. Then $\mathrm{Pic}^0_{C/k}$ is k-isomorphic to a unipotent algebraic subgroup of $G_a^{\ell+1}$ defined as*

$$\mathrm{Spec}\left(k[Y_0, Y_1, \ldots, Y_\ell] / (Y_0 - \sum_{i=0}^{\ell} \binom{\ell}{i} a^i Y_{\ell-i}^p) \right) \ .$$

Furthermore, $\mathrm{Pic}^0_{C/k}$ makes the following sequence exact,

$$0 \longrightarrow G \overset{\tau}{\longrightarrow} \mathrm{Pic}^0_{C/k} \longrightarrow G_a^{\ell-1} \longrightarrow 0 \ ,$$

where G is a k-group of Russell type $G := \mathrm{Spec}\,(k[x,y]/(y^p - x + a^\ell x^p))$ and τ is defined by $\tau^*(Y_0) = x, \tau^*(Y_\ell) = y$ and $\tau^*(Y_i) = 0$ for $1 \leq i \leq \ell - 1$.

For the proof, see [41, 4.7, p.70].

PART II

Purely inseparable and Artin-Schreier coverings

Chapter 1

Vector fields and infinitesimal group schemes

Hereafter throughout the present book, unless otherwise mentioned, the base field k is an algebraically closed field of fixed positive characteristic p. Let A be an affine k-domain, which is by definition a finitely generated integral k-domain. Let $\varphi : k \to k$ be the pth power endomorphism $\lambda \to \lambda^p$. Then there is an injective k-algebra homomorphism $\varphi_A : A \otimes_k (k, \varphi) \to A$ defined by $\varphi_A(a \otimes \lambda) = \lambda a^p$ for $a \in A$ and $\lambda \in k$. The image of φ_A is denoted by $A^{(p)}$ and identified with $A \otimes_k (k, \varphi)$. In short, $A^{(p)}$ is a k-subalgebra of A generated by $\{a^p \mid a \in A\}$, and called the Frobenius image of A. In terms of schemes, let X be an algebraic k-scheme. Then there exists the Frobenius morphism $F_X : X \to X \otimes_k (k, \varphi)$, which is given on an affine open set $\operatorname{Spec} A$ by the above k-algebra homomorphism φ_A. As in Part I, we denote $X \otimes_k (k, \varphi)$ by $X^{(p)}$.

1.1 Differential 1-forms and derivations

Given a ring C and a C-algebra A, let $\Omega^1_{A/C}$ be the A-module of differential 1-forms of A over C which is generated by elements $\{da \mid a \in A\}$ as an A-module such that

(1) $d(a + b) = da + db, \quad d(ab) = adb + bda, \quad a, b \in A$

(2) $d\lambda = 0, \quad \lambda \in C$.

Let $m : A \otimes_C A \to A$ be the multiplication $a \otimes b \mapsto ab$ and let $I = \operatorname{Ker} m$. By the left multiplication $a(a_1 \otimes b) = aa_1 \otimes b$, we view $A \otimes_C A$ as an A-module. Then $\operatorname{Ker} m$ is an A-module and hence so is I/I^2. We set $\Omega^1_{A/C} = I/I^2$. For $a \in A$, set da to be the class of $a \otimes 1 - 1 \otimes a$ in I/I^2. Then $d : A \to \Omega^1_{A/C}$ is a C-module homomorphism and d has the above properties. In particular,

$d(ab) = adb + bda$ follows from

$$ab \otimes 1 - 1 \otimes ab$$
$$= a(b \otimes 1 - 1 \otimes b) + b(a \otimes 1 - 1 \otimes a) - (a \otimes 1 - 1 \otimes a)(b \otimes 1 - 1 \otimes b).$$

It is clear that $\Omega^1_{A/C}$ is generated by the set $\{da \mid a \in A\}$ as an A-module.

We denote the dual A-module $\mathrm{Hom}_A(\Omega^1_{A/C}, A)$ by $\mathrm{Der}_C(A)$ and call it the *module of C-derivations* of A. Let $\alpha : \Omega^1_{A/C} \to A$ be an A-module homomorphism and let $D := \alpha \cdot d : A \to A$. Then D satisfies the following two conditions by the above properties (1) and (2):

$$(1)' \quad D(ab) = aD(b) + bD(a), \quad a, b \in A$$
$$(2)' \quad D(\lambda) = 0, \quad \lambda \in C.$$

We call a C-module endomorphism D of A a C-*derivation* of A if it satisfies (1)' and (2)' above. Conversely, given a C-derivation D of A, define an A-module homomorphism $\widetilde{\alpha} : A \otimes_C A \to A$ by $\widetilde{\alpha}(a \otimes b) = -aD(b)$. Then $\widetilde{\alpha}(a \otimes 1 - 1 \otimes a) = D(a)$ and $\widetilde{\alpha}|_{I^2} = 0$. Hence $\widetilde{\alpha}$ induces an A-module homomorphism $\alpha : \Omega^1_{A/C} \to A$ which satisfies $d \cdot \alpha = D$. Thus the A-module $\mathrm{Der}_C(A)$ is the set of all C-derivations of A.

Here we collect the results on $\Omega^1_{A/C}$ and $\mathrm{Der}_C(A)$ which we make frequent use in the forthcoming arguments (see [EGA, Chap. IV]).

1.1.1

Lemma. *We have the following assertions.*

(1) *Let $f : B \to A$ be a homomorphism of C-algebras. With A viewed as an B-algebra via f, we have an exact sequence of A-modules*

$$A \otimes_B \Omega^1_{B/C} \longrightarrow \Omega^1_{A/C} \longrightarrow \Omega^1_{A/B} \longrightarrow 0.$$

(2) *Let J be an ideal of a C-algebra A. Then there is an exact sequence of A/J-modules*

$$J/J^2 \longrightarrow (A/J) \otimes_A \Omega^1_{A/C} \longrightarrow \Omega^1_{(A/J)/C} \longrightarrow 0.$$

(3) *Suppose that $A = C[a_1, \ldots, a_r]$, i.e., A is a finitely generated C-algebra with a set of generators $\{a_1, \ldots, a_r\}$. Then $\Omega^1_{A/C}$ is generated by $\{da_1, \ldots, da_r\}$ as an A-module.*

(4) *Let S be a multiplicative set of A. Then we have: $\Omega^1_{S^{-1}A/C} \cong S^{-1}\Omega^1_{A/C}$, where $S^{-1}A$ is the ring of fractions of A with respect to S and $S^{-1}M = S^{-1}A \otimes_A M$ for an A-module M.*

(5) *Let C' be a C-algebra which is a flat C-module. Then $C' \otimes_C \Omega^1_{A/C} \cong$* $\Omega^1_{(C' \otimes_C A)/C'}$.

If A is a C-algebra by a ring homomorphism $f : C \to A$, it is clear that $\Omega^1_{A/C} = \Omega^1_{A/f(C)}$. So, only the image $f(C)$ in A matters.

1.1.2

Since $\mathrm{Hom}_A(\,\cdot\,, A)$ is a left-exact contravariant functor on the category of A-modules, we have the following result.

Lemma. *With the respective settings in Lemma 1.1.1, we have:*

(1) *The following is an exact sequence of A-modules:*

$$0 \longrightarrow \mathrm{Der}_B(A) \longrightarrow \mathrm{Der}_C(A) \longrightarrow A \otimes_B \mathrm{Der}_C(B).$$

(2) *The following is an exact sequence of A/J-modules*

$$0 \longrightarrow \mathrm{Der}_C(A/J) \longrightarrow A/J \otimes_A \mathrm{Der}_C(A) \longrightarrow \mathrm{Hom}_{A/J}(J/J^2, A/J).$$

(3) *If $A = C[a_1, \ldots, a_r]$, then an element D of $\mathrm{Der}_C(A)$ is determined by the values $D(a_1), \ldots, D(a_r)$.*

(4) *If $\Omega^1_{A/C}$ is finitely presented as an A-module, i.e., $\Omega^1_{A/C}$ is the cokernel of an A-module homomorphism of free A-modules of finite rank, then $\mathrm{Der}_C(S^{-1}A) \cong S^{-1} \otimes_A \mathrm{Der}_C(A)$.*

1.1.3

Lemma. *Let k be an arbitrary field of k and let A be an affine k-algebra, i.e., A is finitely generated over k. Then $\Omega^1_{A/k}$ is a finitely generated A-module by Lemma 1.1.1, (3). Since A is noetherian, there exists an exact sequence of A-modules*

$$A^{\oplus s} \longrightarrow A^{\oplus r} \longrightarrow \Omega^1_{A/k} \longrightarrow 0.$$

Hence, by taking the dual sequence, we have an exact sequence

$$0 \longrightarrow \mathrm{Der}_{A/k} \to A^{\oplus r} \longrightarrow A^{\oplus s}.$$

So, $\mathrm{Der}_k(A)$ is a finitely generated A-module.

1.2 *p*-Lie algebra and Galois correspondence

On the other hand, with a k-algebra A which is not necessarily finitely generated over k, the A-module $\mathrm{Der}\,_k(A)$ satisfies the following conditions.

(1) Define the *bracket product* $[D, D']$ by $[D, D'] = DD' - D'D$. Namely $[D, D'](a) = D(D'(a)) - D'(D(a))$. Then $[D, D']$ is a k-derivation of A, i.e., $[D, D'] \in \mathrm{Der}\,_k(A)$. We call $[D, D']$ a *bracket product* of D and D'.

(2) Define the *pth power product* D^p of D by setting $D^p(a) = \underbrace{D \circ \cdots \circ D}_{p}(a)$. Then D^p is a k-derivation of A by the following Leibniz rule.

$$D^n(ab) = \sum_{i=0}^{n} \binom{n}{i} D^{n-i}(a) D^i(b), \quad a,\ b \in A.$$

If $n = p$, all the binomial coefficients $\binom{p}{i} = 0$ except for $\binom{p}{0} = \binom{p}{p} = 1$. Hence D^p is a k-derivation.

Hence the A-module $\mathrm{Der}\,_k(A)$ equipped with the bracket product and pth power product is a *p-Lie algebra*. An A-submodule L of $\mathrm{Der}\,_k(A)$ is a p-Lie subalgebra if it is closed under the bracket product and the pth power product. If L is an A-module generated by a single element D, then the condition that L is a p-Lie subalgebra is equivalent to the condition that $D^p = aD$ for some element $a \in A$. We then say that D is *p-closed*. A p-Lie subalgebra L of $\mathrm{Der}\,_k(A)$ is *saturated* if $aD \in L$ for $a \in A$ and $D \in \mathrm{Der}\,_k(A)$ implies $D \in L$.

1.2.1

The following is a renowned analogue of the Galois theory due to Jacobson [39].

Lemma. *Let K be a field extension of k. Then there is a bijective correspondence between the set $\mathcal{M}$ of intermediate field extensions M with $K^{(p)} \subset M \subset K$ and the set $\mathcal{L}$ of p-Lie subalgebras L of $\mathrm{Der}\,_k(K)$. The correspondence is given by*

$$M \mapsto L(M) = \{\delta \in \mathrm{Der}\,_k(K) \mid \delta|_M = 0\}$$

and

$$L \mapsto M(L) = \{\xi \in K \mid \delta(\xi) = 0 \text{ for all } \delta \in L\}.$$

1.2.2

We can generalize this result to the following effect. This result is implicitly written in [77]. For a k-domain A, we denote by $Q(A)$ the quotient field, i.e., the field of fractions, of A.

Lemma. *Suppose that a given affine k-domain A is normal. Then there is a bijective correspondence between the set $\mathcal{B}$ of normal k-subalgebras B with $A^{(p)} \subset B \subset A$ and the set $\mathcal{L}_A$ of saturated p-Lie subalgebras L of $\operatorname{Der}_k(A)$. The correspondence is given by*

$$B \mapsto L(B) := L(Q(B)) \cap \operatorname{Der}_k(A)$$

and

$$L \mapsto A \cap M(L_K), \ L_K = L \otimes_A K, \ K = Q(A).$$

Proof. Given a normal k-subalgebra B of R with $A^{(p)} \subset B$, $L(B)$ is a p-Lie subalgebra of $\operatorname{Der}_k(A)$. We shall show that it is saturated. In fact, if $f\delta \in L(B)$ with $f \in A$ and $\delta \in \operatorname{Der}_k(A)$, then $\delta \in L(Q(B)) \cap \operatorname{Der}_k(A)$, whence $\delta \in L(B)$. Set $B' = \{a \in A \mid \delta(a) = 0 \text{ for all } \delta \in L(B)\}$. Suppose $a \in B'$. Since $L(Q(B)) = L(B) \otimes_A K$, we have $a \in M(L(Q(B))) = Q(B)$ by the Jacobson theory. Since a is integral over B and B is normal, we have $a \in B$. Hence $B' \subseteq B$. The converse $B \subseteq B'$ is clear. So, $B = B'$.

Conversely, let L be a saturated p-Lie subalgebra of $\operatorname{Der}_k(A)$. Then $Q(A \cap M(L_K)) = M(L_K)$. In fact, $Q(A \cap M(L_K)) \subseteq M(L_K)$ is clear. If $\xi = ba^{-1} \in M(L_K)$ with $a, b \in A$, then $\xi = (a^{p-1}b)a^{-p}$ with $a^p, a^{p-1}b \in A \cap M(L_K)$. So, the converse holds. Hence $A \cap M(L_K)$ is a normal subalgebra of A containing $A^{(p)}$. Furthermore, we have

$$
\begin{aligned}
L(A \cap M(L_K)) &= L(Q(A \cap M(L_K))) \cap \operatorname{Der}_k(A) \\
&= L(M(L_K)) \cap \operatorname{Der}_k(A) = L_K \cap \operatorname{Der}_k(A) = L.
\end{aligned}
$$

$\square$

1.3 Unramified extension of affine domains

Let A be a k-domain and let B be a k-subalgebra of A such that A is finitely generated over B. We say that A is *unramified* over B if, for every prime ideal $\mathfrak{p}$ of A and $\mathfrak{q} := \mathfrak{p} \cap B$, we have $\mathfrak{p}A_{\mathfrak{p}} = \mathfrak{q}A_{\mathfrak{p}}$ and $Q(A/\mathfrak{p})$ is a separable extension of $Q(B/\mathfrak{q})$. Furthermore, if A is flat over B, then we say that A

is *étale* over B. We consider the relationship between the unramifiedness of A over B and the vanishing of the module of differential 1-form $\Omega^1_{A/B}$.

Namely, we have the following result.

Theorem. *Assume that B is an affine k-domain. With the above setting, A is unramified over B if and only if $\Omega^1_{A/B} = 0$.*

Since this result is one of the important results used in this book, we give a full proof, which is given step by step in the subsequent subsections.

1.3.1

Lemma. *Let L be a finite algebraic extension of a field K. Then the following assertions hold.*

(1) *If L is purely inseparable over K, then $\Omega^1_{L/K} \neq 0$. More precisely, if $L = K(\alpha)$ with $\alpha^p = a \in K \setminus K^p$, then $\Omega^1_{L/K} = Ld\alpha$. Hence $\mathrm{Der}\,_K(L) = L \cdot D$, where D is a K-derivation of L such that $D(\alpha) = 1$ and $D^p = 0$.*

(2) *If L is separable over K, then $\Omega^1_{L/K} = 0$.*

(3) *L is a separable extension of K if and only if $\Omega^1_{L/K} = 0$.*

Proof. (1) The field L is the residue ring of a polynomial ring $K[x]$ modulo the ideal $J = (x^p - a)$. By Lemma 1.1.1, (2), we have an exact sequence of L-modules

$$ J/J^2 \longrightarrow L \otimes_{K[x]} \Omega^1_{K[x]/K} \longrightarrow \Omega^1_{L/K} \longrightarrow 0 \, , $$

where J/J^2 is mapped to an L-submodule of the middle term generated by $d(x^p - a)$, which is obviously zero. Hence we have an isomorphism $L \otimes_{K[x]} \Omega^1_{K[x]/K} \cong \Omega^1_{L/K}$, where $\Omega^1_{K[x]/K} = K[x]dx$. Hence $\Omega^1_{L/K} = Ld\alpha$, which is nonzero. Then $\mathrm{Der}\,_{L/K} = L \cdot D$ with a K-derivation D of L determined by $D(\alpha) = 1$. It is clear that $D^p = 0$.

(2) It is well-known that L is a simple extension. Namely there exists an element θ such that $L = K(\theta)$. Let $f(x) = 0$ be the (monic) minimal equation of θ over K. Then $d(f(\theta)) = f'(\theta)d\theta = 0$, where $f'(\theta) \neq 0$. Hence $d\theta = 0$. Since $\Omega^1_{L/K} = L \cdot d\theta$, it follows that $\Omega^1_{L/K} = 0$.

(3) Let L_s be the separable closure of K in L. Then Lemma 1.1.1, (1) yields an exact sequence of L-modules

$$ L \otimes_{L_s} \Omega^1_{L_s/K} \longrightarrow \Omega^1_{L/K} \longrightarrow \Omega^1_{L/L_s} \longrightarrow 0 \, , $$

where the left term is zero by (2) above. Hence $\Omega^1_{L/K} \cong \Omega^1_{L/L_s}$. In view of (2) above, it suffices to show that $\Omega^1_{L/K} = 0$ implies that L is a separable extension of K. Otherwise, $L \neq L_s$. Since L is a purely inseparable extension of L_s, there exists a subfield L_1 of L such that $L_s \subseteq L_1$ and $L = L_1(\alpha)$ with $\alpha \in L \setminus L_1^p$. Then we have an exact sequence of L-modules

$$L \otimes_{L_1} \Omega^1_{L_1/K} \longrightarrow \Omega^1_{L/K} \longrightarrow \Omega^1_{L/L_1} \longrightarrow 0 \,,$$

where $\Omega^1_{L/L_1} \neq 0$ by (1) above. Hence $\Omega^1_{L/K} \neq 0$. This is a contradiction. Thus $L = L_s$. $\qquad\qquad\square$

1.3.2

Lemma. *The following assertions hold.*

(1) *Let L/K be a finitely generated field extension. If $\operatorname{tr.deg}_K L > 0$ then $\Omega^1_{L/K} \neq 0$. Hence if $\Omega^1_{L/K} = 0$, then L/K is a finite algebraic extension.*

(2) *Let A be an affine domain over a field K. If $\Omega^1_{A/K} = 0$, then A is a finite algebraic extension field of K.*

Proof. (1) Suppose that $\Omega^1_{L/K} = 0$. Let $d = \operatorname{tr.deg}_K L$, let $\{x_1, \ldots, x_d\}$ be a transcendence basis of L/K and let $L_1 = K(x_1, \ldots, x_d)$. Then we have an exact sequence of L-modules

$$L \otimes_{L_1} \Omega^1_{L_1/K} \longrightarrow \Omega^1_{L/K} \longrightarrow \Omega^1_{L/L_1} \longrightarrow 0 \,,$$

whence $\Omega^1_{L/L_1} = 0$. By Lemma 1.3.1, L/L_1 is a separable algebraic extension. Suppose $d > 0$ and consider the basis $\{x_1^p, x_2, \ldots, x_d\}$. Let $L_2 = K(x_1^p, x_2, \ldots, x_d)$. Then the above argument shows that L/L_2 is also separable extension. Meanwhile, we have inclusions $L_2 \subset L_1 \subseteq L$, where L_1 is a purely inseparable extension, which is a contradiction.

(2) Let $L = Q(A)$. Then L is a finitely generated field extension of K. Let $S = A \setminus \{0\}$. Then S is a multiplicative set of A, and $S^{-1}A = Q(A) = L$. By Lemma 1.1.1, (4), we have $\Omega^1_{L/K} = S^{-1}\Omega^1_{A/K} = 0$ since $\Omega^1_{A/K} = 0$. By the assertion (1), L is a finite algebraic extension of K, whence L is a finite K-module. Since A is a sub-K-module of L, A is a finite K-algebra. In particular, A is integral over K. This implies that A is a finite algebraic field extension of K because A is a domain. $\qquad\qquad\square$

1.3.3

Lemma. *Let A and B be affine domains over a field k such that $B \subseteq A$, where k is not necessarily algebraically closed. If $\Omega^1_{A/B} = 0$ then A is unramified over B.*

Our proof consists of several steps which are developed in two paragraphs below.

1.3.3.1. Let $\mathfrak{p}$ be a prime ideal of A and let $\mathfrak{q} = \mathfrak{p} \cap B$. Let $T = B \setminus \mathfrak{q}$ and let $T^{-1}B = B_{\mathfrak{q}}$. Then $T^{-1}\Omega^1_{A/B} = \Omega^1_{T^{-1}A/B_{\mathfrak{q}}} = 0$. By replacing A and B by $T^{-1}A$ and $B_{\mathfrak{q}}$ respectively, we may assume that $(B, \mathfrak{q})$ is a local ring. Note that A is finitely generated over B.

Let $\overline{A} = A/\mathfrak{q}A$ and $\kappa = B/\mathfrak{q}$, where κ is the residue field of B. By Lemma 1.1.1,(2), $\Omega^1_{\overline{A}/\kappa} = 0$. We shall show that $\overline{A}$ is a product of fields $\prod_{i=1}^d K_i$, where K_i is a finite separable algebraic extension of κ expressed as $K_i = \overline{A}_{\overline{\mathfrak{p}}_i}$. Here $\overline{\mathfrak{p}}_i = \mathfrak{p}_i/\mathfrak{q}A$ if $\mathfrak{p}_1, \ldots, \mathfrak{p}_d$ exhaust all prime ideals of A lying over $\mathfrak{q}$. This implies that $\mathfrak{p}_i = \mathfrak{q}A$ and $A_{\mathfrak{p}_i}/\mathfrak{p}_i A_{\mathfrak{p}_i}$ is a separable algebraic extension of κ. Namely, the assertion of Lemma 1.3.3 follows. Thus we are reduced to the case $B = \kappa$ is a field and A is a finitely generated κ-algebra.

Let $\mathfrak{n} = \sqrt{0}$ be the nilradical of A and let $\mathfrak{n} = \mathfrak{p}_1 \cap \mathfrak{p}_2 \cap \cdots \cap \mathfrak{p}_r$ be the minimal prime decomposition. Let $S = A \setminus (\mathfrak{p}_1 \cup \mathfrak{p}_2 \cup \cdots \cup \mathfrak{p}_r)$. Then S is a multiplicative set of A and $S^{-1}A$ is the total quotient ring. Let $\mathfrak{P}_i = \mathfrak{p}_i S^{-1}A$ for $1 \le i \le r$. Then $\mathfrak{P}_i + \mathfrak{P}_j = S^{-1}A$ if $i \ne j$. In fact, if $\mathfrak{P}_i + \mathfrak{P}_j \ne S^{-1}A$, then there exists a maximal ideal M of $S^{-1}A$ such that $\mathfrak{P}_i + \mathfrak{P}_j \subseteq M$. Let $\mathfrak{m}$ be the inverse image of M by the canonical homomorphism $A \to S^{-1}A$. Then $\mathfrak{p}_i + \mathfrak{p}_j \subseteq \mathfrak{m}$. Since $\mathfrak{p}_1, \ldots, \mathfrak{p}_r$ are minimal prime divisors of A, $\mathfrak{m}$ is different from $\mathfrak{p}_1, \ldots, \mathfrak{p}_r$. Take an element $a_i \in (\mathfrak{m} \cdot \prod_{j \ne i} \mathfrak{p}_j) \setminus \mathfrak{p}_i$ and let $a = \sum_{i=1}^d a_i$. Then $a \in \mathfrak{m}$ and $a \notin \mathfrak{p}_1 \cup \cdots \cup \mathfrak{p}_r$. Hence a is invertible in $S^{-1}A$ and $a \in M$. This is a contradiction. Hence, by Chinese Remainder Theorem, we have $S^{-1}A = \prod_{i=1}^d (S^{-1}A)_{\mathfrak{P}_i}$, where $(S^{-1}A)_{\mathfrak{P}_i}$ is an Artin local ring with maximal ideal $\mathfrak{P}_i(S^{-1}A_{\mathfrak{P}_i})$. Since

$$S^{-1}\Omega^1_{A/\kappa} = \prod_{i=1}^d \Omega^1_{(S^{-1}A)_{\mathfrak{P}_i}/\kappa} \, ,$$

we have $\Omega^1_{(S^{-1}A_{\mathfrak{P}_i})/\kappa} = 0$ for every i. Hence we are reduced to show that if $\Omega^1_{A/\kappa} = 0$ then $\mathfrak{p} = 0$ and $A/\mathfrak{p}$ is a finite separable algebraic extension of κ in the case where $(A, \mathfrak{p})$ is an Artin local ring.

1.3.3.2. By Lemma 1.1.1,(2), $\Omega^1_{(A/\mathfrak{p})/\kappa} = 0$. It is clear by construction that $A/\mathfrak{p}$ is a finitely generated field extension of κ By Lemmas 1.3.1 and 1.3.2, it follows that $A/\mathfrak{p}$ is a separable algebraic extension of κ. Let $\overline{\kappa}$ be an algebraic closure of κ. Then $\overline{\kappa} \otimes_\kappa A$ is a direct product of Artin local rings whose residue fields are isomorphic to $\overline{\kappa}$. If each direct component of $\overline{\kappa} \otimes_\kappa A$ is a field, we know that $\mathfrak{p} = 0$. Hence we may assume that the residue field $A/\mathfrak{p}$ is equal to κ. Then there exists an ideal J of A such that $\mathfrak{p}/J = \varepsilon(A/J)$ and $\varepsilon^2 = 0$. To simplify the notation, by replacing $(A, \mathfrak{p})$ by $(A/J, \mathfrak{p}/J)$, we assume that $\mathfrak{p} = \varepsilon A$ and $\varepsilon^2 = 0$. Then $A = \kappa \oplus \kappa\varepsilon$ with multiplication given by

$$(\alpha + \beta\varepsilon) \cdot (\alpha' + \beta'\varepsilon) = \alpha\alpha' + (\alpha\beta' + \alpha'\beta)\varepsilon, \quad \alpha, \alpha', \beta, \beta' \in \kappa .$$

Define a κ-module homomorphism $D : A \to A$ by $D(\alpha + \beta\varepsilon) = \beta\varepsilon$. Then D is a κ-derivation of A. Since $\operatorname{Der}_\kappa(A) = \operatorname{Hom}_A(\Omega^1_{A/\kappa}, A)$, we know that $\Omega^1_{A/\kappa} \neq 0$, which is a contradiction to the hypothesis. Hence $\mathfrak{p} = 0$. This proves Lemma 1.3.3.

1.3.4

Lemma. *Let A and B be as in Lemma 1.3.3. If A is unramified over B then $\Omega^1_{A/B} = 0$.*

Proof. Since A is an affine k-domain, there exists a surjective k-algebra homomorphism $\sigma : k[x_1, \ldots, x_r] \to A$ with $J = \operatorname{Ker} \sigma$. Then, by Lemma 1.1.1,(3), $\Omega^1_{A/k}$ is a finitely generated A-module. Since there exists a surjection of A-modules $\Omega^1_{A/k} \to \Omega^1_{A/B}$, $\Omega^1_{A/B}$ is a finitely generated A-module. Then $\Omega^1_{A/B} = 0$ if and only if $A_\mathfrak{m} \otimes \Omega^1_{A/B} = 0$ for every maximal ideal $\mathfrak{m}$ of A. Hence we may assume that A is a local ring $(A, \mathfrak{m})$ and $\Omega^1_{A/B}$ is a finitely generated A-module, where B is taken to be a local ring $(B, \mathfrak{n})$ with $\mathfrak{n} = B \cap \mathfrak{m}$. Then, by a lemma of Nakayama, it suffices to show that $A/\mathfrak{m} \otimes_A \Omega^1_{A/B} = 0$. Meanwhile, we have an exact sequence by Lemma 1.1.1,(2),

$$\mathfrak{m}/\mathfrak{m}^2 \longrightarrow A/\mathfrak{m} \otimes_A \Omega^1_{A/B} \longrightarrow \Omega^1_{(A/\mathfrak{m})/(B/\mathfrak{n})} \longrightarrow 0 ,$$

where $\mathfrak{m}/\mathfrak{m}^2 \cong \mathfrak{n}/\mathfrak{n}^2 \otimes_{B/\mathfrak{n}} (A/\mathfrak{m})$ because $\mathfrak{m} = \mathfrak{n}A$ and $\Omega^1_{(A/\mathfrak{m})/(B/\mathfrak{n})} = 0$ because $A/\mathfrak{m}$ is a finite separable algebraic extension of $B/\mathfrak{n}$ (cf. Lemma 1.3.1,(3)). The image of $\mathfrak{n}/\mathfrak{n}^2$ in $A/\mathfrak{m} \otimes_A \Omega^1_{A/B}$ is zero because $\xi \pmod{\mathfrak{n}^2}$ for $\xi \in \mathfrak{n}$ is mapped to $1 \otimes d\xi$, where $d\xi = 0$ because $\xi \in B$. So, $A/\mathfrak{m} \otimes_A \Omega^1_{A/B} = 0$. $\square$

1.4 Sheaf of differential 1-forms and tangent sheaf

Let $X = \operatorname{Spec} A$ be an affine variety defined over k. Then the coherent sheaf $\widetilde{\Omega}^1_{A/k}$ associated with $\Omega^1_{A/k}$ is denoted by $\Omega^1_{X/k}$ and called the *sheaf of differential 1-forms* of X. For a point P of X, the stalk $\Omega^1_{X/k,P}$ of $\Omega^1_{X/k}$ at the point P is $\Omega^1_{A_\mathfrak{p}/k}$, where $\mathfrak{p}$ is a prime ideal of A corresponding to P. Let X be now a (not necessarily affine) algebraic variety defined over k and let $\mathcal{U} = \{U_i\}_{i \in I}$ be an affine open covering, where $U_i = \operatorname{Spec} A_i$. Then the sheaves $\Omega^1_{U_i/k}$ on U_i patch together and form a coherent sheaf $\Omega^1_{X/k}$ on X. The dual sheaf $\mathcal{H}om_{\mathcal{O}_X}(\Omega^1_{X/k}, \mathcal{O}_X)$ is denoted by $\mathcal{T}_{X/k}$ and called the *tangent sheaf* of X. The stalk $\mathcal{T}_{X,P}$ at P is isomorphic to $\operatorname{Hom}_{A_\mathfrak{p}}(\Omega^1_{A_\mathfrak{p}/k}, A_\mathfrak{p}) \cong \operatorname{Der}_k(A) \otimes_A A_\mathfrak{p}$ if $X = \operatorname{Spec} A$ and the point P corresponds to a prime ideal $\mathfrak{p}$. We consider now the local behavior of $\Omega^1_{X/k}$ and $\mathcal{T}_X$ for a smooth affine variety $X = \operatorname{Spec} A$ of dimension n over k. We denote the residue field $\mathcal{O}_{X,P}/\mathfrak{m}_{X,P}$ by $k(P)$.

1.4.1

Lemma. *Let P be a closed point of X which corresponds to a maximal ideal of $\mathfrak{m}$ of A. Let $x_1, \ldots, x_n$ be local parameters of X at P. Then the following assertions hold.*

(1) $\Omega^1_{X/k,P} \otimes k(P) := \Omega^1_{A/k} \otimes_A (A/\mathfrak{m}) \cong \mathfrak{m}/\mathfrak{m}^2 \cong \bigoplus_{i=1}^n k\, dx_i$.

(2) $\Omega^1_{A/k} \otimes_A A_\mathfrak{m} \cong \bigoplus_{i=1}^n A_\mathfrak{m} dx_i$, *which is a free $A_\mathfrak{m}$-module of rank n.*

(3) $\mathcal{T}_{X,P} \otimes k(P) := \operatorname{Der}_k(A) \otimes_A A/\mathfrak{m} \cong \operatorname{Hom}_k(\mathfrak{m}/\mathfrak{m}^2, k) \cong \bigoplus_{i=1}^n k(\partial/\partial x_i)$.

Proof. (1) We have an exact sequence by Lemma 1.1.1,(2),

$$\mathfrak{m}/\mathfrak{m}^2 \xrightarrow{\;d\;} \Omega^1_{A/k} \otimes_A k \longrightarrow \Omega^1_{(A/\mathfrak{m})/k} \longrightarrow 0 \;,$$

where $A/\mathfrak{m} = k$ as k is algebraically closed, whence $\Omega^1_{(A/\mathfrak{m})/k} = 0$. Hence the homomorphism d is surjective, where d is defined as $d(x \pmod{\mathfrak{m}^2}) =$ the residue class of dx in $\Omega^1_{A/k} \otimes_A k$ for $x \in \mathfrak{m}$. We shall show that the homomorphism d is injective. For this end, it suffices to show that the dual homomorphism d^* of d

$$d^* : \operatorname{Hom}_k(\Omega^1_{A/k} \otimes_A k, k) \longrightarrow \operatorname{Hom}_k(\mathfrak{m}/\mathfrak{m}^2, k)$$

is surjective because both $\mathfrak{m}/\mathfrak{m}^2$ and $\Omega^1_{A/k} \otimes_A k$ are finite k-modules. Here $\operatorname{Hom}_k(\Omega^1_{A/k} \otimes_A k, k) \cong \operatorname{Hom}_A(\Omega^1_{A/k}, k) := \operatorname{Der}_k(A, k)$, where $\operatorname{Der}_k(A, k)$ is the k-module of k-derivations $\delta : A \to k$. Namely, δ is a k-module homomorphism $A \to k$ such that $\delta(ab) = \bar{a}\delta(b) + \bar{b}\delta(a)$, where $\bar{a}, \bar{b}$ are the residue classes of a, b in $A/\mathfrak{m}$. Then $\delta|_{\mathfrak{m}}$ induces a k-module homomorphism $\mathfrak{m}/\mathfrak{m}^2 \to k$ because $\delta(ab) = \bar{a}\delta(b) + \bar{b}\delta(a) = 0$ if $a, b \in \mathfrak{m}$. So, $d^*(\delta) = \delta|_{\mathfrak{m}}$. Given a k-module homomorphism $\sigma : \mathfrak{m}/\mathfrak{m}^2 \to k$, define $\delta_\sigma : A \to k$ by $\delta(a) = \sigma((a - \alpha) \pmod{\mathfrak{m}^2})$ with $\alpha = \bar{a}$. Let $b \in A$ with $\beta = \bar{b}$. Since

$$(ab - \alpha\beta) \pmod{\mathfrak{m}^2} = \alpha(b - \beta) + \beta(a - \alpha) \pmod{\mathfrak{m}^2},$$

δ_σ is an element of $\mathrm{Der}_k(A, k)$ such that $d^*(\delta_\sigma) = \sigma$. So, d^* is surjective and d is injective.

(2) Let $K = Q(A)$. Then K is a regular extension of k since $k = \bar{k}$. Hence there exists a separating transcendence basis $\{x_1, \ldots, x_n\}$. Namely, $K_1 := k(x_1, \ldots, x_n)$ is a purely transcendental extension and K/K_1 is a finite separable algebraic extension. Then we have an exact sequence of K-modules

$$\Omega^1_{K_1/k} \otimes_{K_1} K \xrightarrow{\gamma} \Omega^1_{K/k} \longrightarrow \Omega^1_{K/K_1} \longrightarrow 0,$$

where $\Omega^1_{K/K_1} = 0$ by Lemma 1.3.1,(3), and $\Omega^1_{K_1/k} = \bigoplus_{i=1}^n K_1 dx_i$. We shall show that the dual homomorphism γ^* of γ

$$\gamma^* : \mathrm{Hom}_K(\Omega^1_{K/k}, K) \longrightarrow \mathrm{Hom}_K(\Omega^1_{K_1/k} \otimes_{K_1} K, K)$$

or equivalently

$$\gamma^* : \mathrm{Der}_k(K) \longrightarrow \mathrm{Der}_k(K_1, K)$$

is surjective, which implies that γ is injective, hence, an isomorphism. Then $\Omega^1_{K/k} \cong \bigoplus_{i=1}^n K dx_i$. Let $\delta \in \mathrm{Der}_k(K_1, K)$. Since K/K_1 is a finite separable algebraic extension, it is a simple extension, i.e., $K = K_1(\theta)$. Let $f(\theta)$ be a (monic) minimal equation of θ over K_1. Write $f(\theta) = \sum_{i=0}^d a_i \theta^{d-i}$ with $a_i \in K_1$ and $a_0 = 1$. Let $f^\delta(\theta) = \sum_{i=0}^d \delta(a_i)\theta^{d-i}$. Put $\tilde{\delta}(\theta) = -f^\delta(\theta)/f'(\theta)$, where $f'(\theta) = \sum_{i=0}^{d-1} a_i(d-i)\theta^{d-i-1}$. Then, with $\tilde{\delta}$ defined naturally on $K = K_1(\theta)$, $\tilde{\delta}$ is a k-derivation of K such that $\tilde{\delta}|_{K_1} = \delta$. This implies that γ^* is surjective. Now, by [24, II, Lemma 8.9], $\Omega^1_{A/k} \otimes_A A_\mathfrak{m}$ is a free $A_\mathfrak{m}$-module of rank n.

(3) Since $\Omega^1_{A/k} \otimes_A A_\mathfrak{m}$ is a free $A_\mathfrak{m}$-module, we have

$$\mathrm{Der}_k(A) \otimes_A A/\mathfrak{m} \cong (\mathrm{Hom}_A(\Omega^1_{A/k}, A) \otimes_A A_\mathfrak{m}) \otimes_{A_\mathfrak{m}} k$$

$$\cong \mathrm{Hom}_{A_\mathfrak{m}}(\Omega^1_{A_\mathfrak{m}/k}, A_\mathfrak{m}) \otimes_{A_\mathfrak{m}} k \cong \mathrm{Hom}_k(\oplus_{i=1}^n k dx_i, k)$$

$$\cong \mathrm{Hom}_k(\mathfrak{m}/\mathfrak{m}^2, k) \cong \bigoplus_{i=1}^n k(\partial/\partial x_i).$$

$\square$

1.4.2

Let X be an algebraic variety defined over k and let $\mathcal{U} = \{U_i\}_{i \in I}$ be an affine open covering of X. For each $i \in I$, we can define $\Omega^1_{U_i/k}$ as before since we can choose U_i as an affine variety with coordinate ring A_i. For a point $P \in U_i \cap U_j$, the stalks $\Omega^1_{U_i/k,P}$ and $\Omega^1_{U_j/k,P}$ coincide with $\Omega^1_{A_i/k} \otimes_{A_i} (A_i)_{\mathfrak{m}_i} \cong \Omega^1_{A_j/k} \otimes_{A_j} (A_j)_{\mathfrak{m}_j} \cong \Omega^1_{(\mathcal{O}_{X,P})/k}$, where $\mathfrak{m}_i$ (resp. $\mathfrak{m}_j$) is the maximal ideal corresponding to the point $P \in U_i$ (resp. $P \in U_j$). Thus as sheaves on U_i and U_j, $\Omega^1_{U_i/k}$ and $\Omega^1_{U_j/k}$ patch together on the intersection $U_i \cap U_j$. Namely, there exists an $\mathcal{O}_{U_i \cap U_j}$-isomorphism $\theta_{ji} : \Omega^1_{U_i/k}|_{U_i \cap U_j} \xrightarrow{\sim} \Omega^1_{U_j/k}|_{U_i \cap U_j}$ such that, on the intersection $U_i \cap U_j \cap U_\ell$, $\theta_{\ell i} = \theta_{\ell j} \cdot \theta_{ji}$. Hence we obtain a coherent $\mathcal{O}_X$-sheaf $\Omega^1_{X/k}$ such that $\Omega^1_{X/k}|_{U_i} = \Omega^1_{U_i/k}$ for every $i \in I$. We say that $\Omega^1_{X/k}$ is the *sheaf of differential 1-forms* of X. The dual sheaf $\mathcal{T}_{X/k} := \mathcal{H}om_{\mathcal{O}_X}(\Omega^1_{X/k}, \mathcal{O}_X)$ is the *tangent sheaf* of X.

1.4.3

Since an algebraic variety is an integral separated k-scheme of finite type, the diagonal morphism $\Delta : X \to X \times_k X$, $P \mapsto (P, P)$ is a closed immersion. Let $\mathcal{I}$ be the defining ideal of $\Delta(X)$ in $\mathcal{O}_{X \times X}$. Then $\mathcal{I}/\mathcal{I}^2$ has support on $\Delta(X)$ and is viewed as a coherent $\mathcal{O}_X$-sheaf by the ring homomorphism $p_1^* \mathcal{O}_X \to \mathcal{O}_{X \times X}$, where $p_1 : X \times_k X \to X$ is the first projection $(P, Q) \mapsto P$. Then $\Omega^1_{X/k} \cong \mathcal{I}/\mathcal{I}^2$. For any affine open set $U = \operatorname{Spec} A$ of X, $\Omega^1_{X/k}|_U$ is the associated sheaf $\widetilde{\Omega}^1_{A/k}$ and $\mathcal{T}_{X/k}|_U$ is the sheaf associated with $\operatorname{Der}_k(A)$.

1.4.4

Lemma. *Let X be a smooth algebraic variety of dimension n over k. Then both $\Omega^1_{X/k}$ and $\mathcal{T}_{X/k}$ are locally free $\mathcal{O}_X$-Modules.*

Proof. Let P be a closed point of X. By Lemma 1.4.1, $\Omega^1_{X/k,P}$ is a free $\mathcal{O}_{X,P}$-module of rank n. Hence $\mathcal{T}_{X/k,P} = \mathcal{H}om_{\mathcal{O}_X}(\Omega^1_{X/k}, \mathcal{O}_X) \otimes \mathcal{O}_{X,P} \cong \operatorname{Hom}_{\mathcal{O}_{X,P}}(\Omega^1_{X/k,P}, \mathcal{O}_{X,P})$ is also a free $\mathcal{O}_{X,P}$-module of rank n. Hence follows the assertion. $\square$

1.4.5

Let X be an algebraic variety defined over k. An element $D \in \Gamma(X, \mathcal{T}_{X/k})$ is called a *vector field* on X. If $U = \operatorname{Spec}(A)$ is an affine open set of X, the restriction $D|_U$ is naturally viewed as a k-derivation of A. In fact, we have

$$D|_U \in \Gamma(U, \mathcal{T}_{X/k}|_U) = \Gamma(U, \mathcal{H}om_{\mathcal{O}_U}(\Omega^1_{U/k}, \mathcal{O}_U))$$
$$= \operatorname{Hom}_A(\Omega^1_{A/k}, A) = \operatorname{Der}_k(A).$$

Suppose that X is smooth. Then, for any closed point $P \in X$, D_P is an element of $\mathcal{T}_{X/k} \otimes k(P)$. By Lemma 1.4.1,(3), D_P is an element of $\operatorname{Hom}_k(\mathfrak{m}/\mathfrak{m}^2, k) \cong \bigoplus_{i=1}^n (\partial/\partial x_i)$, where $\mathfrak{m} = (x_1, \ldots, x_n)$ is the maximal ideal of $\mathcal{O}_{X,P}$. Hence a vector field D assigns a direction $\sum_{i=1}^n a_i(P)(\partial/\partial x_i)$ at P. Since a system of local coordinates $\{x_1, \ldots, x_n\}$ at P becomes, after a suitable translation $\{x_1 - c_1, \ldots, x_n - c_n\}$ with $c_1, \ldots, c_n \in k$, a system of local coordinates of a neighboring point P', the system of functions $P \mapsto \{a_1(P), \ldots, a_n(P)\}$ defines a system of regular functions near P. Thus the direction D_P is continuous near the point P. So, if $\{x_1, \ldots, x_n\}$ is a local system of parameters of P, a vector field D is written as

$$D = a_1 \frac{\partial}{\partial x_1} + \cdots + a_n \frac{\partial}{\partial x_n},$$

where $a_1, \ldots, a_n \in \mathcal{O}_{X,P}$. Hence D is a k-derivation of the local ring $\mathcal{O}_{X,P}$ for every $P \in X$. We denote $\Gamma(X, \mathcal{T}_{X/k})$ by $\mathcal{T}(X)$. Then $\mathcal{T}(X)$ is a p-Lie algebra with the bracket product $[D_1, D_2]$ and the pth power product (see 1.2).

1.5 Actions of affine group schemes

Let $G = \operatorname{Spec} A$ be an affine k-group scheme. The multiplication $m : G \times_k G \to G$, the inverse $i : G \to G$ and the unit $e : \operatorname{Spec} k \to G$ are defined respectively by k-algebra homomorphisms

$$\Delta : A \to A \otimes_k A, \quad \iota : A \to A, \quad \varepsilon : A \to k,$$

which satisfy the following conditions [1]:

(1) Two composites of k-algebra homomorphisms $A \xrightarrow{\Delta} A \otimes A \xrightarrow{\Delta \otimes \operatorname{id}_A}$ $A \otimes A \otimes A$ and $A \xrightarrow{\Delta} A \otimes A \xrightarrow{\operatorname{id}_A \otimes \Delta} A \otimes A \otimes A$ coincide.
(2) The composite of k-algebra homomorphisms $A \xrightarrow{\varepsilon} k \xrightarrow{\nu} A$ coincides with the composites $A \xrightarrow{\Delta} A \otimes A \xrightarrow{\operatorname{id}_A \otimes \iota} A \otimes A \xrightarrow{\pi} A$ and $A \xrightarrow{\Delta}$ $A \otimes A \xrightarrow{\iota \otimes \operatorname{id}_A} A \otimes A \xrightarrow{\pi} A$, where $\nu : k \to A$ is the natural inclusion as a k-algebra and $\pi : A \otimes A \to A$ is the algebra multiplication.

---

[1] Δ is the k-algebra homomorphism associated to the multiplication m. So, the Greek letter μ is better than Δ. But the μ is kept for the infinitesimal, multiplicative group scheme μ_p. The readers must be careful of not confusing Δ with the diagonal morphism of a scheme.

(3) The composites $A \xrightarrow{\Delta} A \otimes A \xrightarrow{\mathrm{id}_A \otimes \varepsilon} A \otimes k \cong A$ and $A \xrightarrow{\Delta} A \otimes A \xrightarrow{\varepsilon \otimes \mathrm{id}_A} A \otimes k \cong A$ are equal to id_A.

The readers are requested to draw commutative diagrams from the above conditions. Then it will be understood that the diagrams (or those converted to schemes and morphisms) describe the associativity law, existence of the inverse and existence of the unit of the group scheme G. The k-algebra homomorphisms Δ, ι and ε are respectively called the *comultiplication*, the *coinverse* and the *augmentation* (or *counit*) of the group scheme G.

An (algebraic) action of a k-group scheme G on an affine variety $X = \operatorname{Spec} B$ is given by a k-morphism $\sigma : G \times X \to X$ such that $\sigma \circ (m \times \mathrm{id}_X) = \sigma \circ (\mathrm{id}_G \times \sigma)$, both of which are morphisms $G \times G \times X \to X$, and $\sigma \circ ((e \cdot p) \times \sigma) = \mathrm{id}_X$ which maps $X \to X$, where $p : X \to \operatorname{Spec} k$ is the natural projection and $(e \cdot p) \times \mathrm{id}_X : X \to G \times X$ is given by $P \mapsto (e, P)$ with the unit e of G. These conditions are given in terms of the associated k-algebra homomorphisms as follows, where $\sigma^* : B \to A \otimes B$ is a k-algebra homomorphism which determines σ:

(i) $(\Delta \otimes \mathrm{id}_B) \circ \sigma^* = (\mathrm{id}_A \otimes \sigma^*) \circ \sigma^*$.

(ii) $(\varepsilon \otimes \mathrm{id}_B) \circ \sigma^* = \mathrm{id}_B$, where $\varepsilon \otimes \mathrm{id}_B : A \otimes B \xrightarrow{\varepsilon \otimes \mathrm{id}_B} k \otimes B \cong B$.

1.5.1

The additive group scheme G_a has $A, \mu, \iota, \varepsilon$ defined by $A = k[t]$, $\Delta(t) = t \otimes 1 + 1 \otimes t$, $\iota(t) = -t$ and $\varepsilon(t) = 0$. The multiplicative group G_m has $A, \Delta, \iota, \varepsilon$ defined by $A = k[t, t^{-1}]$, $\Delta(t) = t \otimes t$, $\iota(t) = t^{-1}$ and $\varepsilon(t) = 1$. With the notations in Part I, (2.2), $\Theta^n(G_a) = \operatorname{Spec}((k, \varphi^n) \otimes_k k[t]) \cong G_a$ and $\Theta^n(G_m) \cong G_m$ similarly. Hence the n-times iterated Frobenius endomorphism $F_{G_a}^n$ and $F_{G_m}^n$ are defined by the k-algebra homomorphism $k[t] \to k[t]$, $t \mapsto t^{p^n}$ and $k[t, t^{-1}] \to k[t, t^{-1}]$, $t \mapsto t^{p^n}$. We denote $F_{G_a}^n$ and $F_{G_m}^n$ simply by F^n.

Let $\sigma : G_a \times X \to X$ be an action on $X = \operatorname{Spec} B$. Then the k-algebra homomorphism $\sigma^* : B \to k[t] \otimes_k B$ is written as

$$\sigma^*(b) = \sum_{i=0}^{\infty} \delta_i(b) t^i \tag{1.1}$$

with k-linear endomorphisms $\delta_i : B \to B$ satisfying the conditions

(0) $\delta_\ell(ab) = \sum_{i+j=\ell} \delta_i(a) \delta_j(b)$ for $i \geq 0$, $j \geq 0$.

(i) $\delta_j \delta_i = \binom{i+j}{i} \delta_{i+j}$ for $i \geq 0$, $j \geq 0$, where $\binom{i+j}{i}$ is the binomial coefficient [2].

(ii) $\delta_0 = \mathrm{id}_B$.

The following result is well-known (see [54, Chapter I, (1.1)]).

Lemma. (1) *If k has characteristic zero, then δ_1 is a locally nilpotent k-derivation of B, which is by definition a k-derivation of B such that $\delta_1^n(b) = 0$ for $n \gg 0$ for every $b \in B$, and $\delta_i = \frac{1}{i!}\delta_1^i$ for $i \geq 1$.*

(2) *If k has characteristic $p > 0$, we have*

$$\delta_i = \frac{(\delta_1)^{i_0}(\delta_p)^{i_1} \cdots (\delta_{p^r})^{i_r}}{(i_0)!(i_1)! \cdots (i_r)!},$$

where $i = i_0 + i_1 p + \cdots + i_r p^r$ is the p-adic expansion of i.

The set $\delta := \{\delta_i \mid i \geq 0\}$ is called an *iterative, locally finite higher derivation*. By (1.1), $\sigma^*(b)$ is a polynomial in $B[t]$, whence $\delta_i(b) = 0$ if $i \gg 0$. This is why δ is called *locally finite*. The above condition (i) and the above lemma explain how δ_i is generated *iteratively* by lower δ_js. In the case of characteristic zero, δ is determined by δ_1, which is called a *locally nilpotent derivation* (lnd, in short).

1.5.2

An *infinitesimal group scheme* over k is a k-group scheme $G = \mathrm{Spec}\, A$ with Artin local ring A. In what follows, we consider only commutative group schemes. We consider only typical examples in later sections. Namely, we give the following definition.

Definition. (1) For an integer $n \geq 1$, $\alpha_{p^n} = \mathrm{Ker}\, F_{G_a}^n = \mathrm{Spec}\, k[t]/(t^{p^n})$ with

$$\Delta(t) = t \otimes 1 + 1 \otimes t, \quad \iota(t) = -t, \quad \varepsilon(t) = 0,$$

where t in α_{p^n} is abused as the residue class $t + (t^{p^n})$.

(2) $\mu_p = \mathrm{Ker}\, F_{G_m} = \mathrm{Spec}\, k[t, t^{-1}]/(t^p - 1) \cong \mathrm{Spec}\, k[\tau]/(\tau^p)$ with

$$\Delta(\tau) = \tau \otimes 1 + 1 \otimes \tau + \tau \otimes \tau,$$
$$\iota(\tau) = -\tau + \tau^2 - \cdots + (-\tau)^{p-1}, \quad \varepsilon(\tau) = 0,$$

where τ in μ_p is abused as the residue class $t - 1 + (t^p - 1)$.

---

[2]For a precise interpretation of the binomial coefficients in the case of positive characteristic, see the section 5.1, where higher derivations are treated from the beginning.

1.5.3

We have the following result on the actions of α_{p^n} and μ_p on algebraic varieties.

Lemma. *Let X be an algebraic variety defined over k. Then the following assertions hold.*

(1) *There is a one-to-one correspondence between α_p-actions (resp. μ_p-actions) on X and vector fields $D \in \Gamma(X, \mathcal{T}_{X/k})$ such that $D^p = 0$ (resp. $D^p = D$).*

(2) *Let $X = \operatorname{Spec} B$ be an affine variety. Then an α_{p^n}-action on X corresponds bijectively to a (truncated) iterative, locally finite higher derivation $\delta = \{\delta_i \mid 0 \le i < p^n\}$, which is by definition an iterative, locally finite higher derivation such that $\delta_i = 0$ for every $i \ge p^n$.*

Proof. (1) Since $(\alpha_p)_{\mathrm{red}}$ is the unit group, the variety X is set-theoretically fixed by α_p. Hence we may replace X by an affine open set $U = \operatorname{Spec} B$. Then the α_p-action on $X = \operatorname{Spec} B$ is given by a k-algebra homomorphism $\sigma^* : B \to B[t]$ such that

$$\sigma^*(b) = \sum_{i=0}^{p-1} \delta_i(b) t^i, \quad b \in B , \tag{1.2}$$

where $\{\delta_i \mid 0 \le i < p\}$ satisfies the conditions (0), (i), and (ii) of 1.5.2. In particular, δ_1 is a k-derivation and $\delta_i = \frac{1}{i!}\delta_1^i$. Since $\delta_i \delta_{p-i} = \binom{p}{i}\delta_p = 0$, we have $\delta_1^p = 0$. We set $D = \delta_1$. Conversely, if we are given a k-derivation D of B with $D^p = 0$, set $\delta_i = \frac{1}{i!}D^i$. Then the formula (1.2) gives an α_p-action on X. In general, let $\mathcal{U} = \{U_i\}_{i \in I}$ be an affine open covering of X. Then a vector field D on X with $D^p = 0$ gives a k-derivation D_i on $\Gamma(U_i, \mathcal{O}_{U_i})$ such that $D_i^p = 0$ and $D_i|_{U_i \cap U_j} = D_j|_{U_i \cap U_j}$ for $i \neq j$, the associated α_p-actions on the open sets U_i patch together and gives an α_p-action on X. This process can be reversed. Hence an α_p-action on X gives a k-derivation D with $D^p = 0$.

In the case of μ_p-actions, by the same reason as for α_p-actions, we may assume that $X = \operatorname{Spec} B$. Then the k-algebra homomorphism $\sigma^* : B \to B[\tau]$ is given by

$$\sigma^*(b) = \sum_{i=0}^{p-1} \delta_i(b) \tau^i, \quad b \in B \tag{1.3}$$

such that

(0) $\delta_\ell(ab) = \sum_{i+j=\ell} \delta_i(a)\delta_j(b)$ for $0 \leq i < p$ and $0 \leq j < p$.

(i) For $0 \leq \alpha, \beta < p$, we have

$$\delta_\alpha \delta_\beta = \sum_{k=0}^{\min(\alpha,\beta)} \frac{(\alpha+\beta-k)!}{(\alpha-k)!(\beta-k)!k!} \delta_{\alpha+\beta-k} \,. \tag{1.4}$$

(ii) $\delta_0 = \mathrm{id}_B$.

Let $t = 1 + \tau$. Then $\sigma^* : B \to B[t]$ is written as

$$\sigma^*(b) = \sum_{i=0}^{p-1} \gamma_i(b)t^i, \quad b \in B \,, \tag{1.5}$$

where $t^p = 1$. Then the formulas (i) and (ii) before 1.5.1 are expressed in terms of the γ_i as

(i') $\gamma_i\gamma_j = \delta_{ij}\gamma_i$, where δ_{ij} is Kronecker's delta.
(ii') $\mathrm{id}_B = \gamma_0 + \gamma_1 + \cdots + \gamma_{p-1}$.

Namely, $\mathrm{id}_B = \gamma_0 + \gamma_1 + \cdots + \gamma_{p-1}$ is the idempotent decomposition of id_B. The comparison of the formulas (1.3) and (1.5) via $t = 1 + \tau$ yields the following relations:

$$\delta_0 = \mathrm{id}_B = \gamma_0 + \gamma_1 + \cdots + \gamma_{p-1}$$
$$\delta_1 = \gamma_1 + 2\gamma_2 + \cdots + (p-1)\gamma_{p-1}$$
$$\cdots\cdots$$
$$\delta_i = \gamma_i + \binom{i+1}{i}\gamma_{i+1} + \cdots + \binom{p-1}{i}\gamma_{p-1}$$
$$\cdots\cdots$$

$$\delta_{p-1} = \gamma_{p-1}$$

The formula (1.4) implies that $\delta_\alpha\delta_\beta = \delta_\beta\delta_\alpha$ for $0 \leq \alpha, \beta < p$ and δ_α is expressed as an polynomial of δ_1. For example, $2\delta_2 = \delta_1^2 - \delta_1$, $3\delta_3 = \delta_1\delta_2 - 2\delta_2, \ldots$. Furthermore, δ_1 is a k-derivation of B. We show that $\delta_1^p = \delta_1$. For this end, from the expression of δ_i in terms of γ_j, we can compute

$$\delta_1\gamma_0 = 0, \; \delta_1\gamma_1 = \gamma_1, \; \delta_1\gamma_2 = 2\gamma_2, \; \cdots \; \delta_1\gamma_{p-1} = (p-1)\gamma_{p-1}$$
$$\delta_1^2\gamma_0 = 0, \; \delta_1^2\gamma_1 = \gamma_1, \; \delta_1^2\gamma_2 = 2^2\gamma_2, \; \cdots \; \delta_1^2\gamma_{p-1} = (p-1)^2\gamma_{p-1}$$

$$\vdots \qquad \vdots \qquad \vdots$$

$$\delta_1^i\gamma_0 = 0, \; \delta_1^i\gamma_1 = \gamma_1, \; \delta_1^i\gamma_2 = 2^i\gamma_2, \; \cdots \; \delta_1^i\gamma_{p-1} = (p-1)^i\gamma_{p-1}$$

$$\vdots \qquad \vdots \qquad \vdots$$

$$\delta_1^p\gamma_0 = 0, \; \delta_1^p\gamma_1 = \gamma_1, \; \delta_1^p\gamma_2 = 2^p\gamma_2, \; \cdots \; \delta_1^p\gamma_{p-1} = (p-1)^p\gamma_{p-1}$$

Hence we have

$$\delta_1^p = \delta_1^p \gamma_0 + \delta_1^p \gamma_1 + \delta_1^p \gamma_2 + \cdots + \delta_1^p \gamma_{p-1}$$
$$= 0 + \gamma_1 + 2^p \gamma_2 + \cdots + (p-1)^p \gamma_{p-1}$$
$$= \gamma_1 + 2\gamma_2 + \cdots + (p-1)\gamma_{p-1} = \delta_1 .$$

We set $D = \delta_1$. In order to get back a μ_p-action on $X = \operatorname{Spec} B$ from a given k-derivation D on B with $D^p = D$, we need to express δ_i ($0 < i < p$) in terms of $\delta_1 = D$ by making use of the formula (1.4) or to express γ_i ($0 \leq i < p$) in terms of D by making use of the previous formula between δ_is and γ_js. The computation is straightforward but tedious. So, we omit the computation. The readers are advised to conduct computations in the case of small characteristic p.

(2) The correspondence between α_{p^n}-actions and truncated iterative locally finite higher derivations $\delta = \{\delta_i \mid 0 \leq i < p^n\}$ is given as in 1.5.1.

□

A vector field D on an algebraic variety X is said to be of *additive type* or of *multiplicative type* if $D^p = 0$ or $D^p = D$ respectively.

1.5.4

For later use, we consider also $\mathbb{Z}/p\mathbb{Z}$-actions. We denote by G the k-group scheme $\mathbb{Z}/p\mathbb{Z}$ over k. As the kernel of the homomorphism $F - \operatorname{id} : G_a \to G_a$ given by $t \mapsto t^p - t$, $G \cong \operatorname{Spec} k[t]/(t^p - t)$. We set $A = k[t]/(t^p - t)$ and identify t with the residue class $t + (t^p - t)$. The comultiplication Δ, the coinverse ι and the augmentation ε are defined respectively by $\Delta(t) = t \otimes 1 + 1 \otimes t$, $\iota(t) = -t$ and $\varepsilon(t) = 0$.

To a G-action σ on an affine variety $X = \operatorname{Spec} B$ over k we can associate a k-algebra homomorphism $\sigma^* : B \to B[t]$ which is written as

$$\sigma^*(b) = \sum_{i=0}^{p-1} \delta_i(b) t^i, \quad b \in B .$$

As in the case of α_p-action, $\delta_0 = \operatorname{id}_B$, δ_1 is a k-module endomorphism of B, and $\delta_i = \frac{1}{i!}\delta_1^i$ for $0 \leq i < p$. Note that δ_1 is not a k-derivation. In fact, it satisfies a relation for $a, b \in B$,

$$\delta_1(ab) = (a\delta_1(b) + b\delta_1(a)) + \sum_{i=1}^{p-1} \frac{1}{i!(p-i)!} \{\delta_1^i(a)\delta_1^{p-i}(b) + \delta_1^i(b)\delta_1^{p-i}(a)\} .$$

Then $\delta_1^p = 0$ as δ_i is considered to be zero for $i \geq p$. Set $\delta := \delta_1$. For the computations below, see Takeda [91] and Miyanishi [59]. The G-action on

B is given as follows when G is identified with $\mathbb{Z}/p\mathbb{Z} = \{0, 1, \ldots, p-1\}$ and γ is a generator 1 of G:

$$\gamma^i(b) = \sigma^*(b)|_{t=i}.$$

On the other hand, we define the trace T by

$$T = 1 + \gamma + \gamma^2 + \cdots + \gamma^{p-1}.$$

Then we have the following result.

Lemma. *The k-module endomorphisms T and $\mathrm{Im}\,(\gamma - 1)$ of B are given by the following formulas in terms of δ.*

$$(1) \quad T = \delta^{p-1}$$

$$(2) \quad \gamma - 1 = \delta + \frac{1}{2!}\delta^2 + \cdots + \frac{1}{(p-1)!}\delta^{p-1}.$$

Proof. (2) is straightforward. To prove (1), we use the following congruence relations

$$\frac{1}{i!}\left(1^i + 2^i + \cdots + (p-1)^i\right) \equiv 0 \pmod{p} \text{ for } 1 \leq i < p-1$$

$$\frac{1}{(p-1)!}\left(1^{p-1} + 2^{p-1} + \cdots + (p-1)^{p-1}\right) \equiv 1 \pmod{p}$$

$\square$

1.6 Cartier dual of a commutative finite group scheme

Let $G = \mathrm{Spec}\,A$ be a commutative finite group scheme over k, where A is a finite k-module. Then A is given a set of k-module homomorphisms

$$\pi : A \otimes A \to A, \quad \nu : k \to A, \quad \Delta : A \to A \otimes A, \quad \iota : A \to A, \quad \varepsilon : A \to k,$$

where π is the algebra multiplication, ν is the natural inclusion of k into the k-algebra A, Δ is the comultiplication, ι is the coinverse, and ε is the augmentation.

Let A^* be the dual k-vector space. Then $A^* \otimes A^*$ is identified with $(A \otimes A)^*$. In fact, if $\{v_1, \ldots, v_d\}$ is a k-basis of A and $\{v_1^*, \ldots, v_d^*\}$ is the dual basis of A^*, then $\{v_i^* \otimes v_j^* \mid 1 \leq i, j \leq d\}$ is the basis of $(A \otimes A)^*$.

Dualize the above k-module homomorphisms to obtain $\{\Delta^*, \varepsilon^*, \pi^*, \iota^*, \nu^*\}$. Then Δ^* and ε^* give the algebra multiplication and the natural inclusion of k into A^* to give A^* a k-algebra structure. Furthermore, π^*, ι^*

and ν^* give the comultiplication, the coinverse and the augmentation on A^*. Thus $G^* := \operatorname{Spec} A^*$ becomes a commutative finite group scheme over k, which is called the *Cartier dual* of G. The following relations will be clear from the above argument:

$$\pi_{A^*} = (\Delta_A)^*, \quad \nu_{A^*} = (\varepsilon_A)^*, \quad \Delta_{A^*} = (\pi_A)^*, \quad \iota_{A^*} = (\iota_A)^*, \quad \varepsilon_{A^*} = (\nu_A)^*.$$

Hence it is an easy exercise to show that $(G^*)^* \cong G$. We have a well-known result (cf. [12]).

1.6.1

Lemma. *The following assertions hold.*

(1) *If a natural number n is prime to the characteristic p, then $(\mathbb{Z}/n\mathbb{Z})^* \cong \mathbb{Z}/n\mathbb{Z}$.*
(2) *$(\mathbb{Z}/p\mathbb{Z})^* \cong \mu_p$ and $\mu_p^* \cong \mathbb{Z}/p\mathbb{Z}$.*
(3) *$\alpha_p^* \cong \alpha_p$.*

Proof. (1) Suppose that $G \cong \mathbb{Z}/n\mathbb{Z}$. Then, as the kernel of the nth power product of G_m, $t \mapsto t^n$, the coordinate ring A of G is isomorphic to $k[t]/(t^n-1)$, hence to the group algebra $k[G] = \bigoplus_{i=0}^{n-1} kx_i$, where $x_0 = 1$ and $x_i \cdot x_j = x_{i+j}$ for $0 \le i, j < n$. In fact, x_i is identified with $t^i \pmod{(t^n-1)}$. Since $\Delta(t) = t \otimes t$ and $\varepsilon(t) = 1$, we have

$$\Delta_A(x_i) = x_i \otimes x_i, \quad \varepsilon_A(x_i) = 1 \quad (0 \le i < n).$$

Then the dual space A^* is given as $A^* = \bigoplus_{i=0}^{n-1} kz_i$, where $\{z_0, z_1, \ldots, z_{n-1}\}$ is the dual basis of $\{x_0, x_1, \ldots, x_{n-1}\}$. Then we have for $0 \le i, j, \ell < n$

$$z_i \cdot z_j = \Delta_A^*(z_i \otimes z_j) = \delta_{ij} z_i$$

$$\Delta_{A^*}(z_i) = \sum_{j+\ell=i} z_j \otimes z_\ell$$

$$\varepsilon_{A^*}(z_i) = \delta_{i0}\ .$$

In fact, $(z_i \cdot z_j)(x_\ell) = (z_i \otimes z_j)(\Delta_A(x_\ell)) = (z_i \otimes z_j)(x_\ell \otimes x_\ell) = \delta_{i\ell}\delta_{j\ell}$. Hence $z_i \cdot z_j = \delta_{ij} z_i$, where δ_{ij} is Kronecker's delta. As for the second formula, $\Delta_{A^*}(z_i)(x_j \otimes x_\ell) = z_i(x_{j+\ell}) = \delta_{i,j+\ell}$. The formula follows from this remark. The last formula follows from $\varepsilon_{A^*}(z_i) = z_i(\nu_A(1)) = z_i(x_0) = \delta_{i0}$. Since $(z_0 + z_1 + \cdots + z_{n-1})(x_i) = 1$ for every $0 \le i < n$, we have $1 = z_0 + z_1 + \cdots + z_{n-1}$, which is, by the first formula, the idempotent decomposition of the unity. Let ζ be a primitive nth root of unity. Set

$$y_i = \frac{1}{n}(x_0 + \zeta^i x_1 + \zeta^{2i} x_2 + \cdots + \zeta^{(n-1)i} x_{n-1}), \quad 0 \le i < n.$$

Then it is easy to see that $x_0 = 1 = y_0 + y_1 + \cdots + y_n$ is the idempotent decomposition of the unity in A and that $\Delta(y_i) = \sum_{j+\ell=i} y_j \otimes y_\ell$, $0 \le i < n$. Hence, by assigning z_i to y_i, we know that A is isomorphic to A^* as the coordinate ring of the finite group scheme G. So, the Cartier dual of $\mathbb{Z}/n\mathbb{Z}$ is isomorphic to itself.

(2) Suppose that $G \cong \mathbb{Z}/p\mathbb{Z}$. Then the group algebra $k[G] \cong k[x]/(x^p - 1)$, which is the kernel of the pth power product of G_m. Let $A = k[G]$. Hence $\operatorname{Spec} A = \mu_p$. As in the above calculation, $A^* = \bigoplus_{i=0}^{p-1} kz_i$, whose algebra structure is given by the idempotent decomposition of the unity, i.e., $1 = z_0 + z_1 + \cdots + z_{p-1}$ and $z_i z_j = \delta_{ij} z_i$. Hence $\operatorname{Spec} A^* \cong \mathbb{Z}/p\mathbb{Z}$.

(3) Let $A = k[x]/(x^p)$. Identify x with the residue class $x + (x^p)$, whence $x^p = 0$. Let $\{z_0, z_1, \ldots, z_{p-1}\}$ be the dual basis of $\{1, x, \ldots, x^{p-1}\}$. Then we can show that $z_1^i = (i!)z_i$ and $\Delta_{A^*}(z_1) = z_1 \otimes 1 + 1 \otimes z_1$. Thence we know that $z_1^p = 0$ and $\operatorname{Spec} A^* \cong \alpha_p$. It is indeed an easy task to show that $\iota_{A^*}(z_1) = -z_1$ and $\varepsilon_{A^*}(z_1) = 0$. $\square$

1.7 Invariant subrings

Let $G = \operatorname{Spec} A$ be a commutative finite group scheme over k acting on an affine variety $X = \operatorname{Spec} B$ by $\sigma : G \times X \to X$. Then an element $b \in B$ is a *G-invariant* element if $\sigma^*(b) = 1_A \otimes b$. Let B^G be the set of G-invariant elements. Then it is easy to see that B^G is a k-subalgebra of B.

1.7.1

Example. (1) Suppose that $G = \alpha_p$. Then the action σ corresponds to a k-derivation D of B such that $D^p = 0$. Then $B^{\alpha_p} = \{b \in B \mid D(b) = 0\}$. With the notations in Part I, (2.2), since $D(b^p) = 0$, it is clear that $\operatorname{Im}(F_B)$ is a k-subalgebra of B contained in B^{α_p}. We denote $\operatorname{Im}(F_B)$ by $k[B^p]$. This implies that B^{α_p} is an affine k-algebra because B is a finite $k[B^p]$-module and $k[B^p]$ is a noetherian ring.

(2) Similarly, a μ_p-action corresponds to a k-derivation D of B such that $D^p = D$. The ring of μ_p-invariants is given as $B^{\mu_p} = \{b \in B \mid D(b) = 0\}$. It also holds that $k[B^p] \subseteq B^{\mu_p} \subseteq B$ and B^{μ_p} is an affine k-domain.

(3) Suppose that $G = \mathbb{Z}/n\mathbb{Z}$ with $p \nmid n$. Write $\sigma^*(b) = \sum_{i=0}^{n-1} \gamma_i(b)t^i$ for $b \in B$, where G is identified with the kernel of the nth power product of G_m. Then $\gamma_i \gamma_j = \delta_{ij}\gamma_i$ and $\operatorname{id}_B = \sum_{i=0}^{n-1} \gamma_i$. Then $\sigma^*(b) = 1_A \otimes b$ if and only if $\gamma_i(b) = 0$ for $1 \le i < n$. The G-invariant subalgebra B^G is an affine k-domain. The standard argument to prove this fact is as follows. Write

$B = k[b_1, \ldots, b_n]$ with a generating system $\{b_1, \ldots, b_n\}$ of B. For each i, consider a polynomial $f_i(x) = \prod_{j=0}^{n-1}(x - g^j b_i)$, where g is a generator of G and $g^j b_i$ is the translate of b_i by the action of g^j on B. Then $f_i(x)$ is a monic polynomial in $B^G[x]$ such that $f_i(b_i) = 0$. Hence B is integral over a k-subalgebra B_0 which is generated over k by all the coefficients of the polynomial $f_i(x)$ for $1 \le i \le n$. Then B_0 is an affine k-algebra and B is a finite B_0-module. Since B^G is a B_0-submodule of B, it is a finite B_0-module, and hence B^G is an affine k-domain.

(4) Suppose that $G = \mathbb{Z}/p\mathbb{Z}$. With the notations in 1.5.4, $b \in B$ is a G-invariant element if and only if $\delta_i(b) = 0$ for $1 \le i < p$. As in the case (3), B^G is an affine k-domain.

We denote $\operatorname{Spec} B^G$ by $X /\!/ G$ and call it the quotient variety of X by G.

1.7.2

In the foregoing arguments, the field k is always G-invariant with respect to a finite group scheme G. If k is not necessarily algebraically closed, we can consider a *twisted* action of G where G acts also on the field. In the next examples, the coordinate ring or the function field of forms of $\mathbb{A}^1$ is obtained as the invariant subring or subfield of $k'[t]$ or $k'(t)$, where k' is a purely inseparable extension of k.

Example. (1) Let $X := \{y^p = x + ax^p\}$ be a k-group of Russell type with $a \in k \setminus k^p$ (cf. Part I, Chap. 5). Let $\alpha^p = a$ and let $k' = k(\alpha)$. Then $X \otimes_k k' \cong \mathbb{A}^1_{k'}$ with coordinate $t = y - \alpha x$. Let D be a k-derivation of k' such that $D(\alpha) = 1$. Then D extends naturally to $k' \otimes_k A$ so that $D|_{1 \otimes A} = 0$, where A is the coordinate ring of X. In fact, $x = t^p$, $y = t + \alpha t^p$, $D(x) = D(y) = 0$ and $D(t) = -t^p$. Hence $D^p = 0$, and we can consider D as obtained from an α_p-action on $\operatorname{Spec} k'[t]$, which is twisted in the sense that α_p acts also on the field k'. It is clear that the α_p-quotient of $\mathbb{A}^1_{k'}$ is X.

(2) Consider a hyperelliptic k-form of $\mathbb{A}^1$ defined by $y^2 = x^p - a$, where $p = 2\ell + 1$. Let $K = k(x, y)$ be the function field. Let $a = \alpha^p$ and let $k' = k(\alpha)$. Then $K' := k' \otimes_k K = k'(t)$, where $t = y/(x - \alpha)^\ell$. Define a k-derivation D of k' as above and extend it to K' by setting $D|_{1 \otimes K} = 0$. Then $D(t) = -1/(2t)$, $D(x) = D(y) = 0$ and $D^p = 0$. Thus α acts on the field K' in a twisted way and the α_p-invariant subfield of K' is K.

1.8   Frobenius sandwiches

Let X be a smooth algebraic variety of dimension n, and let $F_X : X \to X^{(p)} \cong X \otimes_k (k, \varphi)$ be the Frobenius morphism. A normal algebraic variety Y is called a *Frobenius sandwich* of X if there exist finite purely inseparable morphisms $f_1 : X \to Y$ and $f_2 : Y \to X^{(p)}$ such that $F_X = f_2 \cdot f_1$ and $\deg f_1 = p^r$ with $0 < r < n$.

Let $U = \operatorname{Spec} A$ be an affine open set of X. Then $F_X(U) = U^{(p)} = \operatorname{Spec} A^{(p)}$, and $V := f_2^{-1}(U^{(p)})$ is also an affine open set $\operatorname{Spec} B$ such that B is integrally closed in the quotient field $Q(B)$. By Lemma 1.2.2 which is generalized to the case where A and B are integrally closed in the respective quotient fields, B corresponds bijectively to a saturated p-Lie subalgebra L of $\operatorname{Der}_k(A)$, where

$$L = \{\delta \in \operatorname{Der}_k(A) \mid \delta|_B = 0\}$$
$$B = \{a \in A \mid \delta(a) = 0, \forall \delta \in L\}.$$

If we consider an affine open covering $\mathcal{U} = \{U_\lambda\}_{\lambda \in \Lambda}$ of X with $U_\lambda = \operatorname{Spec} A_\lambda$, we can extend the above Galois correspondence between Frobenius sandwiches of X and sheaves of saturated p-Lie algebras which are $\mathcal{O}_X$-submodules of $\mathcal{T}_{X/k} := \mathcal{H}om_{\mathcal{O}_X}(\Omega^1_{X/k}, \mathcal{O}_X)$.

1.8.1   *Some auxiliary results*

Let A be an affine k-domain and let δ be a k-derivation of A. We denote by $I(\delta)$ the ideal of A generated by the set $\{\delta(a) \mid a \in A\}$. Suppose that A is a regular ring of $\dim A = n$. For a maximal ideal $\mathfrak{m}$ of A, let $\{x_1, \ldots, x_n\}$ be a regular system of parameters of $A_{\mathfrak{m}}$. Then a k-derivation of $A_{\mathfrak{m}}$ is written as $\delta = \sum_{i=1}^n f_i \frac{\partial}{\partial x_i}$ with $f_i := \delta(x_i) \in A_{\mathfrak{m}}$ (see Lemma 1.4.1). Then $I(\delta; A_{\mathfrak{m}}) = I(\delta) \otimes_A A_{\mathfrak{m}} = (f_1, \ldots, f_n)$. Since $A_{\mathfrak{m}}$ is factorial, there exists an element $h \in A_{\mathfrak{m}}$ such that the principal ideal (h) of $A_{\mathfrak{m}}$ is the intersection of prime divisors of height one of $I(\delta; A_{\mathfrak{m}})$. In fact, (h) is obtained by localization $\otimes_A A_{\mathfrak{m}}$ from the intersection of prime divisors of height one of $I(\delta)$, which defines an effective divisor (δ) on X. We call (δ) the *divisorial part* of the k-derivation δ. If $(\delta) = (0)$, we say that δ is *reduced*.

A k-derivation δ of A is *p-closed* if $\delta^p = f\delta$ with $f \in A$.

Lemma.   *With the above notations, let $R = A_{\mathfrak{m}}$. Then the following assertions hold.*

(1) $\operatorname{Der}_k(R) = \operatorname{Der}_k(A) \otimes_A R$ *is a free R-module of rank n.*

(2) *Let L be a saturated p-Lie subalgebra of rank one of $\operatorname{Der}_k(R)$, i.e., $\dim L \otimes_R Q(R) = 1$. Then there exists a k-derivation δ of R such that $L = R\delta$, where δ is a reduced, p-closed k-derivation of R.*

(3) *Conversely, if δ is a reduced, p-closed k-derivation of R, the R-submodule $L = R\delta$ is a saturated p-Lie subalgebra of $\operatorname{Der}_k(R)$.*

Proof. (1) If $\{x_1, \ldots, x_n\}$ is a regular system of parameters of R then $\operatorname{Der}_k(R) = \oplus_{i=1}^n R\delta_i$ with $\delta_i = \frac{\partial}{\partial x_i}$.

(2) $L \otimes_R Q(R)$ is a $Q(R)$-subspace of $\oplus_{i=1}^n Q(R)\delta_i$. Hence $L \otimes_R Q(R) = Q(R)\delta$ for $\delta \in L$. If we write $\delta = \sum_{i=1}^n f_i \delta_i$ with $f_i \in R$, we may assume that $\gcd(f_1, \ldots, f_n) = 1$. We show that $L = R\delta$. In fact, let $\delta' \in L$. Then $a\delta' = b\delta$ for $a, b \in R$ with $\gcd(a, b) = 1$. This implies $a \mid bf_i$ for every i. So, a is invertible and $\delta' \in R\delta$. Thus $L = R\delta$. Since L is a p-Lie subalgebra, we have $\delta^p = f\delta$ for $f \in R$. Namely, δ is p-closed.

(3) We show that $L = R\delta$ is saturated. For a k-derivation δ' of R, assume that $a\delta' \in L$. Then $a\delta' = b\delta$ with $b \in R$. We may assume that $\gcd(a, b) = 1$. By the same argument as in (2), it follows that a is invertible in R. So, $\delta' \in L = R\delta$. $\qquad\square$

If $\dim X = 2$, then a Frobenius sandwich Y of X has degrees $[k(X) : k(Y)] = [k(Y) : k(X^{(p)})] = p$. Hence, for a closed point $y \in Y$, the local ring $\mathcal{O}_{Y,y} = k(Y) \cap \mathcal{O}_{X,x}$ is obtained as the kernel of a reduced, p-closed k-derivation δ of $\mathcal{O}_{X,x}$.

Corollary. *Let X be a smooth algebraic variety and let $\mathcal{L}$ be a saturated p-Lie algebra sheaf of rank one of $\mathcal{T}_{X/k}$. Then $\mathcal{L}$ is an invertible shead, locally generated by reduced, p-closed k-derivations.*

1.8.2 Singularity of Frobenius sandwiches

We look into a local ring. A local ring $(\mathcal{O}, \mathfrak{m})$ is said to be a *geometric local ring* if it is isomorphic to the local ring $\mathcal{O}_{X,P}$ of an algebraic variety X at a closed point P.

Lemma. *Let $(\mathcal{O}, \mathfrak{m})$ be a geometric local ring and let $K = Q(\mathcal{O})$. Let δ be a reduced, p-closed k-derivation of $\mathcal{O}$, i.e., $\delta^p = \lambda\delta$ with $\lambda \in \mathcal{O}$ and $\operatorname{ht} I(\delta) \geq 2$. Let $\mathcal{O}_0 = \operatorname{Ker}\delta$ and $\mathfrak{m}_0 = \mathcal{O}_0 \cap \mathfrak{m}$. Then the following assertions hold.*

(1) *$(\mathcal{O}_0, \mathfrak{m}_0)$ is a geometric local ring over k and $Q(\mathcal{O}_0) = \operatorname{Ker}\delta_K$, where*

δ_K is the natural extension of δ to K.

(2) $\mathcal{O}$ is a finite $\mathcal{O}_0$-module. If $\mathcal{O}$ is normal, so is $\mathcal{O}_0$.

(3) Let $\widehat{\mathcal{O}}$ be the $\mathfrak{m}$-adic completion of $\mathcal{O}$. Then δ extends naturally to a reduced, p-closed k-derivation $\widehat{\delta}$ of $\widehat{\mathcal{O}}$. Furthermore, $\operatorname{Ker}\widehat{\delta}$ is the $\mathfrak{m}_0$-adic completion of $\mathcal{O}_0$.

(4) If $\mathcal{O}_0$ is regular then $I(\delta) = \mathcal{O}$. If $\dim\mathcal{O} = 2$, $I(\delta) = \mathcal{O}$ implies that $\mathcal{O}_0$ is regular.

Proof. (1) and (2). Let $\mathcal{O}^{(p)}$ be the subalgebra of $\mathcal{O}$ generated over k by $\{a^p \mid a \in \mathcal{O}\}$. Then $\mathcal{O}^{(p)} \subseteq \mathcal{O}_0$. Hence $\mathcal{O}$ is integral over $\mathcal{O}_0$. Write $\mathcal{O} = k[a_1, \ldots, a_n]_{\mathfrak{p}}$ for some prime ideal $\mathfrak{p}$. Then it is clear that $\mathcal{O} = \mathcal{O}^{(p)}[a_1, \ldots, a_n]$. Hence $\mathcal{O}$ is a finite $\mathcal{O}^{(p)}$-module. Let $b \in \mathcal{O}_0 \setminus \mathfrak{m}_0$. Then $b \in \mathcal{O} \setminus \mathfrak{m}$ and $b^{-1} \in \operatorname{Ker}\delta \cap \mathcal{O} = \mathcal{O}_0$. Hence $(\mathcal{O}_0, \mathfrak{m}_0)$ is a geometric local ring. If $\xi = \frac{a}{b} \in \operatorname{Ker}\delta_K$ then $\xi = \frac{ab^{p-1}}{b^p}$ and $\delta(ab^{p-1}) = 0$. Hence $\xi \in Q(\mathcal{O}_0)$. The inclusion $Q(\mathcal{O}_0) \subseteq \operatorname{Ker}\delta_K$ is clear. If $\xi \in Q(\mathcal{O}_0)$ is integral over $\mathcal{O}_0$ then $\xi \in \mathcal{O}$ because $\mathcal{O}$ is normal. Hence $\xi \in \mathcal{O} \cap \operatorname{Ker}\delta_K = \mathcal{O}_0$, and $\mathcal{O}_0$ is normal.

(3) Since $\delta(\mathfrak{m}^r) \subseteq \mathfrak{m}^{r-1}$ for every $r > 0$, δ is continuous in the $\mathfrak{m}$-adic topology of $\mathcal{O}$. So, δ induces a k-derivation $\widehat{\delta}$ of $\widehat{\mathcal{O}}$ such that $\widehat{\delta}|_{\mathcal{O}} = \delta$. It is clear that $\widehat{\delta}$ is p-closed because so is δ. Since $I(\widehat{\delta}; \widehat{\mathcal{O}}) = I(\delta; \mathcal{O}) \otimes_{\mathcal{O}} \widehat{\mathcal{O}}$, in order to show that $\widehat{\delta}$ is reduced, it suffices to show that any prime divisor of $I(\widehat{\delta})$ has height ≥ 2. Suppose that $\mathfrak{P}$ is a prime divisor of $I(\widehat{\delta})$ of height one. Let $\mathfrak{p} = \mathfrak{P} \cap \mathcal{O}$. Since $\widehat{\mathcal{O}}$ is $\mathcal{O}$-flat, $\operatorname{ht}(\mathfrak{P}) \geq \operatorname{ht}(\mathfrak{p})$. Hence $\mathfrak{p}$ is a prime divisor of height one of $I(\delta)$. This is a contradiction.

On the other hand, $\widehat{\mathcal{O}} = \mathcal{O} \otimes_{\mathcal{O}_0} \widehat{\mathcal{O}_0}$. Hence an exact sequence

$$0 \longrightarrow \mathcal{O}_0 \longrightarrow \mathcal{O} \overset{\delta}{\longrightarrow} \mathcal{O}$$

yields an exact sequence

$$0 \longrightarrow \widehat{\mathcal{O}_0} \longrightarrow \widehat{\mathcal{O}} \overset{\widehat{\delta}}{\longrightarrow} \widehat{\mathcal{O}}.$$

Hence $\widehat{\mathcal{O}_0} = \operatorname{Ker}\widehat{\delta}$.

(4) Suppose that $\mathcal{O}_0$ is regular. Then $\widehat{\mathcal{O}} = k[[x_1, \ldots, x_n]]$ and $\widehat{\mathcal{O}_0} = k[[x_1, \ldots, x_s, x_{s+1}^p, \ldots, x_n^p]]$ after a suitable choice of regular system of parameters of $\widehat{\mathcal{O}}$ (see Lemma 2.2 in the next chapter). Hence we can identify $\widehat{\delta}$ with

$$\frac{\partial}{\partial x_{s+1}} + \cdots + \frac{\partial}{\partial x_n}.$$

Then $I(\widehat{\delta}) = \widehat{\mathcal{O}}$. Since $I(\widehat{\delta}) \cap \mathcal{O} = I(\delta)$, it follows that $I(\delta) = \mathcal{O}$.

If $\dim \mathcal{O} = 2$, the converse holds. In fact, Rudakov-Shafarevich [77, Theorem 1 (p.1208)] shows that $\mathcal{O}_0$ is regular if δ has only divisorial singularity, which follows from $I(\delta) = \mathcal{O}$. Further, we will observe a similar but particular case in Lemma 2.3. □

The assertion (3) of the lemma implies that the analytic behavior of singularity of $\mathcal{O}_0$ can be explored via $\operatorname{Ker} \widehat{\delta}$, where $\widehat{\delta}$ is a reduced, p-closed k-derivation on $\widehat{\mathcal{O}} = k[[x_1, \ldots, x_n]]$.

1.8.3 *Plane sandwiches*

If $A = k[x, y]$ is a polynomial ring in two variables, i.e., X is the affine plane, we call a Frobenius sandwich an *affine Zariski surface* and state relevant results in Chapter 2. Here we call X a *plane sandwich* and give several examples. The first one makes use of Theorem 3.3.1 in Chapter 3.

1.8.3.1

Example. Let $p = 3$ and let $B = k[x, y, t]/(y^2 = x^3 + \varphi(t))$, where $\varphi(t) \in k[t]$. Let $d = \deg \varphi(t)$ and $m = \left[\frac{d}{6}\right]$. If $d \geq 7$, we assume the following two conditions.

(i) For every root α of $\varphi'(t) = 0$, $v_\alpha(\varphi(t) - \varphi(\alpha)) \leq 5$, where v_α is the valuation of $k(t)$ with $v_\alpha(t - \alpha) = 1$.
(ii) If
$$\varphi(t) - \varphi(\alpha) = a(t - \alpha)^3 + \text{terms of degree} > 3 \text{ in } t - \alpha$$
for some root α of $\varphi'(t) = 0$ and $a \in k^*$ then $v_\alpha(\varphi(t) - \varphi(\alpha) - a(t-\alpha)^3) \leq 5$.

Then the following assertions hold.

(1) Let $Y = \operatorname{Spec} B$. Then Y is normal, and a point $(x, y, t) = (\alpha, \beta, \gamma)$ is a singular point of Y if and only if $\beta = 0$, $\varphi'(\gamma) = 0$ and $\alpha^3 + \varphi(\gamma) = 0$.
(2) Y is k-rational iff $m = 0$, birational to a K3-surface iff $m = 1$ and birational to a surface with Kodaira dimension 1 iff $m > 1$.
(3) Write $t = \tau^3$ and $\varphi(t) = \psi(\tau)^3$. Set $u = y(x + \psi(\tau))^{-1}$. Then $y = u^3$, $x = u^2 - \psi(\tau)$ and $B = k[\tau^3, u^3, u^2 - \psi(\tau)]$. Hence B is a plane sandwich, and the corresponding derivation is
$$\delta = u\frac{\partial}{\partial \tau} - \psi'(\tau)\frac{\partial}{\partial u},$$
where $\delta^3 = -\psi''(\tau)\delta$. Hence δ is additive iff $\psi''(\tau) = 0$ and multiplicative iff $\psi''(\tau) = -1$.

1.8.3.2

The above example shows that a normal algebraic surface Y is not necessarily rational even if there exists a finite morphism $f : X \to Y$ from a rational surface X. This is one of peculiar phenomena in positive characteristic. Similar examples of plane sandwiches are given in [60, Theorems 4.1 and 4.12] in the case $p > 2$ and [*ibid.*; Theorem 5.1] in the case $p = 2$. We only consider the case $p > 2$.

Example. Let $p > 2$. Let $B = k[x^p, y^p, (x^p y^p + 1)x + y^{p+1}]$ and $Y = \operatorname{Spec} B$. Then the following assertions hold.

(1) $k[x^p, y^p] \subset B \subset k[x, y]$. Hence Y is a plane sandwich. Further, B is a regular, factorial, irrational k-affine domain.
(2) Y is a hypersurface in $\mathbb{A}^3 = \operatorname{Spec} k[u, v, w]$ defined by

$$w^p = (u^p v^p + 1)u + v^{p+1},$$

where $u = x^p, v = y^p$ and $w = (x^p y^p + 1)x + y^{p+1}$.
(3) Let W be a minimal smooth completion of Y with the boundary $W \setminus Y$ which is a divisor with simple normal crossings. Then W is a surface of general type.

1.8.3.3

In the case of characteristic zero, the following results are known (see [57, Theorems 2.4.8 and 2.5.1, Chapter 3]).

Theorem. *Let X be the affine plane defined over an algebraically closed field of characteristic zero, and let $f : X \to Y$ be a finite morphism onto a normal algebraic surface Y. Then the following assertions hold.*

(1) *Y is isomorphic to $\mathbb{A}^2/G$ with a finite group G of $\operatorname{GL}(2, k)$. Hence Y has at most one singular point which has quotient singularity.*
(2) *If f is flat, Y is isomorphic to the affine plane.*

In positive characteristic, both assertions do not hold. For instance, let $\varphi(t) = t^{10} - t^4 + 1$ in Example 1.8.3.1. Then Y is birational to a K3-surface, and Y has three singular points, each of which has the same configuration of resolution locus of a rational double point of type E_6 (see Part III, Chapter 1). Hence the assertion (1) does not hold.

In Example 1.8.3.2, Y is smooth, but irrational. If Y is smooth, the morphism $f : X \to Y$ is flat. Hence the assertion (2) does not hold either.

1.8.4   *Abelian sandwiches*

Construction of plane sandwiches can be applied to an algebraic surface of another type. In [58, Theorem 5.1], we have the following example.

Example. Let $p = 2$. Let E_1 be an elliptic curve defined by $u^2 + u = x^3$ and let E_2 be its copy defined by $v^2 + v = y^3$. Let $A = E_1 \times E_2$. Let δ be a k-derivation of the field $k(A) = k(x, y, u, v)$ defined by

$$\delta = (x + x^{2n})\frac{\partial}{\partial x} + (y + y^{2n})\frac{\partial}{\partial y}$$

with $n > 0$. Then $\delta(x) = x + x^{2n}, \delta(y) = y + y^{2n}, \delta(u) = x^2(x + x^{2n})$ and $\delta(v) = y^2(y + y^{2n})$. Then $\delta^2 = \delta$. Let $Y = A/\delta$ be the Frobenius sandwich defined by $\text{Ker } \delta$, which is actually a μ_2-quotient of A (see 2.9). Let $\widetilde{Y}$ be the minimal resolution of singularity. Then $\widetilde{Y}$ is a minimal surface of general type such that $(K_{\widetilde{Y}}^2) = 8n^2 + 8n - 2$ and $q_{\widetilde{Y}} = 2$.

Chapter 2

Zariski surfaces

The purpose of the present chapter is to give some general results and tools with which we can describe the structures of affine Zariski surfaces or Zariski surfaces. In the subsequent chapters, these results will be applied to the study of Zariski surfaces with quasi-elliptic pencils or with quasi-hyperelliptic pencils. All results treated here have sources in Miyanishi-Russell [60].

2.1 Affine Zariski surface

An *affine Zariski surface*, or a *purely inseparable covering of p-exponent one of the affine plane* $\mathbb{A}_k^2 := \operatorname{Spec} k[u, v]$ is, by definition, a normal affine surface $X = \operatorname{Spec} A$ such that

$$k[u, v] \subset A \subset k[x, y] \tag{2.1}$$

and $[Q(A) : k(u, v)] = p$, where $x^p = u$, $y^p = v$ and $Q(A)$ is the quotient field of A. Meanwhile, a *Zariski surface* is a smooth projective surface V defined over k such that the function field $k(V)$ is a simple purely inseparable extension of the rational function field $k(u, v)$. Then V is k-birationally equivalent to an affine hypersurface in $\mathbb{A}_k^3 = \operatorname{Spec} k[u, v, w]$ defined by an equation of the type $w^p = f(u, v) \in k[u, v]$. Define $\varphi(x, y) \in k[x, y]$ by $\varphi(x, y)^p = f(u, v)$, and let $B := k[u, v, \varphi(x, y)]$. Let A be the normalization of B. Then $X := \operatorname{Spec} A$ is an affine Zariski surface, k-birationally equivalent to the Zariski surface V. Conversely, any smoothified completion of an affine Zariski surface is a Zariski surface.

Let $X = \operatorname{Spec} A$ be an affine Zariski surface as defined above. Since the quotient field $K = Q(A)$ is a purely inseparable extension of $k(u, v)$ of degree p, by Lemma 1.2.1, there exists a k-derivation δ' of $k(x, y)$ such that

$\delta'^p = a'\delta'$, $a' \in k(x,y)$, and

$$K = k(x,y)^{\delta'} := \{f \in k(x,y) \mid \delta'f = 0\}.$$

Furthermore, K determines δ' up to a nonzero factor in $k(x,y)$. Hence there exists $b \in k(x,y)$, unique up to a nonzero constant in k, such that

$$\delta = b\delta' = f_1 \frac{\partial}{\partial x} - f_2 \frac{\partial}{\partial y}, \qquad (2.2)$$

where $f_1, f_2 \in k[x,y]$, $\gcd(f_1, f_2) = 1$ and $K = k(x,y)^{\delta} = k(x,y)^{\delta'}$. Since A is normal, we have

$$A = k[x,y]^{\delta} = K \cap k[x,y].$$

Conversely, if δ' is a nonzero k-derivation of $k(x,y)$ such that $\delta'^p = a'\delta'$ with $a' \in k(x,y)$, $K := k(x,y)^{\delta'}$ is a purely inseparable extension of $k(u,v)$ of degree p, and $A := K \cap k[x,y]$ is a normal k-subalgebra of $k[x,y]$ containing $k[u,v]$. Hence $X := \operatorname{Spec} A$ is an affine Zariski surface. Note that by the chain rule, the derivation δ is independent of the choice of coordinate system $\{x,y\}$ (and hence depends only on the ring $k[x,y]$) up to a nonzero constant in k. Namely, if $k[x,y] = k[x',y']$, determine δ and δ' by

$$\delta = f_1 \frac{\partial}{\partial x} - f_2 \frac{\partial}{\partial y}, \quad f_1, f_2 \in k[x,y], \ \gcd(f_1,f_2) = 1$$

$$\delta' = f_1' \frac{\partial}{\partial x'} - f_2' \frac{\partial}{\partial y'}, \quad f_1', f_2' \in k[x',y'], \ \gcd(f_1',f_2') = 1.$$

Then $\delta' = c\delta$ for some $c \in k^*$. In fact, since $\delta = b\delta'$ with $b \in k(x,y)$, we can write $b_1\delta = b_2\delta'$ for $b_1, b_2 \in k[x,y]$ such that $\gcd(b_1,b_2) = 1$. Applying δ to x and y, we have

$$b_1 f_1 = b_2 \left(f_1' \frac{\partial x}{\partial x'} - f_2' \frac{\partial x}{\partial y'} \right)$$

$$-b_1 f_2 = b_2 \left(f_1' \frac{\partial y}{\partial x'} - f_2' \frac{\partial y}{\partial y'} \right).$$

Hence $b_2 \mid b_1 f_1$ and $b_2 \mid b_1 f_2$. Since $\gcd(f_1, f_2) = 1$ and $\gcd(b_1, b_2) = 1$, it follows that $b_2 \in k^*$. Exchanging the roles of δ and δ', we have $b_1 \in k^*$. So, $\delta' = c\delta$ for $c \in k^*$.

2.2   A theorem of Kimura-Niitsuma

We use the following result due to Kimura-Niitsuma [44], also due to Ganong [19] in the case $n = 2$.

Lemma. *Let k be a field of characteristic p and let $k[[x_1, \ldots, x_n]]$ be the formal power series ring in n variables over k. Let A be a local k-algebra such that $k[[x_1, \ldots, x_n]] \supseteq A \supseteq k[[x_1^p, \ldots, x_n^p]]$. Then A is regular if and only if, after a suitable change of variables, A is written in the form $A = k[[x_1, \ldots, x_s, x_{s+1}^p, \ldots, x_n^p]]$.*

2.3 A regularity criterion

Write the derivation δ in the form (2.2). Let $I(\delta)$ be the ideal of $k[x, y]$ generated by f_1 and f_2. Then we have the following:

Lemma. *With the notations as above and in 2.1, A is regular if and only if $I(\delta) = k[x, y]$.*

Proof. Let $P = (x = \alpha, y = \beta)$ be a closed point of $\operatorname{Spec} k[x, y]$, let $\mathcal{O}$ be the local ring at P, and let $\widehat{\mathcal{O}}$ be the completion of $\mathcal{O}$ with respect to the maximal ideal $\mathfrak{M}$. Let $\mathfrak{o} = \mathcal{O} \cap K$ and let $\widehat{\mathfrak{o}}$ be the completion of $\mathfrak{o}$ with respect to the maximal ideal $\mathfrak{m} := \mathfrak{M} \cap \mathfrak{o}$. Note that $\mathfrak{o}$ is the local ring of $\operatorname{Spec} A$ at the closed point $A \cap \mathfrak{M}$. It is clear that the derivation δ is extended naturally to the derivations of $\mathcal{O}$ and $\widehat{\mathcal{O}}$, which we denote by the same letter δ, and that $\mathfrak{o} = \mathcal{O}^\delta$ and $\widehat{\mathfrak{o}} = \widehat{\mathcal{O}}^\delta$.[1] We have only to show that $\widehat{\mathfrak{o}}$ is regular if and only if $I(\delta)\widehat{\mathcal{O}} = \widehat{\mathcal{O}}$ for every closed point P of $\operatorname{Spec} k[x, y]$.

Obviously we may assume that $\alpha = \beta = 0$. Suppose $I(\delta)\widehat{\mathcal{O}} \subseteq \widehat{\mathfrak{M}} = \mathfrak{M}\widehat{\mathcal{O}}$ and $\widehat{\mathfrak{o}}$ is regular. Then we can find a regular system of parameters (ξ, η) of $\widehat{\mathcal{O}}$ such that $\widehat{\mathfrak{o}} = k[[\xi, \eta^p]]$ and $\delta(\xi) = 0$ (see Lemma 2.2). Since $\delta(\xi) = 0$, we have

$$f_1 \xi_x = f_2 \xi_y, \quad \xi_x = \frac{\partial \xi}{\partial x}, \quad \xi_y = \frac{\partial \xi}{\partial y}.$$

---

[1] For the convenience of readers, we show that $\mathfrak{o} = \mathcal{O}^\delta$ and $\widehat{\mathfrak{o}} = \widehat{\mathcal{O}}^\delta$. It is clear that $\mathfrak{o} \subseteq \mathcal{O}^\delta$. Let $a/s \in \mathcal{O}^\delta$, where $a, s \in k[x, y]$ and $s \notin \mathfrak{M}$. Then $a/s = (as^{p-1})/s^p$, where $as^{p-1}, s^p \in A$ and $s^p \notin \mathfrak{m}$. Hence $a/s \in \mathfrak{o}$. To prove the second equality, note that $\mathcal{O}^{(p)} := \{\xi^p \mid \xi \in \mathcal{O}\} \subseteq \mathfrak{o}$ and $\mathcal{O}$ is a finite $\mathcal{O}^{(p)}$-module. In fact, since any element $\xi \in \mathcal{O}$ is written as $\xi = (as^{p-1})/s^p$ with $a, s \in k[x, y]$ and $s \notin \mathfrak{M}$, it suffices to show that $k[x, y]$ is a finite $k[x^p, y^p]$-module. But this is clear. Hence $\mathcal{O}$ is a finite $\mathfrak{o}$-module. We have an exact sequence of $\mathfrak{o}$-modules

$$0 \longrightarrow \mathfrak{o} \longrightarrow \mathcal{O} \xrightarrow{\delta} \mathcal{O}.$$

Since $\widehat{\mathfrak{o}}$ is a flat $\mathfrak{o}$-module, by taking the tensor product, we have an exact sequence

$$0 \longrightarrow \widehat{\mathfrak{o}} \longrightarrow \mathcal{O} \otimes_{\mathfrak{o}} \widehat{\mathfrak{o}} \xrightarrow{\delta} \mathcal{O} \otimes_{\mathfrak{o}} \widehat{\mathfrak{o}},$$

where $\mathcal{O} \otimes_{\mathfrak{o}} \widehat{\mathfrak{o}} \cong \widehat{\mathcal{O}}$ since $\mathcal{O}$ is a finite $\mathfrak{o}$-module. Hence we have an isomorphism $\widehat{\mathfrak{o}} \cong \widehat{\mathcal{O}}^\delta$.

Since $k[[\xi, \eta]] = k[[x, y]]$, we can write ξ as follows:

$$\xi = \alpha x + \beta y + \rho$$

with $\alpha, \beta \in k$, $\alpha \neq 0$ or $\beta \neq 0$, and $\rho \in \mathfrak{M}^2$. Then we have

$$f_1 \cdot (\alpha + \rho_x) = f_2 \cdot (\beta + \rho_y).$$

Assume that $\alpha \neq 0$. Then $f_1 = f_2 \cdot (\beta + \rho_y) \cdot (\alpha + \rho_x)^{-1}$, which implies that $I(\delta)\widehat{\mathcal{O}}$ has a prime divisor of height 1. This implies in turn that $I(\delta)\mathcal{O}$, hence $I(\delta)$, has a prime divisor of height 1, which contradicts the choice of f_1 and f_2. Hence, if $I(\delta)\widehat{\mathcal{O}} \subseteq \widehat{\mathfrak{M}}$, then $\hat{\mathfrak{o}}$ is not regular. This proves the *only if* part of the assertion. We shall prove the *if* part. We may assume that $f_1 \in (\widehat{\mathcal{O}})^*$ since $I(\delta)\widehat{\mathcal{O}} = \widehat{\mathcal{O}}$. Let

$$\delta' := \frac{\partial}{\partial x} - f_1^{-1} \cdot f_2 \frac{\partial}{\partial y}.$$

Since $\hat{\mathfrak{o}} = (\widehat{\mathcal{O}})^\delta = (\widehat{\mathcal{O}})^{\delta'}$, we know that $(\delta')^p = a'\delta'$ with $a' \in Q(\widehat{\mathcal{O}})$. Since $\delta'(x) = 1$ and $(\delta')^p(x) = 0$, we have $a' = 0$. Then we show that $\widehat{\mathcal{O}}$ is a free $\hat{\mathfrak{o}}$-module with a free basis $\{1, x, x^2, \ldots, x^{p-1}\}$. Let z be an element of $\widehat{\mathcal{O}}$, not in $\hat{\mathfrak{o}}$. Suppose $(\delta')^\ell(z) \neq 0$ and $(\delta')^{\ell+1}(z) = 0$. Such an integer ℓ exists because $(\delta')^p = $ and $1 \leq \ell < p$. Then

$$(\delta')^\ell(z - (\ell!)^{-1}(\delta')^\ell(z)x^\ell) = 0.$$

By induction on ℓ, z belongs to $\hat{\mathfrak{o}}[x]$. Therefore, $\widehat{\mathcal{O}} = \hat{\mathfrak{o}}[x]$. Since $Q(\widehat{\mathcal{O}})$ is purely inseparable over $Q(\hat{\mathfrak{o}})$, we see easily that $\{1, x, \ldots, x^{p-1}\}$ is a free $\hat{\mathfrak{o}}$-basis of $\widehat{\mathcal{O}}$. Since $\widehat{\mathcal{O}}$ is a faithfully flat $\hat{\mathfrak{o}}$-module, we know that $\hat{\mathfrak{o}}$ is regular.

$$\square$$

2.4   Generators of $k[x, y]^\delta$

Write $K = k(u, v, \psi\varphi^{-1})$, where $\varphi, \psi \in k[x, y], \gcd(\varphi, \psi) = 1$, and φ and ψ have no irreducible factor whose pth power divides φ or ψ. Set anew

$$\delta' := (\varphi\psi_y - \psi\varphi_y)\frac{\partial}{\partial x} - (\varphi\psi_x - \psi\varphi_x)\frac{\partial}{\partial y}.$$

Then $\delta' = \rho\delta$ with $\rho \in k[x, y]$. Let $A' := k[x^p, y^p, \varphi^{p-1}\psi, \varphi\psi^{p-1}]$. Then $A' \subseteq A$ and $Q(A') = Q(A)$. With the notations of 2.3, let $\mathcal{O}$ be the local ring of $\operatorname{Spec} k[x, y]$ at a closed point P. Let $\mathfrak{o} := \mathcal{O} \cap K$ and let $\mathfrak{o}'$ be the localizations of A' with respect to $A' \cap \mathfrak{M}$. Then we have the following:

2.4.1

Lemma. *Assume that the curve on* $\operatorname{Spec} k[x,y]$ *defined by* $\varphi(P)\psi - \psi(P)\varphi = 0$ *is smooth at* P. *Then* $\mathfrak{o}' = \mathfrak{o}$ *and* $\mathfrak{o}$ *is regular.*

Proof. Let $\alpha = \varphi(P)$ and $\beta = \psi(P)$. We have $(\alpha\psi_x - \beta\varphi_x)(P) \neq 0$ or $(\alpha\psi_y - \beta\varphi_y)(P) \neq 0$ by the assumption. In particular, $\alpha \neq 0$ or $\beta \neq 0$. Suppose $\alpha \neq 0$. Then we may choose a regular system of parameters (ξ, η) of $\mathcal{O}$ such that

$$\xi = \frac{\psi}{\varphi} - \frac{\beta}{\alpha} = \frac{\alpha\psi - \beta\varphi}{\alpha\varphi}.$$

Since $\varphi^p \in k[x^p, y^p]$ and $\xi\varphi^p = \psi\varphi^{p-1} - \beta\alpha^{-1}\varphi^p \in A'$, we have

$$\widehat{\mathcal{O}} = k[[\xi, \eta]] \supset \widehat{\mathfrak{o}} \supseteq \widehat{\mathfrak{o}'} \supseteq k[[\xi, \eta^p]] .$$

Since $\widehat{\mathfrak{o}}$ is normal [66, Th. 37.5] and $Q(\widehat{\mathfrak{o}}) = k((\xi, \eta))^{\delta'} = k((\xi, \eta^p))$, we find that $\widehat{\mathfrak{o}} = \widehat{\mathfrak{o}'} = k[[\xi, \eta^p]]$. The argument is similar if $\beta \neq 0$. $\square$

2.4.2

Corollary. *Let* $I(\delta')$ *be the ideal of* $k[x,y]$ *generated by the coefficients of* δ' *defined in 2.4. Suppose* $I(\delta') = k[x,y]$. *Then* $A = A'$ *and* A *is regular.*

Proof. If $I(\delta') = k[x,y]$ then the assumption in Lemma 2.4.1 holds for any closed point P of $\operatorname{Spec} k[x,y]$. Hence $A' = \cap\mathfrak{o}' = \cap\mathfrak{o} = A$ and A is regular. $\square$

2.4.3

Lemma. *The following assertions hold true.*

(1) *If* $p = 2$ *then there exists an element* ψ *of* $k[x,y]$ *such that* $A = k[x^2, y^2, \psi]$.

(2) *If* $p = 3$ *then there exist elements* φ, ψ *of* $k[x,y]$ *such that* $\gcd(\varphi, \psi) = 1$ *and* $A = k[x^3, y^3, \varphi^2\psi, \varphi\psi^2]$.

Proof. (1) Since $K = k(x^2, y^2, \psi\varphi^{-1}) = k(x^2, y^2, \varphi\psi)$, we can write $K = k(x^2, y^2, \psi)$ with $\psi \in k[x,y]$. We can choose $\psi \in k[x,y]$ in such a way that if we write $\psi = a_0 + a_1 x + a_2 y + a_3 xy$ with $a_i \in k[x^2, y^2]$, then $a_0 = 0$ and $\gcd(a_1, a_2, a_3) = 1$ in $k[x^2, y^2]$. If it happens that $d := \gcd(a_1, a_2, a_3) \notin k$ then ψ is divisible by d. Since we may replace ψ by $d^{-1}\psi$, we can assume that $\gcd(a_1, a_2, a_3) = 1$ as well as $a_0 = 0$.

Let $b_0 + b_1 x + b_2 y + b_3 xy \in A$. Since $k(x^2, y^2, \psi) = k(x^2, y^2) \cdot 1 + k(x^2, y^2) \cdot \psi$, there exists an element $t \in k[x^2, y^2]$ such that

$$t \cdot (b_0 + b_1 x + b_2 y + b_3 xy) = c_0 + c_1 (a_1 x + a_2 y + a_3 xy), \quad c_0, c_1 \in k[x^2, y^2] .$$

Hence we have

$$t b_0 = c_0, \quad t b_1 = c_1 a_1, \quad t b_2 = c_1 a_2, \quad t b_3 = c_1 a_3 .$$

By the additional assumption, t divides c_1. Since t divides c_0, it follows that $b_0 + b_1 x + b_2 y + b_3 xy \in k[x^2, y^2, \psi]$.

(2) We choose φ, ψ of $k[x, y]$ so that both φ and ψ have only irreducible factors with multiplicity one. This is possible since $p = 3$. For example, if ψ has an irreducible factor ψ_1 with multiplicity two, then replace $\psi \varphi^{-1}$ by $(\psi_1)^{-3} \psi \varphi^{-1}$ to shift ψ_1 to denominator. Let $X = \varphi^2 \psi$ and $Y = \varphi \psi^2$. Then we have

$$A' := k[x^3, y^3, \varphi^2 \psi, \varphi \psi^2]$$
$$= k[u, v, X, Y]/(X^3 - (\varphi^{(3)})^2 \psi^{(3)}, Y^3 - \varphi^{(3)} (\psi^{(3)})^2) ,$$

where $(\varphi(x, y))^3 = \varphi^{(3)}(u, v)$ and $(\psi(x, y))^3 = \psi^{(3)}(u, v)$. We show that A' is normal. Since $Q(A') = Q(A)$, it then follows that $A = A'$. Note that, for any closed point P' of Spec A', the local ring $\mathcal{O}_{P'}$ has a regular sequence $(u - u(P'), v - v(P'))$, hence depth $\mathcal{O}_{P'} = 2$. By Serre criterion of normality, it suffices to show that, for any prime ideal $\mathfrak{P}'$ of height 1 of A', the local ring $A'_{\mathfrak{P}'}$ is a discrete valuation ring. Since $k[u, v] \subset A' \subset k[x, y]$, let $\mathfrak{P}$ be a prime ideal of $k[x, y]$ lying over $\mathfrak{P}'$. Since $k[x, y]$ is a UFD, $\mathfrak{P}$ is a principal ideal (τ). If τ is not an irreducible factor of φ nor ψ, then $\mathfrak{P}' A'_{\mathfrak{P}'}$ is generated by a principal ideal $\mathfrak{P}'' := \mathfrak{P} \cap k[u, v]$, whence $\mathfrak{P}' A'_{\mathfrak{P}'}$ is a principal ideal. Suppose that τ is an irreducible factor of ψ. In the case where τ is an irreducible factor of φ, we change the role of φ, ψ and consider $\varphi \psi^{-1}$ instead of $\psi \varphi^{-1}$. By the Jacobian criterion of regularity, it suffices to show that, in $k[x, y]_{\mathfrak{P}}$,

$$\operatorname{rank} \begin{pmatrix} \varphi(\varphi \psi_x - \psi \varphi_x) & \varphi(\varphi \psi_y - \psi \varphi_y) \\ \psi(\varphi \psi_x - \psi \varphi_x) & \psi(\varphi \psi_y - \psi \varphi_y) \end{pmatrix} \geq 1 .$$

Suppose that the rank is zero. Namely, suppose that all four matrix components are divisible by τ. Then, since $\tau \mid \psi$ and $\tau \nmid \varphi$, we have $\tau \mid d(\psi \varphi^{-1})$, where

$$d\left(\frac{\psi}{\varphi}\right) = \frac{(\varphi \psi_x - \psi \varphi_x)}{\varphi^2} dx + \frac{(\varphi \psi_y - \psi \varphi_y)}{\varphi^2} dy.$$

This implies that $\tau^2 \mid \psi \varphi^{-1}$ in $k[x, y]_{\mathfrak{P}}$, whence $\tau^2 \mid \psi$. This is a contradiction to the choice of ψ. $\qquad\square$

Remark. If $p \geq 3$, A is not necessarily a simple extension of $k[x^p, y^p]$ as shown by the following example. Let $f(x) \in k[x]$ be a non-constant polynomial free from square factors. Let $\varphi = f(x), \psi = yf(x) + 1$ and

$$A := k[x, y] \cap k(x^p, y^p, \psi\varphi^{-1}) = k[x^p, y^p, \varphi^{p-1}\psi, \varphi\psi^{p-1}] .$$

Then A is not a simple extension of $k[x^p, y^p]$.

In fact, $\delta = \delta' = f(x)^2 \frac{\partial}{\partial x} - f'(x)\frac{\partial}{\partial y}$. Since $\gcd(f(x)^2, f'(x)) = 1$, we have $I(\delta) = k[x, y]$. By Corollary 2.4.2, $A = k[x^p, y^p, \varphi^{p-1}\psi, \varphi\psi^{p-1}]$. Suppose $A = k[x^p, y^p, g]$ for some $g \in k[x, y]$. Then $c\delta = g_y \frac{\partial}{\partial x} - g_x \frac{\partial}{\partial y}$ for some $c \in k^*$. Since $g_x = cf'(x)$, we have $g(x, y) = cf(x) + h(y)$ with $h(y) \in k[y]$. Then $g_y = h'(y) = cf(x)^2$. Hence $f(x)$ is a constant, and this is a contradiction to the choice of $f(x)$.

2.5 Divisor class group of affine Zariski surface

Let $X = \mathrm{Spec}\,(A)$ be an affine Zariski surface. Write $A = k[x, y]^\delta$ with the derivation δ as in 2.1. We shall compute the divisor class group $\mathrm{C}\ell\,(X)$. See Samuel [79, Chap.III].

2.5.1

Lemma. (1) $\mathrm{C}\ell\,(X)$ *is isomorphic to the abelian group L of all logarithmic differentials $\delta f/f$ such that $\delta f/f \in k[x, y]$, where $f \in k(x, y)$.*

(2) *Write $\delta^p = a\delta$. Then $a \in A$. Moreover, for $h \in k[x, y]$, $h \in L$ if and only if*

$$(\delta^{p-1} + F - a)h = 0,$$

where $Fh = h^p$.

Proof. We only show that $a \in A$. The other results are well-known and proved in the reference cited above. Since δ^p is a k-trivial derivation on $k[x, y]$ such that $\delta^p = 0$ on $A := k[x, y]^\delta$, we can write $\delta^p = a\delta$ with $a \in k[x, y]$ by the definition of δ. Since $\delta^{p+1} = \delta(a\delta) = a\delta^2$, we have $\delta(a)\delta + a\delta^2 = a\delta^2$, whence $\delta(a) = 0$. Hence we obtain $a \in A$. $\square$

2.5.2

Lemma. (1) $\mathrm{C}\ell\,(X)$ *is a finite direct sum of cyclic groups of order p, i.e., a finite elementary p-group.*

(2) *Write* $\delta = f_1 \frac{\partial}{\partial x} - f_2 \frac{\partial}{\partial y}$ *as in (2) of 2.1. Letting here* deg *stand for the total degree with respect to* x *and* y, *set* $d_1 = \max\,(\deg f_1, \deg f_2)$ *and* $d_2 = \deg a$ *for a satisfying* $\delta^p = a\delta$, *where* $a \in A$. *Let* $\rho :=$ $\max\,(d_1 - 1, [d_2/(p-1)])$. *Then the order* $|C\ell\,(X)|$ *is bounded by* $p^{\rho(\rho+1)/2}$.

Proof. Suppose we have $(\delta^{p-1} + F - a)h = 0$ for $h \in k[x,y]$. Let $d = \deg_{x,y} h$. Then we have

$$\deg_{x,y} \left(\delta^{p-1}h - ah\right) \leq \max\,(d + (p-1)(d_1 - 1), d_2 + d)\,.$$

Hence we have

$$pd \leq \max(d + (p-1)(d_1 - 1), d_2 + d),$$

or equivalently

$$d \leq \max\,(d_1 - 1, [d_2/(p-1)]) = \rho.$$

Let V be the k-vector space generated by all polynomial of total degree $\leq \rho$. Then we know that $L \subset V$ by Lemma 2.5.1. Note that L is a $\mathbb{F}_p$-vector space, where $\mathbb{F}_p$ is the prime field of characteristic p. We shall show that if $\{h_1, \ldots, h_m\}$ is a set of $\mathbb{F}_p$ linearly independent elements of L, then $\{h_1, \ldots, h_m\}$ is k-linearly independent. Suppose we have a nontrivial relation

$$\lambda_1 h_1 + \cdots + \lambda_m h_m = 0, \quad \lambda_i \in k.$$

Let $\{\omega_1, \ldots, \omega_n\}$ be a linearly independent basis of the $\mathbb{F}_p$-vector subspace of k spanned by $\lambda_1, \ldots, \lambda_m$. Write

$$\lambda_i = \sum_{j=1}^{n} \mu_{ij}\omega_j, \quad \mu_{ij} \in \mathbb{F}_p, \quad 1 \leq i \leq m$$

and set

$$v_j = \sum_{i=1}^{m} \mu_{ij}h_i, \quad 1 \leq j \leq n.$$

Then we have $v_j \in L$ and

$$\omega_1 v_1 + \cdots + \omega_n v_n = 0, \quad \omega_j \in k.$$

Hence we have

$$0 = \delta^{p-1}\left(\sum_j \omega_j v_j\right) = \sum_j \omega_j \left(\delta^{p-1} v_j\right)$$

$$= -\sum_j \omega_j v_j^p + a\sum_j \omega_j v_j = -\sum_j \omega_j v_j^p,$$

whence
$$\omega_1^{1/p} v_1 + \cdots + \omega_n^{1/p} v_n = 0.$$

Similarly,
$$\omega_1^{p^{-i}} v_1 + \cdots + \omega_n^{p^{-i}} v_n = 0, \quad 0 \le i \le n - 1.$$

Let Δ' be the determinant of the $(n \times n)$-matrix whose (i,j)-entry is $\omega_j^{p^{-i}}$, where $0 \le i \le n - 1$ and $1 \le j \le n$. Note that
$$(\Delta')^{p^{n-1}} = (-1)^{n(n-1)/2} \Delta ,$$

where Δ is the determinant of the $(n \times n)$-matrix
$$\begin{pmatrix} \omega_1 & \cdots & \cdots & \omega_n \\ \omega_1^p & \cdots & \cdots & \omega_n^p \\ \cdots & \cdots & \cdots & \cdots \\ \omega_1^{p^{n-1}} & \cdots & \cdots & \omega_n^{p^{n-1}} \end{pmatrix}$$

It is then not hard to show that
$$\Delta^{p-1} \equiv (-1)^n \prod_{(\varepsilon_1, \ldots, \varepsilon_n)} (\varepsilon_1 \omega_1 + \cdots + \varepsilon_n \omega_n) \pmod{p},$$

where $(\varepsilon_1, \ldots, \varepsilon_n)$ moves over all elements of $(\mathbb{F}_p)^n - (0, \ldots, 0)$. By the definition of $\{\omega_1, \ldots, \omega_n\}$, Δ, and hence Δ', is nonzero. This implies that $v_1 = \cdots = v_n = 0$. Since $\mu_{ij} \in \mathbb{F}_p$, and since $\{h_1, \ldots, h_m\}$ is linearly independent over $\mathbb{F}_p$, we know that $\mu_{ij} = 0$ for every pair (i, j). Hence $\lambda_1 = \cdots = \lambda_m = 0$. This is a contradiction. Thus, we have shown that L is a sublattice of the k-vector space V. Then, our assertion follows immediately from this fact. $\qquad\square$

2.6 Resolution of singularities

In this section, we are interested in the resolution of singularities of an affine Zariski surface.

2.6.1

Let d be an integer such that $0 < d < p$.

(1) We consider the continued fraction expansion
$$\frac{p}{d} = [b_1, \ldots, b_s] = b_1 - \cfrac{1}{b_2 - \cfrac{1}{\ddots - \cfrac{1}{b_{s-1} - \cfrac{1}{b_s}}}},$$

where b_i is an integer ≥ 2.

(2) The integers $b_1, \ldots, b_s$ are determined by the Euclidean algorithm

$$\lambda_0 = p, \ \lambda_1 = d, \ \lambda_{i-1} = b_i \lambda_i - \lambda_{i+1} \ \text{with} \ 0 \le \lambda_{i+1} < \lambda_i$$
$$\text{for} \ 1 \le i \le s, \ \lambda_s = 1, \ \lambda_{s+1} = 0.$$

Note that $\lambda_{i+1} = b_i \lambda_i - \lambda_{i-1}$ for $1 \le i \le s$.

(3) We define integers $\mu_0, \mu_1, \ldots, \mu_{s+1}$ by

$$\mu_0 = 0, \ \mu_1 = 1, \ \mu_{i+1} = b_i \mu_i - \mu_{i-1} \ \text{for} \ 1 \le i \le s \ .$$

Then $0 < \mu_i < \mu_{i+1}$ for $1 \le i \le s$ and $\mu_{s+1} = p$. Suppose $dd' \equiv 1$ (mod p) with $0 < d' < p$. Put $\lambda_i' = \mu_{s+1-i}$. Then

$$\lambda_{i-1}' = b_{s+1-i}\lambda_i' - \lambda_{i+1}', \ \ 0 \le \lambda_{i+1}' < \lambda_i' \ \text{for} \ 1 \le i \le s,$$
$$\lambda_s' = 1, \ \lambda_{s+1}' = 0 \ ,$$

is the Euclidean algorithm for p and d'. In particular,

$$\frac{p}{d'} = [b_s, \ldots, b_1].$$

(4) Let α, β be integers. We define integers $\gamma_i = \gamma_i(\alpha, \beta)$ $(0 \le i \le s)$ by

$$\gamma_0 = p\alpha, \ \gamma_1 = \beta + \alpha d, \ \gamma_{i+1} = b_i \gamma_i - \gamma_{i-1} \ \text{for} \ 1 \le i \le s \ .$$

Then we have

$$\gamma_i = \alpha\lambda_i + \beta\mu_i \ \text{for} \ 0 \le i \le s+1 \ .$$

(5) Suppose $\beta + \alpha d \equiv 0$ (mod p). Then $\gamma_i(\alpha, \beta) \equiv 0$ (mod p), and we put

$$\nu_i(\alpha, \beta) := \frac{1}{p}\gamma_i(\alpha, \beta) \ \text{for} \ 0 \le i \le s+1 \ .$$

For the proof, see [31, p.16].

2.6.2

Lemma. *Let* $A := k[x, y] \cap k(x^p, y^p, y/x^d)$ *and* $\delta = x\frac{\partial}{\partial x} + dy\frac{\partial}{\partial y}$. *With the notations of 2.6.1, we have the following assertions.*

(1) $A = k[x, y]^\delta$, $X := \operatorname{Spec} A$ *has an isolated singular point* P *defined by* $x = y = 0$, *and* $X - \{P\}$ *is smooth.*

(2) *The monomials* $x^\beta y^\alpha$ *with* $\beta + d\alpha \equiv 0$ (mod p) *form a* k-*basis of* A. *Furthermore,*

$$A = k[x^p, x^{p-d}y, \ldots, x^{r_t}y^t, \ldots, y^p] \ ,$$

where for $0 \le t \le p$, *we write*

$$dt = q_t p - r_t \ \text{with} \ q_t \ge 1 \ \text{and} \ 0 \le r_t < p \ .$$

(3) *The minimal resolution of singularity at P has the following dual graph of exceptional curves:*

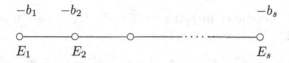

where " $\circ$ " stands for a smooth rational curve. Hence the singularity at P is a cyclic singularity. The proper transform of the curve on X defined by $y^p = 0$ (resp. $x^p = 0$) meets transversally the curve E_1 (resp. E_s) and does not meet any other curves E_i.

Proof. (1) This follows from 2.1 and Lemma 2.4.1.

(2) Assign weights $1, d$ to x, y, respectively. Then δ is a homogeneous derivation. Namely, we have $\delta(x^\beta y^\alpha) = (\beta + d\alpha)x^\beta y^\alpha$. This proves the first part of the assertion (2). Suppose $\beta + d\alpha = np$. Write $\alpha = p\ell + t$ with $0 \le t < p$. Then $\beta = np - d\alpha = p(n - d\ell - q_t) + r_t$ with q_t, r_t as in the statement and $n - d\ell - q_t \ge 0$. This proves the second part of (2).

(3) Consider a rational variety $\widetilde{X}$ which is a union of $(s + 1)$ copies of the affine plane $\mathbb{A}_k^2$ patched together by the following data:

$$\widetilde{X} = \bigcup_{i=0}^{s} W_i, \quad W_i = \operatorname{Spec} k[u_i, v_i] \cong \mathbb{A}_k^2,$$

where

$$u_i = \frac{1}{v_{i-1}}, \quad v_i = (v_{i-1})^{b_i} u_{i-1} \quad (1 \le i \le s). \tag{2.3}$$

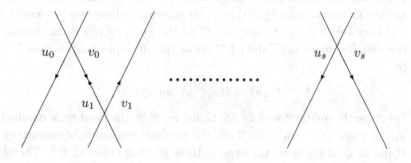

Then, setting

$$u_0 = x^p, \quad v_0 = \frac{y}{x^d}, \tag{2.4}$$

we have a birational mapping $\pi : \widetilde{X} \to X$. If $\beta + d\alpha \equiv 0 \pmod{p}$, we find, using (2.3), (2.4) and 2.6.1, that

$$x^\beta y^\alpha = u_0^{\nu_1} v_0^{\nu_0} = \cdots = u_i^{\nu_{i+1}} v_i^{\nu_i} = \cdots = u_s^{\nu_{s+1}} v_s^{\nu_s}, \tag{2.5}$$

with $\nu_i := \nu_i(\alpha, \beta)$ for $0 \le i \le s$ as in 2.6.1, (5). Furthermore, $\nu_i \ge 0$ by 2.6.1, (3) and (4). By the assertion (2) above, $A \subset k[u_i, v_i]$ for $0 \le i \le s$, and hence π is a morphism. Define the curve E_i $(1 \le i \le s)$ on $\widetilde{X}$ by

$$u_{i-1} = 0 \quad \text{on} \quad W_i \quad \text{and} \quad v_i = 0 \quad \text{on} \quad W_{i+1}.$$

It is then easily verified that $\pi^{-1}(P) = E_1 \cup \cdots \cup E_s$ and that $\widetilde{X} - \pi^{-1}(P) \cong X - \{P\}$. Moreover, $(E_i^2) = -b_i$ by (2.3). The proper transform on $\widetilde{X}$ of the curve on X defined by $y^p = 0$ (resp. $x^p = 0$) is defined by $v_0 = 0$ on W_0 (resp. $u_s = 0$ on W_s). All the assertions of (3) are thus verified. $\quad\square$

2.6.3

Remark. (1) Let $\pi_1 : \mathbb{A}_k^2 = \operatorname{Spec} k[x, y] \to X$ be the finite morphism induced by the canonical inclusion $A \hookrightarrow k[x, y]$. Then the rational mapping $T := \pi^{-1} \cdot \pi_1 : \mathbb{A}_k^2 \to \widetilde{X}$ is given by

$$T_i : (x, y) \mapsto (x^{\lambda_i} y^{-\nu_i}, x^{-\lambda_{i+1}} y^{\mu_{i+1}})$$

on W_i.

(2) The procedure of obtaining the minimal resolution of singularity at P and the dual graph of the exceptional curves are the same as those for the cyclic quotient singularity of type (p, d) in characteristic zero (see Fujiki [18] and Hirzebruch [31]). Namely, let G be the group of pth roots of unity in the complex numbers $\mathbb{C}$, and let G act on the affine plane $\mathbb{A}_\mathbb{C}^2 = \operatorname{Spec} \mathbb{C}[x, y]$ by

$$(x, y) \mapsto (\xi x, \xi^d y) \quad \text{for} \quad \xi \in G.$$

Let X be the quotient variety $\mathbb{A}_\mathbb{C}^2 / G$ and let P be the point of X dominated by the point $(x = 0, y = 0)$. Then the minimal resolution of singularity at P and its dual graph are the same as those given in Lemma 2.6.2. Therefore we will say that $\operatorname{Spec} A$ has a *cyclic quotient singularity of type* (p, d) at P.

2.7 Resolution data

Let V be a smooth surface defined over k, let K be a purely inseparable extension of degree p of the function field $k(V)$ and let S be the normalization of V in K. Let C_1, C_2 be closed curves on V meeting transversally at $Q \in V$. Suppose there exists an element $w_1 \in K$ such that $K = k(V)(w_1)$ and $w_1^p = \tau' u^{b'} v^{a'} \in \widehat{\mathcal{O}}_{V,Q}$, where $u = 0$ and $v = 0$ are local equations for C_2 and C_1, respectively, a' and b' are integers and $\tau' \in \widehat{\mathcal{O}}_{V,Q}^*$. Note that (u, v) is a regular system of parameters of V at Q, We write $a' = q_1 p + a$, $b' = q_2 p + b$ with $0 \le a, b < p$ and put $w = w_1 u^{-q_2} v^{-q_1}$. Then $w^p = \tau u^b v^a$ with $\tau \in \widehat{\mathcal{O}}_{V,Q}^*$,

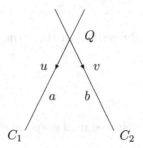

Suppose $0 < a, b < p$ and let $d = d(a, b)$ be the integer defined by $0 < d < p$ and $b + ad \equiv 0 \pmod{p}$. Note that $d(a, b)d(b, a) \equiv 1 \pmod{p}$. Let P be the point of S lying over Q. Then we have the following result.

2.7.1

Lemma. (1) S has a cyclic quotient singularity of type (p, d) at P.

(2) Let $\widetilde{S}$ be the normal surface obtained from S by resolving minimally the singularity of S at P and let $\sigma : \widetilde{S} \to V$ be the canonical morphism. Let $\widetilde{C}_1, \widetilde{C}_2$ be the set-theoretic proper inverse images on $\widetilde{S}$ of C_1, C_2, respectively (i.e., the closures in $\widetilde{S}$ of the inverse images of σ of dense open sets of C_1, C_2). Let α, β be integers. Then, with the notations of Lemma 2.6.2, we have

$$\sigma^*(\alpha C_1 + \beta C_2) = \gamma_0 \widetilde{C}_1 + \gamma_1 E_1 + \cdots + \gamma_s E_s + \gamma_{s+1} \widetilde{C}_2$$

with $\gamma_i = \gamma_i(\alpha, \beta)$ as in 2.6.1, (4). In particular,

$$\sigma^*(a C_1 + b C_2) = p(a \widetilde{C}_1 + \nu_1 E_1 + \cdots + \nu_s E_s + b \widetilde{C}_2)$$

with $\nu_i = \nu_i(a, b)$ as in 2.6.1, (5).

Proof. (1) Let $R_1 := \hat{\mathcal{O}}_{V,Q} = k[[u,v]]$ and let $R := \hat{\mathcal{O}}_{S,P}$. Let $u = x^p$ and $v = y^p$, and let $\tilde{R} := k[[x,y]]$. Then $R_1 = k[[x^p, y^p]]$, and R is isomorphic to the normalization of $k[[x^p, y^p, x^b y^a]]$. Let

$$\delta := ax\frac{\partial}{\partial x} - by\frac{\partial}{\partial y} = a\left(x\frac{\partial}{\partial x} + dy\frac{\partial}{\partial y}\right).$$

Then it is clear that $R = k[[x,y]]^\delta = k[[x,y]] \cap k((x^p, y^p, \frac{y}{x^d}))$. Hence S has a cyclic quotient singularity of type (p,d) at P.

(2) In an open neighborhood of $\sigma^{-1}(Q)$, $\sigma^*(\alpha C_1 + \beta C_2)$ is defined by $t = u^\beta v^\alpha = 0$. It follows from 2.6.1, (4) and (2.3), (2.4) in the proof of Lemma 2.6.2 that

$$t = u_0^{\gamma_1} v_0^{\gamma_0} = \cdots = u_i^{\gamma_{i+1}} v_i^{\gamma_i} = \cdots = u_s^{\gamma_{s+1}} v_s^{\gamma_s}.$$

The last assertion is already verified in the formula (2.5) of the proof of Lemma 2.6.2. □

2.7.2

We shall use the following weighted dual graph to represent $\sigma^*(\alpha C_1 + \beta C_2)$:

We also use the next graph to represent $\frac{1}{p}\sigma^*(aC_1 + bC_2)$:

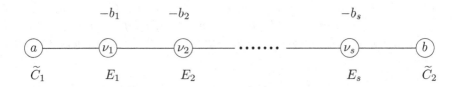

2.7.3

Lemma. *Assume that p is odd and write $p = 2\ell + 1$. With assumptions and notations as in Lemma 2.7.1, we have the following assertions.*

(1) *Assume $a = b$. Then $d = s = p - 1$, $b_1 = b_2 = \cdots = b_s = 2$ and $\nu_1 = \nu_2 = \cdots = \nu_s = a$ ($\nu_i = \nu_i(a, b)$). Also we have*

$$\sigma^*(C_1 + C_2) = p(\widetilde{C}_1 + E_1 + \cdots + E_{2\ell} + \widetilde{C}_2)$$
$$\sigma^*(C_2) = E_1 + 2E_2 + \cdots + 2\ell E_{2\ell} + p\widetilde{C}_2.$$

(2) *Assume $a = p-1$ and $b = \ell$. Then $d = s = \ell$, $b_1 = 3$, $b_2 = \cdots = b_\ell = 2$ and $\nu_1 = \cdots = \nu_\ell = \ell$ ($\nu_i = \nu_i(a, b)$). Moreover,*

$$\sigma^*(C_1 + (\ell + 1)C_2) = p(\widetilde{C}_1 + E_1 + 2E_2 + \cdots + \ell E_\ell + (\ell + 1)C_2).$$

Proof. Straightforward from Lemma 2.7.1.

2.7.4

Lemma. *Assume $p = 2$. With assumptions and notations of Lemma 2.7.1, we have $a = b = 1$, $d = s = 1$, $b_1 = 2$ and $\nu_1 = 1$. Also $\sigma^*(C_1 + C_2) = 2(\widetilde{C}_1 + E_1 + \widetilde{C}_2)$ and $\sigma^*(C_2) = E_1 + 2\widetilde{C}_2$.*

2.7.5

Remark. (1) With the notations in 2.4, assume that the curves $\varphi = 0$ and $\psi = 0$ meet transversally at a point $\widetilde{P}$. Let P be the point of $X := \operatorname{Spec} A$ dominated by $\widetilde{P}$. Then X has a cyclic quotient singularity of type $(p, 1)$ at P. This follows immediately from Lemma 2.7.1.

(2) Assume now that the curves $\varphi = 0$ and $\psi = 0$ are smooth at $\widetilde{P}$ and meet each other with order of contact $\mu > 1$, where $\gcd(\mu, p) = 1$. Let $\mu \equiv d \pmod{p}$ with $0 < d < p$. Though it is expected that X has a cyclic quotient singularity of type (p, d) at the point P dominated by $\widetilde{P}$, this might or might not be the case as shown by the following examples.

(i) $p = 3$, $\varphi = y$, $\psi = y + xy + x^2$ and $\widetilde{P} = (0, 0)$. With the notations in the proof of Lemma 2.7.1, R is the normalization of $k[[x^3, y^3, y^2(y + xy + x^2)]]$. The singularity of X at P has the following weighted graph for the exceptional curves of the resolution of singularity.

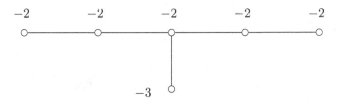

(ii) $p = 3$, $\varphi = y$, $\psi = y + x^2$ and $\widetilde{P} = (0,0)$. Then R is the normalization of $k[[x^3, y^3, y^2(y+x^2)]]$ and hence $R = k[[x^3, y^3, xy, x^2y^2]]$. So, X has a cyclic quotient singularity of type $(3, 2)$ at P and the minimal resolution has the following weighted dual graph.

$$\overset{-2}{\circ}\!\!-\!\!-\!\!-\!\!-\!\!-\!\!-\!\!-\!\!-\!\!\overset{-2}{\circ}$$

2.8   Nice ramification with given data

Let V and K be as in 2.7. Let C be a smooth complete curve on V and $D_1, \ldots, D_r$ the curves on V such that D_i meets C transversally at a point Q_i and no points else. We allow the case with no D_i, too. Let $a, b_1, \ldots, b_r$ be positive integers less than p. We assume that for each point $Q \in C$ there exist $w \in K$ and a regular system of local parameters (u, v) for V at Q such that $K = k(V)(w)$, C has a local equation $v = 0$ at Q and

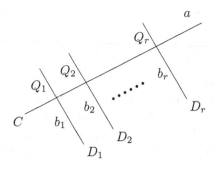

(i) if $Q \neq Q_i$ $(1 \leq i \leq r)$, then

$$w^p = \tau v^a \quad \text{with} \quad \tau \in \widehat{\mathcal{O}}^*_{V,Q},$$

(ii) if $Q = Q_i$, then $u = 0$ is a local equation for D_i at Q and

$$w^p = \tau u^{b_i} v^a \text{ with } \tau \in \widehat{\mathcal{O}}_{V,Q}^*.$$

We refer to this situation more briefly by saying that K *ramifies nicely above C with ramification data* $(a, b_1, \ldots, b_r)$. We call the integer a the *multiplicity of ramification* of K over C.

Let S be the normalization of V in K, $\widetilde{S}$ the surface obtained from S by resolving minimally all singular points of S lying over C, and let $\sigma : \widetilde{S} \to V$ be the canonical morphism.

2.8.1

Lemma. *With the assumptions and notations as above, we have the following assertions.*

(1) *S has a cyclic quotient singularity of type (p, d_i) with $d_i = d(a, b_i)$ (see 2.7) at the point P_i on S lying above Q_i $(1 \leq i \leq r)$ and S is smooth at any point P lying above a point $Q \neq Q_i$ of C.*
(2) *Let $\widetilde{C}$ be the set-theoretic proper transform of C on $\widetilde{S}$. Then $\widetilde{C}$ is a smooth curve isomorphic to C, and $[k(\widetilde{C}) : k(C)] = 1$.*
(3) *$(\widetilde{C})^2 = \frac{1}{p}\left((C)^2 - d_1 - \cdots - d_r\right)$.*

Proof. (1) This follows from Lemmas 2.7.1 and 2.3.

(2) We have proved in Lemma 2.6.2 that $\widetilde{C}$ is smooth at the point $\widetilde{Q}_i$ lying above Q_i. Let P be the point of S lying above a point $Q \neq Q_i$ of C. Since $K = k(V)(w)$ with $w^p = \tau v^a$ at Q, there exists $t \in \widehat{\mathcal{O}}_{S,P}$ such that $\widehat{\mathcal{O}}_{S,P} = k[[u, t]]$ with $w = \tau^m t^a$ and $v = \tau^n t^p$, where m, n are integers satisfying $mp - na = 1$. Thus $\widetilde{C}$ is defined by $t = 0$ at P and $\widetilde{C}$ is smooth at P. Furthermore, $\widetilde{C}$ is ramified with index p over C and hence $[k(\widetilde{C}) : k(C)] = 1$. In particular, $\widetilde{C}$ is isomorphic to C.

(3) Let $\alpha := \left(\sigma^*(aC + b_1 D_1 + \cdots + b_r D_r) \cdot \widetilde{C}\right)$. By the projection formula, we have

$$\alpha = a(C^2) + b_1 + \cdots + b_r.$$

By Lemmas 2.7.1 and 2.6.2, we have

$$\alpha = p\left(a(\widetilde{C}^2) + \nu_{11} + \cdots + \nu_{r1}\right),$$

where $\nu_{i1} = \nu_1(a, b_i) = \frac{1}{p}(b_i + a d_i)$. Solving for $(\widetilde{C}^2)$ from these equations, we obtain the stated formula. $\qquad\square$

2.8.2

Lemma. *Suppose K ramifies nicely over C with data $(a, b_1, \ldots, b_r)$. Put $\alpha = -(C^2)$ and $\gamma = -(\widetilde{C}^2)$.*

(1) *Suppose $r = 0$. Then $\alpha \equiv 0 \pmod{p}$ and $\gamma = \frac{\alpha}{p}$.*

(2) *Suppose $r = 1$. Then $b_1 \equiv a\alpha \pmod{p}$ and $\gamma = \left[\frac{\alpha}{p}\right] + 1$. (Note that $\alpha \not\equiv 0 \pmod{p}$.)*

(3) *Suppose $r = 2$ and $\alpha \equiv \alpha' \pmod{p}$ with $0 \le \alpha' \le 2$ and $\alpha' < p$. Then one of the following cases holds.*

   (i) *$\alpha' = 0$, $b_1 + b_2 = p$ and $\gamma = \frac{\alpha}{p} + 1$.*

   (ii) *$\alpha' = 1$, $b_1 + b_2 \equiv a \pmod{p}$ and $\gamma = \left[\frac{\alpha}{p}\right] + 1$.*

   (iii) *$\alpha' = 2$, $b_1 + b_2 \equiv 2a \pmod{p}$, $b_1 \ne b_2$ and $\gamma = \left[\frac{\alpha}{p}\right] + 1$.*

   (iv) *$\alpha' = 2$, $b_1 = b_2 = a$ and $\gamma = \left[\frac{\alpha}{p}\right] + 2$.*

Proof. We have $p\gamma = \alpha + d_1 + \cdots + d_r$ by Lemma 2.8.1, (3). If $r = 0$, we have the assertion (1). Suppose $r = 1$. We have $b_1 + ad_1 \equiv 0 \pmod{p}$ with $0 < d_1 < p$. Hence the assertion (2) holds. Suppose the assumptions of the assertion (3) hold. We have

$$\gamma = \left[\frac{\alpha}{p}\right] + \frac{d_1 + d_2 + \alpha'}{p} \quad \text{with} \ \ 0 < d_1 + d_2 \le 2p - 2.$$

If $0 \le \alpha' \le 1$, then $d_1 + d_2 + \alpha' = p$. Furthermore, $-b_1 - b_2 + \alpha'a \equiv a(d_1 + d_2 + \alpha') \equiv 0 \pmod{p}$. If $\alpha' = 2 < p$, then $-b_1 - b_2 + 2a \equiv 0 \pmod{p}$ and either $d_1 + d_2 + 2 = p$ and hence $d_1 \ne d_2$ (since p is odd) and $b_1 \ne b_2$, or $d_1 + d_2 + 2 = 2p$ and hence $d_1 = d_2 = p - 1$. Since then $b_i + (p-1)a \equiv 0 \pmod{p}$, we find $b_1 = a = b_2$. Thus the assertion (3) is verified. $\qquad \square$

2.9 μ_p-quotients

We considered a μ_p-action on an affine variety $X = \operatorname{Spec} B$ and observed that the μ_p-action is described in terms of a k-derivation δ on B such that $\delta^p = \delta$. Our objective is to prove first a fundamental result of Rudakov-Shafarevich [77, Theorem 2] and apply it to produce interesting examples of affine normal surfaces sandwiched by the affine plane.

2.9.1 A result of Rudakov-Shafarevich

We assume that X is smooth. Since μ_p is an infinitesimal group scheme, i.e., a k-group scheme with one pointed underlying space, every point of X is fixed set-theoretically. However, we say that a point $P \in X$ is *fixed* if $\delta(\mathfrak{m}_P) \subset \mathfrak{m}_P$ and *non-fixed* otherwise, i.e., $\delta(\mathfrak{m}_P) \not\subset \mathfrak{m}_P$, where $\mathfrak{m}_P$ is the maximal ideal of B corresponding to P. The following result due to Rudakov-Shafarevich is important.

Theorem. *Let k be an algebraically closed field of positive characteristic p and let $R = k[[x_1, \ldots, x_n]]$ be a formal power series ring in n variables. let δ be a k-derivation of R such that $\delta^p = \delta$ and $\delta(\mathfrak{m}) \subseteq \mathfrak{m}$, where $\mathfrak{m}$ is the maximal ideal $(x_1, \ldots, x_n)$ of R. Then, after a suitable change of variables, δ is written as*

$$\delta = \sum_{i=1}^{n} \alpha_i x_i \frac{\partial}{\partial x_i}, \quad \alpha_i \in \mathbb{F}_p.$$

Proof. Let L_0 be the p-Lie algebra $\operatorname{Der}_k(R)$ (see 1.2), which is an R-module, and let $L_r = \sum_{i=1}^{r} \mathfrak{m}^r \frac{\partial}{\partial x_i}$, which is an R-submodule of L_0. Then we have a descending chain

$$\operatorname{Der}_k(R) = L_0 \supset L_1 \supset \cdots \supset L_r \supset \cdots$$

satisfying the conditions:

(1) $\cap_{r \geq 0} L_r = (0)$,
(2) $[L_r, L_s] \subseteq L_{r+s-1}$,
(3) $L_r^p \subseteq L_{pr}$.

Write $\delta = \sum_{i-1}^{n} f_i \frac{\partial}{\partial x_i}$ with $f_i \in R$. Since $\delta(\mathfrak{m}) \subseteq \mathfrak{m}$, we have $\delta \in L_1$. Since $\delta^p = \delta$, we have $\delta \notin L_2$. Suppose otherwise that $f_i \in \mathfrak{m}^s$, $s \geq 2$ for every i and some $f_i \notin \mathfrak{m}^{s+1}$. Then $\delta^p(x_j) \in \mathfrak{m}^{ps-p+1}$ for every j. Since $ps - p + 1 > s$, this is a contradiction.

After a change of coordinates $x_1, \ldots, x_n$, we may write

$$f_i = \alpha_i x_i + P_i, \quad P_i \in \mathfrak{m}^2, \quad \alpha_i \in \mathbb{F}_p \tag{2.6}$$

To see this, write the linear part of f_i as

$$\sum_{j=1}^{n} a_{ij} x_j, \quad a_{ij} \in k.$$

Let $A = (a_{ij})$ be the coefficient matrix. Note that, by a linear base change

$$\begin{pmatrix} y_1 \\ \vdots \\ y_n \end{pmatrix} = P \begin{pmatrix} x_1 \\ \vdots \\ x_n \end{pmatrix}, \quad P \in \mathrm{GL}(n, k)$$

the linear part of the y_i in $\delta = \sum_{i=1}^{n} g_i \frac{\partial}{\partial y_i}$ becomes

$$A \begin{pmatrix} x_1 \\ \vdots \\ x_n \end{pmatrix} \mapsto PAP^{-1} \begin{pmatrix} y_1 \\ \vdots \\ y_n \end{pmatrix}.$$

We change A into the Jordan canonical form B. Then $B^p = B$. Considering a Jordan cell B_1 of size r

$$B_1 = \begin{pmatrix} \beta_1 & 1 & \cdots & \cdots & 0 \\ 0 & \beta_2 & 1 & \cdots & 0 \\ \vdots & 0 & \ddots & \ddots & \vdots \\ \vdots & \vdots & \ddots & \beta_{r-1} & 1 \\ 0 & 0 & \cdots & 0 & \beta_r \end{pmatrix},$$

B_1^p has the second diagonal consisting of all 0. This implies that B is a diagonal matrix with all diagonal entries lying in the finite field $\mathbb{F}_p$. Hence, after a linear change of parameters, we may write f_i as in (2.6).

We prove by induction on ℓ that, for every i, we have

$$\delta(x_i) = \sum_{i=1}^{n} \alpha_i x_i \frac{\partial}{\partial x_i} + \sum_{i=1}^{n} P_i \frac{\partial}{\partial x_i}, \quad P_i \in \mathfrak{m}^\ell. \tag{2.7}$$

We look for a change of coordinates $x_i' = x_i + \xi_i$ with $\xi_i \in \mathfrak{m}^2$ such that

$$\delta(x_i + \xi_i) \equiv \alpha_i(x_i + \xi_i) \pmod{\mathfrak{m}^{\ell+1}}. \tag{2.8}$$

For this, set $\delta_0 = \sum_{i=1}^{n} \alpha_i x_i \frac{\partial}{\partial x_i}$ and $\delta_1 = \sum_{i=1}^{n} P_i \frac{\partial}{\partial x_i}$. Then our requirement (2.8) is

$$\alpha_i \xi_i - \delta_0 \xi_i \equiv P_i \pmod{\mathfrak{m}^{\ell+1}} \tag{2.9}$$

because $\delta_1(\xi_i) = \sum_{j=1}^{n} P_j \frac{\partial \xi_i}{\partial x_j} \in \mathfrak{m}^{\ell+1}$ since $\xi_i \in \mathfrak{m}^2$.

Meanwhile, we have by the Jacobson identity [38]

$$\delta^p = (\delta_1 + \delta_0)^p = \delta_1^p + \delta_0^p + \sum_{i=1}^{p-1} s_i(\delta_1, \delta_0) = \delta_1 + \delta_0,$$

where $(p-i)s_i(\delta_1, \delta_0)$ is the coefficient of λ^{p-i-1} in the λ-extension of

$$\mathrm{ad}(\lambda\delta_1 + \delta_0)^{p-1}(\delta_1) = [(\lambda\delta_1 + \delta_0)[\cdots[(\lambda\delta_1 + \delta_0), \delta_1]\cdots]].$$

Since $\delta_0 \in L_1$ and $\delta_1 \in L_\ell$ by the hypothesis, any term in $s_i(\delta_1, \delta_0)$, which contains in some order $(p-i)$-copies of δ_1 and i-copies of δ_0, belongs to $L_{(p-i)\ell-(p-i-1)}$. Further, $\delta_1^p \in L_{p\ell-(p-1)}$. Note that $\delta_0^p = \delta_0$, $(p-i)\ell - (p-i-1) \geq \ell+1$ for $1 \leq i < p-1$ and $p\ell - (p-1) \geq \ell+1$. Hence we have

$$(\mathrm{ad}\,\delta_0)^{p-1}(\delta_1) \equiv \delta_1 \pmod{L_{\ell+1}}.$$

Since we have

$$(\mathrm{ad}\,\delta_0)\left(\sum_{i=1}^{n} Q_i \frac{\partial}{\partial x_i}\right) = \sum_{i=1}^{n}(\delta_0 - \alpha_i \mathrm{id}\,_R)Q_i\frac{\partial}{\partial x_i},$$

the last two equations show that

$$(\delta_0 - \alpha_i\mathrm{id}\,_R)^{p-1}\left(P_i\frac{\partial}{\partial x_i}\right) = (\mathrm{ad}\,\delta_0)^{p-1}\left(P_i\frac{\partial}{\partial x_i}\right) \equiv P_i\frac{\partial}{\partial x_i} \pmod{L_{\ell+1}}.$$

Hence we have

$$(\delta_0 - \alpha_i\mathrm{id}\,_R)^{p-1}P_i \equiv P_i \pmod{\mathfrak{m}^{\ell+1}}.$$

Let $\xi_i = -(\delta_0 - \alpha_i\mathrm{id}\,_R)^{p-2}P_i$. Then ξ_i satisfies the requirement (2.9). $\quad\square$

Remark. An alternative proof of Theorem 2.9.1 was suggested by Yuya Matsumoto. We use the notations in Lemma 1.5.3. Note that γ_i for $0 \leq i \leq p-1$ is a linear combination of $D^0 = \mathrm{id}, D, D^2, \ldots, D^{p-1}$, where $D = \delta_1$ is a k-derivation of R defining the μ_p-action. Since $D(\mathfrak{m}) \subseteq \mathfrak{m}$, it follows that $\gamma_i(\mathfrak{m}) \subseteq \mathfrak{m}$. For any element a of $\mathfrak{m}$, $\gamma_i(a)$ is an eigenvector of a k-linear endomorphism D of $\mathfrak{m}$ with eigenvalue in $\mathbb{F}_p$. In fact, $D = \sum_{j=1}^{p-1} j\gamma_j$ and $\gamma_i\gamma_j = \delta_{ij}\gamma_i$, which implies that $D(\gamma_i(a)) = \sum_{j=1}^{p-1} j\gamma_j\gamma_i(a) = i\gamma_i(a)$. For every $1 \leq i \leq n$, $x_i = \gamma_0(x_i) + \gamma_1(x_i) + \cdots + \gamma_{p-1}(x_i)$. Hence the maximal ideal $\mathfrak{m} = (x_1, \ldots, x_n)$ is generated by eigenvectors $\gamma_i(x_j)$ of D for $0 \leq i \leq p-1$ and $1 \leq j \leq n$. Hence $\mathfrak{m}$ is generated by n elements of these eigenvectors. So, we may assume that $x_1, \ldots, x_n$ are eigenvectors of D. Now write $D = \sum_{i=1}^{n} f_i\frac{\partial}{\partial x_i}$. Then $D(x_i) = f_i = \alpha_i x_i$ with $\alpha_i \in \mathbb{F}_p$.

2.9.2 *Case of algebraic surfaces*

Let $X = \operatorname{Spec} A$ be a smooth affine surface with a μ_p-action. Let δ be the corresponding k-derivation on A such that $\delta^p = \delta$. A closed point P of X is called a *zero* of δ if $\delta(\mathfrak{m}_P) \subseteq \mathfrak{m}_P$, where $\mathfrak{m}_P$ is the maximal ideal of A defining the point P. If P is a zero of δ, the $\mathfrak{m}_P$-adic completion of A is the completion $\widehat{\mathcal{O}}_{X,P}$ of the local ring $\mathcal{O}_{X,P}$, and is isomorphic to a formal power series ring $k[[x, y]]$. By Theorem 2.9.1, we may choose local coordinates x, y in such a way that

$$\delta = x\frac{\partial}{\partial x} + dy\frac{\partial}{\partial y}, \quad d \in \mathbb{F}_p.$$

Let $B = \operatorname{Ker}\delta = A^{\mu_p}$ be the μ_p-invariant subalgebra of A and let $Y = \operatorname{Spec} B$. We call Y the *quotient* of X by μ_p and denote it by $X/\!/\mu_p$ or X/μ_p. The inclusion $B \hookrightarrow A$ induces a k-morphism $q : X \to Y$ called the *quotient morphism*. Since A is a finite B-module, q is a finite morphism.

Suppose that $d \neq 0$. By the arguments in 2.6, Y has analytically a cyclic quotient singularity of type (p, d) at the point $Q = q(P)$. Namely, $\widehat{\mathcal{O}}_{Y,Q}$ is isomorphic to the completion of $k[x, y]^\delta$ by the ideal induced by (x, y) in Lemma 2.6.2. The minimal resolution of singularity has the exceptional locus consisting of a linear chain of smooth rational curves

$$E_1 \relbar\joinrel\relbar E_2 \relbar\joinrel\relbar \cdots \relbar\joinrel\relbar E_s$$

such that the self-intersection numbers $(E_i^2) = -b_i$ are obtained by a continued fraction expansion

$$\frac{p}{d} = [b_1, \ldots, b_s]$$

(see 2.6.1). If $d = 0$, then the invariant subring under δ is equal to the kernel of $\frac{\partial}{\partial x}$, hence Y is smooth at the point Q.

Suppose next that P is not a zero of δ, i.e., $\delta(\mathfrak{m}_P) \not\subseteq \mathfrak{m}_P$. The arguments in 2.3 can be applied to $R = \widehat{\mathcal{O}}_{X,P} \cong k[[x, y]]$, and the quotient is smooth at Q if and only if $I(\delta) = R$, where $I(\delta)$ is the ideal generated by the coefficients f, g of δ when δ is written as $\delta = f\frac{\partial}{\partial x} + g\frac{\partial}{\partial y}$.

We thus obtain the following result.

Theorem. *Let $X = \operatorname{Spec} A$ be a smooth affine surface with a μ_p-action, let δ be the corresponding k-derivation of A and let $q : X \to Y$ be the quotient morphism. Let P be a closed point of X and let $Q = q(P)$. Then the following assertions hold.*

(1) *Q is a smooth point of Y if and only if P is not a zero of δ.*

(2) *If P is a zero of δ, Y has a cyclic quotient singularity of type (p, d) for $d \in \mathbb{F}_p^*$.*

2.9.3 μ_p-actions on the affine plane

Note that μ_p is a subgroup of the multiplicative group scheme G_m as $\mu_p = \text{Ker } F_{G_m}$. In characteristic zero, it is well-known that G_m and its subgroup scheme $\mathbb{Z}/n\mathbb{Z}$ are diagonalizable when they act on the affine plane $\mathbb{A}^2$. Namely, there exists a system of coordinates (x, y) of $\mathbb{A}^2$ such that $^t(x, y) = (t^m x, t^n y)$ for $t \in G_m$ and $m, n \in \mathbb{Z}$ (Case of G_m), or $^g(x, y) = (\zeta x, \zeta^m y)$ for a generator g of $\mathbb{Z}/n\mathbb{Z}$, corresponding to a primitive nth root ζ of unity and $0 \le m < n$ (Case of $\mathbb{Z}/n\mathbb{Z}$). In particular, in the case of $\mathbb{Z}/n\mathbb{Z}$, the quotient surface $\mathbb{A}^2/(\mathbb{Z}/n\mathbb{Z})$ has a unique singular point. We show by an example that this is not the case with μ_p.

Lemma. *Let*
$$\delta = (\alpha x + f(x^p, y^p))\frac{\partial}{\partial x} + (\beta y + g(x^p, y^p))\frac{\partial}{\partial y}$$
be a k-derivation of a polynomial ring $A := k[x, y]$, where $\alpha, \beta \in \mathbb{F}_p^*$ and $f(x^p, y^p), g(x^p, y^p) \in k[x^p, y^p]$. Then the following assertions hold.

(1) *δ satisfies $\delta^p = \delta$.*
(2) *Let C and D be the affine plane curve defined respectively by $\alpha x + f(x^p, y^p) = 0$ and $\beta y + g(x^p, y^p) = 0$. Then C and D intersect transversally at every point $P \in C \cap D$.*
(3) *Let $B = \ker \delta$, $Y = \text{Spec } B$ and $q : X \to Y$ the quotient morphism. Then Y has cyclic quotient singularity at the images of intersection point $C \cap D$.*

Proof. (1) It suffices to show that $\delta^p(x) = \delta(x)$ and $\delta^p(y) = \delta(y)$ because both δ^p and δ are k-derivations. It is then easy to show that $\delta^p(x) = \delta^{p-1}(\alpha x + f(x^p, y^p)) = \alpha^{p-1}(\alpha x + f(x^p, y^p)) = \delta(x)$ because $\alpha^{p-1} = 1$. Similarly, $\delta^p(y) = \delta(y)$ holds.

(2) Let $P = (a, b) \in C \cap D$. Then the tangent directions of C and D at P are respectively $(\alpha, 0)$ and $(0, \beta)$. Hence C and D meet transversally at P.

(3) The result follows from Theorem 2.9.1. $\square$

By taking polynomials $f(x^p, y^p)$ and $g(x^p, y^p)$ more concretely, we can say more about the number of singular points.

Example. Let $\alpha = 1, f(x^p, y^p) = -x^{p^n}$ and $\beta = 1, g(x^p, y^p) = -y^{p^m}$ in the above lemma, where n, m are positive integers. Then $C \cap D$ is exhausted by points $\{(\alpha, \beta) \mid \alpha \in \mathbb{F}_{p^n}, \beta \in \mathbb{F}_{p^m}\}$. Hence the quotient surface Y, which is Frobenius sandwiched by the affine plane, has p^{m+n} singular points, all of which have cyclic singularity of type $(p, p-1)$.

In fact, write

$$x - x^{p^n} = x \prod_{\alpha \in \mathbb{F}_{p^n}^*} (\alpha - x)$$

$$y - y^{p^m} = y \prod_{\beta \in \mathbb{F}_{p^m}^*} (\beta - y).$$

Then, by a change of coordinates $x' = x - \alpha, y' = y - \beta$, δ is written as

$$\delta = \left(x' - x'^{p^n}\right) \frac{\partial}{\partial x'} + \left(y' - y'^{p^m}\right) \frac{\partial}{\partial y'}.$$

Further, by a change of local coordinates $\left(\xi = x' - x'^{p^n}, \eta = y' - y'^{p^m}\right)$ at $P = (\alpha, \beta) \in C \cap D$, δ is written as

$$\delta = \xi \frac{\partial}{\partial \xi} + \eta \frac{\partial}{\partial \eta}.$$

Hence Y has a quotient singularity of type $(p, p-1)$ at the image point of P.

Remark. Not many results are known about μ_p-actions. We make some remarks and pose questions.

(1) A k-derivation δ on $A := k[x, y]$ is considered as a rational[2] vector field on the Hirzebruch surface $F_0 = \mathbb{P}^1 \times \mathbb{P}^1$ which contains $\mathbb{A}^2$ as an open set. Let δ be a k-derivation as in Lemma 2.9.3. Suppose $p > 2$ for simplicity. Then δ as a rational vector field of F_0 is regular everywhere if and only if

$$\frac{\alpha x + f(x^p, y^p)}{x^2}, \quad \frac{\beta y + g(x^p, y^p)}{y^2} \in k\left[\frac{1}{x}, \frac{1}{y}\right].$$

---

[2]Let V be a smooth projective variety and let δ be a k-derivation of the function field $k(V)$. Let $P \in V$ be a closed point of V and let $\{x_1, \ldots, x_n\}$ be a system of local parameters of V at P. Then δ is expressed as a sum

$$\delta = f_1 \frac{\partial}{\partial x_1} + \cdots + f_n \frac{\partial}{\partial x_n},$$

where $f_1, \ldots, f_n \in k(V)$. If every f_i is regular at P, δ is said to be *regular*. In general, δ is said to be *rational*.

In fact, with the affine open covering of F_0 to be given in 3.4, the regularity of δ in the open set U_4 gives the above conditions. With these conditions satisfied, the regularity of δ in the open sets U_2 and U_3 follows.

(2) Let $\delta = x\frac{\partial}{\partial x} + y\frac{\partial}{\partial y}$. Then δ is regular on $V = F_0$. Hence δ defines a μ_p-action on V. Let $W = V/\mu_p$ and let $q : V \to W$ be the quotient surface. Let $Q_i = q(P_i)$, where P_1, P_2, P_3, P_4 are the points of V defined by $(x, y) = (0, 0), (\infty, 0), (0, \infty), (\infty, \infty)$ respectively. Then the points Q_1 and Q_4 have cyclic quotient singularity of type $(p, 1)$, and the points Q_2 and Q_3 have singularity of type $(p, p-1)$. The $\mathbb{P}^1$-fibration $p_1 : F_0 \to \mathbb{P}^1$, i.e., the projection to the first factor $\mathbb{P}^1$, induces a $\mathbb{P}^1$-fibration on the minimal resolution $\widetilde{W}$ of W, which has two singular fibers over $x = 0, \infty$, each of which consists of a (-1)-curve connected to the exceptional $(-p)$-curve coming from the singular point of type $(p, 1)$ and to the terminal component of a linear chain of (-2)-curves of length $(p - 1)$ which is the exceptional locus of the singular point of type $(p, p - 1)$.

(3) If $p = 2$, every k-derivation of $k[x, y]$ satisfying $\delta^2 = \delta$ is written in the form

$$\delta = (x + f(x^2, y^2))\frac{\partial}{\partial x} + (y + g(x^2, y^2))\frac{\partial}{\partial y}$$

provided δ is *irreducible*, i.e., the coefficients $\delta(x)$ and $\delta(y)$ are coprime in $k[x, y]$. In fact, write $\delta = \varphi\frac{\partial}{\partial x} + \psi\frac{\partial}{\partial y}$, where φ, ψ are coprime elements of $k[x, y]$. Then we have

$$\delta(x) = \varphi, \quad \delta^2(x) = \varphi\varphi_x + \psi\varphi_y$$
$$\delta(y) = \psi, \quad \delta^2(y) = \varphi\psi_x + \psi\psi_y.$$

Hence we have $\varphi(1 + \varphi_x) = \psi\varphi_y$. Since $\gcd(\varphi, \psi) = 1$, φ divides φ_y. Since $\deg \varphi_y < \deg \varphi$, it follows that $\varphi_y = 0$. Hence $1 + \varphi_x = 0$. This implies that $\varphi = x + f(x^2, y^2)$. Similarly, $\psi = y + g(x^2, y^2)$.

(4) If $p = 3$, there is a k-derivation of multiplicative type of $k[x, y]$ which is not of the above form. Let

$$\delta = (x + x^2)\frac{\partial}{\partial x} + (y + y^2)\frac{\partial}{\partial y}.$$

Then $\delta^3 = \delta$. It is an open problem to write down all k-derivations of multiplicative type of $k[x, y]$ for the characteristic p in general.

Chapter 3

Quasi-elliptic or quasi-hyperelliptic fibrations

3.0 Introduction

Throughout this chapter the ground field k is assumed to be an algebraically closed field of characteristic $p > 0$. A smooth projective surface V defined over k is called an *elliptic surface* (resp. *quasi-elliptic surface*) if there exists a surjective morphism $f : V \to C$ to a smooth projective curve C such that general fibers of f are elliptic curves (resp. irreducible singular curves of arithmetic genus 1). Such a morphism f is called an *elliptic fibration* (resp. *quasi-elliptic fibration*). The generic fiber $V_{\mathfrak{K}}$ of f defined over the function field $\mathfrak{K} = k(C)$ is a smooth (resp. singular, normal) projective curve of arithmetic genus 1 defined over $\mathfrak{K}$ if f is an elliptic (resp. quasi-elliptic) fibration. By a fundamental result of Tate [92], with a scheme-theoretic proof available in Schröer [81], the genus drop of the curve $V_{\mathfrak{K}}$ caused by a purely inseparable extension of $\mathfrak{K}$ is a multiple of $(p-1)/2$. Hence a quasi-elliptic fibration on V is possible only if $p = 2$ or 3.

In section 3.1, we explain general results on elliptic and quasi-elliptic fibrations which include Weierstrass models of quasi-elliptic fibrations. More general results can be found in Schütt-Shioda [87]. From section 3.2 on, we restrict ourselves to the *unirational* case with quasi-elliptic fibration. Namely, the surface V is dominated by a rational surface, and hence the curve C is k-isomorphic to $\mathbb{P}^1$.

In the latter part of the chapter, we consider the case where $p > 3$ and the generic fiber is a hyperelliptic form of the affine line $\mathbb{A}^1_{\mathfrak{K}}$ with arithmetic genus $(p-1)/2$. Then we call $f : V \to C$ a *quasi-hyperelliptic fibration*.

In general, a surjective morphism from a smooth projective surface V to a smooth projective curve is called a *fibration* if general fibers are integral, i.e., irreducible and reduced. A fiber $V_P := f^*(P)$ lying over a point $P \in C$

is a *reducible fiber*[1] if it has two or more irreducible components. Write $V_P = \sum_i n_i C_i$ as a divisor with irreducible components C_i and positive integers n_i. The greatest common divisor m of the n_i is called the *multiplicity* of the fiber V_P. If $m > 1$, $f^*(P)$ is called a *multiple fiber*. Then $\frac{1}{m}V_P := \sum_i \frac{n_i}{m}C_i$ is called the *reduced form* of the multiple fiber V_P.

A fibration $f : V \to C$ is *relatively minimal* if there are no (-1)-fiber components, i.e., a (-1)-curve which is an irreducible component of a fiber of f. Given a fibration $f : V \to C$, we can always find a relatively minimal fibration $f_0 : V_0 \to C$ such that $f = f_0 \cdot \sigma$, where $\sigma : V \to V_0$ is the contraction of all (-1)-curves appearing in the fibers of f. We call such a (-1)-curve simply a (-1)-*fiber component of f*.

3.1 General results

In this section, we explain general results on elliptic and quasi-elliptic surfaces to give some perspectives of the theory on unirational quasi-elliptic surfaces, which is a main subject of this chapter.

3.1.1

We recall the canonical divisor formula in Bombieri-Mumford [9] and Bombieri-Husemoller [8]. We mostly refer to [57, Chap. 1, sect. 3].

Theorem. *Let $f : V \to C$ be a relatively minimal, elliptic or quasi-elliptic fibration. Then the canonical divisor of V is written as*

$$K_V \sim f^*(K_C - L) + \sum_\lambda a_\lambda Z_\lambda \,,$$

where:

(1) K_C *is the canonical divisor of C and L is a divisor on C defined by $\mathcal{O}_C(L) \cong \mathcal{L}$ with the free part $\mathcal{L}$ of $R^1 f_* \mathcal{O}_V$, i.e., $R^1 f_* \mathcal{O}_V = \mathcal{L} \oplus \mathcal{T}$ with the torsion $\mathcal{O}_C$-Module $\mathcal{T}$.*

(2) $\{m_\lambda Z_\lambda\}$ *ranges over all multiple fibers of f with multiplicity m_λ and the reduced form Z_λ.*

(3) $0 \le a_\lambda < m_\lambda$.

---

[1]In the case of characteristic zero, a fiber which is not isomorphic to a general fiber is called a *singular fiber*. But, for a quasi-elliptic fibration, a general fiber itself is a singular curve. Hence if V_P is not isomorphic to a general fiber, it has either not less than two irreducible components or a single irreducible component with multiplicity $m \ge 2$. Hence we use the calling *reducible fiber* or *non-reduced fiber* to indicate a fiber which is not isomorphic to a general fiber.

(4) $a_\lambda < m_\lambda - 1$ if $m_\lambda Z_\lambda$ is a wild fiber[2], i.e., $V_P := m_\lambda Z_\lambda = f^*(P)$ with $h^1(V_P, \mathcal{O}_{V_P}) \geq 2$, which is equivalent to $P \in \operatorname{Supp}\mathcal{T}$.

(5) $\deg(K_C - L) = 2p_a(C) - 2 + \chi(\mathcal{O}_V) + \operatorname{length}\mathcal{T}$, where $p_a(C)$ is the genus of C.

3.1.2

In the rest of this chapter, we assume that the fibration $f : V \to C$ has a rational cross-section, i.e., a rational mapping $s : C \to V$ such that $f \cdot s = \operatorname{id}_C$. Since s is then a morphism, the image of s is a regular cross-section of f. Let D be the image of s. Then D is a smooth irreducible curve such that $(D \cdot V_P) = 1$ for every fiber V_P of f. Hence D is isomorphic to C by the restriction $f|_D$. The existence of D makes the canonical divisor formula more feasible.

Corollary. *Let $f : V \to C$ be an elliptic or quasi-elliptic fibration with a cross-section D. Then we have:*

(1) $\mathcal{T} = 0$ and $\mathcal{L} = R^1 f_* \mathcal{O}_V$ is an invertible sheaf $\mathcal{O}_C(L)$ on C.

(2) $K_V \sim f^*(K_C - L)$.

(3) $\chi(\mathcal{O}_V) = -\deg(L) = -(D^2)$.

(4) $f_* \mathcal{O}_V \cong f_* \mathcal{O}_V(D) \cong \mathcal{O}_C$.

(5) $f_* \mathcal{O}_V(nD)$ is a locally free $\mathcal{O}_C$-Module of rank n if $n > 0$.

(6) $R^1 f_* \mathcal{O}_V(nD) = 0$ for $n > 0$.

(7) $R^i f_* \mathcal{O}_V(nD) = 0$ for $i > 1$ and every $n \in \mathbb{Z}$.

(8) $\mathcal{L} \cong \mathcal{O}_V(D) \otimes \mathcal{O}_D$.

Proof. (1) Since the cross-section D exists, there are no multiple fibers. In particular, $\mathcal{T} = 0$ by [57, Lemma 3.6.2] and hence $R^1 f_* \mathcal{O}_V \cong \mathcal{L} = \mathcal{O}_C(L)$.

(2) The formula follows from the previous theorem.

(3) Let $g = p_a(C)$. Then we have

$$2g - 2 = (D^2) + (D \cdot K_V),$$

where $(D \cdot K_V) = \deg(K_C - L) = 2g - 2 - \deg L$ by the assertion (2) above. Hence $(D^2) = \deg L$. On the other hand, the assertion (5) of the previous theorem gives

$$2g - 2 - \deg L = 2g - 2 + \chi(\mathcal{O}_V).$$

---

[2]V_P is a *tame* fiber if $h^1(V_P, \mathcal{O}_{V_P}) = 1$.

Hence $\chi(\mathcal{O}_V) = -(D^2)$.

(8) Consider an exact sequence

$$0 \longrightarrow \mathcal{O}_V \longrightarrow \mathcal{O}_V(D) \longrightarrow \mathcal{O}_V(D) \otimes \mathcal{O}_D \longrightarrow 0 \,,$$

whence we obtain $\mathcal{O}_V(D) \otimes \mathcal{O}_D \cong R^1 f_* \mathcal{O}_V$ because $R^1 f_* \mathcal{O}_V(D) = 0$.

For the rest of the assertions, we refer the readers to a detailed account in Mumford-Suominen [64]. □

3.1.3

Lemma. *Let $f : V \to C$ be either a quasi-elliptic fibration or a quasi-hyperelliptic fibration of genus $g = (p-1)/2$ $(p > 2)$. Assume that f has a rational cross-section D. Let $V_{\mathfrak{K}}$ be the generic fiber of f, where $\mathfrak{K}$ is the function field of C over k. Then we have:*

(1) *$V_{\mathfrak{K}}$ has a unique singular point P_∞ which is a one-place point. Furthermore, $V_{\mathfrak{K}} - \{P_\infty\}$ is a $\mathfrak{K}$-form of the affine line. Hence almost all fibers of f are irreducible rational curves with unique one-place singular point, i.e., a cuspidal singular point.*

(2) *If V is a quasi-elliptic surface, we have necessarily $p = 2$ or 3. The generic fiber $V_{\mathfrak{K}}$ is birationally equivalent to one of the following affine plane curves:*

(i) *If $p = 3$, $y^2 = x^3 + \gamma$ with $\gamma \in \mathfrak{K} \setminus \mathfrak{K}^3$.*
(ii) *If $p = 2$, $y^2 = x^3 + \beta x + \gamma$ with $\beta, \gamma \in \mathfrak{K}$ such that $\beta \notin \mathfrak{K}^2$ or $\gamma \notin \mathfrak{K}^2$.*

(3) *If f is a quasi-hyperelliptic fibration, $V_{\mathfrak{K}}$ is birationally equivalent to an affine plane curve:*

$$y^2 = x^p + \gamma \quad \text{with} \quad \gamma \in \mathfrak{K} \setminus \mathfrak{K}^p.$$

Proof. (1) Note that $V_{\mathfrak{K}}$ is an irreducible normal projective curve with arithmetic genus $p_a(V_{\mathfrak{K}}) = 1$ (the quasi-elliptic case) and $p_a(V_{\mathfrak{K}}) = g = (p-1)/2$ (the quasi-hyperelliptic case). By Lemma I-2.4.2, there exists a finite separable extension $\mathfrak{K}'$ of $\mathfrak{K}$ such that all singular points of the irreducible projective curve $V_{\mathfrak{K}} \otimes_{\mathfrak{K}} \mathfrak{K}'$ are only one-place points. Let C' be the normalization of C in $\mathfrak{K}'$ and let V' be a desingularization of the irreducible projective surface $V \times_C C'$. Then there exists a surjective morphism $f' : V' \to C'$ such that the generic fiber of f' is a $\mathfrak{K}'$-curve $V_{\mathfrak{K}} \otimes_{\mathfrak{K}} \mathfrak{K}'$. We shall show, under the above hypothesis, that $V_{\mathfrak{K}} \otimes \mathfrak{K}'$ has only one singular point. For this purpose we may assume that $\mathfrak{K}' = \mathfrak{K}$ and hence all singular points of $C_{\mathfrak{K}}$ are one-place points. This claim holds true apparently if $p_a(V_{\mathfrak{K}}) = 1$.

So, suppose that $\Gamma := V_{\mathfrak{K}}$ is a quasi-hyperelliptic curve of genus $g = (p-1)/2$. Let P_∞ be a singular point of Γ and let $\widetilde{C}$ be the closure of P_∞ in V. Then the restriction $f|_{\widetilde{C}}$ is a one-to-one morphism from $\widetilde{C}$ onto C, hence $f|_{\widetilde{C}}$ has degree p^α. Furthermore, $p > \mu$, where μ is the multiplicity (of singularity) of Γ at the point P_∞ because $\mu(\mu-1)/2 \le (p-1)/2 = g$. Find positive integers $q_0, q_1, \ldots, q_s$ and $r_1, \ldots, r_s$ by the Euclidean algorithm,

$$p^\alpha = q_0 \mu + r_1, \quad 0 < r_1 < \mu$$

$$\mu = q_1 r_1 + r_2, \quad 0 < r_2 < r_1$$

$$\cdots\cdots \qquad \cdots\cdots$$

$$r_{s-2} = q_{s-1} r_{s-1} + r_s, \quad 0 < r_s < r_{s-1}$$

$$r_{s-1} = q_s r_s + 1, \quad 1 < r_s,$$

where we note that $\gcd(p^\alpha, \mu) = 1$. Since P_∞ is a one-place point of Γ, its multiplicity sequence is,

$$\Big(\underbrace{\mu, \ldots, \mu}_{q_0}, \ \underbrace{r_1, \ldots, r_1}_{q_1}, \ \ldots, \ \underbrace{r_s, \ldots, r_s}_{q_s} \Big) .$$

Hence the genus drop δ_{P_∞} contributed by the desingularization of the point P_∞ on $\Gamma \otimes_{\mathfrak{K}} \overline{\mathfrak{K}}$ is :

$$\frac{q_0}{2}\mu(\mu-1) + \frac{q_1}{2}r_1(r_1-1) + \cdots + \frac{q_s}{2}r_s(r_s-1) = \frac{1}{2}(p^\alpha-1)(\mu-1)$$

$$\le \frac{1}{2}(p-1),$$

where $\overline{\mathfrak{K}}$ is an algebraic closure of $\mathfrak{K}$. The equality in the above formula is easily derived from the Euclidean algorithm. Therefore, we know that $\alpha = 1, \mu = 2$ and $\delta_{P_\infty} = g$. This implies that Γ has only one singular point P_∞ and $\Gamma \otimes \overline{\mathfrak{K}}$ is rational over $\overline{\mathfrak{K}}$. Similar results hold in the quasi-elliptic case if $p = 3$. In general, the genus drop of a normal projective curve under purely inseparable extension is a multiple of $(p-1)/2$ by Tate [92]. Hence $p = 2$ case is possible. The argument to show that $\Gamma \otimes \overline{\mathfrak{K}}$ is rational is the same as above.

The assertions (2) and (3) follow from I-2.8.2 and Theorem I-4.3 since $\Gamma = V_{\mathfrak{K}}$ has a $\mathfrak{K}$-rational point. This is one of the places we use the hypothesis that $f : V \to C$ has a rational cross-section.

Remark. With the same notations as in the above proof, $\mu = 2$ and

$$\deg(f|_{\widetilde{C}}) = p = (\Gamma \cdot \widetilde{C}) = i(\Gamma, \widetilde{C}; P_\infty),$$

where $i(\Gamma, \widetilde{C}; P_\infty)$ is the local intersection multiplicity of Γ and $\widetilde{C}$ at P_∞.

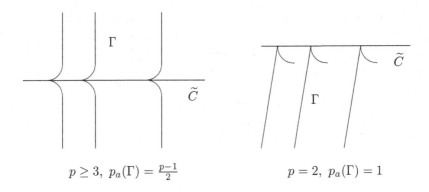

$$p \geq 3, \ p_a(\Gamma) = \tfrac{p-1}{2} \qquad\qquad p = 2, \ p_a(\Gamma) = 1$$

3.1.4

Let $f : V \to C$ be an elliptic or quasi-elliptic fibration with a rational cross-section D. Let $F = \sum_i n_i F_i$ be a reducible fiber of f with two or more irreducible components. For every i, we have

$$(E_i \cdot F) = n_i(E_i^2) + \sum_{j \neq i} n_j(E_i \cdot E_j) = 0 \ .$$

Since F is connected, we have $(E_i^2) < 0$. Since $K_V \sim f^*(K_C - \Delta)$, it also follows that $(E_i \cdot K_V) = 0$. Hence $E_i \cong \mathbb{P}^1$ and $(E_i^2) = -2$.

By a divisor $A \sim nD + f^*(B)$ with $n \gg 0$ and a divisor B on C of sufficiently high degree, the induced rational mapping $\Phi_{|A|}$ is a birational morphism such that every E_i of F is contracted to a point if E_i is different from the irreducible component E_0 of F meeting the cross-section D. Thus we have a singular normal projective surface $\overline{V}$, a birational morphism $\sigma : V \to \overline{V}$ and the induced fibration $\overline{f} : \overline{V} \to C$ such that $f = \overline{f} \cdot \sigma$. As explained above, the restriction of σ on any reducible fiber F is the contraction of $\sum_{i \neq 0} E_i$ to a normal singular point on $\overline{V}$.

The spectral sequence $E_2^{p,q} = R^p \overline{f}_*(R^q \sigma_* \mathcal{O}_V) \Rightarrow R^{p+q} f_* \mathcal{O}_V$ induces an exact sequence [3]

$$0 \longrightarrow R^1 \overline{f}_* \mathcal{O}_{\overline{V}} \longrightarrow R^1 f_* \mathcal{O}_V \longrightarrow \overline{f}_*(R^1 \sigma_* \mathcal{O}_V) \longrightarrow 0 \ ,$$

---

[3] For a spectral sequence $E_2^{p,q} \Rightarrow E^n$ which is biregular [FGA, 0. 11.1.3] and satisfies the condition $E_2^{p,q} = 0$ if $p < 0$ or $q < 0$, there is a long exact sequence of terms of low

where $\sigma_* \mathcal{O}_V = \mathcal{O}_{\overline{V}}$ and $R^1 \overline{f}_* \mathcal{O}_{\overline{V}} \cong R^1 f_* \mathcal{O}_V \cong \mathcal{L}$. In fact, we can apply the same argument as in Corollary 3.1.2 to a fibration $\overline{f} : \overline{V} \to C$ with a cross-section D to show that $R^1 \overline{f}_* \mathcal{O}_{\overline{V}} \cong \mathcal{O}_{\overline{V}}(D) \otimes \mathcal{O}_D$, which is isomorphic to $\mathcal{O}_V(D) \otimes \mathcal{O}_D \cong \mathcal{L}$ because V and $\overline{V}$ are isomorphic to each other in an open neighborhood of D. By the above exact sequence, we have $\overline{f}_*(R^1 \sigma_* \mathcal{O}_V) = 0$, which entails $R^1 \sigma_* \mathcal{O}_V = 0$ because $R^1 \sigma_* \mathcal{O}_V$ is a skyscraper sheaf supported by $\mathrm{Sing}\,(\overline{V})$. This implies that the normal singular points on $\overline{V}$ are all rational singular points [3].

Replacing V by $\overline{V}$, we assume that all closed fibers of $f : V \to C$ are integral though V is only normal. We consider a *Weierstrass model* of f after Lang [48]. Let $\mathcal{U} = \{U_i\}_{i \in I}$ be an affine open covering of C such that $\mathcal{L}|_{U_i} = \mathcal{O}_{U_i} t_i$ with basis t_i. Then $\mathcal{L}^n|_{U_i} = \mathcal{O}_{U_i} t_i^n$. Let $\mathcal{V} = f_* \mathcal{O}_V(2D)$. Then $\mathcal{V}$ is a locally free $\mathcal{O}_C$-Module of rank 2 obtained as an extension [4]

$$0 \longrightarrow \mathcal{O}_C \longrightarrow \mathcal{V} \overset{\rho_2}{\longrightarrow} \mathcal{L}^2 \longrightarrow 0 \;,$$

whence $\mathcal{V}|_{U_i} = \mathcal{O}_{U_i} \cdot 1 + \mathcal{O}_{U_i} \cdot x_i$ with $\rho_2(x_i) = t_i^2$. Similarly, $f_* \mathcal{O}_V(3D)$ is a locally free $\mathcal{O}_V$-Module of rank 3 satisfying an exact sequence

$$0 \longrightarrow \mathcal{V} \longrightarrow f_* \mathcal{O}_V(3D) \overset{\rho_3}{\longrightarrow} \mathcal{L}^3 \longrightarrow 0 \;.$$

Hence $f_* \mathcal{O}_V(3D)|_{U_i} = \mathcal{O}_{U_i} \cdot 1 + \mathcal{O}_{U_i} \cdot x_i + \mathcal{O}_{U_i} \cdot y_i$ with $\rho_3(y_i) = t_i^3$. By a similar argument, we have

$$f_* \mathcal{O}_V(4D)|_{U_i} = \mathcal{O}_{U_i} \cdot 1 + \mathcal{O}_{U_i} \cdot x_i + \mathcal{O}_{U_i} \cdot y_i + \mathcal{O}_{U_i} \cdot x_i^2$$
$$f_* \mathcal{O}_V(5D)|_{U_i} = \mathcal{O}_{U_i} \cdot 1 + \mathcal{O}_{U_i} \cdot x_i + \mathcal{O}_{U_i} \cdot y_i + \mathcal{O}_{U_i} \cdot x_i^2 + \mathcal{O}_{U_i} x_i y_i \;.$$

Consider the sections $1, x_i, y_i, x_i^2, x_i y_i, x_i^3$ of $f_* \mathcal{O}_V(6D)|_{U_i}$. By projecting these elements to $\mathcal{L}^6$, we know that these sections are a basis of $f_* \mathcal{O}_V(6D)|_{U_i}$. On the other hand, y_i^2 is also a section of $f_* \mathcal{O}_V(6D)|_{U_i}$. Hence we have a relation

$$y_i^2 + a_1 x_i y_i + a_3 y_i = a_0 x_i^3 + a_2 x_i^2 + a_4 x_i + a_6 \;,$$

where $a_i \in \Gamma(U_i, \mathcal{O}_C)$. By looking at the images of projection onto $\mathcal{L}^6$ of both sides, we know that $a_0 = 1$. Further, by Lemma 3.1.3, we know that in the case of quasi-elliptic fibrations,

degree

$$0 \to E_2^{1,0} \to E^1 \to E_2^{0,1} \to E_2^{2,0} \to E^2 \;.$$

---

[4] Consider an exact sequence

$$0 \longrightarrow \mathcal{O}_V(D) \longrightarrow \mathcal{O}_V(2D) \longrightarrow \mathcal{O}_V(2D) \otimes \mathcal{O}_D \longrightarrow 0.$$

Since $f_* \mathcal{O}_V(D) = \mathcal{O}_C$ and $R^1 f_* \mathcal{O}_V(D) = 0$, we obtain an exact sequence.

(i) if $p = 3$, $a_1 = a_3 = a_2 = a_4 = 0$;

(ii) if $p = 2$, $a_1 = a_3 = a_2 = 0$.

On the open set U_j, we have a basis $\{1, x_j, y_j, x_j^2, x_j y_j, x_j^3\}$ and a relation

$$y_j^2 + a_1' x_j y_j + a_3' y_j = x_j^3 + a_2' x_j^2 + a_4' x_j + a_6'$$

with $a_j' \in \Gamma(U_j, \mathcal{O}_C)$. Over the intersection $U_i \cap U_j$, we have the following relations:

$$t_i = u_{ij} t_j, \quad x_i = u_{ij}^2 x_j + r_{ij}, \quad y_i = u_{ij}^3 y_j + s_{ij} u_{ij}^2 x_j + t_{ij} \ ,$$

where $u_{ij} \in \Gamma(U_i \cap U_j, \mathcal{O}_C^*)$ and $r_{ij}, s_{ij}, t_{ij} \in \Gamma(U_i \cap U_j, \mathcal{O}_C)$. Then, a part of the *formulas* of Tate [93, p.181] give the following relations:

$$u a_1' = a_1 + 2s$$
$$u^2 a_2' = a_2 - sa_1 + 3r - s^2$$
$$u^3 a_3' = a_3 + ra_1 + 2t$$
$$u^4 a_4' = a_4 - sa_3 + 2ra_2 - (t + rs)a_1 + 3r^2 - 2st$$
$$u^6 a_6' = a_6 + ra_4 + r^2 a_2 + r^3 - ta_3 - rta_1 - t^2 \ ,$$

where $u = u_{ij}$, $r = r_{ij}$, $s = s_{ij}$ and $t = t_{ij}$. Hence, if we *assume* that $a_1 = a_3 = a_2 = 0$, which is the case in the case of quasi-elliptic fibrations, we have $r = s = t = 0$, $a_4 = u^4 a_4'$, $a_6 = u^6 a_6'$, and $x_i = u^2 x_j$, $y_i = u^3 y_j$. Hence, under the assumption that $a_1 = a_2 = a_3 = 0$, we have

$$V \cong \mathcal{O}_C \oplus \mathcal{L}^2 \quad \text{and} \quad f_* \mathcal{O}_V(3D) \cong \mathcal{O}_V \oplus \mathcal{L}^2 \oplus \mathcal{L}^3 \ .$$

If we write $a_4 = a_{4i}$, $a_4' = a_{4j}$, $a_6 = a_{6i}$ and $a_6' = a_{6j}$, we have $a_{4i} = u_{ij}^4 a_{4j}$ and $a_{6i} = u_{ij}^6 a_{6j}$. Hence $\{a_{4i}\}_{i \in I}$ and $\{a_{6i}\}_{i \in I}$ give elements of $\Gamma(C, \mathcal{L}^{-4})$ and $\Gamma(C, \mathcal{L}^{-6})$. Note that a section $a = \{a_i\}_{i \in I}$ of $\mathcal{L}$ is given as satisfying $a_i t_i = a_j t_j$, whence $a_i = u_{ij}^{-1} a_j$. We know that the set of coefficients $\{(a_{6i}, a_{4i})\}_{i \in I}$ gives rise to a section of $V \otimes \mathcal{L}^{-6}$ over C.

Theorem. *Let $f : V \to C$ be a quasi-elliptic fibration with a rational cross-section D. Then we have:*

(1) *There exist a normal projective surface with a surjective morphism $\overline{f} : \overline{V} \to C$ such that*

    (i) *f splits as $V \xrightarrow{\sigma} \overline{V} \xrightarrow{\overline{f}} C$, where σ is the contraction of all irreducible components of reducible fibers of f except for the component meeting D, and*

    (ii) *all closed fibers of $\overline{f}$ are integral.*

(2) Let $\mathcal{L} = \mathcal{O}_V(D) \otimes \mathcal{O}_D$, which is also invertible sheaf on $\overline{V}$. There exists a section of $\Gamma(C, f_*\mathcal{O}_V(2D) \otimes \mathcal{L}^{-6})$ which is represented by $\{(a_{6i}, a_{4i})\}_{i \in I}$ with respect to an affine open covering $\mathcal{U} = \{U_i\}_{i \in I}$ of C such that, for every i, $f^{-1}(U_i)$ is isomorphic to a Weierstrass model

$$y^2 = x_i^3 + a_{4i}x_i + a_{6i}$$

in the projective plane over U_i, $\mathbb{P}^2 \times U_i$, with coordinates $(1, x_i, y_i)$.
(3) If $p = 3$, we may assume that $a_{4i} = 0$ for every i.

3.1.5 Kodaira classification of singular fibers

We recall the reducible (singular) fibers of a relatively minimal elliptic or quasi-elliptic fibration. The following is the list of all possible reduced forms: In the configurations, a line stands for a (-2)-curve, i.e., a smooth rational curve with self-intersection number -2, and the attached integers stand for the multiplicities of the components in a reducible fiber.

Type I_0 : a nonsingular elliptic curve
Type I_1 : an irreducible curve of arithmetic genus 1 with one node
Type I_2 : $C_1 + C_2$ with $C_1 \cap C_2 = \{2 \text{ points}\}$
Type I_n $(n > 2)$:

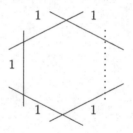

Type I_n^* $(n \geq 0)$:

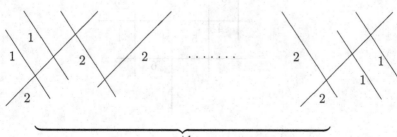

Type II : an irreducible curve of arithmetic genus 1 with one cusp
Type III : $C_1 + C_2$ with $C_1 \cap C_2 = \{1 \text{ point}\}$ and $(C_1 \cdot C_2) = 2$
Type IV :   **Type I_0^*** :

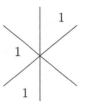

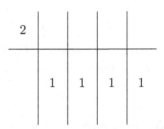

Type IV* :   **Type III*** :

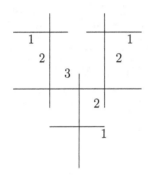

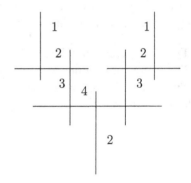

Type II* :

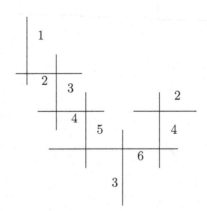

3.1.5.1 *Enumeration of singular fibers in case $p = 3$*

Our naming of fiber types is due to Kodaira [46]. The following result is essentially due to Lang [48, Theorem 1.2].

Theorem. *Assume that $p = 3$. Let $f : V \to C$ be a relatively minimal, quasi-elliptic fibration with a rational cross-section D. Suppose that its Weierstrass model is locally given by an equation $y^2 = x^3 + a_6$, where $a_6 \in \Gamma(C, \mathcal{L}^{-6})$. For a point $P \in C$, let t be a local parameter of C at P and write a_6 near P as $a_6 = c_r t^r + c_{r+1} t^{r+1} + \cdots$ with $c_r \in k^*$ and $r = 1, 2, 4, 5$. Then we have:*

(1) *If the fiber $f^*(P)$ is a reducible fiber, the type of reducible fiber is given as follows:*

r	1	2	4	5
type	II	IV	IV*	II*
$v_P(da_6)$	0	1	3	4

where $v_P(da_6)$ is the order of zero of the differential da_6 at P.

(2) *Let g be the genus of C. Then the Euler Poincaré characteristic $\chi(\mathcal{O}_V)$ is given as follows, where c_j is the number of reducible fibers of f with type B_j and $\nu(T)$ is alternatively the number of singular fibers of Kodaira type T.[5]*

$$6\chi(\mathcal{O}_V) = (2 - 2g) + c_4 + 3c_8 + 4c_{10}$$
$$= (2 - 2g) + \nu(IV) + 3\nu(IV^*) + 4\nu(II^*).$$

Proof. The reduction of a_6 into the given form as well as the classification of possible reducible fibers of f is the subject of the subsequent arguments in this chapter. With the notations of 3.1.4, the differential da_6 is given by a local data $\{da_{6i}\}_{i \in I}$ with respect to an affine open covering $\mathcal{U} = \{U_i\}_{i \in I}$ of C, where $\{da_{6i}\} \in \Gamma(C, K_C \otimes \mathcal{L}^{-6})$ because $da_{6i} = u_{ij}^6 da_{6j}$. Since $\deg \mathcal{L} = -\chi(\mathcal{O}_V)$, the number of zeros of da_6 is $2g - 2 + 6\chi(\mathcal{O}_V)$, which is equal to $c_4 + 3c_8 + 4c_{10}$. The stated formula follows from this observation. $\square$

---

[5]Here is a confusion of the naming of singular fibers by Shafarevich's (see [85]) and Kodaira's. The following is a comparison table, where the top row is Shafarevich's and the bottom row is Kodaira's.

E	A_1'	A_2'	$A_n'(n > 2)$	$A_n''(n \geq 0)$	B_2	B_3	B_4	B_6	B_8	B_9	B_{10}
$_mI_0$	$_mI_1$	$_mI_2$	$_mI_b$	$I_b^*(b>0)$	II	III	IV	I_0^*	IV*	III*	II*

Remark. In Theorems 3.3.1 and 3.3.2, we consider the case where a_6 is a polynomial $\varphi(t)$. The local analysis of the fiber at a point P given by $t = \alpha$ and determination of its type is done by a Taylor expansion of $\varphi(t)$ in terms of $t - \alpha$, which is the same as the expansion of a_6 as a formal power series in $t - \alpha$. But we will see that the degree $\deg \varphi(t)$ affects $\chi(\mathcal{O}_V)$.

3.1.5.2 *Enumeration of singular fibers in case $p = 2$*

We will consider the Mordell-Weil group for a unirational quasi-elliptic fibration (i.e., the base curve is $\mathbb{P}^1$) which is an analogue of the same group defined for an elliptic surface (see [87]). In order to develop the theory in full generality also in the case $p = 2$, we need the data on reducible singular fibers given in a subsequent theorem in this subsection.

Assume that $p = 2$. Let $f : V \to C$ be a relatively minimal quasi-elliptic fibration with a rational cross-section, where we assume that $C \cong \mathbb{P}^1$. By Theorem 3.1.4, V is then birational to a hypersurface in the affine space $\mathbb{A}^3$ defined by

$$y^2 = x^3 + \psi(t)x + \varphi(t) \tag{3.1}$$

with $\psi(t), \varphi(t) \in k[t]$ such that $\psi(t) \notin k[t]^2$ or $\varphi(t) \notin k[t]^2$. Let

$$\Delta(t) = \psi(t)\psi'(t)^2 + \varphi'(t)^2 \in k[t],$$

where $\psi'(t)$ and $\varphi'(t)$ are the derivatives of $\psi(t)$ and $\varphi(t)$, respectively. We call $\Delta(t)$ the *discriminant* of the Weierstrass equation of $f : V \to C$.

Set

$$m = \max\left\{ \left[\frac{\deg \psi(t)}{4}\right], \left[\frac{\deg \varphi(t)}{6}\right] \right\}.$$

By a birational change of variables $\tau = 1/t, \xi = x/t^{2m+2}$ and $\eta = y/t^{3m+3}$, V is birational to an affine hypersurface

$$\eta^2 = \xi^3 + \Psi(\tau)\xi + \Phi(\tau), \tag{3.2}$$

where $\Psi(\tau) = \tau^{4m+4}\psi(1/\tau)$ and $\Phi(\tau) = \tau^{6m+6}\varphi(1/\tau)$. By a straightforward computation, we have

$$\Delta_\infty(\tau) = \tau^{12m+8}\Delta\left(\frac{1}{\tau}\right).$$

By a suitable choice of coordinates (x, y, t), we may assume that the following conditions are satisfied.

(1) $\psi(t)$ has no monomial terms of degree $\equiv 0 \pmod 4$.

(2) $\varphi(t)$ has no monomial terms of even degree.

(3) $\min\{v_\alpha(\psi(t) - \psi(\alpha)) - 4, v_\alpha(\varphi(t) - \varphi(\alpha)) - 6\} < 0$ for all roots α of $\Delta(t) = 0$, where v_α is a normalized valuation of $k(t)$ at $t = \alpha$.

In order to write the Weierstrass equation more precisely, we take a root α of $\Delta(t) = 0$ and assume that $\alpha = 0$. Further, taking the fiber product of $f : V \to C$ by a morphism $\mathrm{Spec}\, k[[t]] \to C$, we consider a Weierstrass equation of a quasi-elliptic fibration $f : V \to C$ in the local case where $C = \mathrm{Spec}\, k[[t]]$.

Lemma. *Let $f : V \to C = \mathrm{Spec}\, k[[t]]$ be a relatively minimal quasi-elliptic fibration with a cross-section defined by*

$$y^2 = x^3 + \psi(t)x + \varphi(t)$$

with $\psi(t), \varphi(t) \in k[[t]]$ and either $\psi(t) \notin k[[t]]^2$ or $\varphi(t) \notin k[[t]]^2$. Then the equation of X can be put into the following local form

$$y^2 = x^3 + (\alpha^4 t^{2s} + \beta^2 t^r)x + \gamma^2 t^k$$

such that

$$\beta \neq 0 \ or \ \gamma \neq 0, \tag{3.3}$$

$$s = 1, \ or \ 1 \leq r \leq 3, \ or \ 1 \leq k \leq 5, \tag{3.4}$$

where each of α, β, γ is either a unit in $k[[t]]$ or a zero, and each of s, r, k is a positive odd integer (resp. zero) if the corresponding coefficient $\alpha^4, \beta^2, \gamma^2$ is not equal to zero (resp. is equal to zero).

Proof. By suitable birational change of coordinates, we may assume that $\psi(t)$ and $\varphi(t)$ satisfy the above conditions (1), (2) and (3). Note that the same conditions hold also for $k[[t]]$ instead of $k[t]$. Hence we can write

$$\psi(t) = \sum_{i \geq 0} \alpha_i t^{4i+2} + \sum_{j \geq 0} \beta_j t^{2j+1}, \quad \varphi(t) = \sum_{k \geq 0} \gamma_k t^{2k+1}.$$

Let

$$i_0 := \begin{cases} \min\{i \mid \alpha_i \neq 0\} & \text{if } \alpha_i \neq 0 \text{ for some } i \\ -1 & \text{otherwise}, \end{cases}$$

let

$$j_0 := \begin{cases} \min\{j \mid \beta_j \neq 0\} & \text{if } \beta_j \neq 0 \text{ for some } j \\ -1 & \text{otherwise}, \end{cases}$$

and let

$$k_0 := \begin{cases} \min\{k \mid \gamma_k \neq 0\} & \text{if } \gamma_k \neq 0 \text{ for some } k \\ -1 & \text{otherwise .} \end{cases}$$

We set

$$\alpha = \alpha'_{i_0} + \alpha'_{i_0+1}t + \cdots \quad \text{and} \quad s = 2i_0 + 1 \text{ if } i_0 \geq 0,$$

$$\beta = \beta'_{j_0} + \beta'_{j_0+1}t + \cdots \quad \text{and} \quad r = 2j_0 + 1 \text{ if } j_0 \geq 0,$$

and

$$\gamma = \gamma'_{k_0} + \gamma'_{k_0+1}t + \cdots \quad \text{and} \quad k = 2k_0 + 1 \text{ if } k_0 \geq 0,$$

where $\alpha'_i = \alpha_i^{1/4}$, $\beta'_j = \beta_j^{1/4}$ and $\gamma'_k = \gamma_k^{1/4}$. Otherwise, we set $\alpha = s = 0$ if $i_0 = -1$, $\beta = r = 0$ if $j_0 = -1$ or $\gamma = k = 0$ if $k_0 = -1$, respectively. Then the assertion follows. $\qquad\square$

By making use of this local form of a Weierstrass equation, we can classify the reducible singular fibers as follows.

Theorem. *Let $f : V \to C$ be a quasi-elliptic surface locally defined so that the conditions* (1) *and* (2) *in the above lemma are satisfied. Then the fiber over $t = 0$ is determined as*

(1) $r \geqslant k \neq 0$ *or* $r = 0$

type	$v(\Delta)$	k	s
II	0	1	$\geqslant 0$
I_0^*	4	3	$\geqslant 0$
II*	8	5	$\neq 1$
I_{2k-6}^*	$2k-2$	$\geqslant 5$	1

(2) $k > r \neq 0$ *or* $k = 0$

type	$v(\Delta)$	r	s
III	1	1	$\geqslant 0$
III*	7	3	$\neq 1$
I_{2r-4}^*	$2r$	$\geqslant 3$	1

Here v denotes the valuation of $k[[t]]$ with $v(t) = 1$ and Δ is the discriminant of the Weierstrass equation.

Regrettably, the proof takes more pages. So we refer the proof to the original paper by Ito [35]. A partial case where $p = 2$ and $\psi(t) = 0$ is treated in subsequent arguments.

Remark. If $\psi(t) = 0$, then we have the following list of reducible fibers

r	1	3	5
type	II	I_0^*	II*
$v_P(d\varphi)$	0	2	4

where $v_P(d\varphi)$ is the order of zero of the differential $d\varphi$ at P. Hence we have an equality

$$6\chi(\mathcal{O}_V) = 2 + 2c_6 + 4c_{10} \ .$$

3.2 Unirational case

Let $f : V \to C$ be a quasi-elliptic fibration with a rational cross-section D.

3.2.1

Lemma. *V is unirational if and only if C is k-rational.*

Proof. If V is unirational, C is rational by Lüroth's theorem. Conversely, suppose that C is k-rational. Then $\mathfrak{K} = k(t)$ and $\Gamma \otimes_{\mathfrak{K}} \mathfrak{K}^{p^{-1}}$ is rational over $\mathfrak{K}^{p^{-1}}$ by Lemma 3.1.3. Hence $\mathfrak{K}^{p^{-1}}(\Gamma) \cong \mathfrak{K}^{p^{-1}}(u)$. Since $\mathfrak{K}^{p^{-1}} = k(t^{p^{-1}})$, we know that $k(V) \hookrightarrow k(t^{p^{-1}}, u)$. Hence V is unirational. $\qquad\square$

3.2.2

Lemma. *Let $f : V \to C$ be a unirational quasi-elliptic surface with a rational cross-section D. Assume that V is relatively minimal. Then the following assertions hold.*

(1) *f has no multiple fibers.*
(2) *$\chi(\mathcal{O}_V) = -(D^2)$.*
(3) *If $\chi(\mathcal{O}_V) \leq 1$, V is rational over k; if $\chi(\mathcal{O}_V) = 2$ then V is a K3-surface; if $\chi(\mathcal{O}_V) \geq 3$ then $p_a(V) = p_g(V)$, $h^1(V, \mathcal{O}_V) = 0$, the r-genus $P_r(V) = r(\chi(\mathcal{O}_V) - 2) + 1$, and the Kodaira dimension $\kappa(V) = 1$.*

Proof. (1) Since f has a cross-section, this is obvious.

(2) This is proved in Corollary 3.1.2.

(3) For a positive integer r, the r-genus $P_r(V)$ is given by

$$P_r(V) := h^0(V, \mathcal{O}(rK_V)) = h^0(C, \mathcal{O}_C(r(\chi(\mathcal{O}_V) - 2))) \ .$$

This follows from Corollary 3.1.2, (2) and Theorem 3.1.1, (5), since $C \cong \mathbb{P}^1$. If $\chi(\mathcal{O}_V) \leq 1$, $P_r(V) = 0$ for every $r > 0$, whence $P_{12}(V) = 0$. This implies that V is ruled (cf. [57, Chap.I, 4.5.1]). Since V is unirational, V is rational. If $\chi(\mathcal{O}_V) = 2$, we have $\mathcal{O}(K_V) \cong \mathcal{O}_V$. Then V is a K3-surface (cf. [8]). If

$\chi(\mathcal{O}_V) \geq 3$, $P_r(V) = r(\chi(\mathcal{O}_V) - 2) + 1$. Hence the Kodaira dimension $\kappa(V)$ equals 1, and $p_g(V) = \chi(\mathcal{O}_V) - 1 := p_a(V)$. Therefore $h^1(V, \mathcal{O}_V) = 0$. $\quad\square$

3.2.3

Corollary. *Let $f : V \to C$ be as in Lemma 3.2.2. Assume that f is relatively minimal and that V is not rational over k. Then V is a minimal smooth model.*

Proof. Set $e := \chi(\mathcal{O}_V) - 2$. Then $K_V \sim ef^*(P)$ (linear equivalence) for a point P on $C \cong \mathbb{P}^1_k$. If V is not rational over k we know by 3.2.2 that $e \geq 0$. Then the canonical linear system $|K_V|$ has no fixed components, which implies that V contains no (-1)-curves. Hence V is a relatively minimal model. Since V is not ruled, V is an absolutely minimal model. $\quad\square$

3.3 Main theorems on unirational quasi-elliptic surfaces

We shall prove the following two theorems. Note that we change from the previous sections the notations for the defining equation of the generic fiber.

3.3.1 *Case $p = 3$*

Theorem. *Let k be an algebraically closed field of characteristic 3. Then any unirational quasi-elliptic surface defined over k with a rational cross-section is birationally equivalent to a hypersurface in $\mathbb{A}^3_k$ defined by $t^2 = x^3 + \varphi(y)$ with $\varphi(y) \in k[y]$ of degree d prime to 3. Conversely, let $K := k(t, x, y)$ be the algebraic function field of the above affine hypersurface, let $m = \left\lceil \frac{d}{6} \right\rceil$ and let V be a smooth minimal model of K if K is not rational over k. Furthermore, if $d \geq 7$ assume that the following conditions hold.*

(1) *For every root α of the discriminant $\Delta(t) := \varphi'(y) = \frac{d\varphi}{dy} = 0$, $v_\alpha(\varphi(y) - \varphi(\alpha)) \leq 5$, where v_α is the $(y-\alpha)$-adic valuation of $k[y]$ with $v_\alpha(y-\alpha) = 1$.*

(2) *If, moreover,*

$$\varphi(y) - \varphi(\alpha) = a(y - \alpha)^3 + (\text{terms of higher degree in } (y - \alpha))$$

for some root α of $\varphi'(y) = 0$ and $a \in k$ then

$$v_\alpha(\varphi(y) - \varphi(\alpha) - a(y - \alpha)^3) \leq 5.$$

Then we have the following assertions.

(i) *If $m = 0$, i.e., $d \leq 5$, then K is rational over k. If $d \geq 7$, K is not rational over k, and the minimal model V exists.*
(ii) *If $m = 1$, i.e., $7 \leq d \leq 11$, then V is a K3-surface.*
(iii) *If $m > 1$, i.e., $d \geq 13$, then $p_a(V) = p_g(V) = m$, $h^1(V, \mathcal{O}_V) = 0$, the r-genus $P_r(V) = r(m - 1) + 1$ for every integer r and the Kodaira dimension $\kappa(V)$ equals 1.*

3.3.2 Case $p = 2$

Theorem. *Let k be an algebraically closed field of characteristic 2. Then any unirational quasi-elliptic surface defined over k with a rational cross-section is birationally equivalent to an affine hypersurface in $\mathbb{A}_k^3$ defined by $t^2 = x^3 + \psi(y)x + \varphi(y)$ with $\varphi(y), \psi(y) \in k[y]$ and either $\psi(y) \notin k[y]^2$ or $\varphi(y) \notin k[y]^2$. Conversely, let $K := k(t, x, y)$ be an algebraic function field of dimension 2 generated by t, x, y over k such that $t^2 = x^3 + \psi(y)x + \varphi(y)$ with $\varphi(y), \psi(y) \in k[y]$ and either $\psi(y) \notin k[y]^2$ or $\varphi(y) \notin k[y]^2$. Define m by*

$$m = \max\left\{ \left[\frac{1}{4}\deg\psi(y)\right], \ \left[\frac{1}{6}\deg\varphi(y)\right] \right\}.$$

Assume moreover that the following conditions are satisfied.

(1) *$\psi(y)$ has no monomial terms of degree congruent to zero modulo 4.*
(2) *$\varphi(t)$ has no monomial terms of even degree.*
(3) *For every root α of the discriminant $\Delta(t) := \psi(y)\psi'(y)^2 + \varphi'(y)^2 = 0$,*

$$\min\{v_\alpha(\psi(y) - \psi(\alpha)) - 4, \ v_\alpha(\varphi(y) - \varphi(\alpha)) - 6\} < 0,$$

where v_α is the $(y - \alpha)$-adic valuation of $k(y)$ such that $v_\alpha(y - \alpha) = 1$.

Then the following assertions hold with $\deg\varphi(y) = 2d_1 + 1$.

(i) *If $m = 0$, i.e., $0 \leq d_1 \leq 2$, then K is rational over k. If $m > 0$, K is not rational over k, and the minimal model V of K over k exists.*
(ii) *If $m = 1$, i.e., $3 \leq d_1 \leq 5$, then V is a (supersingular) K3-surface.*
(iii) *If $m > 1$, i.e., $d_1 \geq 6$, then $p_a(V) = p_g(V) = m$, $h^1(V, \mathcal{O}_V) = 0$, the r-genus $P_r(V) = r(m - 1) + 1$ for every positive integer r and the Kodaira dimension $\kappa(V)$ equals 1.*

In Theorems 3.3.1 and 3.3.2, the first assertion follows from Lemma 3.1.3. So, we shall prove the second (converse) assertion. After a slight change of notations, we start with an affine hypersurface of the following type:

Case: $p = 3$. $z^3 = y^2 + \varphi(x)$ with $\varphi(x) = a_0 x^d + \cdots + a_d \in k[x]$, where $a_0 \neq 0$. We may assume that $\varphi(x)$ contains no terms of degree congruent to zero modulo 3 and that $\varphi(x)$ satisfies the conditions (1), (2) of Theorem 3.3.1.

Case: $p = 2$. $z^2 = y^3 + \varphi(x)$ with $\varphi(x) = a_0 x^d + \cdots + a_d \in k[x]$, where $a_0 \neq 0$. We may assume that $\varphi(x)$ contains no terms of even degree (hence it is of the form $\varphi(x) = x\varphi_1(x)^2$) and that $\varphi(x)$ satisfies the condition that for every α of $\varphi'(x) = 0$, we have $v_\alpha(x - \alpha) < 6$. So, the assumption corresponds to the case $\psi(y) = 0$ in Theorem 3.3.2. The case $\psi(y) \neq 0$ is proved in [35] in full generality.

3.4 Settings toward proofs

Each of the above affine hypersurfaces is a purely inseparable covering of exponent 1 of the affine plane $\mathbb{A}_k^2 = \operatorname{Spec} k[x, y]$. Embed $\mathbb{A}_k^2$ into $F_0 = \mathbb{P}^1 \times \mathbb{P}^1$ as the complement of two transversal generators of F_0, let S_0 be the normalization of F_0 in $K := k(x, y, z)$, the function field of the above hypersurface, and let $\rho_0 : S_0 \to F_0$ be the normalization morphism. Let $U_1 := \rho_0^{-1}(F_0 - (x = \infty) \cup (y = \infty))$, $U_2 := \rho_0^{-1}(F_0 - (u = \infty) \cup (y = \infty))$, $U_3 := \rho_0^{-1}(F_0 - (x = \infty) \cup (v = \infty))$, and $U_4 := \rho_0^{-1}(F_0 - (u = \infty) \cup (v = \infty))$, where $u = 1/x$ and $v = 1/y$. Let B be the curve on F_0 obtained as the closure of the affine plane curve $y^2 + \varphi(x) = 0$ (when $p = 3$) or $y^3 + \varphi(x) = 0$ (when $p = 2$).

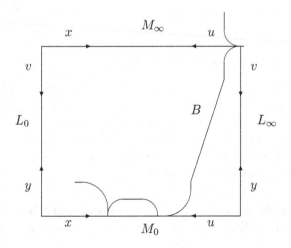

It is straightforward to verify the following:

3.4.1

Lemma. *With the above notations and assumptions, each of the U_i $(1 \le i \le 4)$ is isomorphic to the normalization of a hypersurface V_i in $\mathbb{A}_k^3$ defined by the following equation, where we put $\psi(u) = u^d \varphi(1/u)$:*

(i) $(p = 3)$ (1) $z^3 = y^2 + \varphi(x)$ for V_1; (2) for V_2, $z^3 = u^2(y^2 u^d + \psi(u))$ if $d \equiv 1 \pmod 3$ and $z^3 = u(y^2 u^d + \psi(u))$ if $d \equiv 2 \pmod 3$; (3) $z^3 = v(1 + v^2 \varphi(x))$ for V_3; (4) for V_4, $z^3 = vu^2(u^d + v^2 \psi(u))$ if $d \equiv 1 \pmod 3$ and $z^3 = vu(u^d + v^2 \psi(u))$ if $d \equiv 2 \pmod 3$.

(ii) $(p = 2)$ (1) $z^2 = y^3 + \varphi(x)$ for V_1; (2) $z^2 = u(u^d y^3 + \psi(u))$ for V_2; (3) $z^2 = v(1 + v^3 \varphi(x))$ for V_3; (4) $z^2 = vu(u^d + v^3 \psi(u))$ for V_4, where we note that d is assumed to be odd.

3.4.2

As shown in the figure in 3.4.1, denote $L_0, L_\infty, M_0, M_\infty$ the curves on F_0 defined by $x = 0, u = 0, y = 0, v = 0$ respectively. The following is a straightforward consequence of the Jacobian criterion of smoothness, partly after the one-step normalization by the embedded blowing-up.

Lemma. *The surface S_0 is smooth except for the points lying over $P_\infty :=$ $(u = 0, v = 0)$ and $P_\alpha := (x = \alpha, y = 0)$, where α ranges over all roots of $\varphi'(x) = 0$.*

Proof. Consider the case $p = 3$. With the notations in Lemma 3.4.1, we apply the Jacobian criterion to $z^3 = y^2 + \varphi(x)$ to find singular points P_α in the statement. For the hypersurface $z^3 = u^2(y^2 u^d + \psi(u))$ with $d \equiv 1$ (mod 3), it is singular along the curve $z = u = 0$. Consider a blowing-up $(u', y, z) \mapsto (zu', y, z)$ with $u = zu'$ to find the proper transform

$$z = u'^2 \{y^2 z^d u'^d + \psi(zu')\}$$

which is the normalization of the hypersurface. Since $\psi(0) \ne 0$, this proper transform is smooth over the curve $z = u = 0$. The hypersurface $z^3 = u(y^2 u^d + \psi(u))$ with $d \equiv 2$ (mod 3) is smooth along the locus $u = 0$ because $\psi(u) \ne 0$. The hypersurface $z^3 = v(1 + v^2 \varphi(x))$ is smooth. As for the hypersurface $z^3 = vu^2(u^d + v^2 \psi(u))$ with $d \equiv 1$ (mod 3), it is singular along the locus $u = 0$, and the blowing-up $u = zu'$ changes the equation to $z = vu'^2(z^d u'^d + v^2\psi(zu'))$, which is smooth. As for the hypersurface $z^3 = vu(u^d + v^2\psi(u))$ with $d \equiv 2 \pmod 3$, it is singular only for the point $u = v = 0$.

In the case $p = 2$, the arguments are similar. $\qquad\qquad\qquad\qquad\square$

3.4.3

In order to obtain a smooth projective model of S_0, we apply the arguments explained in the previous chapter. Take up the point P_∞ when $p = 3$. In a neighborhood of $\rho_0^{-1}(P_\infty)$, S_0 (or $k(S_0)$) ramifies over the curves B, L_∞, M_∞ with multiplicities of ramification $1, 2, 1$, respectively, if $d \equiv 1 \pmod 3$ and $1, 1, 1$, respectively, if $d \equiv 2 \pmod 3$, but not nicely over the point P_∞. As shown in the next section 3.5, we can apply a finite number of blowing-ups with centers at P_∞ and its infinitely near points to obtain a smooth projective surface F and a birational morphism $\sigma : F \to F_0$ such that the function field $k(S_0)$ ramifies *nicely over the curves* $\sigma^{-1}(B \cup L_\infty \cup M_\infty)$ (see 2.8). We always assume $\sigma : F \to F_0$ to be the *shortest* series of blowing-ups such that the curve $\sigma^{-1}(B)$ has only normal crossings as singularities and that $k(S_0)$ ramifies nicely not only in a neighborhood of $\sigma^{-1}(P_\infty)$ (over the curves $\sigma^{-1}(B \cup L_\infty \cup M_\infty)$) but also in respective neighborhoods of $\sigma^{-1}(P_\alpha)$ (over the curve $\sigma^{-1}(B)$), where P_α ranges over all singular points of B except P_∞ (cf. Lemma 3.4.2). Then, let S be the normalization of F in $K := k(S_0)$. By the results given in the previous chapter, we know that S has only quotient singularities and we can resolve minimally those isolated singularities to obtain a smooth projective surface $\widetilde{S}$;

$$S_0 \xleftarrow{\;\tau\;} S \xleftarrow{\;\widetilde{\tau}\;} \widetilde{S}$$

$$\downarrow{\rho_0} \qquad \downarrow{\rho}$$

$$F_0 \xleftarrow{\;\sigma\;} F$$

and $\sigma \cdot \rho = \rho_0 \cdot \tau$, where ρ is the normalization morphism and $\widetilde{\tau}$ is the minimal resolution of singularities of S; the morphism τ exists by the definition of S_0. In the case $p = 2$, we can make similar arguments to obtain the above diagram.

Let Λ be the linear pencil $|L_0|$ on F_0. Let L_a be a general member of Λ defined by $x = a$. Then the pull-back $(\sigma \cdot \rho \cdot \widetilde{\tau})^* L_a$ is a quasi-elliptic curve Γ_a birationally equivalent to $z^3 = y^2 + \varphi(a)$ (if $p = 3$) or $z^2 = y^3 + \varphi(a)$ (if $p = 2$) and having a unique singular point lying over the curve M_0.

If the point $x = a$ moves on $C := \mathbb{P}^1_k$, the curve Γ_a moves in a pencil $\widetilde{\Lambda} := (\sigma \cdot \rho \cdot \widetilde{\tau})^* \Lambda$ of quasi-elliptic curves. Hence the pencil $\widetilde{\Lambda}$ defines a surjective morphism $\widetilde{f} : \widetilde{S} \to C$ whose general fibers are members of $\widetilde{\Lambda}$. Thus $\widetilde{f}$ is a quasi-elliptic fibration, though it is not necessarily relatively minimal. By contracting all (-1)-components in the fibers of $\widetilde{f}$, we obtain a relatively minimal quasi-elliptic surface $f : V \to C$ and the contraction $\pi : \widetilde{S} \to V$ such that $\widetilde{f} = f \cdot \pi$. On the other hand, $\widetilde{D} := (\sigma \cdot \rho \cdot \widetilde{\tau})^*(M_\infty)$ defines a regular cross-section of $\widetilde{f}$, hence $D := \pi_*(\widetilde{D})$ is a regular cross-section of f as well. Then, by computing (D^2), we obtain the consequences of Theorems in 3.3 by Lemma 3.2.2. This is an outline of the proof of Theorems 3.3.1 and 3.3.2. We shall consider the details of the proof in the subsequent sections.

3.5 $\sigma^{-1}(B \cup L_\infty \cup M_\infty)$ in the case $p = 3$

In the sections 3.5 and 3.6, we assume $p = 3$. By means of straightforward computations, we have the following result. (See Remark below for a double line in the graph.)

3.5.1

Lemma. *With the same notations and assumptions as in 3.4.3, the curves* $\sigma^{-1}(B \cup L_\infty \cup M_\infty)$ *have one of the following weighted dual graphs in a neighborhood of* $\sigma^{-1}(B)$, *where the negative numbers (resp. the encircled integers) signify the self-intersection numbers (resp. the multiplicities of ramification) of the corresponding curves, the* E_i *are the exceptional curves in the process* $\sigma : F \to F_0$, *and* $\overline{B}, \overline{L}_\infty$ *and* $\overline{M}_\infty$ *are the proper transforms of* B, L_∞ *and* M_∞ *by* σ *respectively (see 2.7.2).*

Case $d = 6m + 1$ $(m > 0)$.

Case $d = 6m + 2$ $(m > 0)$.

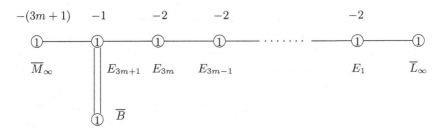

Case $d = 6m + 4$ $(m \geq 0)$.

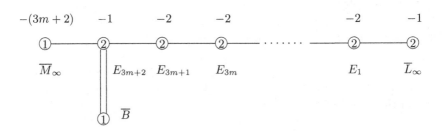

Case $d = 6m + 5$ $(m \geq 0)$.

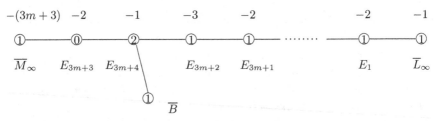

Proof. We indicate the computations only in the cases $d = 6m + 1$ and $d = 6m + 2$. The remaining cases can be treated in similar fashions. If $d = 6m + 1$, the curve $B \cup L_\infty \cup M_\infty$ with weights $1, 2, 1$ is defined by $vu^2(u^d + v^2\psi(u)) = 0$ near the point P_∞, where $\psi(u) \neq 0$. By blowing up P_∞ and its infinitely near points $P_\infty^{(i)}$, we introduce $v_1, \ldots, v_{3m}$ by $v = u^i v_i$ $(1 \leq i \leq 3m)$ and replace z by $z_i = z/u^i$. We thus convolve the u-term with degree a multiple of 3 into z. At the ith step, we have the equation $z_i^3 = v_i u^2(u^{6m+1-2i} + v_i^2\psi(u))$ near the point $P_\infty^{(i)}$ defined by $u = v_{3m} = 0$, where the proper transforms $M_\infty^{(i)}, B^{(i)}$ of M_∞, B and the last exceptional curve E_i are defined by $v_i = 0$, $u^{6m+1-2i} + v_i^2\psi(u) = 0$, and

$u = 0$, respectively. At the $3m$th step, the curve $B^{(3m)}$ is smooth at $P_\infty^{(3m)}$ and touches E_{3m} with degree $i(B^{(3m)}, E_{3m}; P_\infty^{(3m)}) = 2$. Then two more blowing-ups make the total transform of $M_\infty^{(3m)} \cup E_{3m} \cup B^{(3m)}$ a divisor with only normal crossings. The final picture of the curves is given by the weighted graph.

If $d = 6m + 2$, a similar computation gives the equation $z_{3m}^3 = uv_{3m}(u^2 + v_{3m}^2 \psi(u))$. Hence one more blowing-up makes the total transform of $M_\infty^{(3m)} \cup E_{3m} \cup B^{(3m)}$ a divisor with only normal crossings. $\quad\square$

Remark. (1) With the notations of 3.3.1, the hypersurface $t^2 = x^3 + \varphi(y)$ is rational over k if $d \leq 2$. Hence we assume that $m > 0$ if $d = 6m + 1$ or $d = 6m + 2$.

(2) In the above graphs, all irreducible curves possibly except for $\overline{B}$ are smooth rational curves. Also, in the above graph, the subgraph $\circ\!\!=\!\!=\!\!\circ\ \overline{B}$ should be understood as

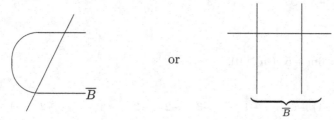

or

3.5.2

Let $\widetilde{L}_\infty, \widetilde{M}_\infty$ and the $\widetilde{E}_i$ be the set-theoretic proper inverse images on $\widetilde{S}$ of L_∞, M_∞ and the E_i, respectively. We have the following:

Lemma. *With the same notations and assumptions as in 3.4.3, $(\sigma \cdot \rho \cdot \widetilde{\tau})^{-1}(L_\infty)$ has one of the following weighted graphs:*

Case $d = 6m + 1$ $(m > 0)$.

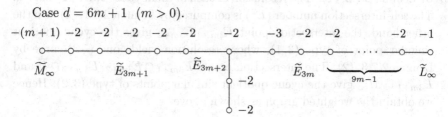

Case $d = 6m + 2$ $(m > 0)$.

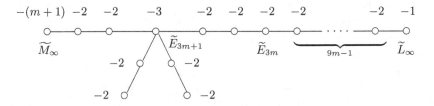

Case $d = 6m + 4$ $(m \geq 0)$.

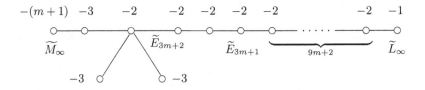

Case $d = 6m + 5$ $(m \geq 0)$.

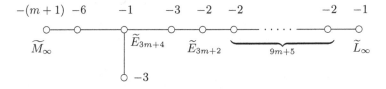

Proof. Except for the last case, we can obtain the required graphs by applying repeatedly Lemmas 2.7.3, 2.8.1 and 2.8.2. We treat the case $d = 6m+1$ as an example. The intersection points $\overline{L}_\infty \cap E_1, E_1 \cap E_2, \ldots, E_{3m-1} \cap E_{3m}$ give on the surface S the cyclic quotient singular points of type $(3, 2)$ by Lemma 2.7.3, (1). The resolution of each of them yields two (-2)-curves. The self-intersection number $(\widetilde{E}_i^2)$ is computed by Lemma 2.8.1, (3). On the other hand, the intersection point $E_{3m} \cap E_{3m+2}$ gives the cyclic quotient singular point of type $(3, 1)$, whose resolution yields one (-3)-curve by Lemma 2.7.3, (2). The intersection points $E_{3m+1} \cap E_{3m+2}, E_{3m+2} \cap \overline{B}$ and $E_{3m+1} \cap \overline{M}_\infty$ give the cyclic quotient singular points of type $(3, 2)$. Hence we obtain the weighted graph as shown above.

We shall verify the last case. The configuration of the curves $\overline{M}_\infty$, E_{3m+3}, E_{3m+4} and $\overline{B}$ is:

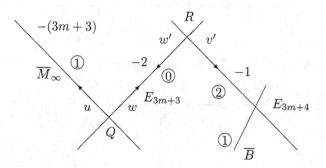

Set $G := E_{3m+3}$ and $H := E_{3m+4}$. Let $\widetilde{G} := \widetilde{E}_{3m+3}$ and $\widetilde{H} := \widetilde{E}_{3m+4}$. It is then easy to see that the surface S is isomorphic to the normalization of a hypersurface defined by

$$\begin{cases} z^3 = w(1 + uw^2\psi(u)) & \text{over } \rho^{-1}(Q) \\ z^3 = w'^2(1 + v'\psi(v'w'^2)) & \text{over } \rho^{-1}(R) \ . \end{cases}$$

We know therefore that S is smooth along $\rho^{-1}(G)$ and $\rho^{-1}(G)$ is a smooth rational curve. Note that $\widetilde{G} \cong \rho^{-1}(G)$, $[k(\widetilde{G}) : k(G)] = p$, and $\widetilde{G}$ meets $\widetilde{M}_\infty$ (resp. $\widetilde{H}$) transversally in one point. Let $\nu := \rho \cdot \widetilde{\tau} : \widetilde{S} \to F$. Then we compute

$$(\nu^*(G + \overline{M}_\infty + H) \cdot \widetilde{G}) = p(G + \overline{M}_\infty + H \cdot G),$$

where $\nu^*(G+\overline{M}_\infty+H) = \widetilde{G}+p\widetilde{M}_\infty+p\widetilde{H}+\cdots$. Hence we obtain $(\widetilde{G}^2)+2p = p(-2+1+1) = 0$, i.e., $(\widetilde{G}^2) = -6$. It is now easy to obtain the last graph.

$\square$

3.5.3

Let Γ_∞ be the fiber of the relatively minimal quasi-elliptic fibration $f : V \to C$, corresponding to the member of L_∞ of Λ and let $D := \pi_*(\widetilde{M}_\infty)$ (cf. 3.4.3). Then we have the following:

Lemma. *The fiber Γ_∞ has one of the following weighted graphs (the configuration of curves if $d = 6m + 5$):*

Case $d = 6m + 1$ $(m > 0)$. *The fiber type is* II*.

$$-(m+1) \quad -2 \quad -2 \quad -2 \quad -2 \quad -2 \quad -2 \quad -2 \quad -2$$

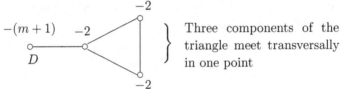

-2

D

Case $d = 6m + 2$ $(m > 0)$. *The fiber type is* IV*.

$$-(m+1) \quad -2 \quad -2 \quad -2 \quad -2 \quad -2$$

D

-2

-2

Case $d = 6m + 4$ $(m \geq 0)$. *The fiber type is* IV.

-2

$-(m+1) \quad -2$

D

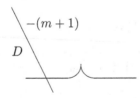

-2

Three components of the triangle meet transversally in one point

Case $d = 6m + 5$ $(m \geq 0)$. *The fiber type is* II.

$-(m+1)$

D

Proof. Straightforward from Lemma 3.5.2. It suffices to make contractions of all possible curves which start with the curve $\tilde{L}_\infty$. $\square$

3.6 $\sigma^{-1}(M_\infty \cup L_\alpha \cup M_0)$ in the case $p = 3$

Let $P_\alpha := (x = \alpha, y = 0)$ be a singular point of the curve B other than P_∞, i.e., α is a root of $\varphi'(x) = 0$. Let $e := v_\alpha(\varphi(x) - \varphi(\alpha))$. The condition (1) of

Theorem 3.3.1 implies that $e = 2, 3, 4$ or 5, while the condition (2) asserts that the case $e = 3$ can be reduced to the case $e = 4$ or 5 by a birational transformation $(x, y, z) \mapsto (x, y, z - a^{1/3}(x - \alpha))$. Let L_α be the member of the pencil Λ on F_0 defined by $x = \alpha$. We can show the following three lemmas in the same fashion as in 3.5.

3.6.1

Lemma. *With the notations and assumptions as above, the curve* $\sigma^{-1}(M_\infty \cup L_\alpha \cup M_0)$ *has one of the following weighted graphs in a neighborhood of* $\sigma^{-1}(P_\alpha)$:

Case $e = 2$.

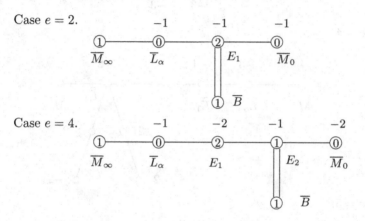

Case $e = 4$.

Case $e = 5$.

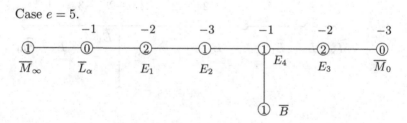

Proof. We indicate the computation in the case $e = 5$. We can set

$$z'^3 = z^3 - \varphi(\alpha) \quad \text{and} \quad \varphi(x) - \varphi(\alpha) = a_5 x'^5 + (x'\text{-terms of degree} \geq 6) ,$$

where $x' = x - \alpha$ and $a_5 \neq 0$. Then, near P_α, the curve B is given by $y^2 + a_5 x'^5 + (x'\text{-terms of degree} \geq 6) = 0$. Then a sequence of blowing-ups at P_α which make $\sigma^{-1}(M_0 \cup L_\alpha \cup B)$ a divisor with normal crossings is indicated by the above weighted graph. $\square$

3.6.2

Lemma. *With the same notations as in 3.4.3, $(\sigma \cdot \rho \cdot \widetilde{\tau})^{-1}(L_\alpha)$ has one of the following weighted graphs.*

Case $e = 2$.

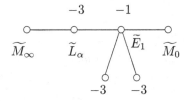

Case $e = 4$.

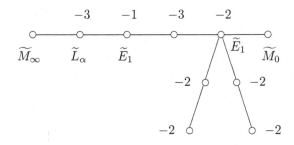

Case $e = 5$.

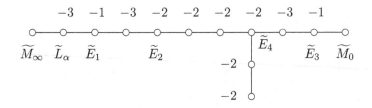

3.6.3

Let Γ_α be the fiber of $f : V \to C$ corresponding to the member L_α of Λ. Then we have the following:

Lemma. *The fiber Γ_α has one of the following weighted graphs:*

Case $e = 2$. *The fiber type is* IV.

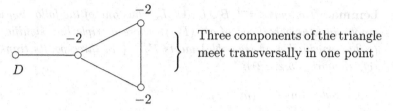

Three components of the triangle meet transversally in one point

Case $e = 4$. *The fiber type is* IV*.

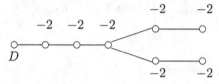

Case $e = 5$. *The fiber type is* II*.

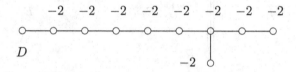

3.6.4

The above construction of Γ_α shows that the regular cross-section D of the relatively minimal quasi-elliptic fibration $f : V \to C$ has self-intersection number $-(m + 1)$. Now Theorem 3.3.1 follows immediately from Lemma 3.2.2 and Corollary 3.2.3.

3.7 $\sigma^{-1}(B \cup L_\infty \cup M_\infty)$ in the case $p = 2$

In the sections 3.7 and 3.8 of this chapter, we assume that $p = 2$. Our goal is to prove Theorem 3.3.2 with $\psi(y) = 0$ by the same method as for Theorem 3.3.1. For the case $\psi(y) \neq 0$, see [35]. *By the condition (2) therein, it suffices to consider the cases $d = 6m + 1, 6m + 3$ and $6m + 5$, and we may further assume that $d \geq 3$. Based on Lemma 3.4.1, (2), we argue as in the case $p = 3$. Since all computations are similar to those in the case $p = 3$, we omit the proof of the subsequent lemmas.*

3.7.1

Lemma. *The curve $\sigma^{-1}(B \cup L_\infty \cup M_\infty)$ has one of the following weighted graphs in a neighborhood of $\sigma^{-1}(P_\infty)$, where a triple line signifies that $\overline{B}$ (either irreducible or reducible) meets E_{2m+1} in three points transversally (cf. Remark 3.5.2, (2)):*

Case $d = 6m + 1$ $(m > 0)$.

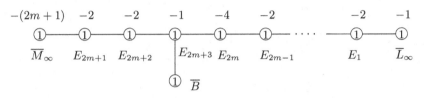

Case $d = 6m + 3$ $(m \geq 0)$.

Case $d = 6m + 5$ $(m \geq 0)$.

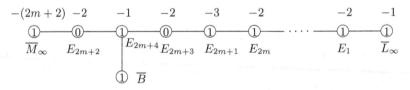

3.7.2

Lemma. *The curve $(\sigma \cdot \rho \cdot \widetilde{\tau})^{-1}(L_\infty)$ has one of the following weighted graphs:*

Case $d = 6m + 1$ $(m > 0)$.

Case $d = 6m + 3$ $(m \geq 0)$.

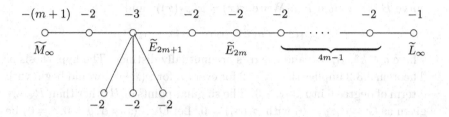

$$
\begin{array}{ccccccccc}
-(m+1) & -2 & -3 & -2 & -2 & -2 & & -2 & -1 \\
\circ & \!\!-\!\!\circ & \!\!-\!\!\circ & \!\!-\!\!\circ & \!\!-\!\!\circ & \!\!-\!\!\circ & \cdots & \!\!-\!\!\circ & \!\!-\!\!\circ \\
\widetilde{M}_\infty & & \widetilde{E}_{2m+1} & \widetilde{E}_{2m} & & & & & \widetilde{L}_\infty
\end{array}
$$

$4m-1$

$$
\begin{array}{ccc}
\circ & \circ & \circ \\
-2 & -2 & -2
\end{array}
$$

Case $d = 6m + 5$ $(m \geq 0)$.

$$
\begin{array}{cccccccccc}
-(m+1) & -6 & -1 & -4 & -2 & -2 & -2 & -2 & -2 & -1 \\
\circ & \circ & \circ & \circ & \circ & \circ & \circ & \cdots & \circ & \circ \\
\widetilde{M}_\infty & \widetilde{E}_{2m+2} & \widetilde{E}_{2m+4} & \widetilde{E}_{2m+3} & \widetilde{E}_{2m+1} & & \widetilde{E}_{2m} & & & \widetilde{L}_\infty
\end{array}
$$

$4m-1$

$$-2 \; \circ$$

3.7.3

Lemma. *The fiber Γ_∞ has one of the following weighted graphs:*

Case $d = 6m + 1$ $(m > 0)$. *The same as the one for the case $d = 6m + 1$ $(m > 0)$ in 3.5.3. Hence its fiber type is* II^*.

Case $d = 6m + 3$ $(m \geq 0)$. *The fiber type is* I_0^*.

$$
\begin{array}{ccccc}
 & & & -2 & \\
 & & & \circ & \\
-(m+1) & -2 & -2 & \mid & -2 \\
\circ & \!\!-\!\!\circ & \!\!-\!\! & \circ & \!\!-\!\!\circ \\
D & & & \mid & \\
 & & & \circ & -2
\end{array}
$$

Case $d = 6m + 5$ $(m \geq 0)$. *The same as the one for the case $d = 6m + 5$ $(d \geq 0)$ in 3.5.3. The fiber type is* II.

3.8 $\sigma^{-1}(M_\infty \cup L_\alpha \cup M_0)$ in the case $p = 2$

As in 3.6, we will look into the singular points $P_\alpha := (x = \alpha, y = 0)$ of the curve B other than P_∞. Write $\varphi(x) = x(\varphi_1(x))^2$ and

$$\varphi_1(x) = a(x - \alpha_1)^{r_1} \cdots (x - \alpha_s)^{r_s},$$

where $a \in k^*$, $\alpha_i \in k$ and the α_i's are mutually distinct. The hypothesis in Theorem 3.3.2 implies that $r_i \le 2$ for every i, for $\varphi(x + \alpha_i)$ would begin with a term of degree 6 in x if $r_i \ge 3$. The singular points of B other than P_∞ are given as $(x = \alpha_i, y = 0)$ with $\varphi_1(\alpha_i) = 0$. Let $Q := (x = \alpha, y = 0, z = 0)$ be a singular point with $\varphi_1(\alpha) = 0$. By a birational transformation $(x, y, z) \mapsto (x + \alpha, y, z)$ which is biregular at Q, the surface S_0 is a hypersurface in $\mathbb{A}^3_k$ defined locally at Q by one of the following equations:

(i)	$z^2 = y^3 + x^2 \delta(x)$	if $\alpha \ne 0$ and $r = 1$
(ii)	$z^2 = y^3 + x^3 \delta(x)$	if $\alpha = 0$ and $r = 1$
(iii)	$z^2 = y^3 + x^4 \delta(x)$	if $\alpha \ne 0$ and $r = 2$
(iv)	$z^2 = y^3 + x^5 \delta(x)$	if $\alpha = 0$ and $r = 2$,

where $\delta(x) \in k[x]$, $\delta(0) \ne 0$, and $\delta(x) \cdot x^{2r+\varepsilon} = \varphi(x + \alpha)$ with $\varepsilon = 0$ or 1 according as $\alpha \ne 0$ or $\alpha = 0$.

Now write

$$\delta(x) = a_0 + a_1 x + a_2 x^2 + a_3 x^3 + \cdots, \qquad a_0 \ne 0.$$

The case (i) above is reduced to the case (ii) or (iv) by a birational transformation

$$(x, y, z) \mapsto \left(x, y, z + a_0^{1/2} x + a_2^{1/2} x^2\right)$$

which is biregular at Q. Namely, if $a_1 \ne 0$ we have the case (ii), and if $a_1 = 0$ and $a_3 \ne 0$ we have the case (iv). Note that the case $a_1 = a_3 = 0$ does not occur by the hypothesis in Theorem 3.3.2. Similarly, the case (iii) is reduced to the case (iv). Thus, in order to look into the singularity of the point Q, we have only to consider the cases (ii) and (iv).

3.8.1

Let L_α be the member of $\Lambda = |L_0|$ defined by $x = \alpha$, where $\varphi_1(\alpha) = 0$. Then we have the following:

Lemma.   *With the notations and assumptions as above, the curve $\sigma^{-1}(M_\infty \cup L_\alpha \cup M_0)$ has one of the following weighted graphs:*

Case (ii).

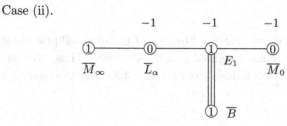

Case (iv).

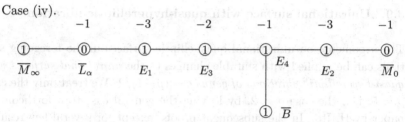

3.8.2

Lemma.   *The curve $(\sigma \cdot \rho \cdot \widetilde{\tau})^{-1}(L_\alpha)$ has one of the following weighted graphs:*

Case (ii). *The fiber type is I_0^*.*

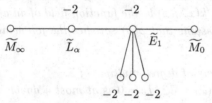

Case (iv). *The fiber type is II^*.*

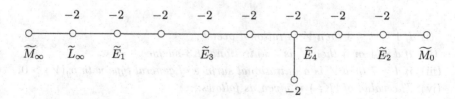

3.8.3

By Lemma 3.8.2, it follows that the fiber Γ_α of the quasi-elliptic fibration $f : V \to C$ has the same weighted graph as above. Thus, the second assertion of Theorem 3.3.2 follows from Lemma 3.2.2 and Corollary 3.2.3.

3.9 Unirational surface with quasi-hyperelliptic fibration

The arguments on unirational quasi-elliptic surfaces in the foregoing sections can be applied with suitable changes to the *unirational surfaces with quasi-hyperelliptic fibrations of genus* $g = (p-1)/2$. We treat only the case $p = 5$, i.e., the case $g = 2$, by leaving the general case to a forthcoming paper by H. Ito. In the subsequent proofs, except for several few results, we only state the results by omitting the proofs which are similar to those in the case of quasi-elliptic surfaces. The goal of the remaining subsections is to prove the following:

Theorem. *Let k be an algebraically closed field of characteristic 5. Let $f : V \to C$ be a quasi-hyperelliptic fibration of genus 2 on a unirational surface V endowed with a rational cross-section. Then V is birationally equivalent to a hypersurface in $\mathbb{A}^3$ defined by $y^2 = z^5 - \varphi(x)$ with $\varphi(x) \in k[x]$. Conversely, let $K := k(x, y, z)$ be the function field of an affine hypersurface of the above type. Assume that $\varphi(x)$ satisfies the following two conditions:*

(1) *$\varphi(x)$ has no terms of degree multiple of 5.*
(2) *Every root of $\varphi'(x) = d\varphi/dx = 0$ is at most a double root.*

Let $d := \deg_x \varphi(x)$, $m = [d/10]$ and V also denote the smooth minimal model of K if K is not rational over k. Then the structure of V is determined as follows:

(i) *If $1 \le d \le 3$ then V is rational over k.*
(ii) *If $d = 4$ or 6 then V is a unirational K3-surface.*
(iii) *If $d \ge 7$ then V is a unirational surface of general type with $p_g(V) > 0$.*
(iv) *The value of (K_V^2) is given as follows:*

   Case $d = 7, 8$ or 9, i.e., $m = 0$. *Then $(K_V^2) = 1$.*

Case $m > 0$.

$$(K_V^2) = \begin{cases} 8m-4 \\ 8m-3 \\ 8m-2 \\ 8m \end{cases} \quad if \quad d = \begin{cases} 10m+1, \ 10m+2 \\ 10m+3 \\ 10m+4, \ 10m+6 \\ 10m+7, \ 10m+8, \ 10m+9 \end{cases}$$

3.10 Canonical divisors for unirational surfaces with quasi-hyperelliptic fibrations

We begin with

Lemma. *Let V be a relatively minimal unirational surface with a quasi-hyperelliptic fibration $f : V \to C \cong \mathbb{P}_k^1$ of genus $g = (p-1)/2$ such that the generic fiber $V_\Re$ is $\Re$-birationally equivalent to an affine plane curve $y^2 = z^p - \varphi$ with $\varphi \in \Re - \Re^p$, where $\Re = k(x)$ (cf. Lemma 3.1.3). Let D be the regular cross-section which meets $V_\Re$ at the unique point lying over the point $(1/y = 0, z/y = 0)$ of the plane curve. Let Γ be a general fiber of f. Then there exist an integer a and an effective divisor Δ such that*

(i) $K_V \sim (p-3)D + a\Gamma + \Delta$.

(ii) Δ *is supported by the union of irreducible components of reducible fibers of f.*

(iii) $|\Delta - \Gamma| = \emptyset$.

Proof. Let $P_\infty := D \cap V_\Re$. Then it is well-known that $K_{V_\Re} \sim (2g - 2)P_\infty = (p-3)P_\infty$ because the curve $V_\Re$ is a hyperelliptic curve $y^2 = z^p - \varphi$ with φ viewed as a constant. Since $(V_\Re + K_V)|_{V_\Re} \sim K_{V_\Re}$ and $V_\Re|_{V_\Re} \sim 0$, we know that

$$(K_V - (p-3)D)|_{V_\Re} \sim 0$$

on $V_\Re$. This implies that $K_V - (p-3)D$ is linearly equivalent to a divisor supported by the union of fibers of f and irreducible components of reducible fibers. It is then easy to find an integer a and an effective divisor Δ satisfying the stated conditions. $\square$

3.11 Settings toward proving Theorem 3.9

Consider an affine hypersurface in $\mathbb{A}_k^3$ defined by

$$z^p = y^2 + \varphi(x), \quad \varphi(x) \in k[x], \quad d := \deg_x \varphi(x),$$

where $\varphi(x)$ contains no terms of degree a multiple of p and $v_\alpha(\varphi(x) - \varphi(\alpha)) \leq 3$ for any root α of $\varphi'(x) = 0$. The second requirement follows from the second assumption of Theorem 3.9. In fact, let α be a root of $\varphi'(x) = 0$. Write $\varphi(x) - \varphi(\alpha) = (x - \alpha)^e \varphi_1(x)$. Then $\varphi'(x)$ is divisible by $(x - \alpha)^{e-1}$, whence $e \leq 3$. View this hypersurface as a purely inseparable covering of exponent one of the affine plane $\mathbb{A}_k^2 = \operatorname{Spec} k[x, y]$. As in the section 3.4, embed $\mathbb{A}_k^2$ into $F_0 = \mathbb{P}^1 \times \mathbb{P}^1$ as the complement of two transversal generators, let S_0 be the normalization of F_0 in $K := k(x, y, z)$ and let $\rho_0 : S_0 \to F_0$ be the normalization morphism. Let B be the closure on F_0 of the curve $y^2 + \varphi(x) = 0$. With the notations similar to the ones in 3.4, we can verify that the following results hold true.

3.11.1

Lemma. *Each of the affine open sets U_i $(1 \leq i \leq 4)$ is isomorphic to the normalization of a hypersurface V_i in $\mathbb{A}_k^3$ defined by the following equation:*

(1) $z^p = y^2 + \varphi(x)$ *for V_1;*
(2) *for V_2, $z^p = u^{p-i}(y^2 u^d + \psi(u))$ if $d \equiv i \pmod{p}$ with $0 < i < p$;*
(3) $z^p = v^{p-2}(1 + v^2 \varphi(x))$ *for V_3;*
(4) *for V_4, $z^p = v^{p-2} u^{p-i}(u^d + v^2 \psi(u))$ if $d \equiv i \pmod{p}$ with $0 < i < p$,*

where $\psi(u) = u^d \varphi(1/u)$.

3.11.2

Lemma. *The surface S_0 is smooth except for the points lying over $P_\infty := (u = 0, v = 0)$ and $P_\alpha := (x = \alpha, y = 0)$, where α ranges over all roots of $\varphi'(x) = 0$.*

3.11.3

In order to prove Theorem 3.9, we use the same idea, arguments and notations as explained and used in 3.4.3. Namely, let $\sigma : F \to F_0$ be the shortest succession of blowing-ups with centers at P_∞, P_α's and their, appropriately chosen, infinitely near points such that the curve $\sigma^{-1}(B)$ has only normal crossings as singularities and that the surface S, which is the normalization of F in $k(S_0)$, ramifies nicely not only in a neighborhood of $\sigma^{-1}(P_\infty)$ (over the curves $\sigma^{-1}(B \cup L_\infty \cup M_\infty)$) but also in respective neighborhoods of $\sigma^{-1}(P_\alpha)$'s (over the curve $\sigma^{-1}(B)$), where P_α ranges over all singular points of B other than P_∞. Then we obtain a smooth projective surface $\widetilde{V}$ from S by resolving minimally all quotient singular points on S. We can endow $\widetilde{V}$ with a fibration $\widetilde{f} : \widetilde{V} \to C \cong \mathbb{P}^1_k$ of quasi-hyperelliptic curves of genus $g = (p-1)/2$ such that the generic fiber is birationally equivalent to the curve $z^p = y^2 + \varphi(x)$ over $\mathfrak{K} := k(C) = k(x)$ and that $\widetilde{f}$ has a regular cross-section $\widetilde{D}$; the section $\widetilde{D}$ meets the generic fiber of $\widetilde{f}$ at the unique point lying over the point $(1/y = 0, z/y = 0)$ of the curve $z^p = y^2 + \varphi(x)$. However, $\widetilde{f}$ is not relatively minimal, in general. By contracting all (-1)-components contained in the fibers of $\widetilde{f}$, we obtain a relatively minimal quasi-hyperelliptic fibration $f : V \to C$ of genus $g = (p-1)/2$ and with a regular cross-section D. We can describe all reducible fibers of f, and after knowing all reducible fibers, we can write down explicitly the canonical divisor K_V by making use of Lemma 3.10.

3.12 Expression of K_V

Lemma. *Let* $P_\alpha := (x = \alpha, y = 0)$ *be a singular point of* B, *let* L_α *be the member of* $\Lambda := |L_0|$ *passing through* P_α, *and let* Γ_α *be the fiber of* $f : V \to C$ *corresponding to* L_α. *Let* $e = v_\alpha(\varphi(x) - \varphi(\alpha))$. *Assume either* $e = 2$ *or* $p = 5$ *and* $e = 3$. *Then we have:*

$$K_V \sim (p-3)D + a\Gamma + \Delta \quad (cf.\ Lemma\ 3.10),$$

where $\mathrm{Supp}(\Gamma_\alpha) \cap \mathrm{Supp}(\Delta) = \emptyset$.

Proof. Consider first the case $e = 2$. By straightforward computations, we can show that:

(i) The curve $\sigma^{-1}(M_\infty \cup L_\alpha \cup M_0)$ has the following weighted graph:

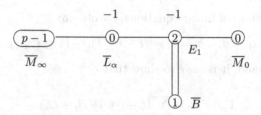

(ii) The curve $(\sigma \cdot \rho \cdot \widetilde{\tau})^{-1}(L_\alpha)$ has the following weighted graph, where $p = 2\ell + 1$:

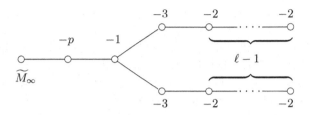

(iii) The fiber Γ_α has the following weighted graph:

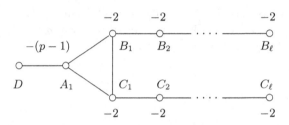

Now write

$$K_V \sim (p-3)D + a\Gamma + a_1 A_1 + \sum_{i=1}^{\ell}(b_i B_i + c_i C_i) + \Delta_1 \,,$$

where a_1, b_i and c_i are non-negative integers and Δ_1 is an effective divisor disjoint from Γ_α. Since A_1, B_i and C_i are smooth rational curves, we obtain relations:

$$(*) \quad \begin{cases} b_1 + c_1 = (p-1)a_1, \quad a_1 - 2b_1 + b_2 + c_1 = 0, \\ a_1 - 2c_1 + c_2 + b_1 = 0, \quad b_i - 2b_{i+1} + b_{i+2} = 0, \\ c_i - 2c_{i+1} + c_{i+2} = 0 \quad (1 \le i \le \ell - 2), \\ b_{\ell-1} = 2b_\ell, \quad c_{\ell-1} = 2c_\ell. \end{cases}$$

Solving this system of linear equations, we obtain:

$$a_1 = \beta, \quad b_i = c_i = (\ell - i + 1)\beta \quad (1 \le i \le \ell).$$

On the other hand, it is easy to show that

$$\Gamma_\alpha = A_1 + \sum_{i=1}^{\ell}(\ell - i + 1)(B_i + C_i) \,.$$

Indeed, since Γ_α is not a multiple fiber, we have only to check that $(\Gamma_\alpha \cdot A_1) = (\Gamma_\alpha \cdot B_i) = (\Gamma_\alpha \cdot C_i) = 0$ for $1 \le i \le \ell$. Since $\Delta_\alpha := a_1 A_1 +$

$\sum_{i=1}^{\ell}(b_i B_i + c_i C_i) = \beta \Gamma_\alpha$ is an effective divisor with $|\Delta_\alpha - \Gamma| = \emptyset$ (cf. Lemma 3.10), we have $\beta = 0$. This proves the assertion in the case $e = 2$.

Consider next the case $p = 5$ and $e = 3$. Then the weighted graphs of the curves $\sigma^{-1}(M_\infty \cup L_\alpha \cup M_0)$, $(\sigma \cdot \rho \cdot \widetilde{\tau})^{-1}(L_\alpha)$ and the fiber Γ_α are respectively given by (i), (ii) and (iii) below:

(i)

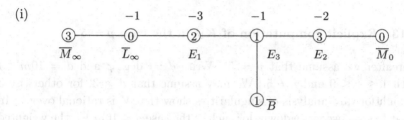

(ii)

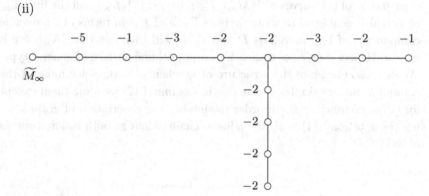

(iii)

Writing $K_V \sim (p-3)D + a\Gamma + \Delta_\alpha + \Delta_1$ and $\Delta_\alpha := bB + \sum_{i=1}^{8} a_i A_i$ as in the previous case, we have

$$\Delta_\alpha = \beta \Gamma_\alpha,$$
$$\Gamma_\alpha = 5B + A_1 + 4A_2 + 7A_3 + 10A_4 + 8A_5 + 6A_6 + 4A_7 + 2A_8 .$$

Hence $\beta = 0$. $\qquad\qquad\square$

This lemma implies that, as far as we consider the case $p = 5$ and assume that $e := v_\alpha(\varphi(x) - \varphi(\alpha)) \le 3$ for every root α of $\varphi'(x)$, there is no contribution to the canonical divisor K_V from the singular points P_α of B other than P_∞. So, we consider the contribution of P_∞.

3.13 Explicit computation of K_V in the case $p = 5$

Hereafter we assume that $p = 5$. Write $d := \deg_x \varphi$ and $d = 10m + i$ with $1 \le i \le 9$ and $i \ne 5$. We may assume that $d \ge 3$, for otherwise a straightforward analysis of singularities show that V is rational over k. In what follows, we write down, for each of the case $d = 10m + i$, the weighted dual graphs of the curves $\sigma^{-1}(M_\infty \cup L_\infty)$, $(\sigma \cdot \rho \cdot \widetilde{\tau})^{-1}(L_\infty)$ and the fiber Γ_∞ of f, which enable us to write divisors Γ_∞ and K_V in terms of irreducible components of Γ_∞ as well as D and Γ. It will turn out that $|K_V| \ne \emptyset$ if $m > 0$. Hence, if $(K_V^2) > 0$ then V is a unirational surface of general type. We can also elucidate the structure of V when $m = 0$. Since most of the computations are similar to the one in Lemma 3.12, we omit them except for only necessary ones. In order to simplify the description of graphs, we use the notation $\Sigma(4)$ to signify a linear chain of four smooth rational curves of weight -2:

3.13.1

Case $d = 10m + 1$ $(m > 0)$.

(i) $\sigma^{-1}(M_\infty \cup L_\infty)$:

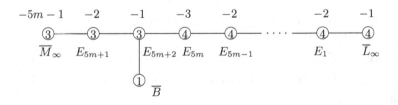

(ii) $(\sigma \cdot \rho \cdot \widetilde{\tau})^{-1}(L_\infty)$:

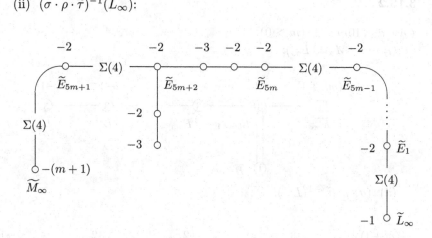

(iii) Γ_∞:

(iv)

$$\Gamma_\infty = 5B + \sum_{i=1}^{10} iA_i + 6A_{11} + 2A_{12}\,,$$

$$K_V \sim 2D + (3m+1)\Gamma + 2\sum_{i=1}^{10} A_i + A_{11} + B\,.$$

The coefficient of Γ is computed from the data: $p_a(D) = 0$, $(D^2) = -(m+1)$ (cf. Lemma 3.12) and $(D \cdot \Gamma) = (D \cdot A_1) = 1$.

(v) $(K_V^2) = 8m - 4$.

3.13.2

Case $d = 10m + 2$ $(m > 0)$.
(i) $\sigma^{-1}(M_\infty \cup L_\infty)$:

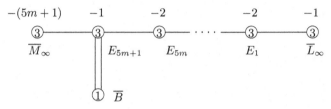

(ii) $(\sigma \cdot \rho \cdot \widetilde{\tau})^{-1}(L_\infty)$:

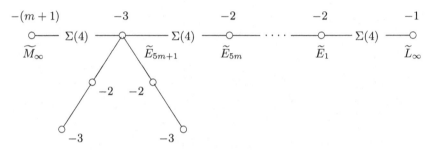

(iii) Γ_∞:

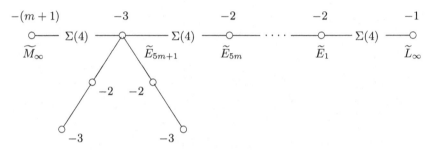

(iv)

$$\Gamma_\infty = \sum_{i-1}^{5} iA_i + 3(B_1 + C_1) + (B_2 + C_2)\,,$$

$$K_V \sim 2D + (3m - 1)\Gamma + 2\sum_{i=1}^{5} A_i + (B_1 + C_1)\,.$$

(v) $(K_V^2) = 8m - 4$.

3.13.3

Case $d = 10m + 3$ $(m \geq 0)$.

(i) $\sigma^{-1}(M_\infty \cup L_\infty)$:

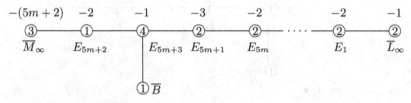

(ii) $(\sigma \cdot \rho \cdot \tilde{\tau})^{-1}(L_\infty)$:

$$-(m+1) \quad -2 \quad -3 \quad -1 \quad -5 \quad -1 \quad -3 \quad -2 \quad -2 \qquad\qquad -2$$

$\widetilde{M}_\infty \qquad\qquad \widetilde{E}_{5m+2} \qquad \widetilde{E}_{5m+3} \qquad \widetilde{E}_{5m+1} \qquad \Sigma(4) \cdots \qquad \widetilde{E}_1$

-5

$\Sigma(4)$

$-1 \quad \widetilde{L}_\infty$

(iii) Γ_∞ :

$$-(m+1) \quad -2 \quad -2 \quad -2 \quad -3$$

$D \qquad\quad A_1 \qquad A_2 \qquad A_3 \qquad A_4$

(iv)

$$\Gamma_\infty = A_1 + 2A_2 + 3A_3 + 2A_4 \, ,$$
$$K_V \sim 2D + (3m-1)\Gamma + 2(A_1 + A_2 + A_3) + A_4 \, .$$

(v) $(K_V^2) = 8m - 3$.

(vi) If $m > 0$, then $p_g(V) > 0, (K_V^2) > 0$ and V is minimal. Hence V is a unirational surface of general type. Suppose $m = 0$. Then D is a (-1)-curve. Let $\mu : V \to W$ be the contraction of D, A_1, A_2 and A_3 in this order. Then it is easy to see that $K_W \sim -G$ with $G = \mu_*(A_4)$. Hence W is rational over k.

3.13.4

Case $d = 10m + 4$ $(m \geq 0)$.

(i) $\sigma^{-1}(M_\infty \cup L_\infty)$:

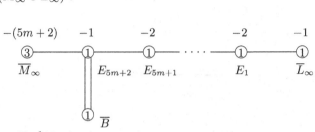

(ii) $(\sigma \cdot \rho \cdot \tilde{\tau})^{-1}(L_\infty)$:

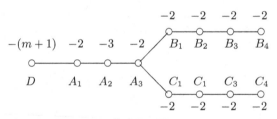

(iii) Γ_∞ :

(iv)

$$\Gamma_\infty = A_1 + 2A_2 + 5A_3 + \sum_{i=1}^{4}(5 - i)(B_i + C_i) \, ,$$

$$K_V \sim 2D + 3m\Gamma + A_1 \, .$$

(v) $(K_V^2) = 8m - 2$.

(vi) If $m > 0$, then $p_g(V) > 0$, $(K_V^2) > 0$ and V is minimal. Hence V is a unirational surface of general type. Suppose $m = 0$. Then D is a (-1)-curve. Let $\mu : V \to W$ be the contraction of D and A_1. Then $K_W \sim 0$. Hence W is a unirational K3-surface.

3.13.5

Case $d = 10m + 6$ $(m \geq 0)$.

(i) $\sigma^{-1}(M_\infty \cup L_\infty)$:

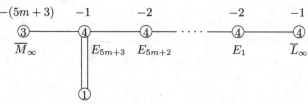

(ii) $(\sigma \cdot \rho \cdot \widetilde{\tau})^{-1}(L_\infty)$:

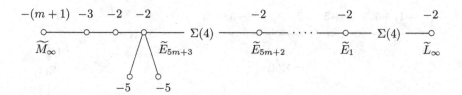

(iii) Γ_∞ :

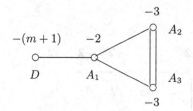

A_1, A_2, A_3 meet in a common point P,
where $i(A_2, A_3; P) = 2$

(iv)

$$\Gamma_\infty = A_1 + A_2 + A_3,$$
$$K_V \sim 2D + 3m\Gamma + A_1.$$

(v) $(K_V^2) = 8m - 2$.

(vi) If $m > 0$, then $p_g(V) > 0$, $(K_V^2) > 0$ and V is minimal. Suppose $m = 0$. Then $(D^2) = -1$. Let $\mu : V \to W$ be the contraction of D and A_1. Then $K_W \sim 0$, and W is a unirational K3-surface.

3.13.6

Case $d = 10m + 7$ $(m \geq 0)$.
  (i) $\sigma^{-1}(M_\infty \cup L_\infty)$:

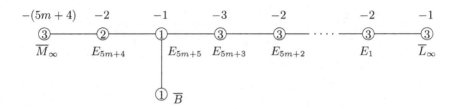

$$-(5m+4) \quad -2 \qquad -1 \qquad -3 \qquad -2 \qquad\qquad -2 \qquad -1$$

$M_\infty \qquad E_{5m+4} \qquad\quad E_{5m+5} \;\; E_{5m+3} \qquad E_{5m+2} \qquad\qquad E_1 \qquad L_\infty$

B

  (ii) $(\sigma \cdot \rho \cdot \widetilde{\tau})^{-1}(L_\infty)$:

$$-(m+1) \;\; -5 \;\; -1 \;\; -3 \;\; -2 \;\; -2 \qquad -3 \;\; -2 \;\; -2 \qquad\qquad\qquad -2$$

$\widetilde{M}_\infty \qquad\quad \widetilde{E}_{5m+4} \qquad\qquad \widetilde{E}_{5m+5} \qquad \widetilde{E}_{5m+3} \qquad \Sigma(4) \; - \cdots - \qquad \widetilde{E}_1$

$\Sigma(4) \qquad\qquad\qquad\qquad\qquad\qquad\qquad\qquad\qquad\qquad \Sigma(4)$

$$-1 \quad \widetilde{L}_\infty$$

  (iii) Γ_∞ :

$$-(m+1) \quad -4 \quad -2 \quad -2 \quad -2 \qquad -2 \quad -2 \quad -2 \quad -2$$

$D \qquad A_1 \quad A_2 \quad A_3 \qquad A_4 \quad A_5 \quad A_6 \quad A_7 \quad A_8$

$$-2 \quad B$$

  (iv)

$$\Gamma_\infty = 5B + A_1 + 4A_2 + 7A_3 + 10A_4 + 8A_5 + 6A_6 + 4A_7 + 2A_8 ,$$
$$K_V \sim 2D + (3m+1)\gamma .$$

  (v) $(K_V^2) = 8m$.

  (vi) If $m > 0$, then $p_g(V) > 0, (K_V^2) > 0$ and V is minimal. Suppose $m = 0$. Then $(D^2) = -1$. Let $\mu : V \to W$ be the contraction of D. Then $K_W \sim G$ with $G := \mu(\Gamma)$. Hence $p_g(W) > 0, (K_W^2) = (G^2) = 1$ and W is minimal. Thus V for $m > 0$ as well as W is a unirational surface of general type for $m \geq 0$.

3.13.7

Case $d = 10m + 8$ $(m \geq 0)$.

(i) $\sigma^{-1}(M_\infty \cup L_\infty)$:

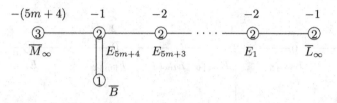

(ii) $(\sigma \cdot \rho \cdot \tilde{\tau})^{-1}(L_\infty)$:

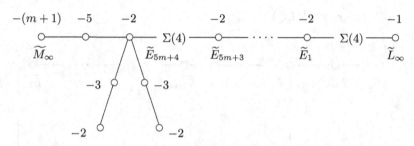

(iii) Γ_∞:

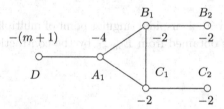

A_1, B_1, C_1 meet transversally in a common point P.

(iv)

$$\Gamma_\infty = A_1 + 2(B_1 + C_1) + (B_2 + C_2),$$
$$K_V \sim 2D + (3m + 1)\Gamma.$$

(v) $(K_V^2) = 8m$.

(vi) If $m > 0$, then $p_g(V) > 0$, $(K_V^2) > 0$ and V is minimal. Suppose $m = 0$. Let $\mu : V \to W$ be the contraction of D. Rhen $K_W \sim G$ with $G := \mu(\Gamma)$. Hence $p_g(W) > 0$, $(K_W^2) = (G^2) = 1$, and W is minimal. Hence V for $m > 0$ as well as W is a unirational surface of general type.

3.13.8

Case $d = 10m + 9$ $(m \geq 0)$.

(i) $\sigma^{-1}(M_\infty \cup L_\infty)$:

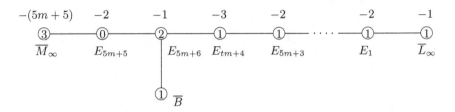

(ii) $(\sigma \cdot \rho \cdot \widetilde{\tau})^{-1}(L_\infty)$:

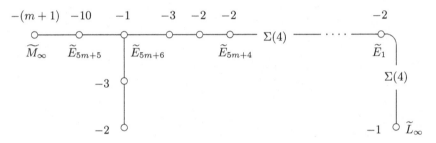

(iii) Γ_∞ is a rational curve A with a cuspidal singular point of multiplicity sequence $(2, 2, 1, \ldots)$ which is obtained from $\widetilde{E}_{5m+5}$ by the contraction of other contractible curves.

(iv)

$$\Gamma_\infty = A,$$
$$K_V \sim 2D + (3m + 1)\Gamma.$$

(v) $(K_V^2) = 8m$.

(vi) If $m > 0$, then $p_g(V) > 0$, $(K_V^2) > 0$ and V is minimal. Suppose $m = 0$. Let $\mu : V \to W$ be the contraction of D. Then $K_W \sim G$ with $G := \mu(\Gamma)$. Hence $p_g(W) > 0, (K_W^2) = (G^2) = 1$ and W is minimal. Thus, V for $m > 0$ as well as W is a unirational surface of general type.

3.14 Final remark

The foregoing observations complete the proof of Theorem 3.9. As witnessed in the above computations, we have in hand techniques necessary to improve Theorem 3.9 to the effect that the restriction $v_\alpha(\varphi(x) - \varphi(\alpha)) \leq 3$ be eliminated for every root α of $\varphi'(x) = 0$ and also to the effect that the same techniques be applied to a unirational surface with a quasi-hyperelliptic fibration $f : V \to C \cong \mathbb{P}^1$ of genus $g = (p-1)/2$ with $p > 5$. But complexity of computations increases accordingly, and we have no concrete results valid without the restriction or for every prime p.

Chapter 4

Mordell-Weil groups of quasi-elliptic or quasi-hyperelliptic surfaces

4.0 Introduction

The theory of Mordell-Weil groups and lattice structures was started by T. Shioda in his works [88], [89] and [90] and developed in his subsequent papers for surfaces with elliptic fibrations over an algebraically closed field of arbitrary characteristic. Since a quasi-elliptic fibration is considered as a degeneration of an elliptic fibration, we can consider the Mordell-Weil group for a surface with a quasi-elliptic fibration by following the main stream of arguments for an elliptic fibration. We will describe the theory in this chapter which shows an enriched geometry behind quasi-elliptic fibrations. In the present chapter, we utilize the results in chapter 3.

4.1 Mordell-Weil groups and Néron-Severi groups of quasi-elliptic surfaces

In this subsection, we define the *Mordell-Weil group* of a quasi-elliptic surface with a section and show that it is a finite group when the surface is unirational.

Let $f : V \to C$ be a quasi-elliptic surface with a section O, which we assume to be relatively minimal, i.e., no (-1) curves in the fibers. Let E be the generic fibre V_η with the unique singular point P_∞ deleted off. Then E is a smooth algebraic curve defined over $K = k(C)$ [1] and it is a quasi-elliptic K-form of the affine line $\mathbb{A}^1$ with a K-rational point O. We here recall that a K-form of the affine line is a smooth affine curve E defined over K such that $E \otimes_K K'$ is K'-isomorphic to the affine line $\mathbb{A}^1_{K'}$ over

---

[1] $k(C)$ is denoted by $\mathfrak{K}$ in the previous chapters. In this chapter, we use, however, the symbol K in accordance with a common use in the theory of Mordell-Weil groups.

K' for some algebraic field extension K' of K and it is quasi-elliptic if a complete normal model $\mathcal{V}$ of $K(E)$ has arithmetic genus one (see I. 2.8).

Let $\mathrm{Pic}^{\circ}_{V_{\eta}/K}$ be the connected component of the Picard scheme $\mathrm{Pic}_{V_{\eta}/K}$ of the curve V_{η}. We now define the Mordell-Weil group of V as the set $E(K)$ of K-rational points of $E = V_{\eta} - \{P_{\infty}\}$ endowed with a group structure via the standard embedding of E into the Picard group $\mathrm{Pic}^{\circ}_{V_{\eta}/K}(K)$, i.e., $P \mapsto (P) - (O)$.

4.1.1 Group structure of $E(K)$

We recall Theorem I. 2.4.3, which is originally [43, Theorem 6.7.9].

Theorem. *Let E be a K-form of the affine line with a K-rational point O and let $\mathcal{V}$ be its normal completion. Then the mapping $i : E \to \mathrm{Pic}^{\circ}_{\mathcal{V}/K}$ defined by $i(Q) = (Q) - [K(Q) : K](O)$ for $Q \in E$ is a closed immersion, and $\mathrm{Pic}^{\circ}_{\mathcal{V}/K}$ is generated as a K-group scheme by the image of i.*

In our case where $E = V_{\eta} - \{P_{\infty}\}$ and $\mathcal{V} = V_{\eta}$, since $p_a(V_{\eta}) = 1$, i is actually an isomorphism. Let Γ be the closure of P_{∞} in V. Hence i induces a group structure on $E(K)$ with the point O as the zero and i is an isomorphism between the group of the rational points: $E(K) \cong \mathrm{Pic}^{\circ}_{V_{\eta}/K}(K)$. Here $\mathrm{Pic}^{\circ}_{V_{\eta}/K}(K)$ is a form of the additive group scheme G_a. So, $E(K)$ is an abelian group such that every element of $E(K)$ is killed by multiplication of the characteristic p, i.e., $pP = O$ for $P \in E(K)$. Namely, $E(K)$ is a vector space over the prime field $\mathbb{F}_p$.

4.1.2 Torsion rank of $E(K)$

Now assume that V is unirational. Then we can show that the group $E(K)$ is finitely generated.

Theorem. *The Mordell-Weil group $E(K)$ of V is a finite-dimensional vector space over the finite field $\mathbb{F}_p$.*

The proof will be given below. The unirationality of V is used only in Proposition 4.1.2.5 which affects Theorems 4.1.2.1 and consequently Theorem 4.1.2. Meanwhile, other results in this subsection hold without the assumption that V be unirational.

The dimension of the Mordell-Weil group $E(K)$ as a vector space over $\mathbb{F}_p$ is called *torsion rank* of the Mordell-Weil group of a unirational quasi-elliptic surface.

4.1.2.1

In order to show that $E(K)$ is finitely generated, we borrow the idea from Shioda-Schütt [87] in the case of the elliptic surfaces and make use of the finite generation of the Néron-Severi group of V. For a proof, we need to fix notations.

Let $F_v = f^{-1}(v)$ denote the fibre over $v \in C$, and let

$$R = \{v \in C \mid F_v \text{ is reducible}\}.$$

For each $v \in R$, write

$$F_v = f^{-1}(v) = \Theta_{v,0} + \sum_{i=1}^{m_v - 1} \mu_{v,i} \Theta_{v,i}$$

be the irreducible decomposition, where $\Theta_{v,i}$ $(0 \leq i \leq m_v - 1)$ with multiplicities $\mu_{v,i}$ are irreducible components of F_v and $\Theta_{v,0}$ is the unique component of F_v meeting the zero section (O). For $P \in E(K)$, we denote by (P) the closure of P in V which is a cross-section of f. Let T denotes the subgroup of the Néron-Severi group NS (V) of V generated by the zero section (O) and all the irreducible components of fibres. We have the following two results.

Theorem. *For a unirational quasi-elliptic surface V with a zero section (O), the Néron-Severi group NS (V) of V is a finitely generated torsion-free $\mathbb{Z}$-module.*

4.1.2.2

In terms of NS (V), we have the following isomorphism of groups.

Theorem. *There is a natural isomorphism*

$$E(K) \cong \text{NS}\,(V)/T,$$

where the isomorphism maps $P \in E(K)$ to (P) (mod T).

It suffices to prove Theorems 4.1.2.1 and 4.1.2.2. In fact, since T is finitely generated, $E(K)$ is finitely generated if and only if so is NS (V).

4.1.2.3

For the finite generation of NS (V), we use a well-known result in Milne [51, Theorem 2.5].

Theorem. *Let ℓ be a prime not equal to p. The ℓ-adic étale cohomology $H^2_{et}(V, \mathbb{Q}_\ell)$ of V is a finite-dimensional vector space over $\mathbb{Q}_\ell$ equipped with a non-degenerate pairing*

$$H^2_{et}(V, \mathbb{Q}_\ell) \times H^2_{et}(V, \mathbb{Q}_\ell) \longrightarrow \mathbb{Q}_\ell$$

and there is a homomorphism (cycle map)

$$\gamma : \mathrm{NS}\,(V) \longrightarrow H^2_{et}(V, \mathbb{Q}_\ell)$$

preserving the pairings. Namely, for divisors D, D' on X, we have

$$(D \cdot D') = (\gamma(D), \gamma(D')).$$

The Néron-Severi group is by definition the free abelian group generated by irreducible divisors modulo numerical equivalence ($\equiv$ by notation), i.e., $D \equiv D'$ if and only if $(D \cdot C) = (D' \cdot C)$ for every curve C, and it is known that the kernel of the cycle map consists of those classes of D which are numerically equivalent to 0 (see Schütt-Shioda [87, §6.3]).

4.1.2.4

As an immediate consequence, we have the following.

Corollary. *The Néron-Severi group $\mathrm{NS}\,(V)$ of V modulo numerical equivalence is finitely generated and torsion free.*

4.1.2.5

This corollary combined with Proposition 4.1.2.5 below implies Theorem 4.1.2.1.

Proposition. *On a unirational quasi-elliptic surface with a section, the following conditions on a divisor D are equivalent:*

(1) *D is linearly equivalent to 0, i.e., $D \sim 0$ by notation.*
(2) *D is algebraically equivalent to 0, i.e., $D \approx 0$ by notation.*
(3) *D is numerically equivalent to 0, i.e., $D \equiv 0$.*

Proof. Obviously, it suffices to show that (3) implies (1). Let D be a divisor on V which is numerically equivalent to 0. Then, by the Riemann-Roch Theorem, we have

$$h^0(V, \mathcal{O}(D)) - h^1(V, \mathcal{O}(D)) + h^2(V, \mathcal{O}(D))$$
$$= \frac{1}{2}((D^2) - (D \cdot K_V)) + \chi(\mathcal{O}_V).$$

In the unirational case where $C \cong \mathbb{P}^1$, $\chi(\mathcal{O}_V)$ is positive by Theorem 3.1.5.1. Hence we have either (a) $h^0(V, \mathcal{O}(D)) > 0$ or (b) $h^2(V, \mathcal{O}(D)) > 0$. In the case (a), D is linearly equivalent to an effective divisor numerically equivalent to 0. Hence $D \sim 0$ and we are done.

Consider the case (b). Then, by the Serre duality

$$h^0(V, \mathcal{O}(K_V - D)) = h^2(V, \mathcal{O}(D)) > 0,$$

which implies that $K_V - D$ is linearly equivalent to some effective divisor, say D'. For each fibre or its component Θ, we have

$$(D' \cdot \Theta) = (K_V \cdot \Theta) - (D \cdot \Theta) = 0,$$

because $K_V \approx (2g(C) - 2 + \chi(\mathcal{O}_V))F$ by Theorem 3.1.1. All the irreducible components of D' are, therefore, contained in some fibres, and we can write

$$D' = \sum_j a_j F_j + \sum_{v \in R} \sum_{i \geq 1} b_{v,i} \Theta_{v,i} \qquad (a_j, b_{v,i} \in \mathbb{Z}).$$

Since $(D' \cdot \Theta_{v,i}) = 0$ for every $v \in R$ and $1 \leq i \leq m_v - 1$ and the intersection matrix $A_v = ((\Theta_{v,i} \cdot \Theta_{v,j}))_{1 \leq i,j \leq m_v - 1}$ is negative-definite for every $v \in R$, we see easily that $b_{v,i} = 0$ for all v, i. It follows that D' is linearly equivalent to some multiple of a fibre F. Since K_V is algebraically equivalent to some multiple of F, D is also linearly equivalent to mF for some $m \in \mathbb{Z}$. Taking the intersection number with zero section (O), we see $m = 0$ by the assumption. Hence $D \sim 0$. □

4.1.2.6

The following result holds.

Proposition. *T is a torsion-free abelian group generated by the divisor classes:*

$$(O), \ F, \ \Theta_{v,i} \ (v \in R, 1 \leq i \leq m_v - 1),$$

where F denote a general fibre of f. In particular, we have

$$\operatorname{rank} T = 2 + \sum_{v \in R} (m_v - 1).$$

Proof. It is clear that T is generated by these divisor classes. In order to show that T is torsion-free, it suffices to show that they are linearly independent over $\mathbb{Z}$. One can easily check it by using the negative-definiteness of the intersection matrices of $\{\Theta_{v,i} \mid (v \in R, 1 \leq i \leq m_v - 1)\}$ and the fact that $((P) \cdot F) = ((O) \cdot F) = 1$ and $(F^2) = 0$. □

4.1.2.7

Proof of Theorem 4.1.2.2 begins with definition of a group homomorphism

$$\psi : \mathcal{D}(V) \longrightarrow E(K),$$

where $\mathcal{D}(V)$ is a free abelian group generated by irreducible curves on V. We denote by $\mathcal{D}_a(V)$ (resp. $\mathcal{D}_\ell(V)$) the subgroup of $\mathcal{D}(V)$ consisting of divisors D such that $D \approx 0$ (resp. $D \sim 0$). We say that a divisor D is *vertical* if D is written as $D = \sum \alpha_i C_i$, where C_i is an irreducible component contained in a fibre of f.

For a divisor D on V, we can define a K-rational divisor $D \cdot V_\eta$ on the generic fibre V_η by

$$\mathcal{O}_{V_\eta}(D \cdot V_\eta) = \mathcal{O}_V(D) \otimes_{\mathcal{O}_V} \mathcal{O}_{V_\eta}.$$

Lemma. *The following assertions hold.*

(1) *A divisor D on V is linearly equivalent to a vertical divisor if and only if $D \cdot V_\eta \sim 0$ on V_η.*
(2) *$\Gamma - p(O)$ is linearly equivalent to a vertical divisor, where Γ is the closure of the singular point of V_η.*

Proof. (1) Suppose that D is linearly equivalent to a vertical divisor. Namely, $D = \sum \alpha_i C_i + (g)_V$ with $g \in k(V)$ and each C_i is contained in a fibre of f. Then $D \cdot V_\eta = (g)_{V_\eta} \sim 0$, where $(g)_{V_\eta}$ is the divisor on V_η of the function $g \in K(V_\eta) = k(V)$. Conversely, suppose that $D \cdot V_\eta \sim 0$. Write $D \cdot V_\eta = (g)_{V_\eta}$ with $g \in K(V_\eta)$. Then $D - (g)_V$ is written as $\sum \alpha_i C_i$, where C_i does not meet V_η. Hence D is linearly equivalent to a vertical divisor.

(2) The generic fibre V_η has a K-rational point O, and the morphism $\Phi_{|pO|}$ associated with a linear system $|pO|$ embeds V_η into a cubic curve $\mathbb{P}^2_K$ defined by an equation

$$y^2 = x^3 + ax + b \qquad \text{with } a, b \in K \text{ such that } a^{\frac{1}{2}} \notin K \text{ or}$$
$$b^{\frac{1}{2}} \notin K \text{ if } p = 2,$$
$$y^2 = x^3 + a \qquad \text{with } a \in K \text{ and } a^{\frac{1}{3}} \notin K \text{ if } p = 3,$$

with respect to a suitable inhomogeneous coordinate system $(x, y, 1)$ on $\mathbb{P}^2_K$, where the point O is sent to $(0, 1, 0)$ and $\Gamma \cdot V_\eta$ is sent to a point $(a^{\frac{1}{2}}, b^{\frac{1}{2}}, 1)$ if $p = 2$, $(-a^{\frac{1}{3}}, 0, 1)$ if $p = 3$ (see Lemma 3.1.3). Since $\Gamma \cdot V_\eta$ is locally defined by $x^{-1}(y^2 - b) = 0$ if $p = 2$ and $a \neq 0$, by $x = 0$ if $p = 2$ and $a = 0$, and by $y = 0$ if $p = 3$, we have

$$\Gamma \cdot V_\eta - pO = \begin{cases} (\frac{y^2 - b}{x})_{V_\eta} & \text{if } p = 2 \text{ and } a \neq 0 \\ (x)_{V_\eta} & \text{if } p = 2 \text{ and } a = 0 \\ (y)_{V_\eta} & \text{if } p = 3. \end{cases}$$

This implies that $\Gamma \sim p(O) +$ a vertical divisor. $\qquad\square$

4.1.2.8 *Proof of Theorem 4.1.2*

On the curve V_η defined over K, we have

$$\text{Pic}^\circ_{V_\eta/K}(K) = \mathcal{D}_a(V_\eta)/\mathcal{D}_\ell(V_\eta),$$

where $\mathcal{D}_a(V_\eta)$ is the group of K-rational divisors with degree zero and $\mathcal{D}_\ell(V_\eta)$ is the group of K-rational divisors linearly equivalent to zero. For the affine curve $E := V_\eta - \Gamma \cdot V_\eta$, we set

$$\mathcal{D}_a(E) := \mathcal{D}(E) \cap \mathcal{D}_a(V_\eta),$$
$$\mathcal{D}_\ell(E) := \mathcal{D}(E) \cap \mathcal{D}_\ell(V_\eta).$$

Since $\Gamma \cdot V_\eta \sim pO$ by Lemma 4.1.2.7, we have

$$\mathcal{D}_a(V_\eta) = \mathcal{D}_a(E) + \mathcal{D}_\ell(V_\eta).$$

Hence we have

$$\mathcal{D}_a(V_\eta)/\mathcal{D}_\ell(V_\eta) = \mathcal{D}_a(E)/\mathcal{D}_\ell(E).$$

Since we have by 4.1.1

$$i : E(K) \cong \text{Pic}^\circ_{V_\eta/K}(K), \qquad (4.1)$$

we have

$$E(K) \cong \mathcal{D}_a(E)/\mathcal{D}_\ell(E). \qquad (4.2)$$

Define a group homomorphism

$$\psi : \mathcal{D}(V) \longrightarrow E(K)$$

by $\psi(D) = D \cdot V_\eta - (D \cdot V_\eta)O$ via the identifications (4.1) and (4.2), where $(D \cdot V_\eta)$ is the degree of a K-rational divisor $D \cdot V_\eta$. The kernel $\text{Ker}\,\psi$ consists of D such that

$$D \cdot V_\eta \sim (D \cdot V_\eta)O,$$

whence we obtain

$$\operatorname{Ker}\psi = \mathcal{D}_\ell(V) + \mathbb{Z}(O) + \mathcal{D}_{ver}(V),$$

where $\mathcal{D}_{ver}(V)$ is the subgroup of vertical divisors on V. Since $\mathcal{D}_\ell(V) = \mathcal{D}_a(V)$ by Proposition 4.1.2.5 and

$$T = \frac{\mathbb{Z}(O) + \mathcal{D}_{ver}(V) + \mathcal{D}_a(V)}{\mathcal{D}_a(V)},$$

we have

$$\operatorname{Im}\psi = \frac{\mathcal{D}(V)/\mathcal{D}_a(V)}{T} = \operatorname{NS}(V)/T.$$

It is clear that ψ is surjective, because $\psi((P)) = P$ holds clearly for $P \in E(K)$. So, ψ induces an isomorphism

$$\overline{\psi} : \operatorname{NS}(V)/T \xrightarrow{\sim} E(K).$$

The inverse of $\overline{\psi}$ is

$$\overline{\varphi} : E(K) \longrightarrow \operatorname{NS}(V)/T$$

defined by $\overline{\varphi}(P) = (P) \pmod{T}$. $\square$

4.1.3 $E(K)^\circ = \{0\}$

We have the following result.

Proposition. *Set*

$$E(K)^\circ = \{P \in E(K) \mid (P) \text{ meets } \Theta_{v,0} \text{ for all } v \in R\}.$$

Then we have

$$E(K)^\circ = \{O\}.$$

Proof. Define a map φ from the Mordell-Weil group $E(K)$ to the Néron-Severi group $\operatorname{NS}(V) \otimes \mathbb{Q}$,

$$\varphi : E(K) \mapsto \operatorname{NS}(V) \otimes \mathbb{Q}$$

by

$$\varphi(P) = (P) - (O) - ((P - O) \cdot (O))F$$
$$- \sum_{v \in R}(\Theta_{v,1}, \cdots, \Theta_{v,m_v - 1})A_v^{-1}\begin{pmatrix} (\Theta_{v,1} \cdot (P)) \\ \vdots \\ (\Theta_{v,m_v - 1} \cdot (P)) \end{pmatrix}$$

for $P \in E(K)$ and $A_v = ((\Theta_{v,i} \cdot \Theta_{v,j}))_{1 \leq i,j \leq m_v - 1}$. Then one can easily check that φ is a homomorphism and makes the following diagram commutative:

$$
\begin{array}{ccc}
E(K) & \xrightarrow{\ \varphi\ } & L^* \\
\uparrow & & \uparrow \\
E(K)^\circ & \xrightarrow{\ \varphi|_{E(K)^\circ}\ } & L,
\end{array}
$$

where L is the orthogonal complement $T^\perp$ of T in $\mathrm{NS}\,(V)_{\mathbb{Q}} := \mathrm{NS}\,(V) \otimes_{\mathbb{Z}} \mathbb{Q}$ and L^* is its dual lattice[2]. Moreover, this map φ is a zero map since $E(K)$ is a torsion group. Meanwhile, $\varphi|_{E(K)^\circ}$ is injective. In fact, if $\varphi(P) = 0$ for $P \in E(K)^\circ$, then $(P) \approx (O) + nF$ for some $n \in \mathbb{Z}$ and

$$(P^2) = (O^2) + 2n,$$

where $(P \cdot Q)$ for $P, Q \in E(K)$ is defined as the intersection number $((P) \cdot (Q))$ of the corresponding sections (P) and (Q). Hence $n = 0$ by Corollary I. 3.1.2 and consequently $(P) \approx (O)$. So, we are done. $\qquad\square$

Note that the map φ in this proof induces the map $\overline{\varphi}$ in 4.1.2.8.

4.1.4 *Generators of* $\mathrm{NS}\,(V)$

The following theorem plays an important role in determining the lattice structure of $NS(V)$.

Theorem. *The Néron-Severi group* $\mathrm{NS}\,(V)$ *of a quasi-elliptic surface* V *with a section is generated by the following divisor classes:*

$$(O),\ F,\ \Theta_{v,i}(v \in R, 1 \leq i \leq m_v - 1),\ D_i(1 \leq i \leq r),$$

where $D_i = (P_i) - (O)$ *with* $P_1, \cdots, P_r$ *representing a basis of the Mordell-Weil group* $E(K)$, *which is a vector space over* $\mathbb{F}_p$ *of torsion-rank* r. *The fundamental relations*[3] *are then exhausted by following* r *relations:*

$$pD_i \approx p(D_i \cdot (O))F + \sum_{v \in R}(\Theta_{v,1}, \cdots, \Theta_{v,m_v-1})pA_v^{-1}\begin{pmatrix} (\Theta_{v,1} \cdot D_i) \\ \vdots \\ (\Theta_{v,m_v-1} \cdot D_i) \end{pmatrix}$$

---

[2]Note that $\mathrm{NS}\,(V)$ is a free $\mathbb{Z}$-module of finite rank by Corollary 4.1.2.4 which is equipped with an indefinite intersection pairing of signature $(1^+, (\rho(V) - 1)^-)$. We then call $\mathrm{NS}\,(V)$ the *Néron-Severi lattice*. Since T is a submodule of $\mathrm{NS}\,(V)$, we set $T^\perp := \{z \in \mathrm{NS}\,(V) \mid (z, x) = 0, \forall x \in T\}$ and call it the *orthogonal complement* of T. The *dual lattice* L^* is defined as

$$L^* = \{w \in \mathrm{NS}\,(V)_{\mathbb{Q}} \mid (w, z) \in \mathbb{Z}, \forall z \in L\}.$$

[3]We call an integral relation among them a *fundamental relation* if the coefficients have no common factors.

for $1 \leq i \leq r$.

Proof. It follows from Lemma 4.1.2.7, (1) that the Picard group $\mathcal{D}(V)/\mathcal{D}_\ell(V)$ is generated by these divisors, where the Picard group coincides with the Néron-Severi group in view of Proposition 4.1.2.5. Indeed, for any divisor D on X, $D \cdot V_\eta - (D \cdot V_\eta)O$ is an element of $E(K)$. Hence $D - (D \cdot V_\eta)O - \sum \alpha_i D_i$ is linearly equivalent to a vertical divisor for $\alpha_i \in \mathbb{Z}$. Suppose that there is a relation:

$$\sum_i \alpha_i D_i + \beta F + \gamma(O) + \sum_j \delta_{v,j} \Theta_{v,j} \approx 0,$$

where $\alpha_i, \beta, \gamma, \delta_{v,j} \in \mathbb{Z}$. We denote by D the left-hand side. By taking intersection with F, (O) and $\Theta_{v,k}$ for $v \in R$ and $1 \leqslant k \leqslant m_v - 1$, we obtain a relation

$$D = \sum_{i=1}^r \alpha_i \left(D_i - (D_i \cdot (O))F - \sum_{v \in R} (\Theta_{v,1}, \cdots, \Theta_{v,m_v-1}) A_v^{-1} \begin{pmatrix} (\Theta_{v,1} \cdot D_i) \\ \vdots \\ (\Theta_{v,m_v-1} \cdot D_i) \end{pmatrix} \right)$$

$$\approx 0.$$

Moreover, by applying the map ψ of Theorem 4.1.2.2, we have

$$\psi(D) = \sum_{i=1}^r \alpha_i P_i = 0 \qquad \text{in } E(K),$$

where $\alpha_i \equiv 0 \pmod{p}$ for each i. Hence $D \approx 0$ modulo the group T if and only if α_i is divisible by p for each i. On the other hand, pD_i, as an element of NS (V), is uniquely expressed as a $\mathbb{Z}$-linear combination of F and $\Theta_{v,i} (v \in R, 1 \leqslant i \leqslant m_v - 1)$. By a straightforward computation, we have

$$R_i := pD_i - p(D_i \cdot O)F - \sum_{v \in R} (\Theta_{v,1}, \cdots, \Theta_{v,m_v-1})pA_v^{-1} \begin{pmatrix} (\Theta_{v,1} \cdot D_i) \\ \vdots \\ (\Theta_{v,m_v-1} \cdot D_i) \end{pmatrix}$$

$$\approx 0.$$

So, D is written as

$$D = \sum_{i=1}^r \beta_i R_i, \quad \text{where} \quad \beta_i = \frac{\alpha_i}{p}.$$

Finally, since $E(K)^\circ = 0$ by Proposition 4.1.3, the gcd of the coefficients in the relations $R_i \approx 0$ must be 1. Thus these are the fundamental relations.
$\square$

Corollary. *We have*

$$\operatorname{rank} \operatorname{NS}(V) = \operatorname{rank} T = 2 + \sum_{v \in R}(m_v - 1).$$

Remark. Let P be a cross-section of $f : V \to \mathbb{P}^1$ which is different from O and let $D = (P) - (O)$. Set $D = (P) - (O)$. Then we have a relation

$$(P) - (O) \approx (D \cdot (O))F + \sum_{v \in R}(\Theta_{v,1}, \cdots, \Theta_{v,m_v-1})A_v^{-1} \begin{pmatrix} (\Theta_{v,1} \cdot D) \\ \vdots \\ (\Theta_{v,m_v-1} \cdot D) \end{pmatrix}.$$

By making use of Corollary 3.1.2, (3), this implies the following formula for $P, Q \in E(K)$,

$$(P \cdot Q) = \chi(\mathcal{O}_V) + (P \cdot O) + (Q \cdot O) - \sum_{v \in R}\operatorname{Cont}_v(P, Q),$$

where

$$\operatorname{Contr}_v(P, Q) = \begin{cases} 0 & \text{if } (P) \text{ or } (Q) \text{ passes through } \Theta_{v,0} \\ (-A_v^{-1})_{ij} & \text{if } ((P) \cdot \Theta_{v,i}) = ((Q) \cdot \Theta_{v,j}) = 1 \text{ for } i, j \geq 1. \end{cases}$$

We omit the proof which is of computational nature (see [34, 35]). In particular, $\operatorname{Contr}_v(P, O) = 0$ for $v \in R$.

4.2 Reducible singular fibers and torsion rank

In this section we consider reducible singular fibers of a quasi-elliptic fibration and give equalities relating with Euler-Poincaré characteristic, the number of the reducible singular fibers of respective type and torsion-rank.

4.2.1 *Discriminant*

Let $\nu(T)$ denote the number of singular fibers of Kodaira type T (see types given in 3.1.5). Let $f : V \to C$ be a relatively minimal, unirational, quasi-elliptic fibration with a cross-section. Its generic fiber V_η has a Weierstrass equation $y^2 = x^3 + \psi(t)x + \varphi(t)$. In order to treat the discriminant free of characteristic, we set the *discriminant* of the Weierstrass model $y^2 = x^3 + \psi(t)x + \varphi(t)$ as a polynomial in t,

$$\Delta(t) = \psi(t)\psi'(t)^2 + \varphi'(t)^2$$

which has the same form as defined in 3.1.5.2 for $p = 2$, but $\Delta(t)$ is the square of the discriminant defined in Theorem 3.3.1. Note that $\Delta(t)$ is defined by the Weierstrass model $y^2 = x^3 + \psi(t) + \varphi(t)$ over the open set $U_0 := \mathbb{A}^1 = \operatorname{Spec} k[t]$ of $\mathbb{P}^1$. For an open set $U_\infty := \operatorname{Spec} k[\tau]$ with $\tau = 1/t$ of the point at infinity $t = \infty$, we consider a Weierstrass model $\eta^2 = \xi^3 + \Psi(\tau)\xi + \Phi(\tau)$ over U_∞ and the discriminant $\Delta_\infty(\tau)$ associated with this model (see 3.1.4 for the Weierstrass models and 3.1.5.2 for a base change of coordinates in the case $p = 2$). Then $\{\Delta(t), \Delta_\infty(\tau)\}$ gives rise to a section of

$$H^0(\mathbb{P}^1, (R^1 f_* \mathcal{O}_V)^{-12} \otimes \omega_{\mathbb{P}^1}^{\otimes 2}),$$

where $R^1 f_* \mathcal{O}_V \cong \mathcal{L}$ by Theorem 3.1.1 because $\mathcal{T} = 0$ as f has a cross-section. We denote the corresponding divisor on $\mathbb{P}^1$ by (Δ) and call it the *discriminant divisor*. Then $\deg(\Delta) = 12\chi(\mathcal{O}_V) - 4$ by the proof of Lemma 3.1.5.1 if $p = 3$, and the same equality holds for $p = 2$ by [35, p.239]. Hence we have the following result.

Proposition. *Let $f : V \to C$ be a unirational quasi-elliptic surface with section. Then $12\chi(\mathcal{O}_V) - 4$ is equal to*

$$\begin{cases} \sum_{k \geqslant 0}(2k + 4)\nu(\mathrm{I}_{2k}^*) + 8\nu(\mathrm{II}^*) + \nu(\mathrm{III}) + 7\nu(\mathrm{III}^*) & \text{if } p = 2 \\ 8\nu(\mathrm{II}^*) + 2\nu(\mathrm{IV}) + 6\nu(\mathrm{IV}^*) & \text{if } p = 3. \end{cases}$$

Proof. The assertion follows from Theorems 3.1.5.1 and 3.1.5.2. □

Remark. If $p = 3$, we have $\psi(t) = 0$ in the Weierstrass equation. Hence $\Delta(t) = \varphi'(t)^2$. In this case we may set $\Delta(t) = \varphi'(t)$. This simplification will make explicit computations easier to handle. We do this simplification in section 4.3.

4.2.2　*Determinant of the trivial lattice T*

For a lattice Z of rank n with a pairing, we denote by $\det Z$ the determinant of the $(n \times n)$-matrix $((e_i, e_j))_{1 \leq i,j \leq n}$, where $\{e_1, \ldots, e_n\}$ is a free basis of Z. Recall that the lattice T is generated by the zero section, one irreducible fiber and all the irreducible components of reducible fibers which do not meet the section O. We call T the *trivial lattice*. Then T has the natural decomposition

$$T = \,<O,\ F>\,\oplus(\bigoplus_{v\in R} T_v),$$

where T_v is the lattice generated by all the irreducible components of a reducible fiber $f^{-1}(v)$ except the irreducible component meeting the zero section (O). Since $\det\langle O,\ F\rangle = -1$ ad since the lattice T_v for the fiber of type II* is unimodular, we have

$$\det T = -\prod_{v\in R}\det T_v = \begin{cases} -2^{\sum_{k\geqslant 0} 2\nu(\mathrm{I}^*_{2k})+\nu(\mathrm{III})+\nu(\mathrm{III}^*)} & p = 2\,, \\ \\ -3^{\nu(\mathrm{IV})+\nu(\mathrm{IV}^*)} & p = 3\,. \end{cases}$$

This implies that the exponent of $\deg T$ is an even integer. Furthermore, we have $E(K) = \mathrm{NS}\,(V)/T$ by Theorem 4.1.2.2. We can take a free basis $\{e_1,\ldots,e_r,e_{r+1},\ldots,e_n\}$ in such a way that $\{pe_1,\ldots,pe_r,e_{r+1},\ldots,e_n\}$ is a free basis of T. In fact, $E(K)$ is a $\mathbb{F}_p$-vector space of rank r. Take elements $e_1,\ldots,e_r \in \mathrm{NS}\,(X)$ so that $e_1 \pmod{T},\ldots,e_r \pmod{T}$ generate $E(K)$ freely. Let $Z = \sum_{i=1}^r \mathbb{Z}e_i$. Then it is clear that $T \cap Z = \langle pe_1,\ldots,pe_r\rangle$ and $\mathrm{NS}\,(V)/Z \cong T/T\cap Z$ is a free $\mathbb{Z}$-module of rank $n-r$. Hence we can choose elements $e_{r+1},\ldots,e_n$ of $\mathrm{NS}\,(V)$ so that $e_{r+1} \pmod{Z},\ldots,e_n \pmod{Z}$ generate $\mathrm{NS}\,(V)/Z$ freely. Then $\{e_1,\ldots,e_r,e_{r+1},\ldots,e_n\}$ is a required free basis. Then the determinant of T with respect to the free basis $\{pe_1,\ldots,pe_r,e_{r+1},\ldots,e_n\}$ shows that

$$\det T = p^{2r}\det \mathrm{NS}\,(V) = |E(K)|^2 \cdot \det \mathrm{NS}\,(V)\,.$$

By virtue of the above remark on the exponent of $\det T$, we can write

$$\det \mathrm{NS}\,(V) = -p^{2\sigma_0},\quad \sigma_0 \in \mathbb{Z},\ \sigma_0 \geq 0.$$

Proposition. *With the above notations, we have*

$$\sigma_0 + r = \begin{cases} \frac{1}{2}\{\sum_{k\geq 0} 2\nu(\mathrm{I}^*_{2k}) + \nu(\mathrm{III}) + \nu(\mathrm{III}^*)\} & p = 2 \\ \\ \frac{1}{2}\{\nu(\mathrm{IV}) + \nu(\mathrm{IV}^*)\} & p = 3\,, \end{cases}$$

where r is the torsion-rank of the Mordell-Weil group $E(K)$.

Proof. By the definition, $2(\sigma_0 + r)$ is the exponent of $\det T$. Hence comparison of exponents gives the equality. □

Remark. The integer σ_0 is called the *Artin invariant* when X is a supersingular K3-surface (then $1 \leq \sigma_0 \leq 10$). It plays a very important role in the theory of moduli space of supersingular K3-surfaces (cf. Artin [6]).

4.2.3 Unboundedness of torsion rank

Torsion rank of a quasi-elliptic surface is unbounded. We show this by an example.

Consider a quasi-elliptic surface over $\mathbb{P}^1$ defined by the equation $y^2 = x^3 + \varphi(t)$, where $\varphi(t)$ is given with $n \in \mathbb{Z}_{\geq 0}$:

$$\varphi(t) = \begin{cases} t^{2^{n+1}+1} + t^3 & p = 2 \\ t^{3^n+1} + t^2 & p = 3. \end{cases}$$

Then the singular fibers are lying over the point at ∞ of $\mathbb{P}^1$ if $p = 2$ and n is even and also if $p = 3$, and at the $\mathbb{F}_{p^n}$-rational points of $\mathbb{P}^1 \setminus \{\infty\}$ determined by $\varphi'(t) = 0$. More precisely, the type of reducible fibers is

I_0^*	over the points $t^{2^{n+1}} + t^2 = 0$	if $p = 2$ and n is odd
I_0^*	over the points $t^{2^{n+1}} + t^2 = 0$ and $t = \infty$	if $p = 2$ and n even
IV	over the points $t^{3^n} - t = 0$ and $t = \infty$	if $p = 3$.

Note that all reducible fibers have the same type. Hence, by Proposition 4.2.1, the Euler-Poincaré characteristic is given as follows:

$$\chi(\mathcal{O}_V) = \begin{cases} \dfrac{4\nu(\mathrm{I}_0^*) + 4}{12} = \dfrac{2^{n+1} + 2}{6} & \text{if } p = 2 \text{ and } n \text{ odd} \\[2ex] \dfrac{4\nu(\mathrm{I}_0^*) + 4}{12} = \dfrac{2^{n+1} + 2^2}{6} & \text{if } p = 2 \text{ and } n \text{ even} \\[2ex] \dfrac{2\nu(\mathrm{IV}) + 4}{12} = \dfrac{3^n + 3}{6} & \text{if } p = 3 \;. \end{cases}$$

Let $\alpha \in \mathbb{F}_{p^n}$, where $\alpha \neq 1$ and $p = 2, 3$, and let σ_α be an element of the automorphism group $\mathrm{Aut}\,(\mathbb{P}^1_{\mathbb{F}_{p^n}})$ defined by

$$(s_0, s_1) \mapsto (s_0 + s_1, s_0 + \alpha s_1),$$

where (s_0, s_1) is a system of homogeneous coordinates of $\mathbb{P}^1$ such that $t = s_1/s_0$. Hence σ_α maps

$$t \mapsto \frac{1 + \alpha t}{1 + t}, \quad \tau \mapsto \frac{\tau + \alpha}{\tau + 1}, \quad \tau = 1/t.$$

It can be readily shown that if $t \in \mathbb{F}_{p^n}$ then $\sigma_\alpha(t) \in \mathbb{F}_{p^n}$. If $f : V \to \mathbb{P}^1$ is the given quasi-elliptic fibration, then $V_{\sigma_\alpha} := V \times_{\mathbb{P}^1} (\mathbb{P}^1, \sigma_\alpha)$ is birationally given by an equation $y^2 = x^3 + \varphi(\sigma_\alpha(t))$. Since $\sigma_\alpha(t) \in \mathbb{F}_{p^n}$ if $t \in \mathbb{F}_{p^n}$, we have $V_{\sigma_\alpha} \cong V$ with all reducible fibers lying over the points of $\mathbb{P}^1$ with t-value in $\mathbb{F}_{p^n}$ permuted by the automorphism σ_α. Thus, we can assume that σ_α extends to an automorphism of V, denoted by the same letter σ_α, such that $\sigma_\alpha \cdot f = f \cdot \sigma_\alpha$ and σ_α stabilizes the section (O).

Consider the case $p = 3$. Let P be a point of $E(K)$ whose (x, y)-coordinates are given by $(x, y) = (t^{2 \cdot 3^{n-1}}, t^{3^n} - t)$. Then $\sigma_\alpha(P), \sigma_\alpha^2(P), \ldots,$ $\sigma_\alpha^{p^n-1}(P)$ are all distinct points of $E(K)$. In fact, σ_α maps the fiber over $t = \infty$ (equivalently $\tau = 0$) to the fiber over $t = \alpha$, and the fiber over $t = \beta$ to the fiber over $t = (1 + \alpha\beta)/(1 + \beta)$. On the other hand (P) intersects the zero section (O) only over the fiber $t = \infty$. These observations show that $P, \sigma_\alpha(P), \sigma_\alpha^2(P), \ldots, \sigma_\alpha^{p^n-1}(P)$ are all distinct p^n points of $E(K)$, where $p = 3$. Hence $p^n \leq |E(K)| = p^r$, where r is the torsion rank of $E(K)$. Since we can take n arbitrarily big, the torsion rank is unbounded.

For the case $p = 2$, the same argument applies with $P = (t^{2^{n-1}} + t, t^{3 \cdot 2^{n-2}} + t^{2^{n-1}+1})$ and the same σ_α gives the unboundedness of torsion rank of $E(K)$.

4.3 Rational quasi-elliptic surfaces

In this section, we consider a rational quasi-elliptic fibration $f : V \to C$ and look into relationships between reducible singular fibers and the Mordell-Weil group. We consider also the configuration of sections and reducible fibers on such a surface.

4.3.1 *Defining equations*

By Theorem 3.3.1 if $p = 3$ and Theorem 3.3.2 if $p = 2$, we may start with the following situation. A rational quasi-elliptic fibration $f : V \to C = \mathbb{P}^1$ is a smooth, relatively minimal, projective model of the affine hypersurface defined by an equation

$$p = 2 \quad y^2 = x^3 + \psi(t)x + \varphi(t), \quad \begin{cases} \psi(t) = a_3 t^3 + a_2 t^2 + a_1 t \\ \varphi(t) = b_5 t^5 + b_3 t^3 + b_1 t, \\ \quad \text{where } \psi'(t) \neq 0 \text{ or } \varphi'(t) \neq 0. \end{cases}$$

$$p = 3 \quad y^2 = x^3 + \varphi(t), \quad \varphi(t) = a_5 t^5 + a_4 t^4 + a_2 t^2 + a_1 t,$$
$$\quad \text{where } \varphi'(t) \neq 0.$$

Reducible singular fibers are determined by the zeros of the discriminant $\Delta(t)$ of V, which is

$$\begin{aligned} p = 2 \quad \Delta(t) &= \psi(t)(\psi'(t))^2 + (\varphi'(t))^2 \\ &= b_5^2 t^8 + a_3^3 t^7 + a_2 a_3^2 t^6 + a_1 a_3^2 t^5 + b_3^2 t^4 \\ &\quad + a_1^2 a_3 t^3 + a_1^2 a_2 t^2 + a_1^3 t + b_1^2 . \end{aligned}$$
$$p = 3 \quad \Delta(t) = \varphi'(t) = -a_5 t^4 + a_4 t^3 - a_2 t + a_1.$$

4.3.2 Case $p = 2$

Set
$$d_1 := a_1^2 b_5 + a_1 a_3 b_3 + a_3^2 b_1, \qquad d_2 := a_1 a_2 a_3 + b_3^2, \qquad d_3 := a_2^2 + a_1 a_3.$$
The following formulas suggest how d_1, d_2 and d_3 are incorporated into concrete computations:

(1) $\Delta'(t) = (a_3 t^2 + a_1)^3 = \psi'(t)^3$.

(2) If $a_3 \neq 0$, $\Delta\left(\sqrt{\dfrac{a_1}{a_3}}\right) = \dfrac{d_1^2}{a_3^4} = \varphi'\left(\sqrt{\dfrac{a_1}{a_3}}\right)^2$.

(3) If $d_1 = 0$ and $a_3 \neq 0$, $\Delta(t) = \left(t^2 + \dfrac{a_1}{a_3}\right)^2 \Delta_1(t) + a_3^3 t \left(t^2 + \dfrac{a_1}{a_3}\right)^3$,

$\Delta_1(t) = b_5^2 t^4 + a_2 a_3^2 t^2 + b_3^2 + \dfrac{a_1^2 b_5^2}{a_3^2}$, and $\Delta_1\left(\sqrt{\dfrac{a_1}{a_3}}\right) = d_2$.

(4) If $d_1 = d_2 = 0$ and $a_3 \neq 0$, $\Delta(t) = \left(t^2 + \dfrac{a_1}{a_3}\right)^3 \Delta_2(t)$

and $\Delta_2(t) = b_5^2 t^2 + a_3^3 t + a_2 a_3^2 + \dfrac{a_1 b_5^2}{a_3}$.

Moreover, $\Delta_2\left(\sqrt{\dfrac{a_1}{a_3}}\right) = 0$ if and only if $d_3 = 0$.

4.3.2.1 Classification of reducible fibers

We have a list of all possible types of reducible singular fibers in the case of rational quasi-elliptic surfaces:

Proposition. *For a rational quasi-elliptic surface, there are seven possible patterns of reducible singular fibers in terms of their types:*

(a) *one* II^*,
(b) *one* I_4^*,
(c) *one* III *and one* III^*,
(d) *two* I_0^**s*,
(e) *one* I_2^* *and two* III*s*,
(f) *one* I_0^* *and four* III*s*,
(g) *eight* III*s*.

Proof. Since V is assumed to be rational, we have $\chi(\mathcal{O}_V) = 1$. Hence, we obtain a possible classification by virtue of Proposition 4.2.1. $\qquad \square$

Remark. The *type* of a reducible singular fiber signifies its shape according to Kodaira classification, while a *pattern* of types means a collection of types of reducible singular fibers on a given quasi-elliptic surface. But we often confuse patterns of types with types.

4.3.2.2

Now we state a classification theorem by means of the given equation for V.

Theorem.

(1) *The types of reducible singular fibers are classified into seven cases (a) ∼ (g) as follows:*

    (i) Suppose $\psi(t) \in K^2$. Then we have necessarily $d_1 = 0$. If $d_2 = d_3 = 0$ then the type is (a); if $d_2 = 0$ and $d_3 \neq 0$ then the type is (b); if $d_2 \neq 0$ then the type is (d).

    (ii) Suppose $\psi \notin K^2$. If $d_1 = d_2 = d_3 = 0$ then the type is (c); if $d_1 = d_2 = 0$ and $d_3 \neq 0$ then the type is (e); if $d_1 = 0$ and $d_2 \neq 0$ then the type is (f); if $d_1 \neq 0$ then the type is (g).

(2) *The torsion-rank r is determined uniquely by the type of reducible singular fibers as in Table 1 in the next page.*

Proof. (1) In order to determine the type of the reducible singular fibers, we look at the discriminant divisor (Δ) on $\mathbb{P}^1$ which corresponds to the polynomial $\Delta(t)$. Noting that $\deg(\Delta) = 8$, by Proposition 4.3.2.1, one can find easily that a possible list of the divisor (Δ) is given by

(T_1)	$8P_1$	$\cdots$ (a), (b)
(T_2)	$7P_1 + P_2$	$\cdots$ (c)
(T_3)	$4P_1 + 4P_2$	$\cdots$ (d)
(T_4)	$6P_1 + P_2 + P_3$	$\cdots$ (e)
(T_5)	$4P_1 + P_2 + P_3 + P_4 + P_5$	$\cdots$ (f)
(T_6)	$P_1 + P_2 + \cdots + P_8$	$\cdots$ (g)

where $P_1, P_2, \cdots P_8$ are distinct points of $\mathbb{P}^1$.

    We start from the equation given in 4.3.2. We consider also the equation
$$\eta^2 = \xi^3 + (a_1\tau^3 + a_2\tau^2 + a_3\tau)\xi + (b_1\tau^5 + b_3\tau^3 + b_5\tau),$$
which is the Weierstrass equation of $f : V \to \mathbb{P}^1$ near the fiber $f^{-1}(t = \infty)$. Moreover, note that b_1, b_3, b_5 cannot be simultaneously zero, for otherwise V is singular.

TABLE 1

pattern of reducible fibers	defining equation	r	nonzero section
(a)	$y^2 = x^3 + t^5$	0	
(b)	$y^2 = x^3 + t^2 x + t^5$	1	$(t^2 + t, t^3)$
(c)	$y^2 = x^3 + t^3 x$	1	$(0,0)$
(d)	$y^2 = x^3 + at^2 x + t^3$ with $a \in k$	2	$(ut, 0)$ with $u^3 + au + 1 = 0$
(e)	$y^2 = x^3 + (t^3 + t)x$	2	$(0,0)$ $(t + 1, t^2 + 1)$ $(t^2 + t, t^3 + t)$
(f)	$y^2 = x^3 + (t^3 + at^2 + t)x$ with $a \in k^*$	3	$(0,0)$ $(a^{1/2}t, a^{1/4}(t^2 + t))$ $(a^{-1/2}(t^2 + at + 1), a^{-3/4}(t^3 + (a+1)t^2 + (a+1)t + 1))$ $(u^{-1}t^2 + ut, u^{-3/2}(t^3 + a^{1/2}ut^2 + u^2 t))$ $(ut + u^{-1}, u^{1/2}(t^2 + a^{1/2}u^{-1}t + u^{-2}))$ with $u^2 + a^{1/2}u + 1 = 0$
(g)	$y^2 = x^3 + (t^3 + at^2 + bt)x + t^3$ with $a \in k$ and $b \in k^*$	4	$(ut, u^{1/2}(t^2 + b^{1/2}t))$ $(u^{-1}t^2 + ut + bu^{-1},$ $u^{-3/2}(t^3 + (b + au^2)^{1/2}t^2 + (b^2 + abu^2)^{1/2}t + b^{3/2}));$ with $u^3 + au + 1 = 0$ $(v^{-1}t^2 + vt, v^{-3/2}(t^3 + a^{1/2}vt^2 + b^{1/2}v^2 t))$ $(vt + bv^{-1}, v^{-3/2}(v^2 t^2 + a^{1/2}b^{1/2}vt + b^{3/2}))$ with $v^4 + av^2 + v + b = 0$ $(t^2/(t^2 + b), (a^{1/2}t^4 + t^3 + a^{1/2}bt^2)/(t + b^{1/2})^3)$

(i) Suppose first $\psi \in K^2$, where $K = k(t)$, i.e., $a_1 = a_3 = 0$. Then $\psi'(t) = 0$ and $\varphi'(t) \neq 0$. Thus $d_1 = 0$ and $\Delta(t) = \varphi'(t)^2 = (\sqrt{b_5}t^2 + \sqrt{b_3}t + \sqrt{b_1})^4$. Clearly the type of (Δ) is (T_1) if and only if $b_3 = 0$, while the type of (Δ) is (T_3) if and only if $b_3 \neq 0$. If $b_3 = 0$, then $(\Delta) = 8P_1$. Hence the reducible singular fiber over P_1 is II* if $a_2 = 0$ and I$_4^*$ if $a_2 \neq 0$ by Theorem 3.1.5.2. Here note that $d_2 = b_3^2$ and $d_3 = a_2^2$ in this case. If $b_3 \neq 0$, then $(\Delta) = 4P_1 + 4P_2$ and the reducible singular fibers over P_1 and P_2 are of type I$_0^*$.

(ii) Next, suppose $\varphi \notin K^2$. The possibilities for the type of (Δ) are (T_2), (T_4), (T_5) and (T_6). We can show that $d_1 \neq 0$ if and only if the discriminant divisor $(\Delta) = 0$ is reduced, i.e., each (distinct) irreducible component has coefficient 1. In fact, if $d_1 \neq 0$ and $a_3 \neq 0$, the formulas (1) and (2) of 4.3.2 show that $\Delta(t)$ and $\Delta'(t)$ have no common roots. Hence either $\Delta(t) = 0$ has eight distinct roots (case $b_5 \neq 0$), or $\Delta(t) = 0$ has seven distinct roots and the point P_∞ with $t = \infty$ is contained in the divisor (Δ) (case $b_5 = 0$). In either case, (Δ) is reduced. We prove the converse by reducing to absurdity. Suppose $d_1 = 0$. If $a_3 \neq 0$ then the formula (3) of 4.3.2 shows that $\Delta(t) = 0$ has a multiple root. If $a_3 = 0$ then $d_1 = a_1^2 b_5 = 0$. Since $a_1 \neq 0$ as $\psi(t) \notin K^2$, we have $b_5 = 0$. Then $\deg \Delta(t) \leq 4$, whence the coefficient of the point P_∞ in the divisor (Δ) is larger than or equal to 4. Hence (Δ) is not reduced. This absurdity shows that the converse holds. Hence, if $d_1 \neq 0$, the type of (Δ) is (T_6). By the formula in 4.3.2, the type of (Δ) is (T_5) if $d_1 = 0$ and $d_2 \neq 0$; more precisely, $P_\infty \notin \mathrm{Supp}(\Delta)$ if $a_3 \neq 0$ and $(\Delta) - 4P_\infty \geq 0$ if $a_3 = 0$. Similarly, (Δ) has type (T_4) if $d_1 = d_2 = 0$ and $d_3 \neq 0$, and (T_2) if $d_1 = d_2 = d_3 = 0$. The point P_∞ is not included (resp. included) in $\mathrm{Supp}(\Delta)$ if $a_3 \neq 0$ (resp. $a_3 = 0$).

(2) When we pick up one standard form among the rational quasi-elliptic surfaces having the same pattern of reducible fibers, we make frequent use of the following lemma.

Lemma. *Suppose V is defined by an affine equation*

$$y^2 = x^3 + (\psi_0(t) + a_k t^k)x + (\varphi_0(t) + b_\ell t^\ell)$$

with $\psi_0(t), \varphi_0(t) \in k[t]$ and $a_k \cdot b_\ell \neq 0$. If $2\ell \neq 3k$, then V is isomorphic to the one defined by the same equation as above with $a_k = b_\ell = 1$ though $\psi_0(t), \varphi_0(t)$ changed.

For the proof of this lemma, one can easily check that two surfaces given in the statement are isomorphic to each other by a suitable change of variables. In fact, for unknown, nonzero constants $c, \gamma \in k$, rewrite the given equation as

$$\left(\frac{y}{c^3}\right)^2 = \left(\frac{x}{c^3}\right)^3 + \frac{(\psi_0(t) + a_k t^k)}{c^4}\left(\frac{x}{c^2}\right) + \frac{(\varphi_0(t) + b_\ell t^\ell)}{c^6}.$$

Consider a change of variables $(x, y, t) \mapsto \left(\frac{x}{c^2}, \frac{y}{c^3}, \frac{t}{\gamma}\right)$. Let $t' = \frac{t}{\gamma}$. Then the coefficients of the t'^k term and the t'^ℓ term are equal to 1 if $\frac{a_k \gamma^k}{c^4} = \frac{b_\ell \gamma^\ell}{c^6} = 1$. Then we can determine c and γ to satisfy the equalities if $2\ell \neq 3k$.

The calculation of rational points in Table 1 is based on the method to be explained in subsection 4.3.4. The second statement of the theorem is clear by Proposition 4.2.2 since $\sigma_0 = 0$ for rational surfaces.   □

4.3.3   Case $p = 3$

Let $f : V \to \mathbb{P}^1$ be a rational, relatively minimal, quasi-elliptic fibration. The Weierstrass equation is the one set in 4.3.1, i.e.,

$$y^2 = x^3 + \varphi(t), \quad \varphi(t) = a_5 t^5 + a_4 t^4 + a_2 t^2 + a_1 t, \quad \varphi'(t) \neq 0.$$

We set

$$d_1 = a_1 a_5 - a_2 a_4, \quad d_2 = a_4 + \sqrt[3]{a_2 a_5^2},$$

and we employ, as in the above cases, Kodaira's notation to indicate the type of reducible fibres. Our results are stated as follows.

Theorem.

(1) *For a given $\varphi(t)$, the reducible fibres of V are completely determined by elements d_1 and d_2. More precisely, we have*

    (a) *If $d_1 = d_2 = 0$, then V has one reducible fibre of type II^*.*
    (b) *If $d_1 = 0$ and $d_2 \neq 0$, then V has two reducible fibres of types IV and IV^*.*
    (c) *If $d_1 \neq 0$, then V has four reducible fibres of type IV.*

(2) *The torsion-rank is determined uniquely by the type of reducible singular fibers as in Table 2 below.*

Proof.   The arguments in the proof are almost parallel to the case $p = 2$. Since we have the equality

$$4 = 6\chi(\mathcal{O}_V) - 2 = \nu(\mathrm{IV}) + 3\nu(\mathrm{IV}^*) + 4\nu(\mathrm{II}^*)$$

by 3.1.5.1, we have one of the following patterns of reducible fibers:

(a) one II*,

(b) one IV* and one IV,

(c) four IVs.

On the other hand, by a simple computation, we have the following relations.

(1) $\Delta'(t) = -(\sqrt[3]{a_5}t + \sqrt[3]{a_2})^3$.

(2) If $a_5 \neq 0$, we have

$$\Delta\left(-\sqrt[3]{\frac{a_2}{a_5}}\right) = \frac{d_1}{a_5}.$$

(3) If $a_5 \neq 0$, we have

$$\Delta(t) = a_5\left(t + \sqrt[3]{\frac{a_2}{a_5}}\right)^4 - d_2\left(t + \sqrt[3]{\frac{a_2}{a_5}}\right)^3 - \frac{d_1}{a_5}.$$

Now we prove the assertion (1). If $d_1 = d_2 = 0$ and $a_5 \neq 0$, then $\Delta(t)$ has a single multiple root of multiplicity 4. Hence the quasi-elliptic fibration $f : V \to \mathbb{P}^1$ has a single reducible fiber of type II* over the point $t = -\sqrt[3]{\frac{a_2}{a_5}}$. If $a_5 = 0$, the fiber over the point $t = \infty$ is a reducible fiber of type II*. If $d_1 = 0$, $d_2 \neq 0$ and $a_5 \neq 0$, $\Delta(t)$ has a root of multiplicity 3 over the point $t = -\sqrt[3]{\frac{a_2}{a_5}}$. If $a_5 = 0$, the reducible fiber lies over the point $t = \infty$. If $d_1 \neq 0$ and $a_5 \neq 0$ then $\Delta(t) = 0$ has four simple roots. Hence f has four reducible fibers of type IV. If $d_1 \neq 0$ and $a_5 = 0$, the fiber over $t = \infty$ is a reducible fiber of type IV. $\square$

4.3.4 Blowing-down of nine $\mathbb{P}^1s$

We describe how reducible singular fibers and sections intersect each other for each of patterns classified in Theorems 4.3.2.2 and 4.3.3.

TABLE 2

pattern of reducible fibers	defining equation	r	nonzero section
(a)	$y^2 = x^3 + t$	0	
(b)	$y^2 = x^3 + t^2$	1	$(0, \pm t)$
(c)	$y^2 = x^3 + t^4 + t^2$	2	$(t^2, \pm(t^3 - t)), (1, \pm(t^2 - 1))$ $(-t, \pm(t^2 + t)), (t, \pm(t^2 - t))$

4.3.4.1 *Case $p = 2$*

For each of the patterns (a), (b), (c), (d) and (e) in Table 1, the configurations are simple to draw, and we can easily find how they are constructed from $\mathbb{P}^2$ by blowing up nine points. In each of the types, we discuss how to obtain the quasi-elliptic fibration $f : V \to \mathbb{P}^1$ from a linear pencil on $\mathbb{P}^2$ by blowing up nine base points which consist of ordinary points and their infinitely near points. This process is uniquely determined and called a *minimal elimination of base points*.

As the reverse process, by blowing down the nine $\mathbb{P}^1$s which are drawn by thick lines in each figure, one can obtain the configuration of curves in $\mathbb{P}^2$ for each of the types (a), (b), (c) and (d) as in the figures as given below, where C_a, C_b are cuspidal cubic curves and C_c is a conic. Further, L_* with subscript $* = a, b_i, c_j, d_k$ and L, Γ are lines. Intersection of these curves are given by

$$(C_b \cdot L_{b_1}) = (C_c \cdot L_{c_1}) = (C_c \cdot L_{c_2}) = 2, \qquad (C_a \cdot L_a) = (C_b \cdot L_{b_2}) = 3.$$

The dotted line Γ gives rise to the locus of moving singular points of fibers of f.

In the case of the type (a), C_a meets the line L_a at a flex Q of C_a, and $C_a \sim 3L_a$. Consider the linear pencil Λ spanned by C_a and $3L_a$. Minimal elimination of base points of the pencil Λ gives rise to the quasi-elliptic fibration $f : V \to \mathbb{P}^1$ which we started with. The first blowin-down is the cross-section O which is (-1)-curve by Corollary 3.1.2, (3) because $\chi(\mathcal{O}_V) = 1$. Further note that all irreducible components of reducible fiber are (-2)-curves.

Figure of type (a)

Figure of type (a) in $\mathbb{P}^2$

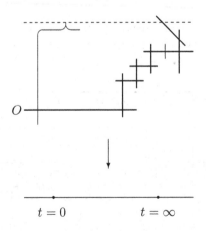

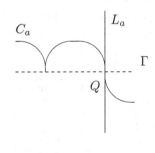

In the case of type (b), the fibration $f : V \to \mathbb{P}^1$ is obtained as a minimal elimination of base points of the linear system $\Lambda = \langle C_b, 2L_{b_1} + L_{b_2} \rangle$, which has two ordinary points Q_1, Q_2 and its infinitely near points as base points, where C_b meets L_{b_1} (resp. L_{b_2}) at Q_1 (resp. Q_2) with multiplicity 2 (resp. 3).

Figure of type (b)   Figure of type (b) in $\mathbb{P}^2$

In the case of type (c), the quasi-elliptic fibration results from a linear pencil $\Lambda = \langle C_c + L_{c_1}, 3L_{c_2} \rangle$, whose base points lie on the ordinary points $Q_1 = L_{c_1} \cap L_{c_2}$ and $Q_2 = C_c \cap L_{c_2}$.

Figure of type (c)   Figure of type (c) in $\mathbb{P}^2$

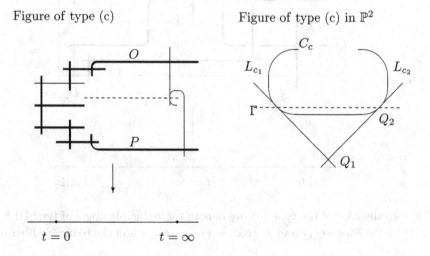

In the case of type (d), the quasi-elliptic fibration is given by the proper transform of a linear pencil $\Lambda = \langle 2L + L_{d_1}, L_{d_2} + L_{d_3} + L_{d_4} \rangle$ by a minimal elimination of base points of Λ, whose base points lie on the ordinary points Q the common point of the L_{d_i} and $Q_i = L \cap L_{d_i}$ for $i = 2, 3, 4$.

Figure of type (d)   Figure of type (d) in $\mathbb{P}^2$

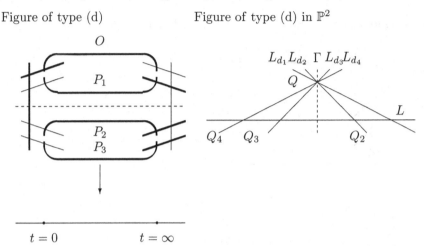

Figure of type (e)

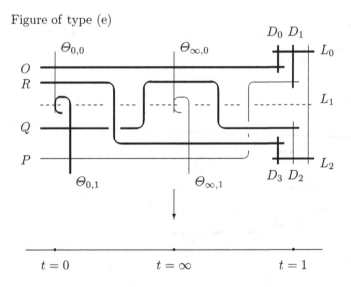

In the case of the type (e), we denote the reducible fibers of type III by $f^{-1}(0) = \Theta_{0,0} + \Theta_{0,1}$ and $f^{-1}(\infty) = \Theta_{\infty,0} + \Theta_{\infty,1}$ and the reducible fiber of type I_2^* by $f^{-1}(1) = D_0 + D_1 + 2(L_0 + L_1 + L_2) + D_2 + D_3$. Let $\sigma : V \to \mathbb{P}^2$ be the contraction of the curves $(O), D_0, L_0, D_1, (R), D_3, L_2, (Q), \Theta_{0,1}$. Then we easily obtain a configuration of four lines $\sigma(L_1), \sigma(D_2), \sigma((P)), \sigma(\Theta_{\infty,1})$, a conic $\sigma(\Theta_{\infty,0})$ and a cubic with cusp $\sigma(\Theta_{0,0})$ in $\mathbb{P}^2$. We, however, give only the configuration of reducible fibers on V, and leave the readers to work out the blowing-down of nine $\mathbb{P}^1$s on V as indicated above and draw the configuration of images on $\mathbb{P}^2$ which we do not give here.

The quasi-elliptic fibration $f : V \to \mathbb{P}^1$ is given by the linear pencil $\Lambda = \langle \sigma(\Theta_{0,0}), 2\sigma(L_1) + \sigma(D_2) \rangle$.

The cases (f) and (g) are more complicated. In the case (f), the reducible fibers are of type III over $t = 0, \infty, \alpha_1, \alpha_2$, where α_1 and α_2 are two solutions of the equation $t^2 + at + 1 = 0$ with $a \neq 0$ and I_0^* over $t = 1$. In fact, since $\Delta(t) = \psi(t)\psi'(t)^2 = t(t - \alpha_1)(t - \alpha_2)(t - 1)^4$, where $\psi(t) = t^3 + at^2 + t$ with $a \neq 0$ (see Table 1). For example, on the fiber $f^{-1}(t = 0)$, there lies an isolated singular point $(x, y) = (0, 0)$ of the hypersurface X defined by $y^2 = x^3 + (t^3 + at^2 + t)x$. An embedded resolution of singularity is obtained by setting $y = y'x$ and $t = t'x$. The proper transform X' of the hypersurface is defined by

$$y'^2 = x + (t'^3 x^2 + at'^2 x + t').$$

Hence the exceptional locus of the blowing-up $X' \to X$ restricted onto the fiber over $t = 0$ is a curve defined by $y'^2 = t'$ in the (y', t')-plane. Meanwhile, the fiber $f^{-1}(t = 0)$ restricted onto the (x, y)-plane is defined by $y^2 = x^3$, whose proper transform on the (x, y')-plane is $x = y'^2$. The former gives the component $\Theta_{0,1}$ and the latter does the component $\Theta_{0,0}$. We can consider also embedded resolutions of singularities to obtain other reducible fibers. We name their components as in the figure below.

Type III fiber over $t = \beta$ Type I_0^* fiber over $t = \gamma$

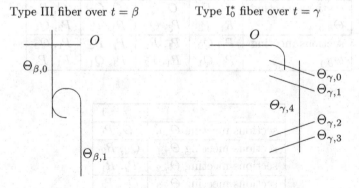

In view of Table 1, we have eight sections which we name as follows:

$$P_1 = (0,0), \ P_2 = \left(\sqrt{a}, \sqrt[4]{a}(t^2 + t)\right),$$

$$P_3 = \left(\frac{t^2 + at + 1}{\sqrt{a}}, \frac{t^3 + (a+1)t^2 + (a+1)t + 1}{\sqrt[4]{a^3}}\right),$$

$$Q_i = \left(\frac{t^2 + u_i^2 t}{u_i}, \frac{t^3 + \sqrt{a}u_i t^2 + u_i^2 t}{\sqrt{u_i^3}}\right),$$

$$R_i = \left(\frac{u_i^2 t + 1}{u_i}, \frac{u_i^2 t^2 + \sqrt{a}u_i t + 1}{\sqrt{u_i^3}}\right) \quad (i = 1, 2),$$

where u_i $(i = 1, 2)$ are the two roots of the equation $u^2 + a^{1/2}u + 1 = 0$ and all sections are represented by K-rational points of $E(K)$ with $K = k(t)$. We can associate each component of a reducible fiber with sections which meet the component. Note that the section (O) is given by the K-rational point at infinity of the generic fiber of f, i.e., $\left(\frac{1}{y}, \frac{x}{y}\right) = (0,0)$. For other sections, we can determine the components $\Theta_{0,0}$ or $\Theta_{0,1}$ of the fiber $f^{-1}(t = 0)$ by setting $t = 0$ in the given expression of sections or their proper transforms by the embedded resolution of singularity. For the fiber $f^{-1}(t = \infty)$, we have to change the coordinates

$$\eta = \frac{y}{t^2}, \ \xi = \frac{x}{t^3}, \ \tau = \frac{1}{t}.$$

Then the defining equation of X is $\eta^2 = \xi^3 + (\tau^3 + a\tau^2 + \tau)\xi$, and the sections O, P_i, Q_j, R_ℓ are accordingly expressed in terms of τ.

Then the configuration is as indicated in Table 3 below. The first table is for reducible fibers of type III, and the second one for type I_0^*.

TABLE 3

	$\beta = 0$	$\beta = \infty$	$\beta = \alpha_1$	$\beta = \alpha_2$
sections meeting	O, R_2	O, Q_2	O, Q_2	O, R_2
$\Theta_{\beta,0}$	R_1, P_3	P_3, Q_1	R_1, P_2	P_2, Q_1
sections meeting	P_1, Q_2	P_1, R_2	P_1, R_2	P_1, Q_2
$\Theta_{\beta,1}$	P_2, Q_1	R_1, P_2	P_3, Q_1	R_1, P_3

	$\gamma = 1$
sections meeting $\Theta_{\gamma,0}$	O, P_1
sections meeting $\Theta_{\gamma,1}$	Q_2, R_2
sections meeting $\Theta_{\gamma,2}$	P_2, R_1
sections meeting $\Theta_{\gamma,3}$	Q_1, P_3

Example. After blowing down the sections and irreducible components of the fibers (for example, $Q_1, R_1, Q_2, R_2, P_2, P_3, P_1, \Theta_{1,0}, \Theta_{1,4}$ in this order), one can obtain a configuration of curves in $\mathbb{P}^2$ which consist of four conics corresponding to $\Theta_{\beta,1}$ for $\beta = 0, \infty, \alpha_1, \alpha_2$, and eight lines corresponding to $\Theta_{\beta,0}$ for $\beta = 0, \infty, \alpha_1, \alpha_2, O, \Theta_{1,1}, \Theta_{1,2}$ and $\Theta_{1,3}$.

In the case of type (g), there are eight reducible fibers of type III over the seven roots of the equation $\Delta(t) = 0$ and $t = \infty$. Since the torsion rank r is 4 in this case, there are sixteen sections which we express concretely by making use of the equations in Table 1 as follows:

$$T = \left(\frac{t^2}{t^2 + b}, \frac{\sqrt{a}\, t^4 + t^3 + \sqrt{a}\, bt^2}{(t + \sqrt{b})^3} \right)$$

$$A_i = \left(\frac{t^2 + u_i^2 t + b}{u_i}, \frac{t^3 + \sqrt{b + au_i^2}\, t^2 + \sqrt{b^2 + abu^2}t + \sqrt{b^3}}{\sqrt{u_i^3}} \right)$$
$$(i = 1, 2, 3)$$

$$A_{j+3} = \left(\frac{v_j^2 t + b}{v_j}, \frac{v_j^2 t^2 + \sqrt{ab}\, v_j t + \sqrt{b^3}}{\sqrt{v_j^3}} \right) \quad (j = 1, \ldots, 4),$$

where u_i $(i = 1, 2, 3)$ are the three roots of the equation $u^3 + au + 1 = 0$ and v_j $(j = 1, \ldots, 4)$ are the four roots of the equation $v^4 + av^2 + v + b = 0$.

Intersection of these sections with reducible fibers of type III, e.g. the fiber $f^{-1}(t = 0) = \Theta_{0,0} + \Theta_{0,1}$, is shown in the figure below, where the seven sections $A_1, A_2, \ldots, A_7$ are those which intersect the same component $\Theta_{0,0}$ of the fiber as the zero section (O). Here T is a unique section intersecting O. In fact, the parametrization of T is also given in the $(\frac{1}{y}, \frac{x}{y})$-coordinates and in a parameter $\tau = \frac{1}{t}$,

$$\frac{1}{y} = \frac{\tau(1 + \sqrt{b}\tau)^3}{\sqrt{a} + \tau + \sqrt{ab}\tau^2}, \quad \frac{x}{y} = \frac{\tau(1 + \sqrt{b}\tau)^3}{(1 + b^2\tau)(\sqrt{a} + \tau + \sqrt{ab}\tau^2)}.$$

Hence T and O meets on the fiber $f^{-1}(t = \infty)$ at $\tau = 0$. Since $T + A_i$ is the addition of T and A_i in $E(K)$, each section A_i intersects only one other section $T + A_i$. One can thus divide sixteen sections into eight pairs (O, T) and $(A_i, T + A_i)$ $(1 \le i \le 7)$, each pair consisting of two mutually intersecting sections.

Figure of type (g) with a fiber $f^{-1}(t = 0)$

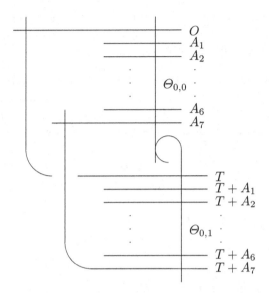

$$O$$
$$A_1$$
$$A_2$$
$$\Theta_{0,0}$$
$$A_6$$
$$A_7$$
$$T$$
$$T + A_1$$
$$T + A_2$$
$$\Theta_{0,1}$$
$$T + A_6$$
$$T + A_7$$

The same situation occurs on the other reducible singular fibers. If one names the fiber components of a reducible fiber of type III as in the case (f) (the fiber on the left hand side figure), one can blow down the following nine sections and components $O, A_1, A_2, \ldots, A_6, T + A_7, \Theta_{0,1}$ in this order.

Remark. The sections are all disjoint except in the case (g).

4.3.4.2 *Case* $p = 3$

For the types (a) and (b) in Theorem 4.3.3, the configurations are exhibited as below. By blowing down the nine $\mathbb{P}^1$s which are drawn by thick lines in each of the figures (a) and (b), we obtain the following configurations of curves in $\mathbb{P}^2$, where C with subscript is a cuspidal rational curve of degree 3 and other curves are lines in $\mathbb{P}^2$. The curve C touches L in a point with intersection multiplicity 3. Reversing the above blowing-down process, we know how to obtain the quasi-elliptic fibrations from the given configurations of curves on $\mathbb{P}^2$. In the case of type (a), the linear pencil $\Lambda = \langle C_a, 3L_a \rangle$ gives the quasi-elliptic fibration by a minimal elimination of base points which lie on the ordinary point $Q = L_a \cap C_a$. The line L_a gives the component in the reducible fiber indicated by D. In the case of type (b), the linear pencil which gives the quasi-elliptic fibration is $\Lambda = \langle L_{b_1} + L_{b_2} + L_{b_3}, 3L \rangle$ and its base points lie over the points Q_1, Q_2 and Q_3.

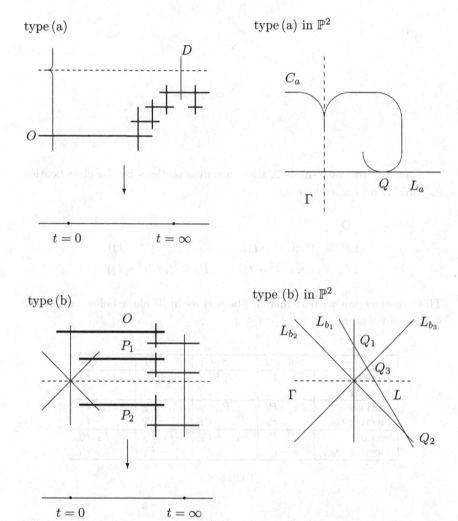

type (a)   type (a) in $\mathbb{P}^2$

type (b)   type (b) in $\mathbb{P}^2$

For the case (c) in Theorem 4.3.3, the situation is more complicated. In this case, there are four reducible fibers of type IV over $t = 0, \pm 1, \infty$. We name their components as in the figure below.

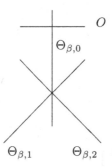

Since the torsion rank is 2, there are nine sections by the classification in Table 1, we name them as

$$O,$$
$$\pm P_1 = (t^2, \pm(t^3 - t)), \quad \pm P_2 = (1, \pm(t^2 - 1)),$$
$$\pm P_3 = (-t, \pm(t^2 + t)), \quad \pm P_4 = (t, \pm(t^2 - t)).$$

Their intersection with reducible fibers is as in Table 4 below, where we write P_i instead of $+P_i$ for $1 \leq i \leq 4$.

	$\beta = 0$	$\beta = 1$	$\beta = -1$	$\beta = \infty$
sections meeting $\Theta_{\beta,0}$	$O, \pm P_2$	$O, \pm P_3$	$O, \pm P_4$	$O, \pm P_1$
sections meeting $\Theta_{\beta,1}$	$P_1, -P_3$ P_4	P_1, P_2 $-P_4$	$P_1, -P_2$ P_3	$-P_2, -P_3$ $-P_4$
sections meeting $\Theta_{\beta,2}$	$-P_1, P_3$ $-P_4$	$-P_1, -P_2$ P_4	$-P_1, P_2$ $-P_3$	P_2, P_3 P_4

TABLE 4

Now we blow down all nine sections as given above. Then the image of twelve fiber components of four reducible fibers is a line arrangement on $\mathbb{P}^2$ shown below. In fact, each fiber component meets exactly three sections by Table 4, whence its image is a line. There are nine quadruple points, i.e., a point where four lines get together, which we put the same name as blown-down sections and mark by black circles.

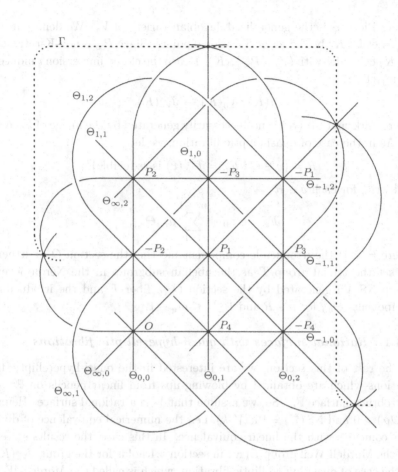

4.4 Case of quasi-hyperelliptic surfaces

Let $f : V \to C$ be a unirational smooth complete surface which has a quasi-hyperelliptic fibration of genus $g = \frac{p-1}{2}$ with a cross-section. Let $K = k(C)$ and let V_η be the generic fiber of f. We assume that $p \geq 5$. Then V_η is birational to a plane curve $y^2 = z^p - \varphi(t)$ over $K = k(t)$ with $\varphi(t) \in k[t]$ (see 3.9 and 3.10). The point at infinity $(\frac{1}{y} = 0, \frac{z}{y} = 0)$ of the plane curve $y^2 = z^p - \varphi(t)$ yields a K-rational smooth point O after normalization. Hence O is a K-rational point of the normal complete curve V_η. Besides, V_η has a unique non-K-rational singular point $P_\infty = (y = 0, z = \sqrt[p]{\varphi(t)})$. By Theorem I.1.8, there is a closed immersion

$$\iota : V_\eta - \{P_\infty\} \longrightarrow \operatorname{Pic}^0_{V_\eta/K}, \quad P \mapsto (P) - (O)$$

where $\mathrm{Pic}_{V_\eta/K}$ is the generalized Jacobian variety of V_η. We denote it by J_{V_η}. Let $V_\eta(K)$ be the set of K-rational points. Since P_∞ is non-K-rational, $V_\eta(K)$ coincides with $(V_\eta \setminus \{P_\infty\})(K)$. Hence the closed immersion ι induces an injection

$$\iota(K) : V_\eta(K) \hookrightarrow J_{V_\eta}(K).$$

We remark that $J_{V_\eta}(K)$ is not necessarily generated by the image of $V_\eta(K)$.

As in the case of quasi-elliptic fibration, we let

$$R = \{v \in C \mid F_v := f^{-1}(v) \text{ is reducible}\}$$

and write for each $v \in R$

$$F_v = \Theta_{v,0} + \sum_{i=1}^{m_v-1} \mu_{v,i}\Theta_{v,i},$$

where $\Theta_{v,0}$ is the irreducible component meeting the section O. Further, we set the trivial group T as the abelian subgroup in the Néron-Severi group $\mathrm{NS}(V)$ generated by the section O, a fiber F and the irreducible components $\Theta_{v,i}$ for $v \in R$ and $1 \le i \le m_v - 1$.

4.4.1   *Rational surfaces with quasi-hyperelliptic fibrations*

In the rest of this section, we are interested in the quasi-hyperelliptic fibrations which are obtained by blowing-ups from linear pencils on $\mathbb{P}^2$ or Hirzebruch surface $\mathbb{F}_n$. So, we assume that V is a rational surface. Hence $\chi(\mathcal{O}_V) > 0$ and $\mathrm{NS}(V) \cong \mathrm{Pic}(V/k)$, i.e., the numerical equivalence of divisors coincides with the linear equivalence. In this case, the results stated for the Mordell-Weil group $E(K)$ in section 4.1 holds for the group $J_{V_\eta}(K)$ in the case of quasi-hyperelliptic fibration, which is called the *Mordell-Weil group* of $f : V \to C$. Namely, we have the following results.

Theorem. *Let V be a rational smooth surface satisfying the above assumptions. Then the following assertions hold.*

(1) *The Néron-Severi group $\mathrm{NS}(V)$ is a free abelian group of finite rank endowed with a non-degenerate intersection pairing.*
(2) $\mathrm{NS}(V)/T \cong J_{V_\eta}(K)$.
(3) $J_{V_\eta}(K)$ *is a finite-dimensional vector space of rank r, and $|J_{V_\eta}(K)|^2 = p^{2r} = \det(T)/\det \mathrm{NS}(V)$.*

Proof.   (1) The arguments in 4.1.2.3 and 4.1.2.4 apply. Furthermore, the assertions in Proposition 4.1.2.5 holds because V is rational. Hence $\mathrm{NS}(V)$

is a finitely generated free abelian group. In particular, the subgroup T is free, and has rank equal to $2 + \sum_{v \in R}(m_v - 1)$.

(2) Let $X = V_\eta - \{P_\infty\}$ and let Γ be the closure of the point P_∞ in V. Then X is a K-form of $\mathbb{A}^1$ with height one. Hence $P_\infty \sim pO$ and $J_{V_\eta}(K) \cong \mathcal{D}_a(X)/\mathcal{D}_\ell(X)$. We can argue exactly as in 4.1.2.8 to prove the assertion (2).

(3) For the first assertion, follow the argument of Theorem 4.1.4, and for the second assertion, follow the argument of 4.2.2. We do not need a precise evaluation of $\det(T)$ in terms of the numbers of reducible fibers. $\square$

4.4.2   Examples of rational quasi-hyperelliptic surfaces

Even for special classes of rational quasi-hyperelliptic surface, there arise interesting reducible fibers, a part of which, in fact, the fiber minus the component Θ_0 meeting the section O induces an interesting isolated surface singularity. We exhibit this by giving several examples in arbitrary characteristic $p \geq 5$. The computation given below allows the case $p = 3$ although the fibration is a quasi-elliptic fibration.

4.4.2.1   Singularity of type $E_8(p)$

Let us consider a quasi-hyperelliptic fibration $f : V \to C = \mathbb{P}^1$ on a uni-rational surface V endowed with a cross-section O, which is birational to a hypersurface in $\mathbb{A}_k^3$ defined by $z^p = y^2 + x$. Then a general fiber has $(2, p)$-cusp, i.e., locally of type $z^p = y^2$, and unique reducible singular fiber is lying over $x = \infty$, where x is a local parameter of the base curve $C \cong \mathbb{P}^1$. We explain how to obtain a smooth complete surface with minimal quasi-hyperelliptic fibration $f : V \to C$. As in 3.4, we consider $F_0 = U_0 \cup U_1 \cup U_2 \cup U_3 \cong \mathbb{P}^1 \times \mathbb{P}^1$ and the curve B which is the closure of the affine curve $y^2 + x = 0$. As in 3.4.3, consider a minimal sequence $\sigma : F \to F_0$ such that $\sigma^{-1}(M_\infty \cup B \cup L_\infty)$ is a divisor of simple normal crossings. It is obtained by blowing up the point $M_\infty \cap B$ and its infinitely near point lying on the proper transform of L_∞. The configuration is given as below. Note that $k(V) = k(x, y, z)$ with $z^p = y^2 + x$. In terms of $u = \frac{1}{x}$ and $v = \frac{1}{y}$, we have $k(V) = k(u, v, z')$ with $z'^p = (zuv)^p = v^{p-2}u^{p-1}(u + v^2)$. Hence, near the point P_1, $k(V) = k(u, v_1, z_1)$ with $z_1^p = u^{p-2}v_1^{p-2}(1 + uv_1^2)$, where $v_1 = \frac{v}{u}$ and $z_1 = \frac{z'}{u}$. Similarly, near the point P_2, $k(V) = k(u_1, v_2, z_2)$ with $z_2^p = u_1^{p-2}v_2^{p-2}(1 + v_2)$, where $z_2 = \frac{z'}{u}$ and $v_2 = \frac{v}{u_1}$. Near the point P_3, $k(V) = k(u_1, v_2', z_2)$ with $z_2^p = v_2'u_1^{p-2}v_2^{p-2}$, where $v_2' = 1 + v_2$ and

v_2 is a unit near P_3. Finally, near the point P_4, $k(V) = k(u_1, v, z_4)$ with $z_4^p = u_2^{p-1} v^{p-2}(1 + u_2)$, where $z_4 = \frac{z'}{v^2}$ and $u_2 = \frac{u_1}{v}$. Now let S be the normalization of F in $k(V)$. Then, Lemma 2.7.1, S has only cyclic quotient singularities lying over the points P_i for $1 = 1, 2, 3, 4$. In order to resolve the singularities, we compute the resolution data by 2.7. We denote the data at the points P_i by d_i. Then $d_1 = d_2 = p - 2, d_3 = 2$ and $d_4 = p - 2$. Let $\overline{S}$ be the resolution of these cyclic quotient singular points.

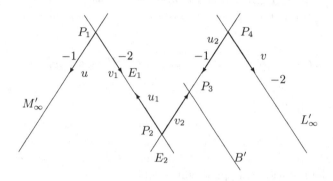

Then $\overline{S}$ has a quasi-hyperelliptic fibration $\overline{f} : \overline{S} \to C$ such that the proper inverse image $\widetilde{M}_\infty$ of M_∞ is a cross-section of $\overline{f}$ and a (-1)-curve. The fiber corresponding to L_∞ has the following dual graph:

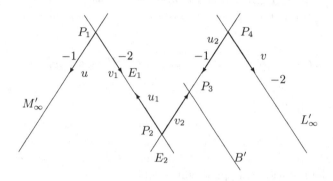

In the graph, $\Sigma(n)$ means a linear chain of length n consisting of (-2)-curves, and $\Sigma(p-1)_i$ $(i = 1, 2)$ signifies that it is the exceptional graph arising from the resolution of singularity at P_i. Furthermore, the chain consisting of Θ_{2p} and Θ_{2p+1} and the chain consisting of Θ_{2p+2} and $\Sigma(p-2)$ are respectively the exceptional loci of the resolution of singularities at P_3 and P_4. Since the proper inverse image $\widetilde{L}_\infty$ of L_∞ is a (-1)-curve, the chain $\widetilde{L}_\infty$ and $\Sigma(p-2)$ are contracted to a smooth point on Θ_{2p+2}. Thus we obtain a smooth complete surface V and a minimal quasi-elliptic fibration $f : V \to C$ with a single reducible fiber. We denote the image of Θ_{2p+2} by the same letter. Name the components of $\Sigma(p-1)_1$ as $\Theta_0, \Theta_1, \ldots, \Theta_{p-2}$, and also the components of $\Sigma(p-1)_2$ as $\Theta_p, \ldots, \Theta_{2p-2}$. Accordingly, name $\widetilde{E}_1$ and $\widetilde{E}_2$ as Θ_{p-1} and Θ_{2p-1}. Then these $(2p+3)$ components form a reducible fiber with the component Θ_0 meeting the cross-section $\widetilde{M}_\infty$, which is the section O in 4.4.1. We say that this fiber has type $\mathrm{II}^*(p)$. In fact, we can make the above computation also in the case $p=3$. Then the obtained reducible fiber in the quasi-elliptic fibration has type II^*.

Let F be a general fiber of the fibration f and let $\Lambda = |aO + bF|$ with $a, b \gg 0$. Then the associated morphism Φ_Λ is a birational morphism $\pi : V \to W$ from V to a normal projective surface such that $\pi(X)$ is a point Q of W and π induces an isomorphism between $V - X$ and $W - \{Q\}$, where $X = \sum_{i=1}^{2p+2} \Theta_i$. Hence X is contractible to an isolated singular point defined by a local equation $z^p + y^2 + x^{2p-1} = 0$[4], say $E_8(p)$, which is in some sense a generalization of the rational double point of type E_8. If the characteristic is equal to 3, this is nothing but a rational double point of type E_8.

This singularity $E_8(p)$ as appeared in the above setting $\pi : V \to W$ has the following properties.

(1) The intersection matrix of this singularity $E_8(p)$, i.e., the one of $X = \sum_{i=1}^{2p+2} \Theta_i$, is unimodular of rank $2p+2$ as given below[5].

$$
\begin{pmatrix}
-2 & 1 & & & & & & & \\
1 & -2 & & & & & & & \\
& & \ddots & & & & & & \\
& & & \ddots & & & & & \\
& & & & -2 & 1 & & & \\
& & & & 1 & -2 & 1 & 0 & 1 \\
& & & & & 1 & -2 & 1 & 0 \\
& & & & & 0 & 1 & -\frac{p+1}{2} & 1 \\
& & & & & 1 & 0 & 1 & -2
\end{pmatrix}
$$

(2) The fundamental cycle is :

$$ Z = 2\Theta_1 + 3\Theta_2 + \cdots + 2p\Theta_{2p-1} + (p+1)\Theta_{2p} + 2\Theta_{2p+1} + p\Theta_{2p+2} $$

(3) $(Z^2) = -2$ and $p_a(Z) = \frac{p-3}{2}$, where $p_a(Z)$ is the arithmetic genus and written also as $p(Z)$. Hence $E_8(p)$ is not a rational singularity if $p > 3$.

---

[4]This local equation is obtained from the initial equation of the surface $z^p + y^2 + x = 0$ by dividing it by x^{2p}.

[5]For the terminology and notation on singularity, the readers are advised to refer to Part III, Chapter 1.

In order to compute (Z^2) and $p_a(Z)$, verify by computation that

$(Z + \Theta_0 \cdot \Theta_i) = 0$ $(i = 0, \ldots, 2p + 2)$, hence $(Z^2) = -2$.

$(K_V \cdot \Theta_i) = 0$ $(i \neq 2p + 1)$, $(K_V \cdot \Theta_{2p+1}) = \dfrac{p-3}{2}$, $(K_V \cdot Z) = p - 3$.

$p_a(Z) = \dfrac{1}{2}(Z^2 + K_V \cdot Z) + 1 = \dfrac{p-3}{2}$.

Remark. There are classifications of singular fibers for genus 2 fibrations by Ogg [71] and Namikawa-Ueno [70]. According to these classifications, our singular fiber with $p = 5$ is named No. 20 in [71] and [VIII-4] in [70].

One can also find this singular fiber (with $p > 5$) in the list of singular fibers of fibrations of genus greater than 2. We mention the structure of the Mordell-Weil group for this example.

Proposition.

(1) *The Mordell-Weil group $J_{V_\eta}(K)$ of the generalized Jacobian of V_η is trivial, thus the cross-section O is a unique section of f.*
(2) *The Néron-Severi group $\mathrm{NS}(V)$ of V is freely generated by*

$$O, F, \Theta_1, \Theta_2, \ldots, \Theta_{2p+2},$$

where F is a general fiber of f.
(3) *Contracting the curves $O = \widetilde{M}_\infty, \Theta_0, \ldots, \Theta_{2p-1}, \Theta_{2p+2}$ in this order, we obtain the Hirzebruch surface $\mathbb{F}_{\frac{p+1}{2}}$ with the image of Θ_{2p+1} the minimal section and the image of Θ_{2p} a fiber of the canonical $\mathbb{P}^1$-fibration.*

4.4.2.2 *Singularity of type $E_6(p)$*

As in the same situation in 4.4.2.1, let $f : V \to C \cong \mathbb{P}^1$ be a unirational quasi-hyperelliptic surface, but which is birational to a hypersurface in $\mathbb{A}_k^3$ defined by $z^p = y^2 + x^2$. Then the reducible singular fibers are located over $x = 0$ and $x = \infty$. Since $z^p = (y + ix)(y - ix)$ with $i^2 = 1$ and the curves $y + ix = 0$ and $y - ix = 0$ intersect normally at the point $(x, y) = (0, 0)$, the normalization S of $\mathbb{F}_0$ in $k(x, y, z)$ has a cyclic quotient singularity of type A_{p-1}, the fiber of f over $x = 0$ is a reducible fiber consisting of p irreducible components, i.e., a linear chain of the irreducible component meeting the section $\widetilde{M}_\infty$ and a linear chain of exceptional curves of the A_{p-1}-singularity. If we assume that the components $\Theta_1, \ldots, \Theta_{2m}$ of the linear chain are arranged in this order, the component Θ_0 of the fiber meeting the section O and the components Θ_m, Θ_{m+1} meet transversally in a single common point (see the proof of Lemma 3.12, the case (iii)).

As for the reducible fiber over $x = \infty$, similar observation as in the case 4.4.2.1 allows us to find a minimal quasi-hyperelliptic fibration $f : V \to C$. The configuration is given by the following dual graph.

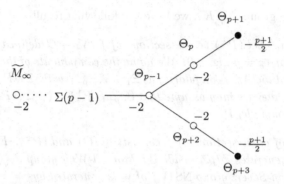

The fiber over $x = \infty$ consists of (-2)-components $\Theta_0, \ldots, \Theta_{p-2}, \Theta_{p-1}$, Θ_p, Θ_{p+2} and two $(-\frac{p+1}{2})$-components $\Theta_{p+1}, \Theta_{p+3}$, where Θ_0 (resp. Θ_{p-2}) meets $\widetilde{M}_\infty$ (resp. Θ_{p-1}). We say that this fiber has type $IV^*(p)$.

Let $X = \sum_{i=1}^{p+3} \Theta_i$. Then X is contractible to a point on a normal projective surface W. After contracting these $(p+3)$ curves, we have an isolated singularity defined by a local equation $z^p + y^2 + x^{2p-2} = 0$, say $E_6(p)$, which is thought of a generalization of a rational double point of type E_6. In fact, if $p = 3$, its type is E_6.

The intersection matrix of this singularity $E_6(p)$ has rank $p + 3$ and is written as follows. Its determinant is $-p$.

$$\begin{pmatrix} -2 & 1 \\ 1 & -2 \\ & & \ddots \\ & & & \ddots \\ & & & & -2 & 1 \\ & & & & 1 & -2 & 1 & 0 & 1 \\ & & & & & 1 & -2 & 1 & 0 & 0 \\ & & & & & 0 & 1 & -\frac{p+1}{2} & 0 & 0 \\ & & & & & 1 & 0 & 0 & -2 & 1 \\ & & & & & 0 & 0 & 0 & 1 & -\frac{p+1}{2} \end{pmatrix}$$

We have the following property of $E_6(p)$ singularity.

(1) The fundamental cycle is given by

$$Z = 2\Theta_1 + 3\Theta_2 + \cdots + p\Theta_{p-1} + \frac{p+1}{2}\Theta_p + \Theta_{p+1} + \frac{p+1}{2}\Theta_{p+2} + \Theta_{p+3}.$$

(2) $(Z^2) = -2$, $(K_V \cdot Z) = p - 3$. Hence $p_a(Z) = \frac{1}{2}(p-3)$.

As for the group $J_{V_\eta}(K)$, we have the following result.

Proposition. *Let* $(P^\pm)$ *be the sections of* $f : V \to C$ *defined respectively by the equations* $z = y \pm ix = 0$. *We name the components of the exceptional components of the* A_{p-1}-*singularity as* $\Xi_1, \ldots, \Xi_{p-1}$ *with* $(\Xi_1 \cdot P^+) = (\Xi_{p-1} \cdot P^-) = 1$. *Further we may assume that* $(\Theta_{p+1} \cdot P^+) = (\Theta_{p+3} \cdot P^-) = 1$. *We denote* P^+ *simply by* P.

(1) *The set of cross-sections of* f *consists of* (O) *and* $(P^\pm)$. *Furthermore, this set generates* $\mathbb{Z}/p\mathbb{Z}$ *inside the Mordell-Weil group* $J_{V_\eta}(K)$.
(2) *The Néron-Severi group* $\mathrm{NS}(V)$ *of* V *is generated by*

$$\Theta_1, \Theta_2, \ldots, \Theta_{p+3}, \Xi_1, \Xi_2, \ldots, \Xi_{p-1}, (O), F, (P)$$

with the following relation.

$$p((P) - (O)) \approx p(((P) - (O)) \cdot (O))F$$
$$+ (\Theta_1, \ldots, \Theta_{p+3}) p A_\infty^{-1} \begin{pmatrix} (\Theta_1 \cdot (P)) \\ \vdots \\ (\Theta_{p+3} \cdot (P)) \end{pmatrix}$$
$$+ (\Xi_1, \ldots, \Xi_{p-1}) p A_0^{-1} \begin{pmatrix} (\Xi_1 \cdot (P)) \\ \vdots \\ (\Xi_{p+3} \cdot (P)) \end{pmatrix},$$

where F *is a general fiber of* f *and* A_∞ *(resp.* A_0*) is the intersection matrix of the* $E_6(p)$-*singularity (resp.* A_{p-1}-*singularity).*

4.4.2.3 *Reducible fibers of* $z^p = y^2 + x^3$

We consider the quasi-hyperelliptic fibration $f : V \to C$ which is obtained by the projection $(x, y, z) \mapsto x$ of the hypersurface $z^p = y^2 + x^3$. We show that the reducible fiber over $x = \infty$ changes its configuration according to the characteristic p. This change makes a sharp contrast to the previous two examples, for which the configurations are generalizations of those of E_8 and E_6 singularities. We observe this in the cases $p = 5$ and 7.

CASE $p = 5$. By Theorem 3.9 the surface V is rational. There are two reducible fibers over $x = 0$ and $x = \infty$. The fiber over $x = 0$ has the following configuration when minimalized, where $\circ$ signifies a (-2)-curve and Ξ_0 is the component meeting the section O.

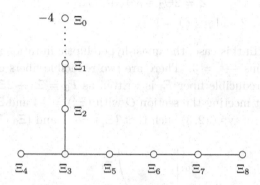

This is a genus 2 singular fiber of Ogg type [7] and Namikawa-Ueno type [VIII-2]. The fiber F_0 is given as

$$\Xi_0 + 4\Xi_1 + 7\Xi_2 + 10\Xi_3 + 5\Xi_4 + 8\Xi_5 + 6\Xi_6 + 4\Xi_4 + 2\Xi_8.$$

However, $F_{0,\mathrm{red}} \setminus \Theta_0$ contracts to a rational double point of type E_8. Hence the fundamental cycle Z is given as

$$2\Xi_8 + 3\Xi_7 + 4\Xi_6 + 5\Xi_5 + 6\Xi_3 + 3\Xi_4 + 4\Xi_2 + 2\Xi_1,$$

and $(Z^2) = -2$ and $p_a(Z) = 0$. If the component Ξ_0 is connected to Ξ_8 instead of Ξ_1, it appears as a minimalized singular fiber of a quasi-elliptic fibration (see Lemma 3.5.2).

On the other hand, the fiber over $x = \infty$ has the following configuration.

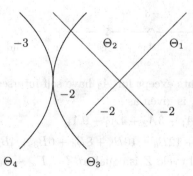

This is a genus 2 singular fiber of Ogg type [16] and Namikawa-Ueno type [VIII-3], where $(\Theta_3 \cdot \Theta_4) = 2$. The fiber F_0 and the fundamental cycle Z are given as follows.

$$F_0 = \Theta_1 + 2\Theta_2 + 3\Theta_3 + 2\Theta_4,$$
$$Z = 2\Theta_2 + 3\Theta_3 + 2\Theta_4.$$

Hence $(Z^2) = -2$ and $p_a(Z) = 1$.

CASE $p = 7$. In this case, the quasi-hyperelliptic fibration $f : V \to C$ has arithmetic genus $\frac{7-1}{2} = 3$. There are two reducible fibers over $x = 0$ and $x = \infty$. The reducible fiber F_0 is written as $F_0 = \Xi_0 + 2\Xi_1$, where Ξ_0 is the component meeting the section O with $(\Xi_0^2) = -4$ and Ξ_1 is a rational cuspidal curve of type $(2, 3)$ such that $(\Xi_1^2) = -1$ and $(\Xi_0 \cdot \Xi_1) = 2$;

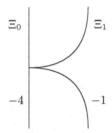

The reducible fiber F_∞ over $x = \infty$ has the following dual graph:

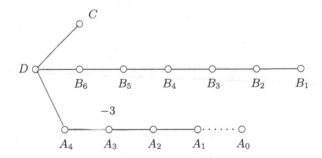

All the components except for A_3 have self-intersection numbers equal to -2. The fiber F_∞ is given as

$$F_\infty = A_0 + 2A_1 + 3A_2 + 4A_3 + 9A_4$$
$$+ 14D + 12B_6 + 10B_5 + 8B_4 + 6B_3 + 4B_2 + 2B_1 + 7C$$

and the fundamental cycle Z is equal to $Z = F_\infty - A_0$. Hence $(Z^2) = -2$ and $p_a(Z) = 2$.

With the above notations, we can compute the canonical divisor K_V as

$$K_V \sim 4(O) + a\Gamma - \Xi_1 + 3A_0 + 2A_1 + A_2$$

for some integer a. On the other hand, V is rational because $z = y'^2 + x'^3$ by the birational change of coordinates $(x, y, z) = (x'z^2, y'z^3, z)$. Hence $a \leq 0$ because $|K_V| \neq \emptyset$ otherwise. In fact, since $\Gamma \sim \Xi_0 + 2\Xi_1$, $|K_V|$ contains an effective divisor if $a > 0$. We can determine (O^2) by making use of resolution data obtained to determine the fiber F_∞ from $(zuv)^7 = u^4v^5(v^2 + u^3)$, which is obtained from $z^7 = y^2 + x^3$ by the change of coordinates $x = \frac{1}{u}$ and $y = \frac{1}{v}$ (see section 2.7). The result is that $(O^2) = -1$. Since the section (O) is a smooth rational curve, we have

$$-2 = (O)^2 + ((O) \cdot K_V) = 5(O)^2 + a + 3 = a - 2.$$

Hence $a = 0$.

4.4.2.4

We consider the following hypersurface to see what kind of reducible fibers appear in the quasi-hyperelliptic fibration

$$z^p = y^2 + x^{p+1} - \frac{1}{2}x^2.$$

Set $\varphi(x) = x^{p+1} - \frac{1}{2}x^2$. Since $\varphi'(x) = x^p - x = \prod_{i=0}^{p-1}$, the corresponding fibration has reducible fibers over the points $x = i$ $(0 \leq i \leq p-1), \infty$.

CASE $x = i$ $(0 \leq i \leq p-1)$. Clearly it suffices to consider the case $x = 0$. Then the equation becomes

$$z^p = y^2 + x^2\left(x^{p-1} - \frac{1}{2}\right),$$

where $x^{p-1} - \frac{1}{2}$ is a unit near the point $(x, y) = (0, 0)$. Hence the reducible fiber of $f : V \to C$ has the dual graph as explained in 4.4.2.2 for the equation $z^p = y^2 + x^2$ over $x = 0$.

CASE $x = \infty$. The equation is written as

$$z'^p = u^{p-1}v^{p-2}\left(v^2\left(1 - \frac{1}{2}u^{p+1}\right) + u^{p+1}\right),$$

where $u = \frac{1}{x}, v = \frac{1}{y}, z' = zu^2v$ and $1 - \frac{1}{2}u^{p+1}$ is a unit near the point $(u, v) = (0, 0)$. Let $p = 2m + 1$. Then, after a change of local coordinates at $(0, 0)$, we may assume that the equation is

$$z^p = u^{p-1}v^{p-2}(v - u^{m+1})(v + u^{m+1}).$$

Then it can be shown by applying the same computation as in the previous cases that the reducible fiber F_∞ is written as

$$F_\infty = \Theta_0 + \Theta_1 + \Theta_2,$$

where $(\Theta_0^2) = -2$, $(\Theta_1^2) = (\Theta_2^2) = -(m+1)$, $(\Theta_1 \cdot \Theta_2) = m$ and $(\Theta_0 \cdot (O)) = 1$. Hence the configuration is as follows.

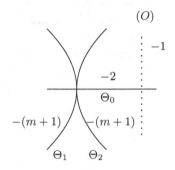

The canonical divisor K_V can be computed easily as

$$K_V \sim (p-3)(O) + (m-2)\Gamma + (m-1)\Theta_0,$$

where Γ is a general fiber of $f : V \to C$. Hence, if $p \geq 5$, V is not rational, and if $p = 3$, V is rational.

Chapter 5

Artin-Schreier coverings

5.1 $\mathbb{Z}/p\mathbb{Z}$-action and quotient morphism

Let B be an affine domain over k, which is an algebraically closed field of characteristic $p > 0$. Let $Y = \operatorname{Spec} B$. Let $G = \mathbb{Z}/p\mathbb{Z}$ be a cyclic group of order p. As a k-group scheme, G is the kernel of a k-group scheme homomorphism $F - \operatorname{id} : G_a \to G_a$ (cf. 1.5.4). Let $\sigma : G \times Y \to Y$ be a G-action. We say that σ is *free* if, for every closed point $Q \in Y$, the G-orbit GQ consists of p points. Let A be the G-invariant subring of B, let $X = \operatorname{Spec} A$ and let $q : Y \to X$ be the quotient morphism which is induced by the natural inclusion $A \hookrightarrow B$.

5.1.1

Lemma. *The following assertions hold.*

(1) A *is an affine domain over* k, *and* B *is a finite* A-*module. If* B *is normal, then so is* A.
(2) *Let* $L = Q(B)$ *and* $K = Q(A)$ *be the quotient fields of* B *and* A. *Then* K *is the* G-*invariant subfield of* L.
(3) *For any closed point* $P \in X$, *the group* G *acts transitively on* $q^{-1}(P)$.
(4) *If* G *acts non-trivially on* $q^{-1}(P)$, *then* q *is unramified over* P.
(5) *If* G *acts freely on* Y, *the quotient morphism* q *is a finite flat morphism. Hence* q *is a finite étale morphism.*
(6) *Assume that* G *acts freely on* Y. *Then the morphism*

$$\Phi := (\sigma, p_2) : \quad G \times Y \longrightarrow Y \times_X Y$$

defined by the action σ *and the second projection* $p_2 : G \times Y \to Y$ *is an isomorphism.*

Proof. (1) Let γ be a generator of G, which we write multiplicatively. Hence $\gamma^p = 1$. For an element $b \in B$, the polynomial

$$\varphi_b(x) = \prod_{0 \leq i < p} (x - \gamma^i(b))$$

has all coefficients invariant under the G-action. Since $\varphi_b(b) = 0$, it follows that b is integral over A. Write $\varphi_b(x)$ as

$$\varphi_b(x) = x^p - \tau_1(b)x^{p-1} + \cdots + (-1)^i \tau_i(b)x^{p-i} + \cdots + (-1)^p \tau_p(b) \ .$$

Since B is finitely generated over k, we can write $B = k[b_1, \ldots, b_n]$. Let $A_0 = k[\tau_i(b_j) \mid 1 \leq i \leq p, \ 1 \leq j \leq n]$. Then $A_0 \subset A \subset B$, and B is a finite A_0-module. Since A_0 is noetherian, A is a finite A_0-submodule of B, whence A is finitely generated over k. We leave the proof of the second assertion of (1) after the proof of (2).

(2) Let L^G be the G-invariant subfield of L. Then the inclusion $K \subset L^G$ is clear. Take an element $\xi \in L^G$ and write $\xi = b_1/b_2$ with $b_1, b_2 \in B$. Let $b_2' = \tau_p(b_2)/b_2$, where $\tau_p(b_2)$ is the norm of b_2. Then $\xi = (b_1 b_2')/\tau_p(b_2)$, where $b_1 b_2' \in A$. Hence $\xi \in K$. So, $K = L^G$. It is a standard fact of the Galois theory that $\mathrm{Aut}_K(L) = G$ and $[L : K] = |G| = p$.

Suppose that B is normal. Since $K \cap B = A$, it follows that A is normal.

(3) Since q is finite and dominating, q is surjective. This implies that $q^{-1}(P) \neq \emptyset$. Let $Q_1, Q_2 \in q^{-1}(P)$. Then P (resp. Q_1, Q_2) corresponds to the maximal ideal $\mathfrak{m}$ (resp. $\mathfrak{M}_1, \mathfrak{M}_2$) of A (resp. B). Clearly, $\mathfrak{M}_i \cap A = \mathfrak{m}$ $(i = 1, 2)$. Since the G-translates of Q_2 belong to $q^{-1}(P)$, let $\mathfrak{M}_2, \ldots, \mathfrak{M}_n$ exhaust all G-translates of $\mathfrak{M}_2$. Suppose that $\mathfrak{M}_1$ is none of $\mathfrak{M}_2, \ldots, \mathfrak{M}_n$. By the lying-over theorem, there are no inclusion relations among $\mathfrak{M}_1, \mathfrak{M}_2, \ldots, \mathfrak{M}_n$. Then, for every $2 \leq i \leq n$, we have

$$\mathfrak{M}_1(\mathfrak{M}_2 \cdots \overset{\vee}{\mathfrak{M}_i} \cdots \mathfrak{M}_n) \not\subset \mathfrak{M}_i \ .$$

Let $b_i \in \mathfrak{M}_1(\mathfrak{M}_2 \cdots \overset{\vee}{\mathfrak{M}_i} \cdots \mathfrak{M}_n) \setminus \mathfrak{M}_i$ and let $b = \sum_{i=2}^n b_i$. Then $b \in \mathfrak{M}_1$ and $b \notin \mathfrak{M}_i$ $(2 \leq i \leq n)$. By the construction, all G-translates of b are not elements of $\mathfrak{M}_i$ $(2 \leq i \leq n)$. Hence the norm $\tau_p(b) = \prod_{i=0}^{p-1} \gamma^i(b) \in \mathfrak{M}_1 \setminus (\cup_{i=2}^n \mathfrak{M}_i)$. Since $\tau_p(b) \in A \cap \mathfrak{M}_1 = \mathfrak{m}$ and $\mathfrak{M}_i \cap A = \mathfrak{m}$ $(2 \leq i \leq n)$, this is a contradiction. Thus the assertion is proved.

(4) The assertion (3) implies that if the G-action is non-trivial on $q^{-1}(P)$, the fiber $q^{-1}(P)$ consists of p points. Let $\mathfrak{m}$ be the maximal ideal of A corresponding to P. Let $\mathfrak{M}_1, \mathfrak{M}_2, \ldots, \mathfrak{M}_p$ be the maximal ideals of B corresponding to the points of $q^{-1}(P)$; there are exactly p points since G acts non-trivially and transitively on $q^{-1}(P)$. We show that $\mathfrak{M}_1 B_{\mathfrak{M}_1} =$

$mB_{\mathfrak{M}_1}$. Then $\mathfrak{M}_i B_{\mathfrak{M}_i} = mB_{\mathfrak{M}_i}$ for $2 \le i \le p$ because $\gamma^i(\mathfrak{M}_1 B_{\mathfrak{M}_1}) = \mathfrak{M}_i B_{\mathfrak{M}_i}$ and $\gamma^i(mB_{\mathfrak{M}_1}) = mB_{\mathfrak{M}_i}$. Since there are no inclusions among $\mathfrak{M}_1, \ldots, \mathfrak{M}_p$, there exists an element $c \in \mathfrak{M}_1 \setminus (\cup_{i=2}^p \mathfrak{M}_i)$. Then $\gamma^i(c) \notin \mathfrak{M}_1$ for $1 \le i < p$. Let $c' = \prod_{i=1}^{p-1} \gamma^i(c)$. Then $c' \notin \mathfrak{M}_1$. Since $cc' = \tau_p(c) \in \mathfrak{M}_1 \cap A = m$, we have $c \in mB_{\mathfrak{M}_1}$. Let $\mathfrak{A} = \mathfrak{M}_2 \cap \cdots \cap \mathfrak{M}_p$. Let c^* be an element of $\mathfrak{M}_1 \cap \mathfrak{A}$. Then $c + c^* \in \mathfrak{M}_1 \setminus (\cup_{i=2}^p \mathfrak{M}_i)$ with the above c, whence $c + c^* \in \mathfrak{M}_1 B_{\mathfrak{M}_1} = mB_{\mathfrak{M}_1}$, and $c^* \in mB_{\mathfrak{M}_1}$. On the other hand, $(\mathfrak{M}_1 \cap \mathfrak{A})B_{\mathfrak{M}_1} = \mathfrak{M}_1 B_{\mathfrak{M}_1}$. In fact, for $i \ne 1$, let b_i be an element of $\mathfrak{M}_i$, not in $\mathfrak{M}_1$. For an element $b \in \mathfrak{M}_1$, $b^* := b \prod_{i \ne 1} b_i \in \mathfrak{M}_1 \cap \mathfrak{A}$ and $b = b^* \cdot (\prod_{i \ne 1} b_i)^{-1} \in (\mathfrak{M}_1 \cap \mathfrak{A})B_{\mathfrak{M}_1}$. This implies that $\mathfrak{M}_1 B_{\mathfrak{M}_1} \subseteq mB_{\mathfrak{M}_1}$. The opposite inclusion is obvious. So, $mB_{\mathfrak{M}_1} = \mathfrak{M}_1 B_{\mathfrak{M}_1}$. This implies that q is unramified over P. The above proof is an adaptation of the proof in [66, Theorem (41.2)].

(5) Let $\mathcal{O} = \mathcal{O}_{X,P}$ and $\mathcal{O}_i = \mathcal{O}_{Y,Q_i}$ $(1 \le i \le p)$, where $q^{-1}(P) = \{Q_1, \ldots, Q_p\}$. Let $\widehat{\mathcal{O}}$ be the m-adic completion of $\mathcal{O}$. Since B is a finite A-module and q is unramified over P, the completion $\widehat{\mathcal{O}}_i$ of $\mathcal{O}_i$ is isomorphic to $\mathcal{O}_i \otimes_{\mathcal{O}} \widehat{\mathcal{O}}$ and $B \otimes_A \widehat{\mathcal{O}}$ decomposes as a product

$$B \otimes_A \widehat{\mathcal{O}} = \widehat{\mathcal{O}}_1 \times \cdots \times \widehat{\mathcal{O}}_p \cong \underbrace{\widehat{\mathcal{O}} \times \cdots \times \widehat{\mathcal{O}}}_{p} .$$

This implies that $B \otimes_A \widehat{\mathcal{O}}$ is a flat $\widehat{\mathcal{O}}$-module. Hence $B \otimes_A \mathcal{O}$ is a flat $\mathcal{O}$-module. Since P is arbitrary, B is a flat A-module. Since q is unramified by the assertion (4), q is, in fact, a finite étale morphism.

(6) The morphism Φ is associated with the B-algebra homomorphism

$$\Phi^* : B \otimes_A B \to k[t]/(t^p - t) \otimes_k B, \quad \Phi^*(b \otimes b') = \sigma^*(b)b', \quad b, b' \in B .$$

See 1.5.4 for the notations. Since B is a locally free A-module of rank p and the G-action σ is free, we infer that Φ^* is an isomorphism. $\quad\square$

5.1.2

Let G be an algebraic k-group scheme, i.e., a group scheme which is of finite type over k. Let Y be an algebraic scheme over k which is given a G-action $\sigma : G \times Y \to Y$. Let $f : Y \to X$ be a G-equivariant morphism with a trivial G-action on X. We say that Y is a *G-torsor* (or a *principal G-homogeneous space*) over X if the following two conditions are satisfied:

(1) f is a faithfully flat morphism.
(2) The morphism $\Phi = (\sigma, p_2) : G \times_k Y \to Y \times_X Y$ is an isomorphism.

By Lemma 5.1.1, an affine variety $Y = \operatorname{Spec} B$ with a free $\mathbb{Z}/p\mathbb{Z}$-action is a $\mathbb{Z}/p\mathbb{Z}$-torsor over X, where $X = \operatorname{Spec} A$ with the $\mathbb{Z}/p\mathbb{Z}$-invariant subring A of B and the role of f is played by the quotient morphism $q : Y \to X$. We also call Y a *free p-cyclic covering* of X.

5.1.3

Theorem. *With the notations of Lemma 5.1.1, assume that Y is a $\mathbb{Z}/p\mathbb{Z}$-torsor over X. Then there exist elements $z \in B$ and $a \in A$ such that $B = A[z]$ with $z^p - z = a$ and $\gamma(z) = z + 1$, where γ is a generator of G and identified with a k-automorphism of B.*

Proof. (I) Let $\mathcal{U} = \{U_\lambda\}_{\lambda \in \Lambda}$ be an affine open covering of X such that $B_\lambda = B \otimes_A A_\lambda$ with $U_\lambda = \operatorname{Spec} A_\lambda$ is a free A_λ-module of rank p. In the subsequent arguments, we take a finer affine open covering of X to make additional A_λ-submodules of B_λ free over A_λ. Since B is a flat A-module of rank p, such an affine open covering of X exists. We fix an open set $U_\lambda = \operatorname{Spec} A_\lambda$ and denote A_λ, B_λ by A, B, respectively.

Now, let $\{e_0, e_1, \ldots, e_{p-1}\}$ be a free basis of B as an A-module, i.e., $B = Ae_0 \oplus \cdots \oplus Ae_{p-1}$. We show that we can take the unity 1 as e_0. In fact, $1 = a_0 e_0 + a_1 e_1 + \cdots + a_{p-1} e_{p-1}$. Then the ideal $(a_0, a_1, \ldots, a_{p-1})$ is unitary, i.e., $A = (a_0, a_1, \ldots, a_{p-1})$. Hence we can write $1 = a_0 a_0' + a_1 a_1' + \cdots + a_{p-1} a_{p-1}'$ with $a_0', \ldots, a_{p-1}' \in A$. Define a surjective A-homomorphism $\theta : B \to A$ by $\theta(e_i) = a_i'$ for $0 \le i \le p-1$. The kernel K of θ is a projective A-module of rank $p - 1$. Hence, by replacing $\mathcal{U}$ by a finer covering, we may assume that K is free. Since $B \cong K \oplus A(a_0 e_0 + a_1 e_1 + \cdots + a_{p-1} e_{p-1})$, the unity 1 is a member of a free basis of B. So, we assume that $e_0 = 1$.

(II) With respect to the free basis, the action of γ is given by

$$
\begin{pmatrix} 1 \\ \gamma(e_1) \\ \vdots \\ \gamma(e_{p-1}) \end{pmatrix} = \begin{pmatrix} 1\,0\,\cdots\,0 \\ * \\ \vdots \quad M \\ * \end{pmatrix} \begin{pmatrix} 1 \\ e_1 \\ \vdots \\ e_{p-1} \end{pmatrix},
$$

where $M \in \operatorname{GL}(p-1, A)$ and $M^p = E$ (the identity matrix). Since the eigenvalues of M are all equal to 1, we can find a non-zero element

$$ e = c_1 e_1 + \cdots + c_{p-1} e_{p-1} \in B, \quad c_i \in A $$

such that $\gamma(e) = e + a$ with $a \in A$. We show that $(c_1, \ldots, c_{p-1}) \subseteq aA$. This is the case if a is a unit of A. Otherwise, let $\overline{A} = A/aA$ and $\overline{B} = B/aB$. Let $\overline{Y} = \operatorname{Spec} \overline{B}$ and $\overline{X} = \operatorname{Spec} \overline{A}$. Then $\overline{Y}$ is a G-torsor over $\overline{X}$ and hence

$\overline{A}$ is the G-invariant subring of $\overline{B}$. If some $c_i \notin aA$, then $\overline{e} = e \pmod{aB}$ is a non-zero element which is G-invariant, not in $\overline{A}$. This is a contradiction. Hence every c_i is divisible by a. Now, setting $y = e/a$, we have $\gamma(y) = y+1$.

(III) We show that $\{1, y, \ldots, y^{p-1}\}$ is a free basis of B. Write the exterior product $1 \wedge y \wedge \cdots \wedge y^{p-1}$ in the form

$$1 \wedge y \wedge \cdots \wedge y^{p-1} = d(1 \wedge e_1 \wedge \cdots \wedge e_{p-1})$$

with $d \in A$. We have only to show that d is a unit in A. For this purpose, it suffices to show that, for every maximal ideal $\mathfrak{m}$ of A, setting $w = y$ $(\mathrm{mod}\ \mathfrak{m}B)$, $1, w, \ldots, w^{p-1}$ are linearly independent over k. Suppose the contrary. Then we have a relation

$$w^r + s_1 w^{r-1} + \cdots + s_r = 0 \ ,$$

where $0 < r < p$ and $s_1, \ldots, s_r \in k$. We may assume that r is the smallest possible. Since $\gamma(w) = w + 1$ and

$$(\gamma(w))^r + s_1(\gamma(w))^{r-1} + \cdots + s_r = 0 \ ,$$

we have a relation of lower degree

$$w^{r-1} + (\text{terms of lower degree}) = 0.$$

This contradicts the choice of r. Therefore $\{1, y, \ldots, y^{p-1}\}$ is a free basis of B. In particular, $B = A[y]$, where y satisfies a monic relation

$$F(y) := y^p + c_1 y^{p-1} + \cdots + c_{p-1} y + c_p = 0$$

with $c_1, \ldots, c_p \in A$. Since $\gamma(y) = y + 1$ and $F(\gamma(y)) = 0$, we infer that

$$c_1 = \cdots = c_{p-2} = 0, \quad c_{p-1} = -1 \ .$$

Hence y satisfies a relation

$$y^p - y = a, \quad a \in A.$$

(IV) Now we come back to an affine open covering $\mathcal{U} = \{U_\lambda\}_{\lambda \in \Lambda}$ of X such that $B_\lambda = A_\lambda[z_\lambda]$ with $z_\lambda^p - z_\lambda = a_\lambda \in A_\lambda$ and $\gamma(z_\lambda) = z_\lambda + 1$ for every $\lambda \in \Lambda$ in the notations A_λ, B_λ as defined in the step (I). Over the intersection $U_\lambda \cap U_\mu$, we have

$$\gamma(z_\mu - z_\lambda) = (z_\mu + 1) - (z_\lambda + 1) = z_\mu - z_\lambda.$$

Hence $a_{\mu\lambda} := z_\mu - z_\lambda \in A_{\lambda\mu} := \Gamma(U_\lambda \cap U_\mu, \mathcal{O}_X)$. It is clear that $\{a_{\mu\lambda}\}$ is a 1-cocycle. Since $H^1(\mathcal{U}, \mathcal{O}_X) = H^1(X, \mathcal{O}_X) = 0$ as X is affine, we have $a_\lambda' \in A_\lambda$ such that $a_{\mu\lambda} = a_\mu' - a_\lambda'$. Then $z_\mu - a_\mu' = z_\lambda - a_\lambda'$ for every pair

(λ, μ). Hence it defines an element $z \in B$ such that $z_\lambda - a'_\lambda$ is the restriction of z onto U_λ. Then $z_\lambda = z + a'_\lambda$ and

$$z^p - z = a_\lambda - (a'^p_\lambda - a'_\lambda) \in \Gamma(U_\lambda, \mathcal{O}_X).$$

It is then clear that $\{a_\lambda - (a'^p_\lambda - a'_\lambda)\}_{\lambda \in \Lambda}$ defines an element a of A such that $a_\lambda - (a'^p_\lambda - a'_\lambda)$ is the restriction of a onto U_λ. Thus we have $B = A[z]$ with $z^p - z = a \in A$ and $\gamma(z) = z + 1$. $\qquad\square$

Remark. The following alternative proof is due to Yuya Matsumoto. As in the step (III) in the above proof, it suffices to show the existence of an element $z \in B$ such that $\gamma(z) = z + 1$. Let $Y = \operatorname{Spec} B$ and $X = \operatorname{Spec} A$. The graph morphism $\Phi : G \times Y \to Y \times_X Y$ is an isomorphism as Y is a G-torsor over X, where $G = \mathbb{Z}/p\mathbb{Z}$.

We first note that $G = \operatorname{Spec}(\oplus_{g \in G} k e_g)$, where $1 = \oplus_{g \in G} e_g$ is the idempotent decomposition, i.e., $e_g e_{g'} = \delta_{gg'} e_g$ for $g, g' \in G$. The comultiplication μ_G, the coinverse ι_G and the counit ε_G are respectively given by k-algebra homomorphisms

$$\mu_G(e_g) = \sum_{g' \in G} e_{g'} \otimes e_{g'^{-1}g}$$

$$\iota_G(e_g) = e_{g^{-1}}$$

$$\varepsilon_G(e_g) = 0 \quad (g \neq 1), \quad \varepsilon_G(e_1) = 1.$$

The coaction σ^* associated to a G-action $\sigma : G \times Y \to Y$ is given by $\sigma^*(b) = \sum_{g \in G} g(b) e_g$.

Let $\delta = \gamma - \operatorname{id}$, where γ is a generator of G identified with the associated k-automorphism of B. Then $\delta^p = (\gamma - \operatorname{id})^p = \gamma^p - \operatorname{id} = 0$ and $\delta^{p-1}(b) = \operatorname{Tr}(b) = \sum_{g \in G} g(b)$ for $b \in B$. Let $I = \operatorname{Im} \operatorname{Tr}(B)$. Then I is an ideal of A. Now consider the composite

$$B \otimes_A B \xrightarrow{\Phi^*} \oplus_{g \in G} B e_g \xrightarrow{\operatorname{Tr}} B,$$
$$b \otimes b' \mapsto \sum_{g \in G} bg(b') e_g \mapsto \sum_{g \in G} bg(b') = b\operatorname{Tr}(b'),$$

where Tr is simply an A-module homomorphism. Then the image of the composite is the ideal IB in B. Meanwhile, since Φ^* is an isomorphism, this image is also the image of $\oplus_{g \in G} B e_g$ by the trace homomorphism. Hence it is equal to B. Namely, we have $IB = B$. Since B is a faithfully flat A-module by Lemma 5.1.1, (5), it follows that $I = A$. Hence $1 \in \operatorname{Im}(\delta)$. This implies that there exists an element $z \in B$ such that $1 = (\gamma - \operatorname{id})(z) = \gamma(z) - z$.

5.1.4

There is a different proof by Kambayashi-Srinivas [42], which is more transparent in ideas than the above proof, but makes use of advanced results on the flat cohomology. The following is an exact sequence of k-group schemes

$$0 \longrightarrow G \longrightarrow G_1 \xrightarrow{q_0} G_2 \longrightarrow 0 , \qquad (5.1)$$

where $G = \mathbb{Z}/p\mathbb{Z}$, $G_1 \cong G_2 \cong G_{a,k}$, and $q_0 := F - \mathrm{id}$, i.e., $q_0(x) = x^p - x$. Then $G_1 \times_{G_2} G_1 \cong G \times G_1$. Hence G_1 is a G-torsor over G_2. Let $X = \mathrm{Spec}\, A$ be an affine variety. Then the set of isomorphism classes of G-torsors over X is represented by the flat cohomology group $H^1_{\mathrm{fl}}(X, G)$, which is equal to the étale cohomology group $H^1_{\mathrm{et}}(X, G)$ in the case $G = \mathbb{Z}/p\mathbb{Z}$. The above exact sequence (5.1) gives rise to a long exact sequence of flat cohomology groups

$$0 \longrightarrow H^0_{\mathrm{fl}}(X, G) \longrightarrow H^0_{\mathrm{fl}}(X, G_1) \xrightarrow{q_0^*} H^0_{\mathrm{fl}}(X, G_2) \xrightarrow{\delta}$$
$$\longrightarrow H^1_{\mathrm{fl}}(X, G) \longrightarrow H^1_{\mathrm{fl}}(X, G_1) \longrightarrow ,$$

where $H^0_{\mathrm{fl}}(X, G) \cong G$ since X is connected, $H^0_{\mathrm{fl}}(X, G_1) \cong H^0_{\mathrm{fl}}(X, G_2) \cong H^0(X, \mathcal{O}_X) = A$, and $H^1_{\mathrm{fl}}(X, G_1) \cong H^1(X, \mathcal{O}_X) = 0$. Hence the long exact sequence becomes

$$0 \longrightarrow \mathbb{Z}/p\mathbb{Z} \longrightarrow A^+ \xrightarrow{q_0} A^+ \xrightarrow{\delta} H^1_{\mathrm{fl}}(X, G) \longrightarrow 0$$

and $H^1_{\mathrm{fl}}(X, G) \cong A^+/q_0(A^+)$. For $a \in A$, the image $\delta(a)$ is given by a G-torsor obtained as a fiber product $Y := (G_1, q_0) \times_{G_2} (X, \alpha)$, where $\alpha : X \to G_2$ is given by a k-algebra homomorphism $\alpha^* : k[t] \to A$, $\alpha^*(t) = a$. Thus $\Gamma(Y, \mathcal{O}_Y) = k[z] \otimes_{k[t]} A \cong A[z]/(z^p - z - a)$.

5.1.5

Let X be a projective variety defined over k. The exact sequence (5.1) of k-group schemes gives rise to an exact sequence

$$0 \longrightarrow H^1_{\mathrm{fl}}(X, G) \longrightarrow H^1(X, \mathcal{O}_X) \xrightarrow{q_0^*} H^1(X, \mathcal{O}_X),$$

where $H^1_{\mathrm{fl}}(X, G) \to H^1(X, \mathcal{O}_X)$ is injective because

$$H^0(X, \mathcal{O}_X) \xrightarrow{q_0^*} H^0(X, \mathcal{O}_X)$$

is surjective as $H^0(X, \mathcal{O}_X) = k$. This exact sequence was observed in detail by Serre [82, §16], where $H^1_{\mathrm{fl}}(X, G)$ is denoted by $\pi(X, G)$ or $\pi^1(X, \mathbb{Z}/p\mathbb{Z})$. Since $H^1(X, \mathcal{O}_X)$ is a k-vector space of finite dimension and $q_0^* = F^* - \mathrm{id}$ with the Frobenius homomorphism F^*, $H^1(X, G)$ is isomorphic to a finite abelian group $(\mathbb{Z}/p\mathbb{Z})^{\sigma(X)}$, where $\sigma(X)$ is the dimension of the *semi-simple* part of the decomposition $H^1(X, \mathcal{O}_X) = H^1(X, \mathcal{O}_X)^{ss} \oplus H^1(X, \mathcal{O}_X)^n$. Here $H^1(X, \mathcal{O}_X)^{ss}$ is the part on which F^* is bijective and $H^1(X, \mathcal{O}_X)^n$ is the part on which F^* is nilpotent.

For example, if X is an elliptic curve, $\sigma(X) = 1$ or 0 according as X is *ordinary* or X is *supersingular*. By definition, X is ordinary (resp. supersingular) if the p-torsion group $\mathrm{Ker}\,(p \cdot \mathrm{id}_X)$ is isomorphic to $\mathbb{Z}/p\mathbb{Z}$ (resp. $\{0\}$).

5.2　$\mathbb{Z}/p\mathbb{Z}$-action and fixed point

$\mathbb{Z}/p\mathbb{Z}$-actions have various aspects. We exhibit them mostly by examples. In this section, we restrict ourselves to the case where $G = \mathbb{Z}/p\mathbb{Z}$ acts on algebraic varieties of dimension ≤ 2.

5.2.1

Lemma. *Let C be a smooth curve isomorphic to $\mathbb{P}^1$ and with a non-trivial G-action σ. Then there exists a unique fixed point P_∞, and G acts on $C \setminus \{P_\infty\}$ by $\gamma(x) = x + 1$, where x is an inhomogeneous coordinate of C such that $x = \infty$ at P_∞.*

Proof. Suppose that σ has no fixed points. Let $D := C/G$ and $q : C \to D$ the quotient morphism. Then $D \cong \mathbb{P}^1$ and q is a finite étale morphism. Hence $K_C = q^* K_D$. Since $C \cong \mathbb{P}^1$ by assumption, we have $-2 = p(-2)$. This is impossible since $p > 1$. Thus σ has a fixed point P_∞. Let x be an inhomogeneous coordinate of $C \setminus \{P_\infty\} \cong \mathbb{A}^1$. Then $\gamma(x) = a(\gamma)x + b(\gamma)$ with $a(\gamma), b(\gamma) \in k$, where γ is a generator of G. Since a is a multiplicative character of G, we have $a(\gamma) = 1$. Since $b(\gamma) \neq 0$ as σ is non-trivial, we replace x by $x/b(\gamma)$, and obtain $\gamma(x) = x + 1$. □

On the other hand, if E is an *ordinary* elliptic curve, its p-torsion group is isomorphic to $\mathbb{Z}/p\mathbb{Z}$. Thus, G acts on E by translations and $E/G \cong E$. Note that if a smooth projective curve C has a free G-action, then the genus g satisfies $g \equiv 1 \pmod{p}$, which follows from the relation $K_C = q^* K_D$.

5.2.2

Theorem. *For a positive integer n with $p \nmid n$, there exists a smooth projective curve C_n with a G-action σ such that σ has a unique fixed point Q_∞ and C_n has genus $(p-1)(n-1)/2$.*

Proof. We consider a connected p-cyclic covering Y of the affine line $\mathbb{A}^1$ defined by

$$y^p - y = c_0 t^n + \cdots + c_{n-1} t,$$

where $c_0, \ldots, c_{n-1} \in k, c_0 \neq 0$ and $c_i = 0$ if $i \equiv 0 \pmod{p}$. Embed $\mathbb{A}^2$ into $\mathbb{P}^2$ as the complement of a line in such a way that, with respect to homogeneous coordinates (X_0, X_1, X_2) of $\mathbb{P}^2$, we have $t = X_1/X_0$ and $y = X_2/X_0$. Suppose $n > p$. Then the closure $\overline{Y}$ of Y in $\mathbb{P}^2$ is defined by an equation

$$X_2^p X_0^{n-p} - X_2 X_0^{n-1} = c_0 X_1^n + \cdots + c_{n-1} X_1 X_0^{n-1}.$$

Let ℓ_∞ be the line defined by $X_0 = 0$. Then $\overline{Y} \cap \ell_\infty$ consists of a single point $P_\infty : (X_0, X_1, X_2) = (0, 0, 1)$. By iterating the embedded blowing-ups with center P_∞ and subsequently appearing singular points over P_∞ on the proper transforms of $\overline{Y}$, we know that the point P_∞ is a one-place point and has the multiplicity sequence

$$(n - p, \underbrace{p, \ldots, p}_{m-1}, \underbrace{r_1, \ldots, r_1}_{m_1}, \ldots, \underbrace{r_a, \ldots, r_a}_{m_a}, 1, \ldots),$$

where the integers $m, m_1, \ldots, m_a, r_1, \ldots, r_a$ are determined by the Euclidean algorithm

$$
\begin{aligned}
n &= mp + r_1, & 0 < r_1 < p, \\
p &= m_1 r_1 + r_2, & 0 < r_2 < r_1, \\
r_1 &= m_2 r_2 + r_3, & 0 < r_3 < r_2, \\
&\cdots\cdots\cdots \\
r_{a-2} &= m_{a-1} r_{a-1} + r_a, & 0 < r_a < r_{a-1}, \\
r_{a-1} &= m_a r_a + 1.
\end{aligned}
$$

Then the genus g is calculated by

$$g = \frac{(n-1)(n-2)}{2} - \frac{1}{2}\{(n-p)(n-p-1) + (m-1)p(p-1)$$

$$+ m_1 r_1(r_1 - 1) + \cdots + m_a r_a(r_a - 1)\} = \frac{(n-1)(p-1)}{2}.$$

Let C_n be the curve obtained from $\overline{Y}$ by resolution of singularities. Then the G-action on Y extends to C_n and the point $Q_\infty := C_n \setminus Y$ is a G-fixed point. The case $p > n$ is also treated in a similar fashion. $\square$

5.2.2.1

We denote by $C(\varphi(t))$ the affine curve $C_n \setminus \{Q_\infty\}$, where $\varphi(t) = c_0 t^n + \cdots + c_{n-1} t + c_n \in k[t]$ as in the proof of Theorem 5.2.2. Then the following result holds.

Corollary. *Let $G = \mathbb{Z}/p\mathbb{Z}$. Then each element ξ of the group of G-torsors $H^1_{\mathrm{fl}}(\mathbb{A}^1, G)$ is represented by an affine curve $C(\varphi(t))$ for some $\varphi(t) \in k[t]$. If $\xi, \eta \in H^1_{\mathrm{fl}}(\mathbb{A}^1, G)$ are represented by $C(\varphi(t)), C(\psi(t))$ with $\varphi(t), \psi(t) \in k[t]$ respectively, then the sum $\xi + \eta$ is represented by $C(\varphi(t) + \psi(t))$. On the other hand, the automorphism group $\mathrm{Aut}_k(\mathbb{A}^1)$ acts on $H^1_{\mathrm{fl}}(\mathbb{A}^1, G)$ as follows. Let α be an automorphism of $\mathbb{A}^1$ given by $\alpha^*(t) = ct + d$ with $c \in k^*$ and $d \in k$ and let $\xi \in H^1_{\mathrm{fl}}(\mathbb{A}^1, G)$ be represented by $C(\varphi(t))$. Then $\alpha(\xi)$ is represented by $C(\varphi(ct + d))$.*

Proof. Let Y be a G-torsor over $\mathbb{A}^1$. If we write $Y = \mathrm{Spec}\, B$ and $\mathbb{A}^1 = \mathrm{Spec}\, k[t]$, by Theorem 5.1.3, B is isomorphic to $k[t][y]/(y^p - y - \varphi(t))$. By Theorem 5.2.2, Y is isomorphic to $C(\varphi(t))$. Hence the isomorphism class of the G-torsor Y is represented by $C(\varphi(t))$. Let Y, Z be G-torsors over $\mathbb{A}^1$ represented by $C(\varphi(t)), C(\psi(t))$ respectively. Since the sum of the isomorphism classes of Y, Z is represented by $(Y \times_{\mathbb{A}^1} Z)/G$, where G acts on $Y \times_{\mathbb{A}^1} Z$ as $(y, z) \mapsto (\gamma(y), \gamma^{-1} z)$ for a generator γ of G when $Y = \mathrm{Spec}\, B$ as above and $Z = \mathrm{Spec}\, k[t][z]/(z^p - z - \psi(t))$. Since $\gamma(y) = y + 1$ and $\gamma^{-1}(z) = z - 1$, $y + z$ is G-invariant. Hence we have

$$(Y \times_{\mathbb{A}^1} Z)/G = \mathrm{Spec}\, k[t][y + z]/((y + z)^p - (y + z) - (\varphi(t) + \psi(t))).$$

So, the sum $Y \times_{\mathbb{A}^1} Z$ is represented by $C(\varphi(t) + \psi(t))$. If $\alpha \in \mathrm{Aut}_k(\mathbb{A}^1)$, α acts on a G-torsor Y by $\alpha(Y) = Y \times_{\mathbb{A}^1} (\mathbb{A}^1, \alpha)$. $\qquad\square$

5.2.2.2

If $W = \mathrm{Spec}\, B$ is a G-torsor of $\mathbb{A}^2 = \mathrm{Spec}\, k[t, u]$, Theorem 5.1.3 implies that B is isomorphic to $k[t, u][z]/(z^p - z - A(t, u))$, where $A(t, u) \in k[t, u]$. We say that Y is a *split G-torsor* of $\mathbb{A}^2$ if $A(t, u) = \varphi(t) + \psi(u)$ with $\varphi(t) \in k[t]$ and $\psi(u) \in k[u]$. Then we have the following result.

Corollary. *Let W be a split G-torsor of $\mathbb{A}^2$. Then W is G-isomorphic to $(Y \times Z)/G$, where Y, Z are G-torsors over $\mathbb{A}^1$ as in 5.2.2.1 with Y for $\varphi(t)$ and Z for $\psi(u)$ and $G = \langle \gamma \rangle$ acts on $Y \times Z$ by $(y, z) \mapsto (\gamma(y), \gamma^{-1}(z))$.*

Proof. Similar to the proof of the above corollary. $\qquad\square$

5.2.3

Theorem. *Let V be a smooth rational projective surface with a G-action. Then V has a G-fixed point.*

Proof. Suppose that V has no G-fixed points. Let $W = V/G$ and $q : V \to W$ the quotient morphism. Note that q is a finite étale morphism, and W is therefore a smooth surface. This implies that $K_V = q^*K_W$. If E is a (-1)-curve on W, we show that $q^*(E)$ is a disjoint union of p curves $q^*(E) = C_1 + \cdots + C_p$, where G acts transitively on the set $\{C_1, \ldots, C_p\}$. Otherwise, $q^*(E) = C$ or $q^*(E) = pC$ for an irreducible curve C. If $q^*(E) = pC$, then $q|_C : C \to E$ is an isomorphism and C is contained in the G-fixed point locus. This contradicts the assumption that the G-action has no fixed points. Hence $C = q^*(E)$ and C is a smooth curve. Then

$$C \cdot K_V = q^*(E) \cdot q^*(K_W) = p(E \cdot K_W) = -p$$

and $(C^2) = (q^*E)^2 = p(E^2) = -p$. The adjunction formula implies

$$C^2 + C \cdot K_V = -2p = 2g - 2 \geq -2,$$

where $g = g(C)$. This is impossible.

Write $q^*(E) = C_1 + \cdots + C_p$, where G acts transitively on $\{C_1, \ldots, C_p\}$. We show that $C_1, \ldots, C_p$ are disjoint from each other. Let $N = C_1 \cdot (C_2 + \cdots + C_p)$. Then the G-transitivity shows that every C_i meets the remaining $(p-1)$ components with multiplicity N. Then we have

$$
\begin{aligned}
p(E^2) = q^*(E^2) &= (C_1 + \cdots + C_p)^2 \\
&= \sum_{i=1}^{p} \{(C_i^2) + C_i \cdot (C_1 + \cdots + C_{i-1} + C_{i+1} + \cdots + C_p)\} \\
&= p(n + N),
\end{aligned}
$$

where $n = (C_1^2)$. Then we have $-p = p(n + N)$, i.e., $-1 = n + N$. Since $N \geq 0$, it follows that $n < 0$. On the other hand, since

$$p(C_1 \cdot K_V) = (q^*(E) \cdot K_V) = p(E \cdot K_W) = -p,$$

we have $(C_1 \cdot K_V) = -1$. Hence C_1 as well as the other C_i is a (-1)-curve. This implies that $N = 0$ because $n = -1$. Hence the curves $C_1, \ldots, C_p$ are mutually disjoint.

Then we can contract $C_1, \ldots, C_p$ simultaneously to obtain a surface V_1 with free G-action, and the quotient $W_1 := V_1/G$ is the contraction of E on W. Repeating contractions of this kind on V and W, we may assume that W is relatively minimal. Since W is a smooth rational surface[1], we have $(K_W^2) = 8$ or 9. Then $(K_V^2) = p(K_W^2) \leq 9$ since V is rational. This is a contradiction. Hence there is a G-fixed point on V. □

In the next section, we show that if V is a smooth projective variety with a nontrivial G-action, the existence of fixed points is related to the vanishing of $H^1(W, \mathcal{O}_W)$ for the quotient variety $W = V/G$.

5.2.4

Here is an easy consequence of *purity of branch loci*. First of all, we recall the result according to [66, Theorem (41.1)].

Theorem. *Let $q : Y \to X$ be a finite surjective morphism from a normal algebraic variety Y onto a smooth algebraic variety X. Suppose that q is unramified over every point of codimension one[2] of X. Then q is unramified.*

Corollary. *Let Y be a normal algebraic variety with a non-trivial $G = \mathbb{Z}/p\mathbb{Z}$-action and let $q : Y \to X$ be the quotient morphism. Suppose that X is smooth. Then either the G-action is free, or there are no irreducible components of codimension two or more in the fixed point locus.*

Proof. Suppose that the G-action is not free and the fixed point locus contains an irreducible component of codimension ≥ 2. Let P be a general point of such component and replace Y by a G-stable affine open neighborhood of P which does not meet any component of the fixed point locus of codimension one. We thus assume that the fixed point locus has codimension ≥ 2. Then, by Lemma 5.1.1, q is unramified over every point of codimension one of X. Hence q is unramified by the purity of branch loci. This is a contradiction because q is ramified at the fixed point P of Y. □

5.2.5

Example. Consider $\mathbb{P}^1$ with G-action $\gamma(x) = x+1$ in 5.2.1. Let $Y = \mathbb{P}^1 \times \mathbb{P}^1$ with the diagonal G-action, i.e., $\gamma(x) = x + 1$ and $\gamma(y) = y + 1$. Then the

---

[1]By Enriques criterion of ruledness, it suffices to show that $H^0(W, 12K_W) = 0$ as the irregularity of W is zero because W is dominated by V. Since $K_V = q^*K_W$, we have $H^0(V, 12K_V) = H^0(W, q_*q^*(12K_W)) = H^0(W, 12pK_W) = 0$. Hence we infer that $H^0(W, 12K_W) = 0$.

[2]Namely, the generic point of an irreducible subvariety of codimension one.

point P_∞ defined by $1/x = 1/y = 0$ is an isolated G-fixed point. By a straightforward computation, we can show the following assertions.

(1) Let $(\mathcal{O}, \mathfrak{m})$ be the local ring $\mathcal{O}_{Y,P_\infty}$. Let $\xi = 1/x$ and $\eta = 1/y$. Then $\mathcal{O} = k[\xi, \eta]_{(\xi,\eta)}$ and $\widehat{\mathcal{O}} = k[[\xi, \eta]]$. Since P_∞ is a fixed point, the action γ maps $\mathcal{O}$ to $\mathcal{O}$ and $\widehat{\mathcal{O}}$ to $\widehat{\mathcal{O}}$. More precisely, $\gamma(\xi) = \xi/(\xi + 1)$ and $\gamma(\eta) = \eta/(\eta + 1)$, and $\widehat{\gamma}(\xi) = \xi\left(1 + \sum_{i \geq 1}(-1)^i \xi^i\right)$ and $\widehat{\eta} = \eta\left(1 + \sum_{i \geq 1}(-1)^i \eta^i\right)$ in $\widehat{\mathcal{O}}$. We shall see below that $\gamma(\xi)$ and $\gamma(\eta)$ cannot be made into polynomials by any change of local coordinates (ξ, η) (cf. Lemma 5.2.6).

(2) Suppose $p = 2$. Then $k[x, y]^G = k[x + y, x(x + 1), y(y + 1)]$. Let $X = x(x + 1), Y = y(y + 1)$ and $Z = x + y$. Then $Z^2 + Z = X + Y$. Hence $k[x, y]^G \cong k[x + y, x(x + 1)]$ which is a polynomial ring.

(3) Suppose also $p = 2$. Set $A = k[\xi, \eta, \gamma(\xi), \gamma(\eta)]$. Then $A^G \supseteq k[u, v, w]$, where

$$u = \frac{\xi^2}{1 + \xi}, \quad v = \frac{\eta^2}{1 + \eta}, \quad w = \frac{\xi^2\eta + \xi\eta^2}{(1 + \xi)(1 + \eta)}$$

with u, v and w satisfying the relation

$$w^2 + uvw = uv(u + v). \tag{$*$}$$

Let S be a hypersurface in $\mathbb{A}^3$ defined by the equation $(*)$. Then S is singular only at the point $(u, v, w) = (0, 0, 0)$. Hence S is a normal surface by Serre's criterion of normality [24, Theorem 8.22A]. [3] Since ξ and η are integral over $k[u, v]$, it follows that A is finite over a normal ring $k[u, v, w]$. Since $Q(A^G) = k(\xi, \eta)^G = k(u, v, w)$, Zariski's Main Theorem shows that $A^G = k[u, v, w]$. Hence $k[[\xi, \eta]]^G = k[[u, v, w]]$. The singular point of S is a rational double point of type D_4^1 (see Part III 1.1.6 for the notation).

(4) Let $p > 0$ in general. Let

$$u = \prod_{i=0}^{p-1} \gamma^i(\xi) = \frac{\xi^p}{1 - \xi^{p-1}}, \quad v = \prod_{i=0}^{p-1} \gamma^i(\eta) = \frac{\eta^p}{1 - \eta^{p-1}}.$$

Since

$$\gamma\left(\frac{1}{\xi} - \frac{1}{\eta}\right) = \frac{1}{\xi} - \frac{1}{\eta},$$

we set

$$w = uv\left(\frac{1}{\xi} - \frac{1}{\eta}\right) = \frac{\xi^{p-1}\eta^p - \xi^p\eta^{p-1}}{(1 - \xi^{p-1})(1 - \eta^{p-1})}.$$

---

[3] The (S_1) condition follows from the Jacobian criterion of smoothness, and the (R_2) condition is satisfied because S is a complete intersection, whence it is Cohen-Macaulay.

Since

$$\frac{w}{uv} = \frac{1}{\xi} - \frac{1}{\eta}, \quad \text{and} \quad \frac{1}{u} - \frac{1}{v} = \left(\frac{1}{\xi^p} - \frac{1}{\xi}\right) - \left(\frac{1}{\eta^p} - \frac{1}{\eta}\right),$$

we have

$$\frac{v - u}{uv} = \left(\frac{1}{\xi} - \frac{1}{\eta}\right)^p - \left(\frac{1}{\xi} - \frac{1}{\eta}\right) = \left(\frac{w}{uv}\right)^p - \frac{w}{uv}.$$

Hence we obtain a relation

$$w^p - (uv)^{p-1}w = u^{p-1}v^p - u^p v^{p-1}.$$

This hypersurface is, however, not normal if $p > 2$. In fact, it is singular along the curve $\{(0, v, 0) \mid v \in k\} \cup \{(u, 0, 0) \mid u \in k\}$. So, $k[[\xi, \eta]]^G$ has not been worked out.

5.2.6

Let Y be anew a smooth algebraic surface with a non-trivial $G = \mathbb{Z}/p\mathbb{Z}$-action σ, and let P be a closed point of Y. A system of local parameters (x, y) of the local ring $\mathcal{O}_{Y,P}$ is called a *system of local coordinates* if $(x - x(Q), y - y(Q))$ is a system of local parameters of the local ring $\mathcal{O}_{Y,Q}$ for every closed point of a Zariski-open neighborhood U of P. Such a system of local coordinates always exists.

Suppose that P is a G-fixed point. We say that the action σ is *locally algebraisable* at P if there exists a system of local coordinates (x, y) such that the induced action $\hat{\gamma}$ on $\widehat{\mathcal{O}}_{Y,P}$ maps x, y to elements of $k[x, y]$, where $k[x, y]$ is viewed naturally as a subring of $\widehat{\mathcal{O}}_{Y,P} = k[[x, y]]$.

Lemma. *We assume that σ is locally algebraisable at P. With the above notations and assumptions, we can take a system of local coordinates (x, y) such that $\gamma(x) = x$ and $\gamma(y) = y + a(x)$, where $a(x) \in k[x]$ and $a(0) = 0$. Hence the curve $x = 0$ coincides with the fixed point locus near the point P. Furthermore, the following assertions hold.*

(1) *Let $q : Y \to X$ be the quotient morphism. Then X is smooth at $Q := q(P)$. In fact, $\widehat{\mathcal{O}}_{X,Q} = k[[y^p - a(x)^{p-1}y, x]]$.*
(2) *The morphism q is flat over Q.*
(3) *$\mathcal{O}_{Y,P} = \mathcal{O}_{X,Q}[y]/(y^p - a(x)^{p-1}y - b)$, where $b \in \mathcal{O}_{X,Q}$.*

Proof. By Theorem 5.2.7 below, after a change of local coordinates x, y, the induced G-action on $k[x, y]$ is normalized to $\gamma(x) = x$ and $\gamma(y) =$

$y + a(x)$ with $a(x) \in k[x]$, and $k[x, y]^G = k[y^p - a(x)^{p-1}y, x]$. In the $\mathfrak{m}$-adic topology, any G-invariant element of $\mathcal{O}_{Y,P}$ is approximated by an element of $k[x, y]^G$. Hence $\widehat{\mathcal{O}}_{X,Q} = k[[y^p - a(x)^{p-1}y, x]]$. Since P is a fixed point, we have $a(0) = 0$, i.e., $x | a(x)$. Then the local curve $x = 0$ coincides with the fixed point locus near P. In fact, $a(x) = x^r a_1(x)$ with $r > 0, a_1(x) \in k[x]$ and $a_1(0) \neq 0$. Then the curve $a_1(x) = 0$ does not contribute to the fixed point locus near the point P. Hence the fixed point locus is purely one-dimensional near P. The assertion (2) follows from the assertion (3). As for the assertion (3), note that $\mathcal{O}_{Y,P}$ is a finite $\mathcal{O}_{X,Q}$-module since P is a fixed point. Then we have

$$\widehat{\mathcal{O}}_{Y,P} = \mathcal{O}_{Y,P} \otimes_{\mathcal{O}_{X,Q}} \widehat{\mathcal{O}}_{X,Q} = \widehat{\mathcal{O}}_{X,Q}[y]/(y^p - a(x)^{p-1}y - b).$$

Since $\widehat{\mathcal{O}}_{X,Q}$ is a faithfully flat $\mathcal{O}_{X,Q}$-module, we can conclude that $\mathcal{O}_{Y,P} = \mathcal{O}_{X,Q}[y]/(y^p - a(x)^{p-1}y - b)$. $\square$

5.2.7

Theorem. *A G-action on the affine plane $\mathbb{A}^2$ is given by $\gamma(x, y) = (x, y + a(x))$ with respect to a suitable system of coordinates (x, y) and $a(x) \in k[x]$. Furthermore, $k[x, y]^G = k[y^p - a(x)^{p-1}y, x]$.*

Proof. This is a result announced first by S. Kuroda, and given a published proof by M. Miyanishi [59, Theorem 3.5 and Corollary 3.6]. We refer the proof to this article, which is an application of the structure theorem of $\mathrm{Aut}_k(\mathbb{A}^2)$. $\square$

5.3 Existence criterion of $\mathbb{Z}/p\mathbb{Z}$-fixed point

As announced earlier, we have the following general result.

Theorem. *Let V be a smooth projective variety with a non-trivial G-action. Let $W = V/G$ be the quotient variety and $q : V \to W$ the quotient morphism. Assume that either W is singular or $H^1(W, \mathcal{O}_W) = 0$. Then there exists a G-fixed point on V.*

By Lemma 5.1.1, (5), it is clear that a G-fixed point exists on V if W is singular. Hence we may assume that W is smooth. The proof consists of Lemmas 5.3.1 and 5.3.2 below.

5.3.1

Lemma. *There exists an affine open covering $\mathcal{V} = \{V_\lambda\}_{\lambda \in \Lambda}$ such that V_λ is a G-stable affine variety.*

Proof. Let $P \in V$ be a closed point. Then there exists an effective ample divisor H_1 of V such that $GP \cap \mathrm{Supp}(H_1) = \emptyset$, where GP is the G-orbit of P. Let $H = \sum_{g \in G} g(H_1)$. Then H is a G-stable ample divisor such that $P \notin \mathrm{Supp}(H)$. Then $V \setminus H$ is a G-stable affine open set of V containing P. By the construction, $V \setminus H$ is an affine variety, i.e., its coordinate ring is an affine k-domain. Cover V by such G-stable affine open sets, and choose a G-stable affine open covering of V from them. $\qquad\square$

5.3.2

Lemma. *Let $\mathcal{V} = \{V_\lambda\}_{\lambda \in \Lambda}$ be a G-stable affine open covering of V as in 5.3.1. Let $V_\lambda = \mathrm{Spec}\, B_\lambda$. Let $A_\lambda = B_\lambda^G$ and $U_\lambda = \mathrm{Spec}\, A_\lambda$. Then the following assertions hold.*

(1) *$\mathcal{U} = \{U_\lambda\}_{\lambda \in \Lambda}$ is an affine open covering of $W := V/G$.*
(2) *Suppose hereafter that the G-action σ is free. Replacing the covering $\mathcal{V}$ by a finer covering, we may assume that, for every λ, $B_\lambda = A_\lambda[z_\lambda]/(z_\lambda^p - z_\lambda - a_\lambda)$ with $a_\lambda \in A_\lambda$ and $\gamma(z_\lambda) = z_\lambda + 1$.*
(3) *On $V_{\lambda\mu} := V_\lambda \cap V_\mu$, $a_{\lambda\mu} := z_\lambda|_{V_{\lambda\mu}} - z_\mu|_{V_{\lambda\mu}}$ defines an element $\alpha \in H^1(W, \mathcal{O}_W)$.*
(4) *The element α is nonzero if the G-action σ on V is free.*

Proof. (1) is clear. (2) follows from Theorem 5.1.3.

(3) Since $\gamma(z_\lambda) = z_\lambda + 1$ and $\gamma(z_\mu) = z_\mu + 1$, $a_{\lambda\mu} \in \Gamma(V_{\lambda\mu}, \mathcal{O}_V)^G = \Gamma(U_{\lambda\mu}, \mathcal{O}_W)$, where $U_{\lambda\mu} := U_\lambda \cap U_\mu$. By construction, $\{a_{\lambda\mu}\}$ is a 1-cocycle, and hence defines an element $\alpha \in H^1(W, \mathcal{O}_W)$.

(4) Suppose that $\alpha = 0$. Then we may assume that $a_{\lambda\mu} = c_\lambda - c_\mu$ for $c_\lambda \in A_\lambda$. Then

$$(z_\lambda - c_\lambda)|_{V_{\lambda\mu}} = (z_\mu - c_\mu)|_{V_{\lambda\mu}}.$$

Hence $\{z_\lambda - c_\lambda\}_{\lambda \in \Lambda}$ defines an element z of $H^0(V, \mathcal{O}_V)$. Since V is projective, it follows that $z \in k$. This contradicts the construction in the assertion (2). Hence $\alpha \neq 0$ if the G-action σ is free. $\qquad\square$

5.3.3

Proposition. *Let V be a smooth projective variety V with a free $G = \mathbb{Z}/p\mathbb{Z}$ action and let $q : V \to W$ be the quotient morphism with $W = V/G$. Then the following assertions hold.*

(1) *There exists an $\mathcal{O}_W$-Module endomorphism $\widetilde{\delta} : q_*\mathcal{O}_V \to q_*\mathcal{O}_V$ such that $\mathrm{Im}\,(\widetilde{\delta})^{p-1} = \mathcal{O}_W$.*

(2) *If $p = 2$, there exists an exact sequence of $\mathcal{O}_W$-Modules*

$$0 \to \mathcal{O}_W \to q_*\mathcal{O}_V \to \mathcal{O}_W \to 0.$$

Hence, if $H^1(V, \mathcal{O}_V) = 0$ then $H^1(W, \mathcal{O}_W) \neq 0$.

Proof. (1) Let $\mathcal{V} = \{V_\lambda\}_{\lambda \in \Lambda}$ be a G-stable affine open covering of V. By 1.5.4, the G-action σ restricted onto V_λ is defined by a k-algebra homomorphism $\varphi_\lambda : B_\lambda \to B_\lambda \otimes_k k[\xi]$ with $\xi^p = \xi$, which is written as

$$\varphi_\lambda(b) = b + \delta_\lambda(b)\xi + \frac{1}{2!}\delta_\lambda^2(b)\xi^2 + \cdots + \frac{1}{(p-1)!}\delta_\lambda^{p-1}(b)\xi^{p-1},$$

where $b \in B_\lambda$ and δ_λ is a k-module endomorphism of B_λ such that $\delta_\lambda^p = 0$ (see [59]). Since φ_λ and φ_μ coincide on $V_{\lambda\mu} = V_\lambda \cap V_\mu$, it follows that $\delta_\lambda = \delta_\mu$ on $V_{\lambda\mu}$. Since $b \in A_\lambda = B_\lambda^G$ if and only if $\delta_\lambda(b) = 0$, δ_λ is an A_λ-module endomorphism of B_λ. Furthermore, if the G-action is free, we have $A_\lambda = \mathrm{Im}\,\delta_\lambda^{p-1}$ [59, Lemma 2.5]. Thus there is an $\mathcal{O}_W$-Module endomorphism $\widetilde{\delta} : q_*\mathcal{O}_V \to q_*\mathcal{O}_V$ such that $(\widetilde{\delta})^{p-1}$ maps $q_*\mathcal{O}_V$ surjectively onto $\mathcal{O}_W$.

(2) Suppose $p = 2$. Then $\widetilde{\delta} : q_*\mathcal{O}_V \to q_*\mathcal{O}_V$ has $\mathcal{O}_W$ as the image and the kernel. Hence we obtain the stated exact sequence. Hence it yields a long exact sequence of k-vector spaces

$$0 \to H^0(W, \mathcal{O}_W) \to H^0(V, \mathcal{O}_V) \to H^0(W, \mathcal{O}_W)$$
$$\to H^1(W, \mathcal{O}_W) \to H^1(V, \mathcal{O}_V) \to H^1(W, \mathcal{O}_W).$$

If $H^1(V, \mathcal{O}_V) = 0$, we have $H^1(W, \mathcal{O}_W) \cong H^0(W, \mathcal{O}_W) \cong k$. Hence $H^1(W, \mathcal{O}_W) \neq 0$. $\qquad\square$

It is shown in [9, Part III] that if $p = 2$ there exists a K3-surface V with a free involution ι so that $W = V/\langle\iota\rangle$ is a (non-classical) singular Enriques surface with $K_W \sim 0$.

5.3.4

Lemma. *Let W be a complete algebraic variety with $H^1(W, \mathcal{O}_W) = 0$. Then every G-torsor over W is trivial, i.e., it is isomorphic to a direct product $G \times W$. Hence there is no connected G-torsor over W.*

Proof. We consider the exact sequence (5.1) in 5.1.4 and derive a long exact sequence

$$0 \longrightarrow G \longrightarrow k^+ \overset{q_0}{\longrightarrow} k^+ \longrightarrow H^1_{\mathrm{fl}}(W, G) \longrightarrow H^1(W, \mathcal{O}_W)$$
$$\overset{q_0^*}{\longrightarrow} H^1(W, \mathcal{O}_W) \,,$$

where $q_0 : k^+ \to k^+, u \mapsto u^p - u$, is surjective. Hence we obtain $H^1_{\mathrm{fl}}(W, G) = 0$ since $H^1(W, \mathcal{O}_W) = 0$ by the assumption. This implies the assertion. $\square$

5.4   Simple Artin-Schreier covering

Let X be a normal algebraic variety. A finite surjective morphism $q : Y \to X$ from a normal variety Y to X is called an *Artin-Schreier covering* of X if the following conditions are satisfied:

(1) The field extension $k(Y)/k(X)$ is an Artin-Schreier extension of degree p, hence $k(Y) = k(X)(z)$ with $z^p - z = a \in k(X)$.
(2) Y is the normalization of X in the field $k(Y)$.

Since the extension $k(Y)/k(X)$ is a Galois extension with group $G \cong \mathbb{Z}/p\mathbb{Z}$ by the condition (1), the group G acts on Y non-trivially, and X is the algebraic quotient variety in the following sense:

Let $\mathcal{U} = \{U_\lambda\}_{\lambda \in \Lambda}$ be an affine open covering of X with $U_\lambda = \operatorname{Spec} A_\lambda$. For every λ, $q^{-1}(U_\lambda)$ is an affine open set of Y because f is a finite morphism. Write $q^{-1}(U_\lambda) = \operatorname{Spec} B_\lambda$. Then A_λ is a subring of B_λ, B_λ is the normalization of A_λ in $Q(B_\lambda) = k(Y)$, G acts non-trivially on B_λ, and $A_\lambda = B_\lambda^G$.

We denote X by Y/G and call $q : Y \to X$ also the quotient morphism.

Let $q : Y \to X$ be an Artin-Schreier covering. We say that q is *free* if the G-action is free. Hence Y is a G-torsor over X. We also say that q is *simple* if there exists an affine open covering $\mathcal{U} = \{U_\lambda\}_{\lambda \in \Lambda}$ of X as above such that, for every λ, B_λ is written as $B_\lambda = A_\lambda[z_\lambda]/(z_\lambda^p - a_\lambda^{p-1} z_\lambda - t_\lambda)$ with $a_\lambda, t_\lambda \in A_\lambda$, and the G-action is given by $\varphi_\lambda(z_\lambda) = z_\lambda + a_\lambda$. We call $Y = \cup_{\lambda \in \Lambda} \operatorname{Spec} B_\lambda$ a *local expression* of Y. For example, if G acts on a smooth algebraic variety Y and the G-action is locally algebraisable at every point of Y, Lemma 5.2.6 shows that the quotient morphism $q : Y \to X := Y/G$ is a simple Artin-Schreier covering.

5.4.1

In order to state Lemma 5.4.2, we have to make a digression and recall basic properties of reflexive sheaves from [25].

Let X be a normal algebraic variety of dimension n. Let $\mathcal{F}$ be a coherent $\mathcal{O}_X$-Module. The *dual* of $\mathcal{F}$ is $\mathcal{F}^\vee := \mathcal{H}om_{\mathcal{O}_X}(\mathcal{F}, \mathcal{O}_X)$, and the *double dual* of $\mathcal{F}$ is $\mathcal{F}^{\vee\vee} := (\mathcal{F}^\vee)^\vee$. Then there exists a natural map of $\mathcal{O}_X$-Modules $\alpha_{\mathcal{F}} : \mathcal{F} \to \mathcal{F}^{\vee\vee}$. The $\mathcal{O}_X$-Module $\mathcal{F}$ is *reflexive* (or $\mathcal{F}$ is a *reflexive sheaf*) if $\alpha_{\mathcal{F}}$ is an isomorphism. We summarize below the basic properties on reflexive sheaves.

Proposition. *With the above notations and assumptions, the following assertions hold.*

(1) $\operatorname{Ker}\alpha_{\mathcal{F}}$ *and* $\operatorname{Coker}\alpha_{\mathcal{F}}$ *are torsion $\mathcal{O}_X$-Modules, and $\operatorname{Coker}\alpha_{\mathcal{F}}$ has support of codimension ≥ 2. In particular, if $\mathcal{F}$ is torsion-free, $\alpha_{\mathcal{F}}$ is injective.*

(2) $\mathcal{F}$ *is reflexive if and only if $\mathcal{F}$ is torsion-free and $\operatorname{depth}\mathcal{F}_x \geq 2$ for every $x \in X$ such that $\dim\mathcal{O}_x \geq 2$.*

(3) *Let $0 \to \mathcal{F}' \to \mathcal{F} \to \mathcal{R} \to 0$ be an exact sequence of coherent $\mathcal{O}_X$-Modules such that $\mathcal{F}$ is reflexive and $\mathcal{R}$ is torsion-free. Then $\mathcal{F}'$ is reflexive.*

(4) *Let $q : Y \to X$ be a surjective finite morphism of normal varieties. If $\mathcal{G}$ is a coherent reflexive sheaf on Y, then $q_*\mathcal{G}$ is a coherent reflexive sheaf on X.*

(5) *Let X be a smooth algebraic surface. Then any reflexive sheaf $\mathcal{F}$ is locally free.*

Proof. (1) Note that any torsion-free finite module M over a Dedekind domain [4] is projective. Then the assertion follows from the definition.

(2) See [25, Prop. 1.3]. (3) follows from [25, Cor. 1.5] by noting that $\operatorname{Ass}\mathcal{R}$ consists of the generic point of X. (4) follows from [25, Cor. 1.7].

(5) Let x be a closed point of X and let $(A, \mathfrak{m})$ be the local ring $\mathcal{O}_{X,x}$. Let $W = \operatorname{Spec} A$ and let $U = W \setminus \{\mathfrak{m}\}$. Let $\mathcal{F}$ be anew the pull-back $j^*\mathcal{F}$ for the natural morphism $j : W \to X$ and let $\mathcal{F}_U = \mathcal{F}|_U$. Let $i : U \to W$ be the open immersion. Since $\mathcal{F}$ is reflexive, $\mathcal{F} \cong i_*\mathcal{F}_U$ by [25, Prop.1.6].

---
[4] It is a normal domain of dimension one.

On the other hand, since $\mathcal{F}_U$ is locally free over U, $\mathcal{F}_U$ is a free $\mathcal{O}_U$-Module by Horrocks [32, Cor. 4.1.1]. Hence $\mathcal{F}$ is a free $\mathcal{O}_X$-Module.   □

5.4.2

To be more precise in Artin-Schreier coverings, we need the following generalization of the theory of Takeda [91].

Lemma. *Let X be a normal algebraic variety of dimension n and let $q : Y \to X$ be an Artin-Schreier covering. Then there is a natural filtration of $\mathcal{O}_X$-sub-Modules of $q_*\mathcal{O}_Y$*

$$\mathcal{O}_X = \mathcal{F}_0 \subset \mathcal{F}_1 \subset \cdots \subset \mathcal{F}_{p-1} = q_*\mathcal{O}_Y$$

satisfying the conditions:

(1) $\mathcal{F}_i$ *is a reflexive sheaf of rank $i + 1$ for $0 \leq i < p$, and $\mathcal{F}_1\mathcal{F}_i \subseteq \mathcal{F}_{i+1}$ if $0 < i < p - 1$.*
(2) $\mathcal{F}_i/\mathcal{F}_{i-1}$ *is a torsion-free $\mathcal{O}_X$-Module of rank 1 for $1 \leq i < p$.*
(3) *Assume that X is smooth and $n = 2$. Then every $\mathcal{F}_i$ is locally free, and $\mathcal{F}_1/\mathcal{F}_0$ is an invertible sheaf.*

Proof. As in the proof of Proposition 5.3.3, there exists an $\mathcal{O}_X$-Module endomorphism $\tilde{\delta} : q_*\mathcal{O}_Y \to q_*\mathcal{O}_Y$ such that $\mathcal{O}_X = \operatorname{Ker}\tilde{\delta}$ and $(\tilde{\delta})^p = 0$. The endomorphism $\tilde{\delta}$ is constructed by patching together local endomorphisms δ_λ, and the assumption that the G-action is free and Y is smooth and projective are not necessary. As δ_λ defines a k-algebra homomorphism $\varphi_\lambda : B_\lambda \to B_\lambda \otimes k[\xi]$ in 5.3.3, we have an $\mathcal{O}_X$-Algebra homomorphism

$$\tilde{\varphi} : q_*\mathcal{O}_Y \longrightarrow (q_*\mathcal{O}_Y)[\xi] = (q_*\mathcal{O}_Y) \oplus (q_*\mathcal{O}_Y)\xi \oplus \cdots \oplus (q_*\mathcal{O}_Y)\xi^{p-1} ,$$

where $\xi^p = \xi$. For $0 \leq i < p$, let $\mathcal{F}_i := \operatorname{Ker}(\tilde{\delta})^{i+1}$. Then $\mathcal{F}_i$ is the pull-back by $\tilde{\varphi}$ of the subsheaf $(q_*\mathcal{O}_Y)[i] := (q_*\mathcal{O}_Y) \oplus (q_*\mathcal{O}_Y)\xi \oplus \cdots \oplus (q_*\mathcal{O}_Y)\xi^i$ of $(q_*\mathcal{O}_Y)[\xi]$. This implies that $\mathcal{F}_1\mathcal{F}_i \subseteq \mathcal{F}_{i+1}$ if $0 < i < p - 1$. Hence we have a filtration of coherent $\mathcal{O}_X$-Modules

$$\mathcal{O}_X = \mathcal{F}_0 \subset \mathcal{F}_1 \subset \cdots \subset \mathcal{F}_{p-1} = q_*\mathcal{O}_Y .$$

Furthermore, $\mathcal{F}_i/\mathcal{F}_{i-1}$ is a subsheaf of $(q_*\mathcal{O}_Y)[i]/(q_*\mathcal{O}_Y)[i-1] \cong q_*\mathcal{O}_Y$. Since $q_*\mathcal{O}_Y$ is a reflexive sheaf by Proposition 5.4.1, (4), $\mathcal{F}_i/\mathcal{F}_{i-1}$ is a torsion-free $\mathcal{O}_X$-Module. More precisely, the coefficient of the ξ^i-term of $\tilde{\varphi}(b)$ for $b \in \mathcal{F}_i$ is annihilated by $\tilde{\delta}$. Hence $\mathcal{F}_i/\mathcal{F}_{i-1}$ is a subsheaf of $\mathcal{O}_X$. Hence the rank of $\mathcal{F}_i/\mathcal{F}_{i-1}$ as an $\mathcal{O}_X$-Module is 1.

Now consider an exact sequence

$$0 \to \mathcal{F}_{p-2} \to \mathcal{F}_{p-1} \to \mathcal{F}_{p-1}/\mathcal{F}_{p-2} \to 0 \ .$$

Since $\mathcal{F}_{p-1} = q_* \mathcal{O}_Y$ is reflexive and $\mathcal{F}_{p-1}/\mathcal{F}_{p-2}$ is torsion-free, it follows from Proposition 5.4.1, (3) that $\mathcal{F}_{p-2}$ is reflexive. By descending induction on i, we infer that every $\mathcal{F}_i$ is a reflexive sheaf of rank $i + 1$.

The assertion (3) follows from Proposition 5.4.1, (5). Assuming that X is a smooth algebraic surface, we show that $\mathcal{F}_1/\mathcal{F}_0$ is invertible. In fact, for every closed point $x \in X$, we have an exact sequence

$$\mathcal{F}_0 \otimes k(x) \xrightarrow{\alpha(x)} \mathcal{F}_1 \otimes k(x) \longrightarrow (\mathcal{F}_1/\mathcal{F}_0) \otimes k(x) \longrightarrow 0 \ ,$$

where, for the inclusion map $\alpha : \mathcal{F}_0 \to \mathcal{F}_1$, the map $\alpha(x) := \alpha \otimes k(x)$ is injective because $\mathcal{F}_0 \otimes k(x) \cong k(x)$ is a $k(x)$-submodule of $\mathcal{F}_{p-1} \otimes k(x)$ as both share the same unity. Hence rank $(\mathcal{F}_1/\mathcal{F}_0) \otimes k(x) = 1$ for every closed point $x \in X$, because $\mathcal{F}_1$ is a locally free sheaf of rank 2. This implies that $\mathcal{F}_1/\mathcal{F}_0$ is an invertible sheaf. $\qquad\square$

5.4.3

According to Takeda [91], an Artin-Schreier covering $q : Y \to X$ of normal algebraic varieties is said to be *of simple type* if, with the notations of 5.4.2,

(1) $\mathcal{F}_1/\mathcal{F}_0$ is an invertible sheaf on X, and
(2) the natural map of $\mathcal{O}_X$-Modules $\mathcal{F}_1^{\otimes i} \to \mathcal{F}_i$ induces an isomorphism $(\mathcal{F}_1/\mathcal{F}_0)^{\otimes i} \cong \mathcal{F}_i/\mathcal{F}_{i-1}$ for $1 \leq i < p$.

We then have the following result.

Lemma. *For an Artin-Schreier covering $q : Y \to X$ of normal algebraic varieties, the following conditions are equivalent.*

(1) *q is of simple type.*
(2) *q is simple.*

Proof. (1) $\Rightarrow$ (2). Let $\mathcal{L}^{-1} := \mathcal{F}_1/\mathcal{F}_0$. Choose an affine open covering $\mathcal{U} = \{U_\lambda\}_{\lambda \in \Lambda}$ of X such that $\mathcal{L}^{-1}|_{U_\lambda} = \mathcal{O}_{U_\lambda} e_\lambda$. By the definition of $\mathcal{L}^{-1}$, there exists an element $z_\lambda \in \mathcal{F}_1(U_\lambda)$ such that $\overline{z}_\lambda := z_\lambda \pmod{\mathcal{F}_0} = e_\lambda$. With the notations of 4.3.2, we have $\varphi_\lambda(z_\lambda) = z_\lambda + \delta_\lambda(z_\lambda)\xi$, where $a_\lambda := \delta_\lambda(z_\lambda) \in A_\lambda := \mathcal{O}_X(U_\lambda)$. Since $\varphi_\lambda(z_\lambda^p) = (\varphi_\lambda(z_\lambda))^p = z_\lambda^p + a_\lambda^p \xi^p = z_\lambda^p + a_\lambda^p \xi$, we have $z_\lambda^p \in \mathcal{F}_1(U_\lambda)$ and hence

$$z_\lambda^p = s_\lambda z_\lambda + t_\lambda, \quad s_\lambda, t_\lambda \in \mathcal{O}_X(U_\lambda) \ .$$

This entails

$$\varphi_\lambda(z_\lambda^p) = \varphi_\lambda(s_\lambda z_\lambda) + t_\lambda = s_\lambda(z_\lambda + a_\lambda \xi) + t_\lambda$$
$$= (s_\lambda z_\lambda + t_\lambda) + s_\lambda a_\lambda \xi = z_\lambda^p + s_\lambda a_\lambda \xi \ ,$$

whence $a_\lambda^p = s_\lambda a_\lambda$. Since $z_\lambda \notin \mathcal{F}_0(U_\lambda)$, it follows that $a_\lambda \neq 0$ and hence $s_\lambda = a_\lambda^{p-1}$. On the other hand, since $\mathcal{F}_i/\mathcal{F}_{i-1} \cong (\mathcal{F}_1/\mathcal{F}_0)^{\otimes i}$ for $1 \leq i < p$, we have $(q_*\mathcal{O}_Y)(U_\lambda) = A_\lambda \oplus A_\lambda z_\lambda \oplus \cdots \oplus A_\lambda z_\lambda^{p-1}$, where $A_\lambda := \mathcal{O}_X(U_\lambda)$. Hence

$$(q_*\mathcal{O}_Y)(U_\lambda) = A_\lambda[z_\lambda]/(z_\lambda^p - a_\lambda^{p-1}z_\lambda - t_\lambda),$$

and the covering q is simple.

We show that $\{a_\lambda\}_{\lambda \in \Lambda}$ gives an element of $H^0(X, \mathcal{L})$. Indeed, let $\{\alpha_{\mu\lambda}\}$ be the set of transition functions of $\mathcal{L}^{-1}$, i.e., $\alpha_{\mu\lambda} \in \mathcal{O}(U_\lambda \cap U_\mu)^*$ and $e_\mu = \alpha_{\mu\lambda}e_\lambda$. Then $z_\mu = \alpha_{\mu\lambda}z_\lambda + \beta_{\mu\lambda}$ with $\beta_{\mu\lambda} \in \mathcal{O}_X(U_\lambda \cap U_\mu)$. Since

$$\varphi_\mu(z_\mu) = \alpha_{\mu\lambda}\varphi_\mu(z_\lambda) + \beta_{\mu\lambda}$$
$$= \alpha_{\mu\lambda}(z_\lambda + a_\lambda \xi) + \beta_{\mu\lambda} = z_\mu + \alpha_{\mu\lambda}a_\lambda \xi$$
$$\varphi_\mu(z_\mu) = z_\mu + a_\mu \xi,$$

we have $a_\mu = \alpha_{\mu\lambda}a_\lambda$. Write $\mathcal{L}|_{U_\lambda} = \mathcal{O}_{U_\lambda}e_\lambda^*$. Then $e_\mu^* = \alpha_{\lambda\mu}e_\lambda^*$. Hence we have

$$a_\mu e_\mu^* = (\alpha_{\mu\lambda}a_\lambda)(\alpha_{\lambda\mu}e_\lambda^*) = a_\lambda e_\lambda^*.$$

Namely, $\{a_\lambda e_\lambda^*\}$ defines an element of $H^0(X, \mathcal{L})$.

(2) $\Rightarrow$ (1). By definition,

$$B_\lambda = (q_*\mathcal{O}_Y)(U_\lambda) = A_\lambda[z_\lambda]/(z_\lambda^p - a_\lambda^{p-1}z_\lambda - t_\lambda)$$

and $\varphi_\lambda(z_\lambda) = z_\lambda + a_\lambda \xi$. This implies that $B_\lambda = A_\lambda \oplus A_\lambda z_\lambda \oplus \cdots \oplus A_\lambda z_\lambda^{p-1}$ and $\mathcal{F}_i(U_\lambda) = A_\lambda \oplus A_\lambda z_\lambda \oplus \cdots \oplus A_\lambda z_\lambda^i$. Hence $(\mathcal{F}_i/\mathcal{F}_{i-1})(U_\lambda) \cong (A_\lambda \overline{z}_\lambda^i)$ with $\overline{z}_\lambda = z_\lambda \pmod{\mathcal{F}_0(U_\lambda)}$. It then follows that $\mathcal{F}_1/\mathcal{F}_0$ is an invertible sheaf on X and $\mathcal{F}_i/\mathcal{F}_{i-1} \cong (\mathcal{F}_1/\mathcal{F}_0)^{\otimes i}$. Namely, q is of simple type. $\square$

Hereafter, we call an Artin-Schreier covering $q : Y \to X$ *simple* if it satisfies the equivalent conditions of Lemma 5.4.3. In the above proof, $\{a_\lambda\}_{\lambda \in \Lambda}$ defines an element of $H^0(X, \mathcal{L})$, hence an effective divisor B[5] on X which is defined by an equation $a_\lambda = 0$. We call B the *branch locus* of a simple Artin-Schreier covering $q : Y \to X$.

---

[5]It is customary to use the letter B to denote the branch locus. The readers must be careful not to confuse with the coordinate ring of Y.

5.4.4

Lemma. *With the notations in the proof of Lemma 5.4.3, the following assertions hold:*

(1) $\beta_{\nu\lambda} = \alpha_{\nu\mu}\beta_{\mu\lambda} + \beta_{\nu\mu}$ *for* $\lambda, \mu, \nu \in \Lambda$.
(2) *Corresponding to* $\{\beta_{\lambda\mu}\}$, *define a 1-cochain* $\widetilde{\beta}$ *in* $C^1(\mathcal{U}, \mathcal{L})$ *by* $\widetilde{\beta}(\mu, \lambda) = \beta_{\mu\lambda}e_\mu^*|_{U_\lambda \cap U_\mu}$. *Then* $\widetilde{\beta}$ *is a 1-cocycle, hence defines an element* $\widetilde{\beta}$ *of* $H^1(X, \mathcal{L})$.
(3) $t_\mu - \alpha_{\mu\lambda}^p t_\lambda = \beta_{\mu\lambda}^p - s_\mu\beta_{\mu\lambda}$. *If* $\widetilde{\beta} = 0$, *i.e.,* $\beta_{\mu\lambda} = \alpha_{\mu\lambda}c_\lambda - c_\mu$ *for* $\lambda, \mu \in \Lambda$, *where* $\{c_\lambda\} \in C^0(\mathcal{U}, \mathcal{L})$, *then* $\{t_\lambda + c_\lambda^p - s_\lambda c_\lambda\}$ *defines an element of* $H^0(X, \mathcal{L}^p)$.

Proof. Straightforward by computations. □

5.4.4.1

The above lemma shows how to construct a simple Artin-Schreier covering under some restrictive conditions. The following important result which guarantees the existence of a simple Artin-Schreier covering is clear by the proof of Lemma 5.4.3 and Lemma 5.4.4.

Theorem. *Let X be a normal algebraic variety and let $\mathcal{L}$ be an invertible sheaf on X such that $H^1(X, \mathcal{L}) = 0$, $H^0(X, \mathcal{L}) \neq 0$ and $H^0(X, \mathcal{L}^p) \neq 0$. Let $\mathcal{U} = \{U_\lambda\}_{\lambda \in \Lambda}$ is an affine open covering of X such that $\mathcal{L}^{-1}|_{U_\lambda} = \mathcal{O}_{U_\lambda}e_\lambda$ for $\lambda \in \Lambda$. Let $\{\alpha_{\mu\lambda}\}$ be the transition functions of $\mathcal{L}^{-1}$ with respect to $\mathcal{U}$. Let $\{a_\lambda\}_{\lambda \in \Lambda}$ and $\{t_\lambda\}_{\lambda \in \Lambda}$ be nonzero elements of $H^0(X, \mathcal{L})$ and $H^0(X, \mathcal{L}^p)$, respectively, such that $a_\mu = \alpha_{\mu\lambda}a_\lambda$ and $t_\mu = \alpha_{\mu\lambda}^p t_\lambda$. Let $A_\lambda = \Gamma(U_\lambda, \mathcal{O}_X)$ and let $B_\lambda = A_\lambda[z_\lambda]/(z_\lambda^p - a_\lambda^{p-1}z_\lambda - t_\lambda)$ for each $\lambda \in \Lambda$. Define a $\mathbb{Z}/p\mathbb{Z}$-action on B_λ by $\varphi_\lambda(z_\lambda) = z_\lambda + a_\lambda$. Then the collection of affine schemes $\{\text{Spec } B_\lambda \mid \lambda \in \Lambda\}$ patch together by the data $z_\mu = \alpha_{\mu\lambda}z_\lambda$ to give an algebraic variety Y and a finite morphism $q : Y \to X$ such that q restricted on $\text{Spec } B_\lambda$ is the quotient morphism of the $\mathbb{Z}/p\mathbb{Z}$-action associated with φ_λ. These local $\mathbb{Z}/p\mathbb{Z}$-actions patch together to give a global $\mathbb{Z}/p\mathbb{Z}$-action, and $q : Y \to X$ is a simple Artin-Schreier covering.*

5.4.4.2

Takeda [91] gives a more general construction of a simple Artin-Schreier covering. Let X be a smooth algebraic variety and let $\mathcal{L}$ be an invertible sheaf on X. Write by $\mathbb{L}$ be the line bundle (as a variety) associated with $\mathcal{L}$. Consider $\mathbb{L}$ and $\mathbb{L}^p$ as smooth X-group schemes, each of which has the additive group G_a as fibers. Let s be an element of $H^0(X, \mathcal{L}^{p-1})$, which we view as a section $s : X \mapsto \mathbb{L}^{p-1}$. Since $F - s : \mathbb{L} \to \mathbb{L}^p$ defined by $(F - s)(x) = x^p - sx$ for $x \in \mathbb{L}$ is a flat surjective morphism by [69], the homomorphism $F - s : \mathbb{L} \to \mathbb{L}^p$ is a surjective homomorphism in the (f.p.q.c)-topology. Namely, for any morphism $h : T \to \mathbb{L}^p$, let $\widetilde{T} = (T, h) \times_{\mathbb{L}^p} (\mathbb{L}, F - s)$ and let $p_1 : \widetilde{T} \to T$ be the first projection. Then p_1 is a flat surjective morphism because so is $F - s$, and the second projection $p_2 : \widetilde{T} \to \mathbb{L}$ satisfies $(F - s) \cdot p_2 = h \cdot p_1$. Let α_s be the kernel of $F - s$. Then there is an exact sequence of X-group schemes

$$0 \longrightarrow \alpha_s \longrightarrow \mathbb{L} \overset{F-s}{\longrightarrow} \mathbb{L}^p \longrightarrow 0 .$$

This yields an exact sequence of cohomology groups

$$0 \to H^0_{\mathrm{fl}}(X, \alpha_s) \to H^0(X, \mathbb{L}) \to H^0(X, \mathbb{L}^p) \overset{\partial}{\longrightarrow} H^1_{\mathrm{fl}}(X, \alpha_s) \to H^1(X, \mathbb{L}) ,$$

where $H^1_{\mathrm{fl}}(X, \alpha_s)$ is the group of isomorphism classes of α_s-torsors[6]. Let $t = \{t_\lambda\}_{\lambda \in \Lambda}$ be an element of $H^0(X, \mathbb{L}^p)$ given locally with respect to an affine open covering $\mathcal{U} = \{U_\lambda\}_{\lambda \in \Lambda}$ of X. Then the image element $\partial(t)$ of $H^1_{\mathrm{fl}}(X, \alpha_s)$ by the coboundary homomorphism ∂ is the fiber product $Z := (\mathbb{L}, F - s) \times_{\mathbb{L}^p} (X, t)$ which is a simple Artin-Schreier covering defined locally by $\{z_\lambda^p - s_\lambda z_\lambda = t_\lambda\}_{\lambda \in \Lambda}$. The second projection $p_2 : Z \to X$ is the quotient morphism of the Artin-Schreier covering.

5.4.4.3

The following result is a restatement of Theorem 5.3 in view of Lemma 5.3.3 and it is proved by the argument used in the proof of Lemma 5.4.4.

Corollary. *Let Y be a smooth projective variety with a free $\mathbb{Z}/p\mathbb{Z}$-action and let $q : Y \to X$ be the quotient morphism. Then $H^1(X, \mathcal{O}_X) \neq 0$.*

Proof. By Lemma 5.3.1, there exists an affine open covering $\mathcal{V} = \{V_\lambda\}_{\lambda \in \Lambda}$ of Y such that each V_λ is a $\mathbb{Z}/p\mathbb{Z}$-stable affine open set. Let B_λ be the coordinate ring of V_λ and let A_λ be the ring of $\mathbb{Z}/p\mathbb{Z}$-invariants of B_λ. Let $U_\lambda = \operatorname{Spec} A_\lambda$. Then $\mathcal{U} = \{U_\lambda\}_{\lambda \in \Lambda}$ is an affine open covering of the quotient variety X which is smooth. We can write $B_\lambda = A_\lambda[z_\lambda]/(z_\lambda^p - z_\lambda - t_\lambda)$ with $t_\lambda \in A_\lambda$, where $\varphi_\lambda(z_\lambda) = z_\lambda + 1$ with the notations in Lemma 5.4.3. On

---

[6] α_s is viewed as a group scheme consisting of $\mathbb{Z}/p\mathbb{Z}$-fibers over points $P \in X$ with $s(P) \neq 0$ and α_p-fibers over points $P \in X$ with $s(P) = 0$.

the intersection open set $V_\lambda \cap V_\mu$, we have $z_\mu - z_\lambda = \beta_{\mu\lambda}$, which is a $\mathbb{Z}/p\mathbb{Z}$-invariant element. By Lemma 5.4.4, $\{\beta_{\mu\lambda}\}$ defines an element $\widetilde{\beta}$ of $H^1(X, \mathcal{O}_X)$. Here we note that $\mathcal{L} \cong \mathcal{O}_X$ because the $\mathbb{Z}/p\mathbb{Z}$-action on Y is free. If $\widetilde{\beta} = 0$, then there exists a coboundary $\{\gamma_\lambda\}$ such that $\beta_{\mu\lambda} = \gamma_\mu - \gamma_\lambda$. Then $z_\mu - \gamma_\mu = z_\lambda - \gamma_\lambda$ for every pair (λ, μ). Hence $\{z_\lambda - \gamma_\lambda\}$ defines an element of $H^0(Y, \mathcal{O}_Y) \cong k$. Since $z_\lambda - \gamma_\lambda$ is a non-constant element as a generator of B_λ over A_λ, this is a contradiction. Hence $\widetilde{\beta} \neq 0$, and $H^1(X, \mathcal{O}_X) \neq 0$. Thus we have proved the contrapositive of the assertion in Theorem 5.3. $\qquad\square$

5.4.5

Hereafter, we assume that X is smooth.

Lemma. *Let $q : Y \to X$ be a simple Artin-Schreier covering, and let $Y = \cup_{\lambda \in \Lambda} \operatorname{Spec} B_\lambda$ be a local expression of Y with $B_\lambda = A_\lambda[z_\lambda]/(z_\lambda^p - a_\lambda^{p-1} z_\lambda - t_\lambda)$. Let Q be a closed point of Y and let $P = q(Q)$. Then the following assertions hold.*

(1) *If $p > 2$, Q is a singular point of Y if and only if $P \in B = V(a_\lambda)$ for some λ and $dt_\lambda = 0$ at P.*

(2) *If $p = 2$, Q is a singular point of Y if and only if $P \in B$ and $d(a_\lambda z_\lambda + t_\lambda) = 0$ at Q.*

Proof. Suppose that the local defining equation of Y at Q is $z_\lambda^p - a_\lambda^{p-1} z_\lambda - t_\lambda = 0$. Let $\{x_1, \ldots, x_n\}$ be a system of local coordinates of $\mathcal{O}_{X,P}$. By the Jacobian criterion, Q is a singular point of Y if and only if the following derivatives of the defining equation vanish at the point Q,

$$-a_\lambda^{p-1}, \quad a_\lambda^{p-2} z_\lambda (a_\lambda)_{x_i} - (t_\lambda)_{x_i} \quad (1 \leq i \leq n) .$$

Hence $a_\lambda(Q) = 0$, $(t_\lambda)_{x_i}(P) = 0$ $(1 \leq i \leq n)$ if $p > 2$ and $z_\lambda(a_\lambda)_{x_i} - (t_\lambda)_{x_i} = 0$ at Q for $1 \leq i \leq n$ if $p = 2$. This implies the stated assertions. $\qquad\square$

5.4.6

Note that, with the notations and assumptions as in 5.4.5, Y is locally a complete intersection, hence Y is a Gorenstein scheme. Namely, Y has the dualizing sheaf ω_Y, which is an invertible sheaf $\mathcal{O}_Y(K_Y)$.

Lemma. $K_Y \sim q^*(K_X + (p-1)B)$.

Proof. We consider the affine open covering $\mathcal{U} = \{U_\lambda\}_{\lambda \in \Lambda}$ as before. For a local piece $q^{-1}(U_\lambda) = \operatorname{Spec} B_\lambda$, the defining equation is $z_\lambda^p - a_\lambda^{p-1} z_\lambda - t_\lambda = 0$, which is viewed as a hypersurface in the $(z_\lambda, x_1, \ldots, x_n)$-space, where $\{x_1, \ldots, x_n\}$ is a system of local coordinates of X (say, at a point P). By differentiating the local equation, we have

$$-a_\lambda^{p-1} dz_\lambda + a_\lambda^{p-2} z_\lambda da_\lambda - dt_\lambda = 0.$$

We may assume that $dz_\lambda \wedge dx_2 \wedge \cdots \wedge dx_n \neq 0$ at a point $Q \in Y$ over the point P. Then we have

$$dz_\lambda \wedge dx_2 \wedge \cdots \wedge dx_n = a_\lambda^{1-p} \left\{ a_\lambda^{p-2} z_\lambda da_\lambda \wedge dx_2 \wedge \cdots \wedge dx_n \right. \\ \left. - dt_\lambda \wedge dx_2 \wedge \cdots \wedge dx_n \right\} \tag{5.2}$$

On the other hand, we have an exact sequence of differential sheaves (cf. Lemma 1.1.1),

$$q^* \Omega_{X/k}^1 \xrightarrow{\psi} \Omega_{Y/k}^1 \longrightarrow \Omega_{Y/X}^1 \longrightarrow 0 . \tag{5.3}$$

This exact sequence induces the map

$$\wedge^n \psi : q^* \Omega_{X/k}^n \longrightarrow \Omega_{Y/k}^n.$$

Since X is smooth and Y is Gorenstein, both $q^* \Omega_{X/k}^n$ and $\Omega_{Y/k}^n$ are invertible sheaves. Furthermore, since q is étale over the open set $X \setminus B$, it follows that $\psi|_{(X \setminus B)}$ is an isomorphism. This implies that $\wedge^n \psi$ is injective, and hence ψ itself is injective because $q^* \Omega_{X/k}^1$ is locally free. We can write $\omega_{Y/k} = q^* \omega_{X/k} \otimes \mathcal{O}_Y(R)$, where $\omega_{X/k} = \Omega_{X/k}^n$, $\omega_{Y/k} = \Omega_{Y/k}^n$ and R is a divisor on Y such that $\operatorname{Supp}(R) \subseteq \operatorname{Supp} q^*(B)$. The relation (5.2) shows that $R \leq (p-1) q^*(B)$. The exact sequence (5.3), where ψ is injective, implies that $R = (p-1) q^*(B)$ because $\Omega_{Y/X}^1|_{q^{-1}(U_\lambda)} \cong \mathcal{O}_{q^{-1}(U_\lambda)} / (a_\lambda^{p-1}) dz_\lambda$. $\quad\square$

5.4.7

Assume that X is a smooth projective surface. Then there exist the following two results due to Takeda [91, Lemmas 1.8 and 1.9], which are useful in constructing various examples of surfaces as simple Artin-Schreier coverings.

5.4.7.1

Lemma. *The following assertions hold.*

(1) $K_Y^2 = p \left(K_X^2 + 2(p-1)(B \cdot K_X) + (p-1)^2 B^2 \right).$

(2) *Let $\chi(\mathcal{O}_X)$ and $\chi(\mathcal{O}_Y)$ be the Euler-Poincaré characteristics of X and Y. Then we have*

$$\chi(\mathcal{O}_Y) = p\left\{\chi(\mathcal{O}_X) + \frac{(p-1)}{4}(B \cdot K_X) + \frac{(p-1)(2p-1)}{12}(B^2)\right\}.$$

(3) *Assume that Y is smooth. Let $e(X)$ and $e(Y)$ be the étale Euler number of X and Y (cf. [57, p.5]). Then we have*

$$e(Y) = p\left\{e(X) + (p-1)(B \cdot K_X) + (p-1)p(B^2)\right\}.$$

(4) *Assume that Y is smooth. Then $\kappa(Y) = \kappa(X, K_X + (p-1)B)$.*

Proof. (1) By Lemma 5.4.6, we have

$$(K_Y^2) = (q^*(K_X + (p-1)B))^2 = p(K_X + (p-1)B)^2$$
$$= p\left(K_X^2 + 2(p-1)(K_X \cdot B) + (p-1)^2 B^2\right).$$

(2) With the filtration $\mathcal{O}_X \subset \mathcal{F}_1 \subset \cdots \subset \mathcal{F}_{p-1} = q_*\mathcal{O}_Y$ in Lemma 5.4.2, we have $\mathcal{F}_i/\mathcal{F}_{i-1} \cong (\mathcal{F}_1/\mathcal{F}_0)^{\otimes i} \cong \mathcal{L}^{-i}$ for $1 \leq i < p$. Since $\mathcal{L} \cong \mathcal{O}_X(B)$, we have $\chi(\mathcal{F}_i) = \chi(\mathcal{F}_{i-1}) + \chi(\mathcal{O}_X(-iB))$. This entails

$$\chi(\mathcal{O}_Y) = \chi(q_*(\mathcal{O}_Y)) = \chi(\mathcal{F}_{p-1}) = \sum_{i=0}^{p-1} \chi(\mathcal{O}_X(-iB)).$$

By the Riemann-Roch theorem on X, we have

$$\chi(\mathcal{O}_X(-iB)) = \frac{1}{2}\left((-iB) \cdot (-iB - K_X)\right) + \chi(\mathcal{O}_X)$$
$$= \frac{1}{2}\left(i^2 B^2 + i(B \cdot K_X)\right) + \chi(\mathcal{O}_X),$$

whence

$$\chi(\mathcal{O}_Y) = \sum_{i=0}^{p-1}\left\{\frac{1}{2}\left(i^2 B^2 + i(B \cdot K_X)\right) + \chi(\mathcal{O}_X)\right\}$$
$$= \frac{1}{12}(p-1)p(2p-1)(B^2) + \frac{1}{4}(p-1)p(B \cdot K_X) + p\chi(\mathcal{O}_X)$$
$$= p\left\{\chi(\mathcal{O}_X) + \frac{(p-1)}{4}(B \cdot K_X) + \frac{(p-1)(2p-1)}{12}(B^2)\right\}.$$

(3) This follows from Noether's formula $12\chi(\mathcal{O}_Y) = K_Y^2 + e(Y)$ and $\chi(\mathcal{O}_X) = K_X^2 + e(X)$.

(4) Since q is a finite morphism, we obtain the equality from $K_Y \sim q^*(K_X + (p-1)B)$. $\square$

5.4.7.2

Lemma. *Let X be a smooth projective surface and let $\mathcal{L}$ be an ample divisor such that $H^1(X, \mathcal{L}^{-1}) = 0$[7]. Let $q : Y \to X$ be a simple Artin-Schreier covering such that $\mathcal{L} = \mathcal{O}_X(B)$ for the branch locus B of q. Then $H^1(X, \mathcal{O}_X) \cong H^1(Y, \mathcal{O}_Y)$.*

Proof. We have an exact sequence

$$0 \longrightarrow \mathcal{L}^{-1} \longrightarrow \mathcal{O}_X \longrightarrow \mathcal{O}_B \longrightarrow 0 , \tag{5.4}$$

which implies that $H^0(B, \mathcal{O}_B) = H^0(X, \mathcal{O}_X) = k$ because $H^1(X, \mathcal{L}^{-1}) = 0$. By induction on i, we show that $H^1(X, \mathcal{L}^{-i}) = 0$ for $0 < i < p$. By taking the tensor product of (5.4) by $\mathcal{L}^{-(i-1)}$, we have an exact sequence

$$0 \longrightarrow \mathcal{L}^{-i} \longrightarrow \mathcal{L}^{-(i-1)} \longrightarrow \mathcal{L}^{-(i-1)} \otimes \mathcal{O}_B \to 0 ,$$

where $H^0(X, \mathcal{L}^{-(i-1)}) = 0$ for $i > 1$ because $\mathcal{L}$ is ample and $H^0(B, \mathcal{L}^{-(i-1)} \otimes \mathcal{O}_B) = 0$ for $i > 1$ because $\mathcal{L} \otimes \mathcal{O}_B$ is ample on B, too. Then $H^1(X, \mathcal{L}^{-(i-1)}) = 0$ implies $H^1(X, \mathcal{L}^{-i}) = 0$.

On the other hand, we have an exact sequence

$$0 \longrightarrow \mathcal{F}_{i-1} \longrightarrow \mathcal{F}_i \longrightarrow \mathcal{L}^{-i} \longrightarrow 0 \quad (0 < i < p)$$

which induces an isomorphism $H^1(X, \mathcal{F}_{i-1}) \cong H^1(X, \mathcal{F}_i)$. Hence we have

$$H^1(X, \mathcal{O}_X) \cong H^1(X, \mathcal{F}_{p-1}) \cong H^1(Y, \mathcal{O}_Y).$$

$\square$

5.5 Geometry of Artin-Schreier coverings

In this section, we take a smooth projective surface X and construct simple Artin-Schreier coverings $q : Y \to X$. The existence is guaranteed by Theorem 5.4.4.1. The covering q has a local expression $\{z_\lambda^p - a_\lambda^{p-1} z_\lambda = t_\lambda\}_{\lambda \in \Lambda}$ with respect to an affine open covering $\mathcal{U} = \{U_\lambda\}_{\lambda \in \Lambda}$ and nonzero sections $\{a_\lambda\}_{\lambda \in \Lambda} \in H^0(X, \mathcal{L})$ and $\{t_\lambda\}_{\lambda \in \Lambda} \in H^0(X, \mathcal{L}^p)$. Let B and T be the curves on X defined locally by $a_\lambda = 0$ and $t_\lambda = 0$. We call Y (or $q : Y \to X$) an *Artin-Schreier covering with sections* (B, T). The surface Y thus constructed is a normal projective surface if B and T share no irreducible components. Since $q^{-1}(U_\lambda)$ is a hypersurface $z_\lambda^p - a_\lambda^{p-1} z_\lambda = t_\lambda$, it is a Gorenstein surface.

---

[7]This corresponds to the Kodaira vanishing theorem in characteristic zero, but it does not hold in positive characteristic.

Suppose that $p > 2$ and T is smooth at every point of $B \cap T$. Then, by replacing $\mathcal{U}$ by a sufficiently fine open covering if necessary, we can take $t_\lambda - t_\lambda(P)$ as one of local coordinates at a point $P \in B \cap U_\lambda$ (see the subsection 5.2.6). Hence, by Lemma 5.4.5, Y is smooth. Suppose that $p = 2$ and the curves B and T intersect transversally. Then Y is smooth. In fact, let $P \in B \cap T \cap U_\lambda$. Since B and T are smooth at P, we can take $\{a_\lambda, t_\lambda - t_\lambda(P)\}$ as a system of local parameters $\{x, y\}$ at $P \in X$. Hence we can assume that the local equation of $q^{-1}(U_\lambda)$ over P is $z^2 + xz + y = 0$, which is a smooth hypersurface.

The condition that T *is smooth at $B \cap T$ if $p > 2$ and that B and T intersect transversally if $p = 2$* suffices to make the covering surface Y smooth. So, *we assume this condition in the subsequent construction.* The morphism $q : Y \to X$ is the quotient morphism under a faithful $\mathbb{Z}/p\mathbb{Z}$-action. A generator σ of $\mathbb{Z}/p\mathbb{Z}$, identified with an X-invariant automorphism of Y of order p is called a *covering automorphism*. If $p = 2$, we call σ the *covering involution*.

5.5.1

The following result is due to Takeda [91, Lemma 3.1].

Lemma. *Assume that the following conditions are satisfied.*

(i) *B is an irreducible curve.*

(ii) *Either B is an ample divisor, or X is relatively minimal, i.e., there are no (-1)-curves on X, and B is semiample, i.e., $|mB|$ is not the empty set and base-point free for some $m > 0$.*

(iii) *K_Y is not nef, i.e., $K_Y \cdot D < 0$ for some irreducible curve D on Y.*

Then either one of the following two cases occurs.

(1) *$X \cong \mathbb{P}^2$ and $p = 2$ or 3.*

(2) *X is a relatively minimal ruled surface. Either p is an arbitrary prime number and $B \sim \ell$, or $p = 2$ and B is a cross-section, where ℓ is a fiber of the associated $\mathbb{P}^1$-fibration of X. If $K_Y^2 > 0$ and $B \sim \ell$, then X is the Hirzebruch surface $\Sigma_n := \mathrm{Proj}\,(\mathcal{O}_{\mathbb{P}^1} \oplus \mathcal{O}_{\mathbb{P}^1}(n))$ [8] and $p = 2$.*

Proof. By the condition (iii), $K_Y \cdot D < 0$ for some irreducible curve D. By Lemma 5.4.6, we have

$$(K_X + (p-1)B) \cdot C < 0,$$

---

[8]In the previous chapters, Hirzebruch surface of degree n is denoted by F_n. We often denote it by Σ_n if there is a fear of confusion of notations.

where $C = q(D)$. By the cone theorem in the Mori theory [47, Theorem 1.24], there exists an extremal rational curve ℓ on X such that

$$(K_X + (p-1)B) \cdot \ell < 0. \tag{5.5}$$

There are three possibilities about ℓ.

(a) ℓ is a line on $\mathbb{P}^2$.
(b) ℓ is a fiber of a relatively minimal $\mathbb{P}^1$-fibration $\rho : X \to C$, where C is a smooth projective curve of genus g.
(c) ℓ is a (-1)-curve.

We first show that the case (c) is impossible. If ℓ is such a (-1)-curve, we have $(p-1)(B \cdot \ell) < 1$. By the condition (ii), B is ample since X is not relatively minimal. Hence $(p-1)(B \cdot \ell) \geq 1$, which is a contradiction.

Consider next the case (a) where $X \cong \mathbb{P}^2$ and ℓ is a line. Write $B \sim d\ell$ with $d > 0$. Then (5.5) implies that $(p-1)d < 3$. Hence $(p, d) = (2, 1), (2, 2), (3, 1)$.

Consider finally the case (b). Write $B \approx aS + b\ell$ (algebraic equivalence), where S is a section of ρ, $a \geq 0$ and $ar + b \geq 0$, where $r = (S^2)$. Since $K_X \approx -2S + (2g - 2 + r)\ell$, the inequality (5.5) implies $(p-1)a < 2$. Hence either $a = 0$ and p is arbitrary, or $p = 2$ and $a = 1$. Suppose that $a = 0$. Since B is irreducible and $B \approx b\ell$, we have $b = 1$. Now suppose that $K_Y^2 > 0$. Since

$$K_X + (p-1)B \approx -2S + (p + 2g + r - 3)\ell,$$

we have by Lemma 5.4.6

$$K_Y^2 = p\{4r - 4(p + 2g + r - 3)\}$$
$$= -4p(p + 2g - 3),$$

whence $p + 2g < 3$. This implies that $p = 2$ and $g = 0$. Hence X is isomorphic to the Hirzebruch surface Σ_n ($n \neq 1$).   ◻

5.5.1.1

Corollary. *Assume the conditions (i), (ii) of Lemma 5.5.1. Then following assertions hold.*

(1) *K_Y is not nef if and only if $\kappa(Y) = -\infty$.*
(2) *If $\kappa(Y) \geq 0$, then K_Y is nef and Y is a minimal surface.*

Proof. (1) By Lemma 5.5.1, X is either $\mathbb{P}^2$ or a relatively minimal ruled surface. Suppose $X \cong \mathbb{P}^2$. Then $K_Y \sim q^*(K_X + (p-1)d\ell)$ with the notations in the proof of Lemma 5.5.1, where $(p, d) = (2, 1), (2, 2), (3, 1)$. Hence $K_Y \sim -q^*(2\ell)$ or $K_Y \sim -q^*(\ell)$. Hence $\kappa(Y) = -\infty$. Suppose that X is a relatively minimal ruled surface. Again with the notations in the proof of Lemma 5.5.1, we have

$$K_X + (p-1)B \sim -2S + (2g-2+r)\ell + (p-1)(aS+b\ell)$$
$$\sim ((p-1)a - 2)S + (2g-2+r+(p-1)b)\ell,$$

where either $a = 0$ or $a = 1$ and $p = 2$. Since $q^*(\ell)$ is a semiample divisor and $K_Y \cdot q^*(\ell) < 0$, $|nK_Y| = \emptyset$ for every $n > 0$. Hence $\kappa(Y) = -\infty$. In order to prove the converse by the contrapositive, assume that $\kappa(Y) = -\infty$ and K_Y is nef. Then Y is relatively minimal. Since Y is then isomorphic to $\mathbb{P}^2$ or a ruled surface, $K_Y \sim -3\widetilde{\ell}$ or $K_Y \approx -2\widetilde{S} + (2g-2+\widetilde{r})\widetilde{\ell}$, where $\widetilde{\ell}$ is either a line on $\mathbb{P}^2$ or a fiber of the ruled surface. Then $K_Y \cdot \widetilde{\ell} < 0$, which contradicts the nefness of K_Y.

(2) Taking the contrapositive of the assertion (1), it follows that K_Y is nef. Hence there is no (-1)-curve on Y. $\qquad\square$

5.5.2

Theorem. *Assume that $p = 2$, $X = \mathbb{P}^2$, $B \sim \ell$ and $T \sim 2\ell$, where ℓ is a line on $\mathbb{P}^2$. Then the following assertions hold.*

(1) $Y \cong \mathbb{P}^1 \times \mathbb{P}^1$ *and* $\widetilde{B} := q^*(B)$ *is the diagonal* $\Delta \sim L + M$, *where L and M are fibers of two distinct $\mathbb{P}^1$-fibrations on $\mathbb{P}^1 \times \mathbb{P}^1$.*
(2) *Let $P \in B$ and let $\widetilde{P}$ be the point of $\widetilde{B}$ lying over P. Let Λ_P be the linear pencil of lines on X passing through P. Then the total transform $q^*\Lambda_P$ consists of curves C on $\mathbb{P}^1 \times \mathbb{P}^1$ such that $C \sim L + M$ and C meets $\widetilde{B}$ at $\widetilde{P}$ with multiplicity 2. There is one special member of $q^*\Lambda_P$ which splits $L_{\widetilde{P}} + M_{\widetilde{P}}$, where $L_{\widetilde{P}}$ and $M_{\widetilde{P}}$ are the members of $|L|$ and $|M|$ passing through the point $\widetilde{P}$. The covering involution associated to $q : Y \to X$ interchanges $L_{\widetilde{P}}$ and $M_{\widetilde{P}}$.*

Proof. (1) Since $K_Y \sim q^*(K_X + B) \sim -2q^*(\ell)$, $-K_Y$ is ample and $K_Y^2 = 8$. Hence $Y \cong \mathbb{P}^1 \times \mathbb{P}^1$. In fact, as a del Pezzo surface [9] of degree 8, Y is isomorphic to the Hirzebruch surface $\Sigma_0 \cong \mathbb{P}^1 \times \mathbb{P}^1$ or Σ_1. Suppose that $Y \cong \Sigma_1$. Then $q^*(\ell) \sim aM + bL$, where M is the minimal section and L is a

---

[9] A reference for del Pezzo surfaces is Manin's textbook [50].

fiber. Then $q^*(\ell)^2 = 2 = -a^2 + 2ab$, whence $a = 2a_1$, and $1 = 2(-a_1^2 + a_1 b)$. This is a contradiction. Hence $Y \cong \Sigma_0$. Further, since $\widetilde{B}^2 = 2$, it follows that $\widetilde{B} \sim L + M$.

(2) Let ℓ be a general line passing through P. Then $C := q^*(\ell)$ meets $\widetilde{B}$ in only one point $\widetilde{P}$ and $C \cdot \widetilde{B} = 2$. Hence C is a smooth curve such that $C \sim L + M$ and C touches $\widetilde{B}$ at $\widetilde{P}$ with multiplicity 2. Such curves C form a pencil on Y [10], $L_{\widetilde{P}} + M_{\widetilde{P}}$ is a member of this pencil. Hence $L_{\widetilde{P}} + M_{\widetilde{P}} = q^*(\ell_0)$ with $\ell_0 \in \Lambda_P$, the covering involution, say σ, interchanges $L_{\widetilde{P}}$ and $M_{\widetilde{P}}$. $\square$

5.5.3

The following example elucidates the situation treated in the above theorem. The readers are requested to verify the assertions as exercises.

Example. Let $p = 2$ and let (X_0, X_1, X_2) be a system of homogeneous coordinates on $\mathbb{P}^2$. Take B to be the line $X_1 = 0$. Let T be a conic defined by an equation

$$F(X_0, X_1, X_2) = aX_1 X_2 + bX_2 X_0 + cX_0 X_1 + dX_0^2 + d' X_1^2 + d'' X_2^2 = 0.$$

Let Y be a hypersurface in $\mathbb{P}^3$ defined by

$$X_3^2 + X_1 X_3 = F(X_0, X_1, X_2),$$

where (X_0, X_1, X_2, X_3) is a system of homogeneous coordinates of $\mathbb{P}^3$ such that $(X_0, X_1, X_2, X_3) \mapsto (X_0, X_1, X_2)$ is the projection $\mathbb{P}^3 \to \mathbb{P}^2$ from the point $(0, 0, 0, 1)$. Verify the following assertions.

(1) By a change of coordinates $X_3' = X_3 + \sqrt{d}X_0 + \sqrt{d'}X_1 + \sqrt{d''}X_2$, the equation F becomes

$$X_3'^2 + X_1 X_3' = (a + \sqrt{d''})X_1 X_2 + bX_2 X_0 + (c + \sqrt{d})X_0 X_1 + \sqrt{d'}X_1^2.$$

Hence we may assume that T is defined by an equation

$$F(X_0, X_1, X_2) = aX_1 X_2 + bX_2 X_0 + cX_0 X_1 + dX_1^2 = 0.$$

Suppose that B is not an irreducible component of T. Then $b \neq 0$. We may assume that $b = 1$. By the change of coordinates

$$X_0' = X_0 + aX_1, \quad X_1' = X_1, \quad X_2' = X_2 + cX_1,$$

the equation of T is reduced to

$$X_0' X_2' + (ac + d)X_1'^2 = 0.$$

Hence the equation of Y is set anew as

$$X_3^2 + X_1 X_3 = X_0 X_2 + aX_1^2, \quad a \in k, \tag{5.6}$$

where T is irreducible if $a \neq 0$ and reducible if $a = 0$.

---

[10]In fact, $\dim |L + M| = 3$. Denote by $|L + M| - 2\widetilde{P}$ the linear subsystem of $|L + M|$ consisting of members which pass through the point $\widetilde{P}$ twice. Then $\dim(|L + M| - 2\widetilde{P}) = 1$.

(2) Suppose $a \neq 0$. For the sake of simplifying computations, we assume that $a = 1$. Let α and β be the roots of $t^2 + t + 1 = 0$. Let (x_0, x_1) (resp. (y_0, y_1)) be a system of homogeneous coordinates of $\mathbb{P}^1$ (resp. a copy of $\mathbb{P}^1$). Let

$$X_0 = x_0 y_0, \quad X_1 = x_0 y_1 + x_1 y_0, \quad X_2 = x_1 y_1, \quad X_3 = \beta x_0 y_1 + \alpha x_1 y_0.$$

Then (X_0, X_1, X_2, X_3) satisfies the equation $X_3^2 + X_1 X_3 = X_0 X_2 + X_1^2$. Let $X_1' = X_3 + \alpha X_1$ and $X_3' = X_3 + \beta X_1$. Then $X_1 = X_1' + X_3', X_3 = \beta X_1' + \alpha X_3', X_1' = x_0 y_1, X_3' = x_1 y_0$ and hence $X_1' X_3' = X_0 X_2$. This implies $(x_0, x_1) = (X_1', X_2)$ and $(y_0, y_1) = (X_0, X_1')$. Hence we have an isomorphism $Y \cong \mathbb{P}^1 \times \mathbb{P}^1$, where Y is the Segre embedding of $\mathbb{P}^1 \times \mathbb{P}^1$ into $\mathbb{P}^3$.

(3) Suppose $a = 1$. The covering morphism $q : Y \to X$ is induced by the projection $(X_0, X_1, X_2, X_3) \mapsto (X_0, X_1, X_2)$. Let $u = X_1/X_0$ and $v = X_2/X_0$. Then the curve T is defined by $v = u^2$, and the branch locus B is defined by $u = 0$ (the v-axis). Hence $\tilde{B} = q^*(B)$ is the diagonal Δ defined by $X_1 = 0$. Let P be the point $(u, v) = (0, c^2)$ with $c \in k$. Let $a_0(v + c^2) = a_1 u$ be a line ℓ passing through P, where $(a_0, a_1) \in \mathbb{P}^1$. It is also defined by $a_0(X_2 + c^2 X_0) = a_1 X_1$. If $a_0 \neq 0$, the pull-back $q^*(\ell)$ is defined by

$$x_1 y_1 = c^2 x_0 y_0 + \frac{a_1}{a_0}(x_0 y_1 + x_1 y_0).$$

Let $x = x_1/x_0$ and $y = y_1/y_0$. Then this equation is written as

$$xy = c^2 + \frac{a_1}{a_0}(x + y).$$

This curve meets the diagonal at $(x, y) = (c, c)$ with multiplicity 2. If $\frac{a_1}{a_0} = c$, the curve is defined by $(x + c)(y + c) = 0$. Hence $q^*(\ell) = L_{\tilde{P}} + M_{\tilde{P}}$.

(4) Suppose $a = 0$. Then T is defined by $X_0 X_2 = 0$. The equation (5.6) is written as

$$X_1' X_3 = X_0 X_2, \quad X_1' = X_1 + X_3.$$

Then $Y \cong \mathbb{P}^1 \times \mathbb{P}^1$ with the identification

$$X_0 = x_0 y_0, \quad X_1 = x_0 y_1 + x_1 y_0, \quad X_2 = x_1 y_1, \quad X_3 = x_1 y_0.$$

Then the rest of the arguments is the same as in the case (3) above.

We do not have clear answers to the following two problems.

5.5.3.1

Problem. In the case $p = 2$, a smooth conic C in $\mathbb{P}^2$ defined by

$$F(X_0, X_1, X_2) = aX_1X_2 + bX_2X_0 + cX_0X_1 + dX_0^2 + d'X_1^2 + d''X_2^2 = 0$$

is a *strange curve* in the sense that the tangent line of all point P of C passes through the same point $P_0 = (a, b, c)$ (see Samuel [80, p.76]). How is the strangeness of the curve C reflected on an Artin-Schreier covering $q : Y \to X$ with sections (B, T), where the notations are the same as above?

Strangeness of a smooth conic in the case $p = 2$ appears in a different settings treating an Artin-Schreier covering of the Hirzebruch surfaces. See Theorem 5.5.7, the assertion (2), (v).

5.5.3.2

Problem. With the notations and assumptions in Theorem 5.5.2, each linear pencil $q^*\Lambda_P$ has a unique split member $L_{\tilde{P}} + M_{\tilde{P}}$. Is there any geometry involved to determine this special member from the sections (B, T)?

It seems that a tangent line to T from a given point $P \in B$ lifted up to Y might split to two lines. But the example 5.5.3 shows that this is not the case.

5.5.4

Theorem. *Assume that $p = 2$, $X = \mathbb{P}^2$, $B \sim 2\ell$ and $T \sim 4\ell$. Then the following assertions hold.*

(1) $K_Y \sim -q^*(\ell)$, $K_Y^2 = 2$ *and hence Y is a del Pezzo surface of degree 2.*
(2) *Suppose that B is smooth. For a line ℓ on $\mathbb{P}^2$, there are two possibilities about $q^*(\ell)$.*

    (i) $q^*(\ell)$ *is a curve of arithmetic genus one.*
    (ii) $q^*(\ell) = E_1 + E_2$, *where E_1 and E_2 are (-1)-curves such that $E_1 \cdot E_2 = 2$.*

(3) *The covering involution σ is the Geiser involution.*
(4) *Let E be a (-1)-curve on Y. Then $\sigma(E) \neq E$ and $E + \sigma(E) = q^*(\ell)$ for a line ℓ on $\mathbb{P}^2$.*

Proof. (1) It follows from Lemma 5.4.6 that $K_Y \sim -q^*(\ell)$. Hence $-K_Y$ is ample and $K_Y^2 = 2$. This implies the assertion.

(2) We consider first the case where $q^*(\ell)$ is irreducible. Since $K_Y \cdot q^*(\ell) = -2$ and $q^*(\ell)^2 = 2$, it follows that $p_a(q^*(\ell)) = 1$. Consider next the case where $q^*(\ell)$ is reducible. We can write $q^*(\ell) = E_1 + E_2$, where $E_2 = \sigma(E_1)$ under the covering involution σ of q. Hence $-2 = K_Y \cdot q^*(\ell) = 2(K_Y \cdot E_1)$, whence $K_Y \cdot E_1 = K_Y \cdot E_2 = -1$. Since q induces a birational morphism from E_1 to ℓ, it follows that $E_1 \cong \mathbb{P}^1$. Similarly, $E_2 \cong \mathbb{P}^1$. By the arithmetic genus formula, we have $E_1^2 = E_2^2 = -1$. Namely, E_1 and E_2 are (-1)-curves. Since $2 = q^*(\ell)^2 = (E_1 + E_2)^2 = -2 + 2(E_1 \cdot E_2)$, it follows that $E_1 \cdot E_2 = 2$.

(3) A smooth del pezzo surface Y of degree 2 is obtained from $Y_0 := \mathbb{P}^2$ by blowing up seven points $P_1, \ldots, P_7$ in general position. Then the linear system of cubic curves on Y_0 has dimension $\dim(|3L| - (P_1 + \cdots + P_7)) = 2$, where L is a line on Y_0. (Since Y_0 is different from our X, we denote a line on Y_0 by a different letter L.) The Geiser involution τ of Y is induced by such a birational automorphism of Y_0 that P and $\tau(P)$ lie on the pencil of elliptic curves $|3L| - (P_1 + \cdots + P_7 + P + \tau(P))$. Namely, the linear pencil $|3L| - (P_1 + \cdots + P_7 + P)$ has still one more base point which we denote by $\tau(P)$. This implies, in particular, that a cubic curve passing through $P_1 + \cdots + P_7$ is stable under the involution τ. Hence, in order to show that $\sigma = \tau$, it suffices to show that

- Let C be an elliptic curve on Y such that $C^2 = 2$, $\sigma(C) = C$ and $\widetilde{P} \in C$ for a given point $\widetilde{P}$. Then $C = q^*(\ell)$ for a line ℓ on X such that $P := q(\widetilde{P}) \in \ell$.
- The linear pencil $|C| - \widetilde{P}$ coincides with the pull-back $q^*\Lambda_P$, where Λ_P is the linear pencil of lines passing through P.

These two claims show that the point $\sigma(\widetilde{P})$ is also the base point of $q^*\Lambda_P$. Let E_i be the exceptional curve of the blowing-up of the point P_i for $1 \leq i \leq 7$. Since $C \cdot E_i = q^*(\ell) \cdot E_i = -K_Y \cdot E_i = 1$, the contraction of $E_1, \cdots, E_7$ brings down σ to an involution of Y_0. Since the involution τ is uniquely defined by the linear pencil $\rho_* q^* \Lambda_P$ as such a map sending $\rho(\widetilde{P})$ to $\rho(\sigma(\widetilde{P}))$ where $\rho : Y \to Y_0$ is the blowing-up of seven points, we have $\sigma = \tau$. The claims follow readily if one notices that $C = q^*(q(C))$ and $2 = C^2 = 2(q(C)^2)$, whence $q(C)$ is a line on X.

(4) Let E be a σ-stable (-1)-curve. Then $E = q^*(A)$ for a curve A on $X = \mathbb{P}^2$. We have $-1 = K_Y \cdot q^*(A) = -q^*(\ell) \cdot q^*(A) = -2(\ell \cdot A) = -2 \deg A$. This is a contradiction. Hence $\sigma(E) \neq E$ and $E + \sigma(E) = q^*(A)$ for some

curve A on $\mathbb{P}^2$. We have then $-2 = K_Y \cdot q^*(A) = -q^*(\ell) \cdot q^*(A) = -2(\ell \cdot A) = -2 \deg A$. Hence A is a line. $\qquad \square$

5.5.4.1

Remark. In the assertion (2), (i) of Theorem 5.5.4, we consider when the irreducible curve $q^*(\ell)$ has a node or a cusp. Suppose that $q^*(\ell)$ has a singular point $\widetilde{P}$. Then the image $P := q(\widetilde{P})$ is a point of B, for otherwise $\widetilde{P}$ is smooth because $q : Y \setminus q^{-1}(B) \to X \setminus B$ is étale. Then either ℓ meets B in distinct points P and P', or ℓ touches B at P.

Consider the first case. Choosing a suitable inhomogeneous coordinate x on ℓ so that $x = 0$ at P, $q^*(\ell)$ viewed locally near $\widetilde{P}$ is isomorphic to a plane curve in the (x, z)-plane defined by

$$z^2 + xz + f_0(x)^2 + f_1(x)^2 x = 0, \tag{5.7}$$

where $f_0(x), f_1(x) \in k[x]$. Since T is a quartic curve, $\deg f_0(x) \leq 2$ and $\deg f_1(x) \leq 1$. We may assume that $f_0(0) = 0$. In fact, let $c = f_0(0)$. The change of variable $z' = z + c$ makes the equation (5.7) in the form

$$z'^2 + xz' + (f_0(x) + c)^2 + (f_1(x)^2 + c)x = 0.$$

Hence we assume that $f_0(0) = 0$. By the Jacobian criterion, the curve $q^*(\ell)$ is singular if and only if $f_1(0) = 0$. In geometric terms, if the local intersection multiplicity $i(\ell, B; P) = 1$, $q^*(\ell)$ is singular if and only if $i(\ell, T; P) \geq 2$. By the blowing-up of the point $(0, 0)$ in the (x, z)-plane, the proper transform of $q^*(\ell)$ becomes

$$z_1^2 + z_1 + x^{-2}(f_0(x)^2 + f_1(x)^2 x) = 0,$$

where $z = xz_1$. Hence the proper transform of $q^*(\ell)$ meets two distinct points of the exceptional curve $x = 0$ whose z_1-values are roots of $z_1^2 + z_1 + a^2 = 0$, where a is the coefficient of the linear term of $f_0(x)$. Namely, the singular point of $q^*(\ell)$ is a node.

Consider next the case $i(\ell, B; P) = 2$. By the same argument as in the first case, $q^*(\ell)$ near the point $\widetilde{P}$ in the (x, z)-plane has equation

$$z^2 + x^2 z + f_0(x)^2 + f_1(x)^2 x = 0, \tag{5.8}$$

where we can also assume that $f_0(0) = 0$. Then the curve $q^*(\ell)$ is singular if and only if $f_1(0) = 0$. In geometric terms, if $i(\ell, B; P) = 2$, $q^*(\ell)$ is singular if and only if $i(\ell, T; P) \geq 2$. The proper transform of $q^*(\ell)$ under the blowing-up of the point P in the (x, z)-plane is defined by

$$z_1^2 + xz_1 + x^{-2}(f_0(x)^2 + f_1(x)^2 x) = 0,$$

which we rewrite as

$$z_1^2 + x z_1 + (a_0 + a_1 x)^2 + a_2^2 x = 0.$$

By the change of variable $z_1' = z_1 + a_0$, we may assume that $a_0 = 0$. If $a_2 = 0$, this curve is reducible, which is not the case. Hence $a_2 \neq 0$ and the above proper transform is smooth. This implies that the singular point of $q^*(\ell)$ is a cusp.

5.5.4.2

Problem. On a del Pezzo surface Y of degree 2 there are 56 (-1)-curves which are paired as $(E, \sigma(E))$ and obtained from 28 lines on $X (\cong \mathbb{P}^2)$ in special position as $q^*(\ell)$. Find geometry to detect these special lines. The conic B and the quartic T meet in 8 points. There are $\binom{8}{2} = 28$ lines connecting two of these 8 points. These lines are possible candidates. However, experiments by examples show that this might not be the case. In fact, over a line ℓ with suitably chosen inhomogeneous coordinate x, $q^{-1}(\ell)$ is a plane curve in the (x, z)-plane defined by a quadratic equation

$$z^2 + a(x)z + f_0(x)^2 + f_1(x)^2 x = 0,$$
$$a(x), f_0(x), f_1(x) \in k[x], \quad \deg f_0(x) \leq 2, \ \deg f_1(x) \leq 1,$$

where we can assume that $a(x) = x(x+1)$ or $a(x) = x^2$ according as ℓ meets B in two distinct points or ℓ touches B at one point. Problem seems to be equivalent to asking whether the quadratic equation splits into two linear factors in z with coefficients in $k[x]$.

In the case of characteristic zero and $X \cong \mathbb{P}^2$, a cyclic $\mathbb{Z}/2\mathbb{Z}$-covering $q : Y \to X$ totally ramified over a smooth quartic curve T is a del Pezzo surface of degree 2 since $K_Y \sim K_X + 2\ell \sim -\ell$, where ℓ is a line on X. A *bitangent* of T is a line which is tangent at two points of T. It is well-known that T has 28 bitangents. Let ℓ_0 be a bitangent. Then $q^*(\ell_0)$ splits to a sum $L_0 + L_0'$, where L_0 and L_0' are (-1)-curves meeting at two points transversally. Hence the 56 of the (-1)-curves on Y are thus obtained.

5.5.5

Theorem. *Assume that $p = 2, X = \mathbb{P}^2, B \sim d\ell$ with $d \geq 3$ and $T \sim 2d\ell$. Let $q : Y \to X$ be an Artin-Schreier covering with sections (B, T). Then the following assertions hold.*

(1) Y *is a minimal surface with invariants*

$$K_Y^2 = 2(d^2 - 6d + 9)$$
$$\chi(\mathcal{O}_Y) = \frac{1}{2}(d^2 - 3d + 4)$$
$$e(Y) = 2(2d^2 - 3d + 3) \ .$$

(2) $q_Y = h^1(Y, \mathcal{O}_Y) = 0$.
(3) *If $d = 3$, then Y is a K3-surface, and if $d \geq 4$, then Y is a surface of general type.*

Proof. (1) Since $K_Y \sim (d - 3)q^*(\ell)$, either $K_Y \sim 0$ (if $d = 3$) or K_Y is ample (if $d \geq 4$). Hence there are no (-1)-curves on Y. Since $\kappa(Y) \geq 0$, Y is then minimal.

(2) The assertion follows from Lemma 5.4.7.2.

(3) If $d = 3$, we have $b_2 = 22$, $e(Y) = 24$, $\chi(\mathcal{O}_Y) = 2$ and $h^1(Y, \mathcal{O}_Y) = 0$. Comparison of these invariants shows that Y is a K3-surface [57, Theorem 4.3.1, p.44]. If $d \geq 4$, it is clear that Y is of general type. $\square$

5.5.5.1

Remark. (1) Assume that $p = 2, X = \mathbb{P}^2$ and B is smooth. Then $\widetilde{B} := q^*(B)$ is a reduced curve, and $q : \widetilde{B} \to B$ is a purely inseparable morphism. In fact, suppose that $q^*(B) = 2B'$ with a reduced irreducible curve B' on Y. Then $q : B' \to B$ is a birational morphism. Since B is smooth, it follows that B' is isomorphic to B. The genera of B' and B are computed as follows:

$$g(B') = p_a(B') = \frac{1}{2}(B'^2 + K_Y \cdot B') + 1$$
$$= \frac{1}{2}\left\{\frac{1}{4}q^*(B)^2 + \frac{1}{2}K_Y \cdot q^*(B)\right\} + 1$$
$$= \frac{1}{4}\left(d^2 + 2d(d - 3)\right) + 1 = \frac{1}{4}(3d^2 - 6d + 4) \ ,$$
$$g(B) = \frac{(d - 1)(d - 2)}{2} \ .$$

Since $g(B') = g(B)$, we have $\frac{1}{4}(3d^2 - 6d + 4) = \frac{1}{2}(d^2 - 3d + 2)$, whence $d = 0$. This is a contradiction.

(2) The curve $\widetilde{B}$ is a reduced irreducible curve, and $q : \widetilde{B} \to B$ is a purely inseparable morphism. Then $\widetilde{B}$ is not a smooth curve if $d > 1$. In fact, if $\widetilde{B}$ is smooth, $\widetilde{B}$ and B have the same genera because $q|_{\widetilde{B}}$ is

the Frobenius morphism of B. We compute $g(\widetilde{B}) = 2d^2 - 3d + 1$ by the arithmetic genus formula on Y and $g(B) = \frac{1}{2}(d^2 - 3d + 2)$. Equating $g(\widetilde{B})$ and $g(B)$, we have $d = 0$ or $d = 1$, which contradicts the assumption.

5.5.6

Assume that $X \cong \mathbb{P}^2$ and $p = 3$. Let $B \sim d\ell$ and $T \sim 3d\ell$. Then $K_Y \sim (2d - 3)q^*(\ell)$. Retaining the initial hypotheses set before the subsection 5.5.1, we have the following consequences:

(1) If $d = 1$, then Y is a del Pezzo surface of degree 3. Let E be a (-1)-curve and let σ be the covering automorphism of Y of order 3. Then E is not σ-stable, hence $E, \sigma(E)$ and $\sigma^2(E)$ are distinct (-1)-curves sharing one and the same point. All (-1)-curves, altogether 27 such curves on Y, are obtained as the splitting inverse image $q^*(\ell_i)$ of a line ℓ_i. But it is not clear how to detect such lines ℓ_i ($1 \le i \le 9$) in connection with sections (B, T).

(2) If $d > 1$, then Y is a minimal, regular surface of general type, i.e., $h^1(Y, \mathcal{O}_Y) = 0$.

We prove the assertion (1). It is immediate that Y is a del Pezzo surface of degree 3 because $-K_Y \sim q^*(\ell)$ is ample and $K_Y^2 = 3$. Let E be a (-1)-curve. Suppose that E is σ-stable. Namely, $E = q^*(A)$ for a curve A on X. Then $-1 = E^2 = (q^*A)^2 = 3A^2$. This is absurd. So, $\sigma(E) \neq E$. Then $E, \sigma(E)$ and $\sigma^2(E)$ are distinct. Write $E + \sigma(E) + \sigma^2(E) = q^*(A)$. Then $3 = -K_Y \cdot q^*(A) = 3(\ell \cdot A)$, whence A is a line. Let $P = B \cap A$. Then $q^{-1}(P) = \{\widetilde{P}\}$ and $\widetilde{P}$ is σ-stable. Thus $\widetilde{P}$ lies on $E, \sigma(E)$ and $\sigma^2(E)$. Since

$$3 = q^*(A)^2 = (E + \sigma(E) + \sigma^2(E))^2$$
$$= 3(E^2) + 2(E \cdot \sigma(E)) + 2(\sigma(E) \cdot \sigma^2(E)) + 2(\sigma^2(E) \cdot E),$$

we have $(E \cdot \sigma(E)) = (\sigma(E) \cdot \sigma^2(E)) = (\sigma^2(E) \cdot E) = 1$. Hence $E, \sigma(E)$ and $\sigma^2(E)$ share the point $\widetilde{P}$. If the inverse image $q^{-1}(\ell)$ splits into three components, each of them is a (-1)-curve passing through the inverse image $q^{-1}(B \cap \ell)$, which is a single point.

The assertion (2) follows from Corollary 5.5.1.1 and Lemma 5.4.7.2.

In the case of characteristic zero and $X \cong \mathbb{P}^2$, a cyclic $\mathbb{Z}/3\mathbb{Z}$-covering $q : Y \to X$ which is totally ramified over a smooth cubic C is a del Pezzo surface of degree 3 because $K_Y \sim q^*(K_X + 2\ell)$, where ℓ is a line on X. Then Y has 27 of (-1)-curves. They are obtained from *inflectional* tangents. A tangent line ℓ_0 to C at a point P is inflectional if ℓ_0 and C meet in one point P with multiplicity 3. Such a point is called a *flex*. The cubic C has 9 flexes. If ℓ_0 is an inflectional tangent of C, then $q^*(\ell_0) = L_1 + L_2 + L_3$ with three (-1)-curves L_1, L_2, L_3 meeting transversally in one common point. Hence 27 of the (-1)-curves on Y are thus obtained.

5.5.7

We will consider as in Lemma 5.5.1 the case where X is a relatively minimal ruled surface with the $\mathbb{P}^1$-fibration $\rho : X \to C$ and a smooth projective curve C of genus g. *We assume here that either B is an ample section or a fiber of ρ.*

Theorem. *With the above notations and assumptions, the following assertions hold.*

(1) *Assume that p is arbitrary and B is a fiber of ρ. Then there exists an Artin-Schreier covering $\alpha : \widetilde{C} \to C$ such that α is ramifying only over the point $\rho(B)$ and $Y \cong (\widetilde{C}, \alpha) \times_C (X, \rho)$. The genus of $\widetilde{C}$ is given by the genus of C as in the following formula:*

$$g(\widetilde{C}) = 1 + p\left(g + \frac{p-3}{2}\right) .$$

(2) *Assume that $p = 2$ and B is a cross-section of ρ. Then $\widetilde{\rho} := \rho \cdot q : Y \to C$ is a $\mathbb{P}^1$-fibration such that the following conditions are satisfied.*

  (i) *$\widetilde{B} := q^*(B)$ is an irreducible curve of arithmetic genus $(B^2) + 2g - 1$.*

  (ii) *If ℓ is a general fiber of ρ, $q^*(\ell)$ is a fiber of $\widetilde{\rho}$ and touches $\widetilde{B}$ with multiplicity 2.*

  (iii) *Every singular fiber of $\widetilde{\rho}$ is of the form $E + \sigma(E)$ with a (-1)-curve E, and there are as many singular fibers as $2(B^2)$.*

  (iv) *If Y is irrational, i.e., $g > 0$, every (-1)-curve of Y is a fiber component of $\widetilde{q}$.*

  (v) *If Y is rational, i.e., $g = 0$, there is a birational mapping $\pi : Y \to \mathbb{P}^2$ such that the proper transform of $\widetilde{B}$ is a smooth conic Q and the general fibers of the $\mathbb{P}^1$-fibration $\widetilde{\rho}$ are mapped to the tangent lines of the conic Q.*

Proof. (1) For a fiber ℓ of ρ which is disjoint from B, $q|_{q^*(\ell)} : q^*(\ell) \to \ell$ is a finite étale covering. Hence either $q^*(\ell)$ is irreducible or $q^*(\ell) = \sum_{i=0}^{p-1} \sigma^i(L)$, where L is an irreducible component of $q^*(\ell)$. If $q^*(\ell)$ is irreducible, $q^*(\ell)^2 = 0$ and $q^*(\ell) \cdot K_Y = p(\ell \cdot (K_X + (p-1)B)) = -2p$, whence

$p_a(q^*(\ell)) = -(p-1) < 0$, a contradiction. If $q^*(\ell)$ is reducible, $q^*(\ell)$ has the irreducible decomposition $q^*(\ell) = \sum_{i=0}^{p-1} \sigma^i(L)$. We have then

$$0 = q^*(\ell)^2 = p(L^2) + \sum_{0 \le i < j < p} 2(\sigma^i(L) \cdot \sigma^j(L)),$$

$$-2p = q^*(\ell) \cdot K_Y = \sum_{i=0}^{p-1} \sigma^i(L) \cdot K_Y = p(L \cdot K_Y),$$

whence $L^2 = 0$ because $L^2 \le 0$, and $L \cdot K_Y = -2$. Hence L is isomorphic to $\mathbb{P}^1$, $L^2 = 0$ and $L \cdot \sigma^i(L) = 0$ for $0 < i \le p-1$. So, $\rho \cdot q : Y \to C$ is composed with a $\mathbb{P}^1$-fibration. We write it by $\tilde{\rho} : Y \to \tilde{C}$, where $\tilde{C}$ is the normalization of C in $k(Y)$. Let S be a cross-section of ρ. Then $q^*(S)$ is a cross-section of $\tilde{\rho}$. In fact, for L as above, we have $q^*(S) \cdot L = S \cdot q_*(L) = S \cdot \ell = 1$. This implies that $q^*(S) \cdot q^*(B) = p(S \cdot B) = p$. Hence $q^*(B) = p\tilde{B}$ as a 1-cycle on Y. Namely, $q^*(B)$ is non-reduced and $q^*(B)_{\mathrm{red}} = \tilde{B}$, where $\tilde{B}$ is a fiber of $\tilde{\rho}$. Then there is a morphism $\tau : Y \to \tilde{C} \times_C X$, which is a birational and finite morphism. Since $\tilde{C} \times_C X$ is smooth as the $\mathbb{P}^1$-fibration $q : X \to C$ is locally trivial in the Zariski topology, τ is an isomorphism by Zariski Main Theorem. Since $k(Y) = k(\tilde{C})(t)$ and $k(X) = k(C)(t)$ for a fiber parameter t of ρ, it follows that the normalization morphism $\nu : \tilde{C} \to C$ is a $\mathbb{Z}/p\mathbb{Z}$-covering and hence an Artin-Schreier covering. The genus $g(\tilde{C})$ is computed by the formula for $\chi(\mathcal{O}_Y)$ in Lemma 5.4.7.1 because $\chi(\mathcal{O}_Y) = 1 - g(\tilde{C})$.

(2) Let ℓ be a fiber of ρ. Since B is a cross-section, the point of Y lying over $\ell \cap B$ is a σ-fixed point. Suppose that $q^*(\ell)$ is decomposable. Then it comprises two components. Write $q^*(\ell) = E_1 + E_2$, where $E_2 = \sigma(E_1)$. Since $q^*(\ell) \cdot K_Y = 2(\ell \cdot (K_X + B)) = 2(-2+1) = -2$, we have $E_1 \cdot K_Y = E_2 \cdot K_Y = -1$. On the other hand, since $0 = q^*(\ell)^2 = E_1^2 + E_2^2 + 2E_1 \cdot E_2 = 2(E_1^2 + E_1 \cdot E_2)$ and $E_1 \cdot E_2 > 0$, it follows that $E_1^2 < 0$. Then the arithmetic genus formula implies that E_1 (hence E_2 as well) is a (-1)-curve and $E_1 \cdot E_2 = 1$. Since there are only finitely many, mutually disjoint (-1)-curves on Y, $q^*(\ell)$ is an irreducible curve isomorphic to $\mathbb{P}^1$ if ℓ is a general fiber of ρ. Namely, $\tilde{\rho} := \rho \cdot q : Y \to C$ is a $\mathbb{P}^1$-fibration. We show the assertion (i). Suppose that $q^*(B) = 2\overline{B}$ with $\overline{B} = q^*(B)_{\mathrm{red}}$. Since $q^*(B)^2 = 2B^2 = 4\overline{B}^2$ and $2(\overline{B} \cdot K_Y) = q^*(B) \cdot K_Y = 2(B \cdot (K_X + B)) = 4(g-1)$ as $B \cong C$, it follows that $\overline{B} \cong B \cong C$ and $2g - 2 = \overline{B} \cdot K_Y + \overline{B}^2 = 2g - 2 + \frac{1}{2}B^2$. Hence $B^2 = 0$. This contradicts the beginning assumption that B is an ample

section. Hence $\widetilde{B} := q^*(B)$ is an irreducible curve of arithmetic genus

$$p_a(\widetilde{B}) = \frac{1}{2}(\widetilde{B}^2 + \widetilde{B} \cdot K_Y) + 1$$

$$= \frac{1}{2}\{q^*(B).(q^*(B) + q^*(K_X + B))\} + 1$$

$$= (B \cdot (K_X + 2B)) + 1 = B^2 + 2g - 1.$$

The assertion (ii) is straightforward because both $\widetilde{B}$ and $q^*(\ell)$ are smooth at the unique intersection point $\widetilde{B} \cap q^*(\ell)$ if ℓ is general and $\widetilde{B} \cdot q^*(\ell) = 2$. We show the second part of the assertion (iii). Suppose that there are altogether N singular fibers of $\widetilde{\rho}$ which are of the form $E_1 + E_2$. By contracting one of two (-1)-curves in each singular fiber, we have a relatively minimal ruled surface Y_0 with base curve of genus g, for which $K_{Y_0}^2 = 8(1 - g)$. Hence we have $K_Y^2 = 8(1 - g) - N$. On the other hand, we have

$$K_Y^2 = 2(K_X + B)^2 = 2\{K_X^2 + 2B \cdot (B + K_X) - B^2\}$$

$$= 2\{8(1 - g) + 4(g - 1) - B^2) = 8(1 - g) - 2B^2.$$

This implies that $N = 2B^2$.

The assertion (iv) is clear because the genus C is positive.

We show the assertion (v). Note that $g = 0$. Since $\widetilde{B} \cdot \ell = 2$ for a general fiber $\widetilde{\ell} = q^*(\ell)$ of $\widetilde{\rho}$, every singular point $\widetilde{P}$ of $\widetilde{B}$ is a cuspidal double point. Namely, if $\ell_{\widetilde{P}}$ is the fiber of $\widetilde{\rho}$ through the point $\widetilde{P}$, by the blowing-up $\theta : Y_1 \to Y$ of $\widetilde{P}$, the proper transform $\theta'(\widetilde{B})$ touches the exceptional curve $\widetilde{E}_{\widetilde{P}}$ with multiplicity 2 and the proper transform $\theta'(\ell_{\widetilde{P}})$ does not meet $\theta'(\widetilde{B})$. Hence the contraction of $\theta'(\ell_{\widetilde{P}})$ resolves the singularity of $\widetilde{B}$ at $\widetilde{P}$ and $\widetilde{E}_{\widetilde{P}}$ gives a smooth fiber of a new $\mathbb{P}^1$-fibration. Let N' be the number of singular points. Then $N' = (B^2) - 1$ by the assertion (i) since the geometric genus of $\widetilde{B}$ is zero. We apply the above elementary transformation at every singular point of $\widetilde{B}$. Furthermore, if $E_1 + E_2$ is a singular fiber of $\widetilde{\rho}$, we contract one of E_1 and E_2. Then the uncontracted (-1)-component gives a smooth fiber, and the self-intersection number of the curve $\widetilde{B}$ increases by 1. By these birational operations, we have a relatively minimal ruled surface $\widehat{Y}$ over C with the smooth proper transform $\widehat{B}$ of $\widetilde{B}$.

Note that there is a purely inseparable morphism $\widehat{B} \to B$ of degree 2. Hence $\widehat{B}$ is isomorphic to $\mathbb{P}^1$. Since $(\widetilde{B}^2) = 2(B^2)$, we compute

$$(\widehat{B}^2) = (\widetilde{B}^2) + 2(B^2) - 4N' = 4(B^2) - 4N' = 4(B^2) - 4((B^2) - 1) = 4.$$

Let $\widetilde{M}$ be the minimal section with $(\widetilde{M}^2) = -\widetilde{n}$. Write $\widehat{B} \sim 2\widetilde{M} + m\widetilde{\ell}$ with $m > 0$. Since $(\widehat{B}^2) = 4$, we have

$$(2\widetilde{M} + m\widetilde{\ell})^2 = -4\widetilde{n} + 4m = 4,$$

whence $m = \widetilde{n} + 1$. Since $(\widetilde{M} \cdot \widehat{B}) \geq 0$, we have $-2\widetilde{n} + \widetilde{n} + 1 \geq 0$. Namely, $\widetilde{n} \leq 1$. If $\widetilde{n} = 1$ then $\widetilde{Y} \cong \Sigma_1$ and $\widehat{B} \cap \widetilde{M} = \emptyset$. By the contraction of $\widetilde{M}$, we have $\mathbb{P}^2$ and the image of $\widehat{B}$ becomes a smooth conic on $\mathbb{P}^2$. Suppose $\widetilde{n} = 0$. Then $\widetilde{Y} \cong \Sigma_0 := \mathbb{P}^1 \times \mathbb{P}^1$, and $\widehat{B}$ meets $\widetilde{M}$ in one point, say $\widetilde{P}$. Let $\widetilde{\ell}_{\widetilde{P}}$ and $\widetilde{M}_{\widetilde{P}}$ be the fiber and the minimal section through $\widetilde{P}$. Then $\widetilde{\ell}_{\widetilde{P}} \cdot \widehat{B} = 2$. Blow up $\widetilde{P}$ and contract the proper transform of $\widetilde{\ell}_{\widetilde{P}}$. Then we have Σ_1 and the same situation as we have treated above. $\qquad\square$

5.5.8

We now consider the case where K_Y is nef. By Corollary 5.5.1.1, $\kappa(Y) \geq 0$ and Y is a minimal surface. We not only inherit the beginning hypotheses of this section, but also assume additionally the condition that B is ample. This is for the sake of simplicity, and the readers can try to relax the ampleness condition to a weaker condition like B being semiample.

Theorem. *Assume that $\kappa(Y) = 0$. Then the following assertions hold.*

(1) *Y is a K3-surface and $p = 2$ or 3.*

(2) *If $p = 2$ then X is $\mathbb{P}^2$, Σ_0 or a del Pezzo surface X_d with $1 \leq d \leq 8$ as specified in the assertion (3) below, and $B \in |-K_X|$. If $p = 3$ then $X \cong \Sigma_0$ and $B \sim M + \ell$ (the diagonal).*

(3) *Suppose $p = 2$. Let O be a point on a smooth cubic C_0 on $\mathbb{P}^2$. For $1 \leq d \leq 8$, let X_d be a del Pezzo surface of degree d which is obtained from $\mathbb{P}^2$ by blowing up $9 - d$ points $P_1 = O, P_2, \dots, P_{9-d}$ on C_0 in general position[11]. Let Λ_d be the proper transform of the linear pencil of lines on $\mathbb{P}^2$ passing through the point O. There are two cases to consider according to how a general line ℓ meets the curve B, which is the proper transform of C_0. Namely, one is the case (i) where ℓ meets B in two distinct points, and the other is the case (ii) where ℓ touches B with degree two (cf. Theorem 5.5.7).*

In the case (i), the linear system $q^* \Lambda_d$ consisting of the inverse images of members of Λ_d is an elliptic linear pencil. Hence $q^* \Lambda_d$ gives an elliptic fibration on Y. A possible singular fiber has type I_1, I_2, I_3, I_4, II, III, IV in the list 3.1.5. Suppose $X \cong \Sigma_0$. Then $q^* \Lambda$ defines an elliptic fibration on Y, where $\Lambda = |\ell|$. A possible singular fiber has type

---

[11]There are no 3 points on a line, no 6 points on a conic and no 8 points on a singular cubic including the singular point. This is equivalent to the condition that no (-2)-curves are produced after blowing up these points.

I_1, I_2, II, III *in the list 3.1.5. In the case (ii), the linear system $q^* \Lambda_d$ gives either an elliptic fibration or a quasi-elliptic fibration. A possible singular fiber has type* II, III, IV.

(4) *Suppose $p = 3$. The inverse image $q^* \Lambda$ defines an elliptic fibration, where $\Lambda = |\ell|$. A possible singular fiber has type* II, IV *in the list 3.1.5.*

Proof. (1) and (2). The assumption $\kappa(Y) = 0$ implies that $K_Y \equiv 0$ (numerical equivalence) by the Enriques classification [57, Theorem 4.3.1, p.44] because Y is minimal. Since $K_Y \sim q^*(K_X + (p-1)B)$, it follows that $-K_X \equiv (p-1)B$. Since B is ample, it follows that X is a del Pezzo surface and $K_X^2 = (p-1)^2 B^2 \le 9$. Hence $p = 2$ or 3. Since the Picard group of a del Pezzo surface has no torsion, we have $K_X + (p-1)B \sim 0$ and hence $K_Y \sim 0$. On the other hand, by Lemma 5.4.7.2, we have $h^1(Y, \mathcal{O}_Y) = 0$. Hence Y is a K3-surface by the Enriques classification (*ibid.*). If $p = 2$, X is $\mathbb{P}^2, \Sigma_0$ or a del Pezzo surface X_d with $1 \le d \le 8$. Since $K_X + B \sim 0$, $B \in |-K_X|$. If $p = 3$, then $K_X + 2B \sim 0$, whence $-1 = K_X \cdot E = 2B \cdot E$ for a (-1)-curve E on X. This is impossible. Hence there are no (-1)-curves on X. The case $X \cong \mathbb{P}^2$ is also excluded because $K_X \sim -3\ell$ and $K_X + 2B \sim 0$. This implies that $X \cong \Sigma_0$ and $B \sim M + \ell$.

(3) X_8 is the blowing-up of $\mathbb{P}^2$ at the point O, and the proper transforms of lines on $\mathbb{P}^2$ through O are the fibers of the $\mathbb{P}^1$-fibration on Σ_1. By abuse of notation, we denote by M the negative section of Σ_1. We consider the case where $X = \Sigma_1$ and $B \sim -K_X \sim 2M + 3\ell$. Let ℓ be a fiber which meets B transversally in two points P, P'. Choose an inhomogeneous coordinate x of ℓ such that $x = 0$ at P. As in Remark 5.5.4.1, we may assume that the curve $q^*(\ell)$ on Y near the point $\widetilde{P} := q^{-1}(P)$ is defined by

$$z^2 + xz + f_0(x)^2 + f_1(x)^2 x = 0, \quad f_0(x), f_1(x) \in k[x], \quad f_0(0) = 0.$$

Then $q^*(\ell)$ is singular if and only if $f_1(0) = 0$. If it is singular, we have

$$z_1^2 + z_1 + x^{-2}(f_0(x)^2 + f_1(x)^2 x) = 0,$$

where $z = xz_1$. This implies that $q^*(\ell)$ has two branches at the point $\widetilde{P}$ lying over P. Since the cusp is the only moving singularity on a smooth projective surface, $q^*(\ell)$ is a smooth curve if ℓ is a general fiber. Since $q^*(\ell)^2 = K_Y \cdot q^*(\ell) = 0$, $q^*(\ell)$ is then an elliptic curve. If ℓ is special, one of the following cases is possible:

(i) $q^*(\ell)$ has a nodal point over one of P, P' and a smooth point over the other.

(ii) $q^*(\ell)$ consists of two irreducible components $L, \sigma(L)$ such that L and $\sigma(L)$ meet transversally in two points over P, P', where σ is the covering involution and $L, \sigma(L)$ are (-2)-curves.

Suppose that $P = P'$. Namely, ℓ touches B at P. Then the defining equation of $q^*(\ell)$ is written as

$$z^2 + x^2 z + f_0(x)^2 + f_1(x)^2 x = 0, \quad f_0(0) = 0.$$

By the same argument as in Remark 5.5.4.1, $q^*(\ell)$ has a cusp if it is singular and irreducible. If $q^*(\ell)$ is reducible, it has type III.

If $d < 8$ we consider the $\mathbb{P}^1$-fibration induced by the one on X_8. Then there happens the case where a fiber ℓ consists of two (-1)-curves $\ell' + E$, where ℓ' is the proper transform of a fiber $\bar{\ell}$ on X_8 and E is the exceptional curve of the blowing-up of some P_i. Then $B \cdot E = 1$, and $B \cdot \ell' = 1$ with $\bar{\ell}$ meeting the image $\overline{B}$ of B on X_8 in points $P = P_i, P'$. If $P \neq P'$, $q^*(\ell)$ is a singular fiber of type I_4 as the inverse images of ℓ' and E decomposing into sums of two (-2)-curves. If $P = P'$, $q^*(\ell)$ is a singular fiber of type III (when both $q^*(\ell')$ and $q^*(E)$ are irreducible) or IV (when either $q^*(\ell')$ or $q^*(E)$ decomposes to two irreducible components).

The case $X = \Sigma_0$ can be treated in a similar fashion only with absence of (-1)-curves taken into consideration.

In the case (ii), we also consider the $\mathbb{P}^1$-fibration on X which is given by Λ_d. A fiber ℓ touches B or $\ell = \ell' + E$ with B, E, ℓ' meeting each other transversally in one common point. In the first case, $q^*(\ell)$ has type II if $q^*(\ell)$ is singular and type III if $q^*(\ell)$ is reducible. In the second case, $q^*(\ell)$ has type III or IV. The arguments are the same as in the corresponding cases in the case (i).

(4) Let ℓ be a fiber of the $\mathbb{P}^1$-fibration $\Sigma_0 \to \mathbb{P}^1$. Since $B \sim M + \ell$, ℓ meets B transversally in one point P. If $q^*(\ell)$ is an irreducible singular curve, the Galois group $G \cong \mathbb{Z}/3\mathbb{Z}$ acts on $q^*(\ell)$ non-trivially and fixes the singular point. Hence G acts on the normalization $\widetilde{q^*(\ell)}$ of $q^*(\ell)$ and stabilizes the inverse set S in $\widetilde{q^*(\ell)}$ of the singular point. Since $\widetilde{q^*(\ell)} \cong \mathbb{P}^1$, S consists of one point or three points by Lemma 5.2.1. Since $q^*(\ell)$ has arithmetic genus 1, $q^*(\ell)$ has at most two branches at the singular point. Hence S consists of a single point. Namely, $q^*(\ell)$ is a cuspidal curve. If $q^*(\ell)$ is reducible, it has three distinct branches over the point of Y lying over P. By the classification of singular fibers of an elliptic fibration (see the subsection 3.1.5), $q^*(\ell)$ consists of three (-2)-curves meeting in one common point (a singular fiber of type IV). Hence the assertion follows. $\square$

5.5.9

Suppose that $\kappa(Y) = 1$. Then the linear system $|nK_Y|$ is composed of a pencil for $n \gg 0$. Namely, the linear system $|nK_Y|$ is free from base points and the morphism $\Phi_{|nK_Y|} : Y \to \mathbb{P}^{\dim |nK_Y|}$ is a composite of a fibration $\rho : Y \to C$ and the closed immersion $C \subset \mathbb{P}^{\dim |nK_Y|}$. It is known that ρ is either an elliptic fibration or a quasi-elliptic fibration (cf. [57, Theorem 4.1.1, (3)]). Let γ be a generator of the Galois group $G \cong \mathbb{Z}/p\mathbb{Z}$. Then γ stabilizes the divisor K_Y upto linear equivalence, hence the linear system $|nK_Y|$. This implies that γ preserves the fibration ρ. Namely, if F is a fiber of ρ, so is the translate $\gamma(F)$.

Theorem. *Let $q : Y \to X$ be an Artin-Schreier covering such that $\kappa(Y) = 1$ and B is an ample divisor on X. With the notations as above, the following assertions hold.*

(1) *We have either $p = 2$ or $p = 3$. The fibration ρ is decomposed as $\rho = \overline{\rho} \cdot q$, where $\overline{\rho} : X \to C$ is a $\mathbb{P}^1$-fibration.*
(2) *Suppose that $p = 2$. The B is a 2-section of the $\mathbb{P}^1$-fibration $\overline{\rho}$. A singular fiber of $\overline{\rho}$ is of the form $\ell = E + E'$, where E, E' are (-1)-curves meeting B in points P, P' (possibly $P = P'$). Distinction of the fibration ρ being elliptic or quasi-elliptic depends in the same way as in Theorem 5.5.8 on how a general fiber ℓ of $\overline{\rho}$ meets B (the cases (i) and (ii)). Classification of a possible singular fiber of ρ is the same as in Theorem 5.5.8, the assertion (3).*
(3) *Suppose that $p = 3$. Then B is a cross-section of $\overline{\rho}$, and the fibration $\overline{\rho}$ is relatively minimal. Furthermore, the fibration ρ is an elliptic fibration. Classification of a possible singular fiber of ρ is the same as in Theorem 5.5.8, the assertion (4).*

Proof. (1) Let F be a fiber of the fibration ρ. Since $F^2 = K_Y \cdot F = 0$, we have $K_Y \cdot F = (K_X + (p-1)B) \cdot q(F) = 0$, whence $K_X \cdot q(F) = -(p-1)B \cdot q(F) < 0$ because B is ample. By the remark before the statement of theorem, $q^*(q(F)) = F$ or $q^*(q(F)) = \sum_{i=0}^{p-1} \gamma^i(F)$ if F is a general fiber. In both cases, $q(F)^2 = 0$, which implies that $q(F) \cong \mathbb{P}^1$. Since F is either an elliptic curve or a rational cuspidal curve, the case $q^*(q(F)) = \sum_{i=0}^{p-1} \gamma^i(F)$ cannot occur, for otherwise $F \cong q(F)$. So, G acts on F and $F/G = q(F)$. Namely, $q|_F : F \to q(F)$ is an Artin-Schreier covering. Furthermore, ρ splits as $\rho = \overline{\rho} \cdot q$, where $\overline{\rho} : X \to C$ is the $\mathbb{P}^1$-fibration whose general fibers are of the form $q(F)$. Since $K_X \cdot q(F) = -2$, we have $-(p-1)B \cdot q(F) = -2$. This implies that $p - 1 \leq 2$, whence $p = 2$ or $p = 3$.

(2) and (3). If a fiber ℓ of $\overline{\rho}$ is reducible, then the fiber $F := q^*(\ell)$ of ρ is reducible, too. Let G be an irreducible component of F. By the classification of reducible singular fibers of an elliptic or quasi-elliptic fibration in 3.1.5, G is a (-2)-curve, i.e., $G^2 = -2$ and $K_Y \cdot G = 0$. Let $E := q(G)$. Then $K_Y \cdot G = (K_X + (p-1)B) \cdot E = 0$ and $E \cdot B > 0$, whence $K_X \cdot E < 0$. On the other hand, $E^2 < 0$ as an irreducible component of ℓ. By the arithmetic genus formula, this implies that E is a (-1)-curve. If $p = 3$, we then have $1 = -K_X \cdot E = 2B \cdot E$, which is a contradiction. So, $\overline{\rho}$ has possibly a reducible fiber only in the case $p = 2$. If $p = 2$, every irreducible component of ℓ is a (-1)-curve, hence $\ell = E + E'$ with (-1)-curves E, E' satisfying $E \cdot E' = 1$. The rest of the assertions (2) and (3) is proved by the same arguments as in Theorem 5.5.8. $\qquad\square$

5.6 Local $\mathbb{Z}/p\mathbb{Z}$-actions near fixed points

Given a $\mathbb{Z}/p\mathbb{Z}$-action on a smooth algebraic variety X, it is important to look into local behaviors of the action near fixed points in order to study the singularity of the image of a fixed point in the quotient variety Y. There are several references by M. Artin [4], B. Peskin [72, 73], J. Fogarty [16], *et al.* In the present section, we extract some of their results by restricting ourselves to the case where $G = \mathbb{Z}/p\mathbb{Z}$ and $\dim X = 2$ ultimately.

5.6.1 *General observations*

Let X be a normal algebraic variety with a nontrivial G-action and let P be a fixed point. Then the local ring $\mathcal{O}_{X,P}$ and its completion $\widehat{\mathcal{O}}_{X,P}$ have the induced G-actions. In order to study the singularity of the (geometric) quotient variety [12] Y at the image point Q of P, it is necessary to look into

---

[12] Suppose for simplicity that $X = \operatorname{Spec} A$ is an affine variety with a G-action. Let $Y = \operatorname{Spec} A^G$ and $q : X \to Y$ the quotient morphism induced by the inclusion $A^G \hookrightarrow A$. Then the following properties are known (cf. [GIT] and [SGA1]).

(1) Y is an affine variety and q is a finite surjective morphism.

(2) The fibers of q are precisely the orbits of the G-action on X.

(3) The topology on Y is the quotient topology, i.e., a subset $V \subset Y$ is an open set if and only if $U = q^{-1}(V)$ is an open set of X.

(4) There is a natural isomorphism $\mathcal{O}_Y \cong (q_*\mathcal{O}_X)^G$. Namely, for any open set V of Y, $\Gamma(V, \mathcal{O}_Y)$ is isomorphic to the G-invariant subring of $\Gamma(q^{-1}(V), \mathcal{O}_X)$.

The variety Y is called the *geometric quotient* of X by G and denoted by X/G.

the G-action on the completion $\widehat{\mathcal{O}}_{X,P}$ and the invariant subring $(\widehat{\mathcal{O}}_{X,P})^G$.

We denote $\mathbb{Z}/p\mathbb{Z}$ by G and $\mathcal{O}_{X,P}$ by R which is a normal local ring of dimension $n = \dim X$. We assume that $n \geq 2$. Define the *trace* Tr : $R \to R^G$ and the *norm* N : $R \to R^G$ by $\operatorname{Tr}(a) = \sum_{g \in G} g(a)$ and $\operatorname{N}(a) = \prod_{g \in G} g(a)$. Then Tr is additive and N is multiplicative as operators. For any element $a \in R$, consider a polynomial $\varphi_a(x) = \prod_{g \in G}(x - g(a))$ which is a polynomial in $R^G[x]$. Since $\varphi_a(a) = 0$, the element a is integral over R^G. As a terminology, a noetherian local ring $(R, \mathfrak{m})$ is a geometric local ring. The local rings R, S appearing in the subsequent arguments are tacitly assumed to be geometric local rings.

5.6.1.1

Lemma. *With the above notations and assumptions, the following assertions hold.*

(1) R^G *is a normal noetherian local ring of dimension n.*
(2) R *is a finite R^G-module.*
(3) *The completion $\widehat{R^G}$ is isomorphic to $(\widehat{R})^G$.*

Proof. (1) and (2). We may assume that $R = A_{\mathfrak{p}}$ for an affine k-domain A and a maximal ideal $\mathfrak{p}$. In fact, there exists an affine open set U of X such that $P \in U$ and the coordinate ring of U is an affine k-domain, i.e., an integral domain which is finitely generated over k. Then $\cap_{g \in G} g(U)$ is an affine open neighborhood of P such that the coordinate ring A is a G-stable affine k-domain. Write $A = k[a_1, \ldots, a_m]$, and let A' be a k-subalgebra of A generated by the coefficients of $\varphi_{a_i}(x)$ for $1 \leq i \leq m$. Then A' is an affine domain, and $A' \subseteq A^G \subset A$. Since A is integral over A', A^G is a submodule of a finite A'-module A. Since A' is noetherian, A^G is a finite A'-module. In particular, A^G is an affine domain over k. Let $S = A \setminus \mathfrak{p}$, and let $\operatorname{N}(S)$ be the multiplicative set consisting of $\operatorname{N}(s)$ for $s \in S$. Then $\operatorname{N}(S) \subset A^G$ and $R = A_{\mathfrak{p}} = \operatorname{N}(S)^{-1}A$. In fact, if $s \in S$ then $g(s) \in S$ for all $g \in S$. This implies that $R^G = \operatorname{N}(S)^{-1}A^G$. In fact, if $\frac{a}{s} \in R^G$ with $a \in A$ and $s \in S$ then $\frac{a}{s} = \frac{a(\operatorname{N}(s)/s)}{\operatorname{N}(s)} \in \operatorname{N}(S)^{-1}A^G$. The inclusion $\operatorname{N}(S)^{-1}A^G \subseteq R^G$ is clear. Since A is a finite A^G-module, $R = \operatorname{N}(S)^{-1}A$ is a finite module over $R^G = \operatorname{N}(S)^{-1}A^G$.

Let $\mathfrak{m}$ be the maximal ideal of R and let $\mathfrak{m}_0 = \mathfrak{m} \cap R^G$. To show that $(R^G, \mathfrak{m}_0)$ is a local ring, it suffices to show that $a \in R^G \setminus \mathfrak{m}_0$ is invertible in R^G. Since a is invertible in R, $au = 1$ for some element u of R. Then $u \in R^G$. Hence a is invertible in R^G. The assertions (1) and (2) are thus verified.

(3) By Nagata [66, (37.5)], $\widehat{R^G}$ is a normal local ring. Since R is a finite R^G-module, $\widehat{R} \cong R \otimes_{R^G} \widehat{R^G}$. Hence $[Q(\widehat{R}) : Q(\widehat{R^G})] = p$. Since $\widehat{R^G} \subset (\widehat{R})^G$ and $[Q(\widehat{R}) : Q((\widehat{R})^G)] = p$, we have $Q(\widehat{R^G}) = Q((\widehat{R})^G)$. Then $\widehat{R^G} = (\widehat{R})^G$ by Zariski's Main Theorem (cf. Hartshorne [24, Cor. 11.4]). $\qquad\square$

5.6.2 *Fogarty's result on* depth R^G

In this subsection, we follow Fogarty's result [16] only for $G = \mathbb{Z}/p\mathbb{Z}$.

5.6.2.1

In the following lemma, R and S are not necessarily geometric local rings.

Lemma. *Let R be a noetherian semilocal domain of characteristic $p > 0$ with the p-group G acting transitively on the maximal ideals. If R is a finite étale extension of $S := R^G$ then there exists an element $f \in R$ such that $\{g(f) \mid g \in G\}$ is a free basis of R over S.*

Proof. Since $S \hookrightarrow R$ is faithfully flat, the ring S is noetherian. In fact, if $\mathfrak{a}$ is an ideal of S, then $\mathfrak{a}R$ is generated by finitely many elements $a_i \in \mathfrak{a}$ since R is noetherian. Let $\mathfrak{b} = \sum_i a_i S$ be an ideal of S contained in $\mathfrak{a}$. Then $\mathfrak{b}R = \mathfrak{a}R$. Since R is faithfully flat over S, $(\mathfrak{a}/\mathfrak{b}) \otimes_S R = 0$. Hence $\mathfrak{a}/\mathfrak{b} = 0$, i.e., $\mathfrak{a} = \mathfrak{b}$, which is finitely generated as an ideal of S. Let $\mathfrak{m}_1, \ldots, \mathfrak{m}_r$ be the maximal ideals of R, and let $\mathfrak{n} = \mathfrak{m}_i \cap S$. This holds because G acts transitively on the set $\{\mathfrak{m}_1, \ldots, \mathfrak{m}_r\}$. Let $a \in S \setminus \mathfrak{n}$. Then $a \notin \mathfrak{m}_i$ for all i. Hence a is a unit in R. So, $au = 1$ for $u \in R$. It is then clear that $u \in S$ and hence a is a unit in S. This implies that S is a local ring with maximal ideal $\mathfrak{n}$.

Since $S \hookrightarrow R$ is a finite étale extension by hypothesis, $R/\mathfrak{n}R$ is a direct sum $\oplus_{i=1}^r R/\mathfrak{m}_i$, where $R/\mathfrak{m}_i$ is a finite separable extension of $S/\mathfrak{n}$. Since G acts transitively on the set $\{\mathfrak{m}_1, \ldots, \mathfrak{m}_r\}$, we have $p = r[R/\mathfrak{m}_1 : S/\mathfrak{n}]$. This implies that either $r = p$ and $R/\mathfrak{m}_i \cong S/\mathfrak{n}$ or $r = 1$ and $[R/\mathfrak{m}_1 : S/\mathfrak{n}] = p$ with $\mathfrak{m}_1 = \mathfrak{n}R$. In the latter case $(R/\mathfrak{n}R)^G = S/\mathfrak{n} = k$.

Let $K = Q(R)$. Then $K^G = Q(S)$ because $\xi \in K^G$ is written as $\xi = \frac{a}{b} = \frac{a(\mathrm{N}(b)/b)}{\mathrm{N}(b)}$, where $a, b \in R, \mathrm{N}(b) = \prod_{g \in G} g(b) \in S$ and $a(\mathrm{N}(b)/b) \in R^G = S$. Note that $K = R \otimes_S Q(S)$ because R is integral over S. Choose an element

δ of $R/\mathfrak{n}R$ such that $\{g(\delta) \mid g \in G\}$ is a k-basis of $R/\mathfrak{n}$. In fact, if $r = p$, then $R/\mathfrak{n}R \cong k[x]/(x^p - x)$, and we can take the residue class $\overline{x}$ as δ. If $r = 1$, then $R/\mathfrak{n}R$ is a Galois extension of k. Here k is not assumed to be algebraically closed. Then $R/\mathfrak{n}R = k(\delta)$ is a simple extension. Then $\{g(\delta) \mid g \in G\}$ is a k-basis of $R/\mathfrak{n}R$. Choose an element $f \in R$ such that $\delta = f \pmod{\mathfrak{n}R}$. Then $\{g(f) \mid g \in G\}$ is a free basis of R over S by Nakayama's lemma.   □

5.6.2.2

Retain the initial notations and assumptions in 5.6.1. Assume further that the point P is an isolated fixed point. By replacing X by a G-stable affine open neighborhood of P in X, we assume that X is affine. Hence Y is a normal affine variety which is smooth outside the point $Q = q(P)$. Let $\widehat{X} = X \setminus \{P\}$ and $\widehat{Y} = Y \setminus \{Q\}$. The restriction of q onto $\widehat{X}$ induces a finite étale morphism $\widehat{q} : \widehat{X} \to \widehat{Y}$ with the group G acting freely on $\widehat{X}$. The following result shows that the singularity of the point Q is bad if $n = \dim X > 2$.

Lemma. depth $R^G = 2$. If $n = \dim R = 2$, R^G is Cohen-Macaulay.

Proof.   The proof due to J. Fogarty [16] depends on a clever use of group cohomologies of G. The first claim is to show that $H^1(G, R) \neq 0$. In fact, a 1-cocycle φ of G with values in R is a collection $\{\varphi(g) \mid g \in G\}$ such that

$$\varphi(g_1 g_2) = g_1(\varphi(g_2)) + \varphi(g_1), \quad g_1, g_2 \in G,$$

and a 1-coboundary is a collection $\{g(a) - a \mid g \in G\}$ for an element $a \in R$. Let σ be a generator of G. If φ is a 1-cocycle, then we have for $1 \le i \le p$

$$\varphi(\sigma^i) = \sigma^{i-1}(\varphi(\sigma)) + \sigma^{i-2}(\varphi(\sigma)) + \cdots + \sigma(\varphi(\sigma)) + \varphi(\sigma).$$

Since $\sigma^p = e$ and $\varphi(e) = 0$, we have $\mathrm{Tr}(\varphi(\sigma)) = 0$. Conversely, if $\mathrm{Tr}(a) = 0$, define a 1-cocyle $\{\varphi(\sigma^i) \mid 0 \le i < p\}$ by

$$\varphi(\sigma^i) = \sigma^{i-1}(a) + \cdots + \sigma(a) + a.$$

Then $\varphi(\sigma) = a$. Thus we have an identification

$$H^1(G, R) = \{a \in R \mid \mathrm{Tr}(a) = 0\}/\{\sigma(a) - a \mid a \in R\}.$$

Since G acts trivially on $R/\mathfrak{m}$, we have $\sigma(a) - a \in \mathfrak{m}$ for all $a \in R$. Hence $1 \notin \{\sigma(a) - a \mid a \in R\}$. Meanwhile, $1 \in \{a \mid \mathrm{Tr}(a) = 0\}$. So, $H^1(G, R) \neq 0$.

The second claim is the existence of a long exact sequence $(*)$ below. By [SGA2, Exposé III, Exemple III-1], depth $R = d$ for a noetherian local ring $(R, \mathfrak{m})$ if $H^d_{\mathfrak{m}}(R) \neq 0$ and $H^i_{\mathfrak{m}}(R) = 0$ for all $i < d$, where $H^{\bullet}_{\mathfrak{m}}(R)$

denotes the local cohomology of R with support $\mathfrak{m}$. There are two spectral sequences $E_2^{p,q}, E'^{p,q}_2$ with the same abutment (see [23, Prop. 5.2.4]);

$$E_2^{p,q} = H^p(G, H^q(\widehat{X}, \mathcal{O}_{\widehat{X}}))$$
$$E'^{p,q}_2 = H^p(\widehat{Y}, \mathcal{H}^q(G, \mathcal{O}_{\widehat{X}})),$$

where $\mathcal{H}^q(G, \mathcal{O}_{\widehat{X}})$ is the sheaf associated to a presheaf on Y,

$$V \mapsto H^q(G, H^0(q^{-1}(V), \mathcal{O}_{\widehat{X}})).$$

Since $\widehat{q} : \widehat{X} \to \widehat{Y}$ as the quotient morphism of a free G-action on $\widehat{X}$ is locally free (hence flat) in the sense of Zariski topology on $\widehat{Y}$ by Lemma 5.6.2.1, it follows from [53, Lemma 1] that $\mathcal{H}^q(G, \mathcal{O}_{\widehat{X}}) = 0$ for $q > 0$. Hence the spectral sequence $E'^{p,q}_2$ degenerates, and the spectral sequence $E_2^{p,q}$ gives rise to a spectral sequence

$$E_2^{p,q} = H^p(G, H^q(\widehat{X}, \mathcal{O}_{\widehat{X}})) \Longrightarrow H^{p+q}(\widehat{Y}, \mathcal{O}_{\widehat{Y}}).$$

We may (and shall) replace X by $\operatorname{Spec} R$ and Y by $\operatorname{Spec} R^G$. The exact sequence $(*)$ of lower degrees of this spectral sequence is written as

$$0 \to H^1(G, R) \to H^1(\widehat{Y}, \mathcal{O}_{\widehat{Y}}) \to H^1(\widehat{X}, \mathcal{O}_{\widehat{X}})^G \to H^2(G, R) \to H^2(\widehat{Y}, \mathcal{O}_{\widehat{Y}}).$$

Since $H^1(G, R) \neq 0$, it follows that $H^1(\widehat{Y}, \mathcal{O}_{\widehat{Y}}) \neq 0$.

The final step is to use an isomorphism

$$H^i(\widehat{Y}, \mathcal{O}_{\widehat{Y}}) \cong H_\mathfrak{n}^{i+1}(R^G), \quad i > 0$$

to obtain $H_\mathfrak{n}^2(R^G) \neq 0$. This implies that $\operatorname{depth} R^G \leq 2$. Meanwhile, since R^G is a normal local ring of dimension $n \geq 2$, $\operatorname{depth} R^G \geq 2$ by Serre criterion of normality. Thus $\operatorname{depth} R^G = 2$. If $n = 2$, then R^G is Cohen-Macaulay. □

Remark. For a finite cyclic group $G = \langle \sigma \rangle$ and a G-module R as above, the following results are known about group cohomologies $H^i(G, R)$ (see Serre [84, p.141] and Peskin [72]).

$$H^0(G, R) = R^G, \quad H^1(G, R) = \{a \in r \mid \operatorname{Tr}(a) = 0\}/\{\sigma(a) - a \mid a \in R\},$$
$$H^2(G, R) = R^G/\operatorname{Tr}(R), \quad H^i(G, R) \cong H^{i-2}(G, R) \ (i > 2).$$

5.6.3 Peskin's criterion for rational singularity

Here we follow a result of B. Peskin [72] on the rationality of singularity of R^G. The field k is assumed to be algebraically closed, and $G = \langle \sigma \rangle \cong \mathbb{Z}/p\mathbb{Z}$.

5.6.3.1

Lemma. *Let $(R, \mathfrak{m})$ be a normal noetherian local ring with a cyclic group G-action such that G acts freely on $\operatorname{Spec} R \setminus \{\mathfrak{m}\}$. Then the cohomology groups $H^i(G, R)$ are isomorphic to k-vector spaces of dimension > 0.*

Proof. Note that $H^i(G, R)$ is, by definition, a finitely generated R-module. By periodicity of order 2 of $H^i(G, R)$ (cf. Remark in 5.6.2.2, the last isomorphism), it suffices to show that $H^i(G, R)$ $(i = 1, 2)$ is isomorphic to a k-vector space of positive dimension. In the long exact sequence of lower degree terms $(*)$ attached to the spectral sequence $E_2^{p,q}$ in the proof of Lemma 5.6.2.2, cohomology groups $H^i(\widehat{Y}, \mathcal{O}_{\widehat{Y}})$ $(i = 1, 2)$ and $H^1(\widehat{X}, \mathcal{O}_{\widehat{X}})$ have supports at the closed points $\mathfrak{n}$ and $\mathfrak{m}$. Hence $H^i(G, R)$ $(i = 1, 2)$ has support at $\mathfrak{m}$, i.e., each element is killed by some power of $\mathfrak{m}$. Hence $H^i(G, R)$ is isomorphic to a k-vector space of finite dimension.

Let $S = R^G$ and $\mathfrak{n} = \mathfrak{m} \cap S$. Since $S \hookrightarrow R$ is totally ramified at $\mathfrak{m}$, $\sigma \equiv \operatorname{id} \pmod{\mathfrak{m}}$. Hence the images of $\sigma - \operatorname{id}$ and Tr are contained in $\mathfrak{m}$. Meanwhile, $\operatorname{Tr}|_k$ is zero and $k \subset R^G$, whence $H^i(G, R) \neq 0$ for $i = 1, 2$. $\quad\square$

5.6.3.2

We assume that $\dim R = 2$ and hence $\dim S = 2$. Let $g : W \to \operatorname{Spec} S$ be a resolution of singularity of S. Such a resolution exists for any characteristic p since $\operatorname{Spec} S$ is geometric and $\dim S = 2$. The singularity of $\operatorname{Spec} S$ at the closed point $\{\mathfrak{n}\}$ is *rational* if $R^1 g_* \mathcal{O}_W = 0$ (see Part III. 1.1.4). Since $H^0(\operatorname{Spec} S, R^1 g_* \mathcal{O}_W) \cong H^1(W, \mathcal{O}_W)$ by the spectral sequence $H^p(\operatorname{Spec} S, R^q g_* \mathcal{O}_W) \Longrightarrow H^{p+q}(W, \mathcal{O}_W)$, the singularity is rational if and only if $H^1(W, \mathcal{O}_W) = 0$.

Let V be the normalization of W in the function field $Q(R)$, and let $f : V \to \operatorname{Spec} R$ be the induced morphism, which exists because R is integral over S. We have a commutative diagram

$$
\begin{array}{ccc}
V & \xrightarrow{\ f\ } & \operatorname{Spec} R \\
{\scriptstyle \nu}\downarrow & & \downarrow \\
W & \xrightarrow{\ g\ } & \operatorname{Spec} S\ ,
\end{array}
$$

where ν is the normalization morphism. Note that g acts on V and W is the geometric quotient of V by G, though V is not necessarily a resolution of $\operatorname{Spec} R$.

Lemma. *The following assertions hold.*

(1) $f : V \to \operatorname{Spec} R$ *is a proper birational morphism.*
(2) *If R has a rational singularity, then $H^i(V, \mathcal{O}_V) = 0$ for all $i > 0$.*

Proof. (1) Since g is birational, it is clear that f is birational. The morphism g is proper as a resolution of singularity, and the morphism ν is finite, hence proper. Then the composite $g \cdot \nu$ is proper, whence the morphism f is proper too, since $\operatorname{Spec} R \to \operatorname{Spec} S$ is separated.

(2) There exist a smooth surface Z and a birational proper morphism $h : Z \to V$ such that $f \cdot h : Z \to \operatorname{Spec} R$ is a resolution of singularity of $\operatorname{Spec} R$. Then $h_* \mathcal{O}_Z = \mathcal{O}_V$ because V is normal. Now $H^1(Z, \mathcal{O}_Z) = 0$ because $\operatorname{Spec} R$ has a rational singularity by hypothesis. The spectral sequence

$$E_2^{p,q} = H^p(V, R^q h_* \mathcal{O}_Z) \Longrightarrow H^{p+q}(Z, \mathcal{O}_Z)$$

yields an exact sequence as a part of a long exact sequence of lower degree terms

$$0 \longrightarrow H^1(V, h_* \mathcal{O}_Z) \longrightarrow H^1(Z, \mathcal{O}_Z) \longrightarrow H^0(V, R^1 f_* \mathcal{O}_Z) .$$

Hence $H^1(V, \mathcal{O}_V) \cong H^1(V, h_* \mathcal{O}_V) = 0$.

Consider a spectral sequence

$$E_2^{p,q} = H^p(\operatorname{Spec} R, R^q f_* \mathcal{O}_V) \Longrightarrow H^{p+q}(V, \mathcal{O}_V) .$$

Since $R^q f \mathcal{O}_V$ is a coherent sheaf on $\operatorname{Spec} R$, $E_2^{p,q} = 0$ for all $p > 0$. Hence

$$H^q(V, \mathcal{O}_V) \cong H^0(\operatorname{Spec} R, R^q f_* \mathcal{O}_V),$$

where $R^q f_* \mathcal{O}_V = 0$ for $q > 1$ because f has relative (fiber) dimension ≤ 1. Hence $H^i(V, \mathcal{O}_V) = 0$ for $i \geq 2$. $\qquad\square$

5.6.3.3

As in 5.6.2.2, there are two spectral sequences with the same abutment:

$$E_2^{p,q} = H^p(G, H^q(V, \mathcal{O}_V))$$
$$E_2'^{p,q} = H^p(W, \mathcal{H}^q(G, \mathcal{O}_V)) ,$$

where $\mathcal{H}^q(G, \mathcal{O}_V)$ is the sheaf on W associated to a presheaf

$$U \mapsto H^q(G, \mathcal{O}_V|_{\nu^{-1}(U)}).$$

Suppose that R has a rational singularity. Since $H^q(V, \mathcal{O}_V) = 0$ for $q > 0$ by Lemma 5.6.3.2, the spectral sequence $E_2^{p,q}$ degenerates, and the second spectral sequence becomes

$$E'^{p,q} = H^p(W, \mathcal{H}^q(G, \mathcal{O}_V)) \implies H^{p+q}(G, R).$$

The associated long exact sequence of lower degree terms yields an exact sequence

$$0 \to H^1(W, \mathcal{H}^0(G, \mathcal{O}_V)) \to H^1(G, R) \xrightarrow{\varphi} H^0(W, \mathcal{H}^1(G, \mathcal{O}_V)) \to$$
$$\to H^2(W, \mathcal{H}^0(G, \mathcal{O}_V)) \to \cdots$$

where $\mathcal{H}^0(G, \mathcal{O}_V) \cong \mathcal{O}_V^G \cong \mathcal{O}_W$. Note that the homomorphism φ is injective if and only if the first term $H^1(W, \mathcal{O}_W) = 0$. Hence we have the following result.

Lemma. *Suppose that R has a rational singularity. Then $S = R^G$ has rational singularity if and only if φ is injective.*

5.6.3.4

An application of Lemma 5.6.3.3 is the following criterion of rational singularity due to B. Peskin.

Theorem. *With the above notations, assume that G acts freely on $\operatorname{Spec} R \setminus \{\mathfrak{m}\}$, $\{\mathfrak{m}\}$ is a rational singularity of $\operatorname{Spec} R$, and that $H^1(G, R) = k$. Then $\operatorname{Spec} S$ with $S = R^G$ has a rational singularity at the closed point $\{\mathfrak{n}\}$.*

Proof. Let T be the ramification divisor of the finite morphism $\nu : V \to W$. Suppose $T = 0$. Then ν is an étale finite morphism and $f : V \to \operatorname{Spec} R$ is a resolution of singularity. The spectral sequence

$$E_2^{p,q} = H^p(W, R^q \nu_* \mathcal{O}_V) \implies H^{p+q}(V, \mathcal{O}_V)$$

yields an isomorphism $H^p(W, \nu_* \mathcal{O}_V) \cong H^p(V, \mathcal{O}_V)$ for all $p \geq 0$ since $R^q \nu_* \mathcal{O}_V = 0$ for $q > 0$. Hence $H^1(W, \mathcal{O}_W) \cong H^1(V, \mathcal{O}_V)$ because $\nu_* \mathcal{O}_V = \mathcal{O}_W$. Since $H^1(V, \mathcal{O}_V) = 0$ by hypothesis, it follows that $H^1(W, \mathcal{O}_W) = 0$, hence the point $\{\mathfrak{n}\}$ of $\operatorname{Spec} S$ has a rational singularity.

Suppose that $T \neq 0$. In fact, it has pure codimension 1 by purity of branch locus. Since $H^1(G, R) = \operatorname{Ker} \operatorname{Tr}/\operatorname{Im}(\sigma - \operatorname{id})$, $H^1(G, R)$ contains the ground field k, and coincides with it by the hypothesis. We show that $\mathcal{H}^1(G, \mathcal{O}_V)$ contains also the ground field k. Then the homomorphism φ is then injective. Since $\nu : V \setminus \nu^{-1}(\nu(T)) \to W \setminus \nu(T)$ is locally trivial in the Zariski topology of W by Lemma 5.6.2.1, $\mathcal{H}^1(G, \mathcal{O}_V) = 0$ on $W \setminus \nu(T)$. Hence $\mathrm{Supp} \mathcal{H}^1(G, \mathcal{O}_V) \subseteq \nu(T)$. Let $\mathcal{I}(T)$ be the ideal sheaf of $\mathcal{O}_W$ defining the closed set $\nu(T)$. Since ν is totally ramified, $\mathrm{Im}\,(\sigma - \mathrm{id})$ is contained in $\mathcal{I}(T)$. Hence k lies outside of $\mathrm{Im}\,(\sigma - \mathrm{id})$. Hence φ is injective on k. $\qquad\square$

5.6.4 *Partial linearization*

From now on we consider the case where R a formal power series ring $k[[u_1, u_2, \ldots, u_n]]$ with the maximal ideal $\mathfrak{m} = (u_1, \ldots, u_n)$. Let $G = \langle \sigma \rangle$ be a p-group isomorphic to $\mathbb{Z}/p\mathbb{Z}$ such that $\sigma(\mathfrak{m}) \subset \mathfrak{m}$ and G acts trivially on $R/\mathfrak{m} = k$. We further assume that G acts freely on $\mathrm{Spec}\,R \setminus \{\mathfrak{m}\}$.

5.6.4.1

A G-action is *linearizable* if there exists a system of local coordinates $\{v_1, \ldots, v_n\}$ such that $\mathfrak{m} = (v_1, \ldots, v_n)$ and $\sigma(v_i) \in \sum_{j=1}^{n} k v_j$. The following remark is due to B. Peskin [73, Prop. 2.1].

Lemma. *With the above notations, assume that the G-action on R is non-trivial and linearizable. Then the following assertions hold.*

(1) *The morphism $\mathrm{Spec}\,R \to \mathrm{Spec}\,R^G$ ramifies over the locus of dimension ≥ 1.*

(2) *If $n = 2$ the ring R^G is regular. If $n > p$, R^G is not necessarily regular.*

Proof. (1) Since the G-action is linearizable and $\sigma^p = 1$, we may assume that the local parameters are chosen in such a way that

$$(\sigma(u_1)\,\sigma(u_2)\,\cdots\,\sigma(u_n)) = (u_1\,u_2\,\cdots\,u_n)A\,,$$

where A is a Jordan matrix

$$A = \begin{pmatrix} 1 & \varepsilon_1 & 0 & \cdots & 0 \\ 0 & 1 & \varepsilon_2 & \cdots & 0 \\ 0 & 0 & 1 & & \vdots \\ \vdots & \vdots & & \ddots & \varepsilon_{n-1} \\ 0 & 0 & \cdots & 0 & 1 \end{pmatrix}\,, \qquad \varepsilon_i = 0,\, 1.$$

Then the fixed point locus of $\operatorname{Spec} R$ contains a dimension one locus $\{u_1 = \cdots = u_{n-1} = 0\}$.

(2) The action is given by $\sigma(u_1) = u_1$ and $\sigma(u_2) = u_2 + u_1$. Then $x = u_1$ and $y = \prod_{i=0}^{p-1}(u_2 - iu_1) = u_2^p - u_1^{p-1}u_2$ generate the invariant subring R^G as a formal power series ring, hence R^G is regular. Suppose that $p = 2$ and $n = 4$. Write $R = k[[u_1, v_1, u_2, v_2]]$ with a G-action $\sigma(u_i) = u_i$ and $\sigma(v_i) = v_i + u_i$ for $i = 1, 2$. Then $R^G = k[[x_1, y_1, x_2, y_2, z]]$, where $x_i = u_i, y_i = v_i(v_i + u_i)$ for $i = 1, 2$ and $z = u_1 v_2 + u_2 v_1$. Then there is a relation $z^2 + x_1 x_2 z + x_1^2 y_2 + x_2^2 y_1 = 0$. Hence R^G is a hypersurface defined by this equation. This hypersurface has the singular locus $\{x_1 = x_2 = z = 0\}$. So, R^G is not regular. $\qquad\qquad\square$

5.6.4.2

Suppose that the G-action is not linearizable. Then $\sigma(u_i)$ for $1 \le i \le n$ is written as follows.

$$\sigma(u_i) = \sum_{j=1}^{n} a_{ij} u_j + h_i(u_1, \ldots, u_n), \quad a_{ij} \in k, \quad \deg h_i(u_1, \ldots, u_n) \ge 2,$$

where the matrix $A = (a_{ij})$ is an invertible matrix such that $A^p = E_n$, where E_n is the identity matrix of size n. By a linear base change of $\{u_1, \ldots, u_n\}$, A is conjugate to a Jordan canonical form consisting of Jordan blocks

$$J = \begin{pmatrix} 1 & 1 & & 0 \\ & \ddots & \ddots & \\ & & \ddots & 1 \\ 0 & & & 1 \end{pmatrix}.$$

The following result is a partial linearization due to B. Peskin [73, Theorem 2.4].

Theorem. *The following assertions hold.*

(1) *The maximal size of a Jordan block J is p.*
(2) *There exists a choice of local coordinates for R so that σ is composed of Jordan blocks of the form*

$$\sigma(u_i) = u_i + f_i(u_1, \ldots, u_n),$$
$$\sigma(u_{i+1}) = u_{i+1} + u_i,$$
$$\vdots$$
$$\sigma(u_{i+s}) = u_{i+s} + u_{i+s-1}.$$

Proof. (1) Let r be the size of a Jordan block J. Since $\sigma^p = 1$, it follows that $J^p = E_r$, hence $(J - E_r)^p = 0$. Meanwhile, if $1 \leq j < r$, the matrix $(J - E_r)^j$ has matrix entries all 1 on the $(j+1)$-st diagonal[13] and 0 elsewhere. If $r > p$, $(J - E_r)^p \neq 0$ by the above remark, which is a contradiction.

(2) The original proof in [73] is convincing. We reproduce it here. Let $r = s + 1$. The σ-action corresponding to the Jordan block J is written as

$$\sigma(u_i) = u_i + h_i(u_1, \ldots, u_n),$$
$$\sigma(u_{i+1}) = u_{i+1} + u_i + h_{i+1}(u_1, \ldots, u_n),$$
$$\vdots$$
$$\sigma(u_{i+s}) = u_{i+s} + u_{i+s-1} + h_{i+s}(u_1, \ldots, u_n),$$

where $\deg h_j(u_1, \ldots, u_n) \geq 2$.

Set

$$\overline{u}_{i+s} = u_{i+s},$$
$$\overline{u}_{i+s-1} = (\sigma - 1)u_{i+s} = u_{i+s-1} + h_{i+s},$$
$$\overline{u}_{i+s-2} = (\sigma - 1)^2 u_{i+s} = u_{i+s-2} + h_{i+s-1} + (\sigma - 1)h_{i+s},$$
$$\vdots$$
$$\overline{u}_i = (\sigma - 1)^s u_{i+s} = u_i + h_{i+1} + (\sigma - 1)h_{i+2} + \cdots + (\sigma - 1)^{s-1}h_{i+s}.$$

Note that this is a change of local coordinates because the Jacobian of this change at $(0, \ldots, 0)$ is the identity matrix. Set

$$f_i = (\sigma - 1)\overline{u}_i = h_i + (\sigma - 1)h_{i+1} + \cdots + (\sigma - 1)^s h_{i+s}.$$

Then we have $\sigma(\overline{u}_{i+j}) = \overline{u}_{i+j} + \overline{u}_{i+j-1}$ for $0 < j \leq s$ and $\sigma(\overline{u}_i) = \overline{u}_i + f_i$. $\square$

5.6.4.3 *Artin's normal form in the case $n = p = 2$*

As a corollary of Theorem 5.6.4.2, we prove the following result due to M. Artin [4].

Theorem. *Let $R = k[[u, v]]$ with a G-action such that $\sigma(\mathfrak{m}) \subseteq \mathfrak{m}$ and G acts freely on $\mathrm{Spec}\, R \setminus \{\mathfrak{m}\}$. Then, after a change of coordinates u, v, there*

---

[13]The diagonal of a matrix is the first diagonal. The parallel line right next to the first diagonal is the second diagonal, so on.

exist non-units a, b of R^G[14] such that a and b are relatively prime in R and that

$$\sigma(u) = u + a, \quad \sigma(v) = v + b.$$

Furthermore, $R^G = k[[x, y, z]]$, where $x = \mathrm{N}(u), y = \mathrm{N}(v)$ and $z = u\sigma(v) + v\sigma(u)$ and x, y, z satisfy a relation

$$z^2 + abz + a^2y + b^2x = 0.$$

Proof. Suppose that $\sigma(u) = u + f(u, v)$ and $\sigma(v) = v + u$ with $\deg f \geq 2$. Since $\mathrm{Tr}(v) = v + v + u = u$, it follows that $u \in R^G$. Hence $f = 0$. This implies that the G-action is linearizable. Then $\mathrm{Spec}\, R^G$ has the fixed point locus $\{u = 0\}$ of dimension one (see Lemma 5.6.4.1). This contradicts the assumption. Hence we can write $\sigma(u) = u + a$ and $\sigma(v) = v + b$, where a, b belong to R^G because $\mathrm{Tr}(u) = a$ and $\mathrm{Tr}(v) = b$. Then $\gcd(a, b) = 1$ in R because otherwise the G-action has the fixed point locus along the dimension one subset. The rest of the assertion is clear because the hypersurface defined by the above equation is normal. $\qquad\square$

Proposition. *Let $G = \mathbb{Z}/2\mathbb{Z}$. In the above theorem, assume that $a, b \in k[[x, y]]$. Then, we have $H^1(G, R) \cong k[[x, y]]/(a, b)$.*

Proof. By Remark 5.6.2.2, $H^1(G, R) \cong (\mathrm{Ker}\,\mathrm{Tr})/(\mathrm{Im}\,(\sigma - \mathrm{id}\,))$. First, we show that $\mathrm{Ker}\,\mathrm{Tr} = R^G$. Since $a \in R^G$, the G-action extends naturally to $R[a^{-1}]$, and $\mathrm{Ker}\,(\mathrm{Tr}|_{R[a^{-1}]}) \cap R = \mathrm{Ker}\,\mathrm{Tr}$. Since $v = a^{-1}(bu + z)$, we can write $R[a^{-1}] = R^G[a^{-1}] + R^G[a^{-1}]u$ as an $R^G[a^{-1}]$-module. Then $\mathrm{Tr}(\alpha + \beta u) = \beta a$ for $\alpha, \beta \in R^G[a^{-1}]$. Hence $\mathrm{Ker}\,\mathrm{Tr}|_{R[a^{-1}]} = R^G[a^{-1}]$, and $\mathrm{Ker}\,\mathrm{Tr} = R^G$.

Since $p = 2$, $\sigma - \mathrm{id} = \mathrm{Tr}$. Note that Tr is an R^G-linear endomorphism of R. Since $u^2 + au + x = 0$ and $v^2 + bv + y = 0$ with $a, b \in k[[x, y]]$, it follows that $\{1, u, v, uv\}$ is a free basis of a $k[[x, y]]$-module $k[[u, v]]$. Then, for an element $\alpha_0 + \alpha_1 u + \alpha_2 v + \alpha_3 uv$ of $k[[u, v]]$ with $\alpha_i \in k[[x, y]]$, we have

$$\mathrm{Tr}(\alpha_0 + \alpha_1 u + \alpha_2 v + \alpha_3 uv) = \alpha_1 a + \alpha_2 b + \alpha_3(z + ab)$$

$$= (\alpha_1 + \alpha_3 b)a + \alpha_2 b + \alpha_3 z.$$

This implies that $\mathrm{Im}\,\mathrm{Tr} \cong k[[x, y, z]]/(a, b, z) \cong k[[x, y]]/(a, b)$. $\qquad\square$

---

[14]Artin asserts a stronger result that $a, b \in k[[x, y]]$ for elements x, y defined below. For this assertion, it is proved that $k[[x, y]] = R_1 \otimes_{k[[x,y]]} R_2$, where $R_1 = k[[x, y]][u]/(u^2 + au + x)$ and $R_2 = k[[x, y]][v]/(v^2 + bv + y)$ and $k((u, v))$ is a Galois extension of $k((x, y))$ with Galois group $\mathbb{Z}/2\mathbb{Z} \times \mathbb{Z}/2\mathbb{Z}$, where $k((u, v))$ and $k((x, y))$ are respectively the quotient fields of $k[[u, v]]$ and $k[[x, y]]$.

5.6.4.4

Giving a normal form of a $\mathbb{Z}/p\mathbb{Z}$-action on $R = k[[u_1, \ldots, u_n]]$ is an unsolved, rather complicated problem. We note the following easy result.

Lemma. *Let $R = k[[u_1, \ldots, u_p]]$, where $p \geq 3$. Assume that $G = \mathbb{Z}/p\mathbb{Z}$ acts on R. If the Jordan block of linear parts has size p, then the action is linearizable.*

Proof. Suppose that the coordinates are normalized so that $\sigma(u_i) = u_i + u_{i-1}$ for $2 \leq i \leq p$ and $\sigma(u_1) = u_1 + f$, with $\deg f \geq 2$. Then we can compute

$$\sigma^i(u_p) = u_p + \binom{i}{1} u_{p-1} + \cdots + \binom{i}{j} u_{p-j} + \cdots + \binom{i}{i} u_{p-i} \quad (0 \leq i < p)$$

$$\sigma^p(u_p) = u_p + \binom{p}{1} u_{p-1} + \cdots + \binom{p}{p-1} u_1 + f = u_p + f.$$

Since $\sigma^p = 1$, we must have $\sigma^p(u_p) = u_p$, whence $f = 0$. $\qquad\square$

Peskin [73, Corollary 5.15] proved the following result when $p = 3$.

Theorem. *Let $R = k[[u_1, u_2]]$, where $p = 3$, and let σ be any automorphism of R of order 3 whose linear terms consist of a single Jordan block. If the extension $R^G \hookrightarrow R$ is unramified off the maximal ideal $\mathfrak{m}$ of R, then*

$$R^G \cong k[[x, y, z]]/(z^3 + y^{2j} z^2 - y^{3j+1} - x^2),$$

for some $j > 0$. If the extension $R^G \hookrightarrow R$ is ramified over a locus of dimension one, then R^G is regular.

5.6.5 *Recent developments*

Let $(R, \mathfrak{m})$ be anew a geometric regular local ring defined over k. Assume that the group $G = \mathbb{Z}/p\mathbb{Z}$ acts on $X = \operatorname{Spec} R$ in such a way that the closed point $\{\mathfrak{m}\}$ is fixed and G acts freely on $\widehat{X} = X \setminus \{\mathfrak{m}\}$. Let $(S, \mathfrak{n}) = (R^G, \mathfrak{m} \cap R^G)$. Let $Y = \operatorname{Spec} S$ and $\widehat{Y} = Y \setminus \{\mathfrak{n}\}$, which is equal to the smooth locus of Y. The algebraic local fundamental group $\pi_1^{alg}(\widehat{Y})$ is defined in [SGA1]. We recall the definition briefly. Set $Z = \widehat{Y}$. We consider a projective system $\{Z_i \to Z \mid i \in I\}$ of finite étale irreducible coverings of Z such that $\operatorname{Aut}_Z(Z_i)$ is a finite group of degree equal to $\deg(Z_i/Z)$, i.e., $Z_i \to Z$ is a Galois covering. If there exists a Z-morphism $\rho_{ij} : Z_j \to Z_i$, it induces a group homomorphism $\rho_{ij_*} : \operatorname{Aut}_Z(Z_j) \to \operatorname{Aut}_Z(Z_i)$. With respect to these induced group homomorphisms, the collection $\{\mathrm{Aut}\,_Z(Z_i), \rho_{ij_*}\}$ forms a projective system of finite groups. The projective limit $\varprojlim_{i\in I} \mathrm{Aut}\,_Z(Z_i)$ is called the *algebraic local fundamental group* and denoted by $\pi_1^{alg}(\widehat{Y})$.

Let $g : \widetilde{Y} \to Y$ be a minimal resolution of singularity at $\mathfrak{n}$ such that the exceptional locus consists of smooth irreducible components meeting each other transversally at the intersection points. A *minimal* resolution means that the contraction of a possible (-1)-curve breaks down the above condition. The following result is one of main results of Ito-Schröer [37, Theorem 2.1].

Theorem. *With the above notations and assumptions, assume that the exceptional locus of $g : \widetilde{Y} \to Y$ is a linear chain of $\mathbb{P}^1 s$. Then the algebraic local fundamental group $\pi_1^{alg}(\widehat{Y})$ is trivial.*

Since the natural morphism $\widehat{X} \to \widehat{Y}$ is a finite étale covering of degree p (Galois covering) and the Galois group G must be the quotient of $\pi_1^{alg}(\widehat{Y})$, the above theorem has the following outstanding conclusion.

Corollary. *With the above notations and assumptions, the exceptional locus of $g : \widetilde{Y} \to Y$ is not a linear chain of $\mathbb{P}^1 s$, i.e., it contains an irreducible component meeting at least three other irreducible components.*

Chapter 6

Higher derivations

6.1 Basic properties

In this chapter we treat the actions of the additive group scheme and its infinitesimal subgroup schemes α_{p^r}. In the case of characteristic zero, G_a-actions on an affine scheme Spec A defined over a field k correspond bijectively to locally nilpotent derivations δ of A in such a way that the action is given as the exponential form $\exp(t\delta) = \sum_{i \geq 0}(1/i!)t^i\delta^i$, where if one puts a value $t = c$ then $\exp(c\delta)$ maps $a \in A$ to $\sum_{i \geq 0}(1/i!)c^i\delta^i(a)$. Local nilpotence of δ guarantees the sum $\sum_{i \geq 0}(1/i!)c^i\delta^i(a)$ to be a finite sum. Since a derivation is one of the simplest algebraic operations on a k-algebra, it has been used to clarify the structure of affine algebraic varieties (cf. [17]).

In the case of positive characteristic, G_a-actions are described in terms of *locally finite iterative higher derivation* which, as we will see below, is a natural translation of G_a-actions, but turn out to be a complex object to handle. We will describe basic properties and applications of higher derivation.

Let k be an algebraically closed field of positive characteristic. Most of properties hold without the condition that k is algebraically closed, but we assume this condition for the sake of coherency throughout the book. Let A be a k-algebra. A *higher derivation*[1] is a collection of k-linear endomorphisms

$$\delta = \{\delta_i \mid i = 0, 1, 2, \ldots\}$$

of A as a k-vector space satisfying the following conditions:

$$\delta_0 = \text{id}, \quad \delta_i(ab) = \sum_{j+\ell=i} \delta_j(a)\delta_\ell(b), \quad a, b \in A. \tag{6.1}$$

---

[1]The definition and partial results are stated in the subsection 1.5.2.

A higher derivation δ is *nonzero* or *non-trivial* if $\delta_i \neq 0$ for some $i > 0$, and is *locally finite* if, for all $a \in A$, there exists an integer N such that

$$\delta_n(a) = 0 \text{ for every } n \geq N. \tag{6.2}$$

A higher derivation δ is *iterative* if δ satisfies the condition

$$\delta_j \delta_{i-j} = \begin{cases} \binom{i}{j} \delta_i & \text{if } i \geq_p j \\ 0 & \text{if } i \not\geq_p j \end{cases} \tag{6.3}$$

For integers $i, j \geq 0$, let $i = i_0 + i_1 p + \cdots + i_r p^r$ with $0 \leq i_0, i_1, \ldots, i_r < p$ and $j = j_0 + j_1 p + \cdots + j_r p^r$ with $0 \leq j_0, j_1, \ldots, j_r < p$ be the p-adic expansions of i, j. We define $i \geq_p j$ if $i_s \geq j_s$ for all $0 \leq s \leq r$. If $i \geq_p j$, we define the binomial coefficient

$$\binom{i}{j} = \binom{i_0}{j_0} \cdots \binom{i_r}{j_r}. \tag{6.4}$$

6.1.1

Theorem. *Let δ be a locally finite, iterative, higher derivation on A. The following assertions hold.*

(1) *δ_1 is a k-derivation of A and $(\delta_{p^r})^p = 0$ for $r \geq 0$.*
(2) *For $i > 0$ with the p-adic expansion $i = i_0 + i_1 p + \cdots + i_r p^r$, we have*

$$\delta_i = \frac{(\delta_1)^{i_0} (\delta_p)^{i_1} \cdots (\delta_{p^r})^{i_r}}{(i_0)!(i_1)! \cdots (i_r)!}. \tag{6.5}$$

(3) *Local finiteness of δ is equivalent to the condition that, for all $a \in A$, $\delta_{p^r}(a) = 0$ for every $r \gg 0$.*

Proof. By (6.3), the commutativity $\delta_i \delta_j = \delta_j \delta_i$ for all $i, j \geq 0$ follows. Define a k-linear homomorphism $\varphi : A \to k[t] \otimes_k A$ by

$$\varphi(a) = \sum_{i=0}^{\infty} t^i \otimes \delta_i(a), \tag{6.6}$$

where t is a variable. Since δ is locally finite, the homomorphism φ is well-defined. Then the condition (6.1) is equivalent to the condition that $\varphi(a)|_{t=0} = a$ and $\varphi(ab) = \varphi(a)\varphi(b)$ for $a, b \in A$. On the other hand, the condition (6.3) is equivalent to the commutativity of the following diagram

$$
\begin{array}{ccc}
A & \xrightarrow{\ \varphi\ } & k[t] \otimes_k A \\
{\scriptstyle \varphi}\downarrow & & \downarrow{\scriptstyle \Delta \otimes \text{id}_A} \\
k[t] \otimes_k A & \xrightarrow[\text{id}_{k[t]} \otimes \varphi]{} & k[t] \otimes_k k[t] \otimes_k A
\end{array}
$$

where $\Delta : k[t] \to k[t] \otimes_k k[t]$ is the comultiplication of G_a defined by $\Delta(t) = t \otimes 1 + 1 \otimes t$. In fact, we have

$$(1 \otimes \varphi) \cdot \varphi = \sum_{j=0}^{\infty} \sum_{\ell=0}^{\infty} t^j \otimes t^\ell \otimes \delta_\ell \delta_j$$

$$(\Delta \otimes 1) \cdot \varphi = \sum_{i=0}^{\infty} (t \otimes 1 + 1 \otimes t)^i \otimes \delta_i$$

$$= \sum_{i=0}^{\infty} \sum_{j, i \geq pj} \binom{i}{j} t^j \otimes t^{i-j} \otimes \delta_i \ ,$$

where we note that $(t \otimes 1 + 1 \otimes t)^i$ is expanded as follows if $i = i_0 + i_1 p + \cdots + i_r p^r$ is the p-adic expansion of i

$$(t \otimes 1 + 1 \otimes t)^i = \prod_{s=0}^{r} \left(t^{p^s} \otimes 1 + 1 \otimes t^{p^s} \right)^{i_s}$$

$$= \sum_{j_0=0}^{i_0} \cdots \sum_{j_r=0}^{i_r} \binom{i_0}{j_0} \cdots \binom{i_r}{j_r} (t^j \otimes t^{i-j}) \ ,$$

for $j = j_0 + j_1 p + \cdots + j_r p^r$ and $i - j = (i_0 - j_0) + (i_1 - j_1)p + \cdots + (i_r - j_r)p^r$. Hence if the term $t^j \otimes t^\ell \otimes \delta_\ell \delta_j$ appears in $(1 \otimes \varphi) \cdot \varphi$ for $j + \ell \not\geq_p j$, then $\delta_\ell \delta_\ell = 0$ because no terms correspond to it in $(\Delta \otimes 1) \cdot \varphi$.

We prove the formula (6.5) by induction on i. We assume that the formula holds for j with $i >_p j$. Since $i >_p (i - j)$, we have

$$\delta_j = \frac{(\delta_1)^{j_0} (\delta_p)^{j_1} \cdots (\delta_r)^{j_r}}{(j_0)!(j_1)! \cdots (j_r)!}$$

$$\delta_{i-j} = \frac{(\delta_1)^{i_0-j_0} (\delta_p)^{i_1-j_1} \cdots (\delta_{p^r})^{i_r-j_r}}{(i_0 - j_0)!(i_1 - j_1)! \cdots (i_r - j_r)!} \ .$$

By the formula (6.3), we have the required formula for i. The repeated use of (6.3) shows that

$$\delta_{i_r p^r} = \frac{1}{(i_r)!}(\delta_{p^r})^{i_r} \quad (0 \leq i_r < p).$$

Since $(p-1)p^r + p^r = p^{r+1} \not\geq_p p^r$, (6.3) implies that $(\delta_{p^r})^p = 0$. $\quad\square$

6.1.2

Let $A^{(p^r)} = k[a^{p^r} \mid a \in A]$ be the k-subalgebra of A generated by the p^rth powers a^{p^r} for all elements $a \in A$. Define a k-algebra homomorphism $\varphi_{p^s} : A \to k[t] \otimes_k A$ for $s \geq 0$ by

$$\varphi_{p^s} = 1 \otimes \mathrm{id} + t^{p^s} \otimes \delta_{p^s} + \frac{1}{2!} t^{2p^s} \otimes (\delta_{p^s})^2 + \cdots + \frac{1}{(p-1)!} t^{(p-1)p^s} \otimes (\delta_{p^s})^{p-1} \ .$$

Identifying $k[t] \otimes_k A$ with a polynomial ring $A[t]$, we view φ_{p^s} (and φ as well) as a k-algebra endomorphism of $A[t]$ by setting $\varphi_{p^s}(t) = t$ (or $\varphi(t) = t$). Then φ_{p^s} and φ_{p^r} are commutative, and $\varphi = \prod_{s=0}^{\infty} \varphi_{p^s}$, which is well-defined as operators by the local-finiteness condition.

Lemma. *The following assertions hold.*

(1) $\delta_{p^r}(a^{p^s}) = \begin{cases} 0 & (s > r) \\ \delta_1(a)^{p^r} & (s = r) \\ \delta_{p^{r-s}}(a)^{p^s} & (s < r) . \end{cases}$

(2) δ_{p^r} *is a k-derivation on* $A^{(p^r)}$.

Proof. We proceed by induction on r. Suppose $r = 1$. Since φ is a k-algebra homomorphism, we have $\varphi(a^p) = \varphi(a)^p$. Since δ_1 is a k-derivation of A, we have $\delta_1(a^p) = p\delta_1(a^{p-1}) = 0$, hence $\varphi_1(a^p) = a^p$. This implies that the t-expansion of $\varphi(a^p)$ has nonzero coefficients only for the t-terms of degree divisible by p. Comparison of the t^pth terms in $\varphi(a^p)$ and $\varphi(a)^p$ shows that $\delta_p(a^p) = \delta_1(a)^p$. In particular, δ_p is a k-derivation of $A^{(p)}$. Suppose that the formulas hold for $1, 2, \ldots, r-1$. Then $\varphi_{p^i}(a^{p^r}) = a^{p^r}$ for $0 \leq i < r$. We have

$$\varphi(a^{p^r}) = \left(\prod_{s=r}^{\infty} \varphi_{p^s} \right)(a^{p^r})$$

$$= \sum_{i=0}^{\infty} \sum_{i_r=0}^{p-1} \cdots \sum_{i_u=0}^{p-1} \frac{(\delta_{p^r})^{i_r} \cdots (\delta_{p^u})^{i_u}}{(i_r)! \cdots (i_u)!}(a^{p^r})t^i$$

$$\varphi(a)^{p^r} = \prod_{j=0}^{\infty} \sum_{j_0=0}^{p-1} \cdots \sum_{j_v=0}^{p-1} \left(\frac{(\delta_1)^{j_0}(\delta_p)^{j_1} \cdots (\delta_{p^v})^{j_v}}{(j_0)!(j_1)! \cdots (j_v)!}(a) \right)^{p^r} t^{p^r j} ,$$

where $i = i_r p^r + \cdots + i_u p^u$ and $j = j_0 + j_1 p + \cdots + j_v p^v$ are the p-adic expansions of i and j, and where we may assume that $i \equiv 0 \pmod{p^r}$. Comparison of the coefficients of t^{p^r}th terms in $\varphi(a^{p^r})$ and $\varphi(a)^{p^r}$ shows that $\delta_{p^r}(a^{p^r}) = \delta_1(a)^{p^r}$. Hence δ_{p^r} is a k-derivation of $A^{(p^r)}$. Other two of the stated formulas are trivial or contained in the induction hypothesis. $\square$

6.1.3

Let α_{p^n} be the infinitesimal subgroup scheme $\mathrm{Ker}\,(F_{G_a}^n)$ of the additive group scheme G_a. An action of α_{p^n} on $\mathrm{Spec}\,A$ is given by a *truncated* iterative higher derivation $\delta = \{\delta_i \mid 0 \leq i < p^n\}$ (cf. Lemma 1.5.3). If $i = i_0 + i_1 p + \cdots + i_n p^n$ is the p-adic expansion of i with value zero allowed for higher p-adic coefficients, then the same arguments as in Theorem 6.1.1 shows that

(1) $\delta_i = \dfrac{(\delta_1)^{i_0}(\delta_p)^{i_1} \cdots (\delta_{p^{n-1}})^{i_{n-1}}}{(i_0)!(i_1)! \cdots (i_{n-1})!}$ for $0 \leq i < p^n$ and $\delta_i = 0$ for $i \geq p^n$.

(2) $(\delta_{p^s})^p = 0$ for $0 \leq s < n$.

We say that the above truncated higher derivation δ has *p-exponent n* if $\delta_{p^{n-1}} \neq 0$. Thus a nontrivial α_p-action is given by an iterative higher derivation of p-exponent 1. We note that Lemma 6.1.2 holds for an iterative higher derivation δ of exponent n if $r < n$.

6.2 Invariant subrings

Let δ be a locally finite iterative higher derivation (*lfihd*, for short) on an affine k-domain A. Then the ring of δ-invariants is a subring A_0 of A consisting of elements a such that $\delta_i(a) = 0$ for every $i > 0$. By Theorem 6.1.1, $A_0 = \{a \in A \mid \delta_{p^s}(a) = 0, \ s = 0, 1, \ldots\}$. If δ is an iterative higher derivation (*ihd*, for short) of finite exponent, we can also define the ring of invariants A_0. We denote A_0 also by A^δ or $\operatorname{Ker} \delta$.

6.2.1

Let A be an affine k-domain and let $K = Q(A)$ be the quotient field. Let δ be a lfihd on A. We extend δ to the higher derivation of K by setting

$$\Phi(\xi) := \frac{\varphi(b)}{\varphi(a)}, \quad \xi = \frac{b}{a}, \quad a, b \in A.$$

Since $\varphi(a), \varphi(b) \in A[t]$, the fraction $\Phi(\xi)$ is an element of the power series ring $K[[t]]$. It is independent of the expression $\xi = b/a$. Hence we can write

$$\Phi(\xi) = \xi + \sum_{i=1}^{\infty} \bar{\delta}_i(\xi) t^i, \quad \bar{\delta}_i(\xi) \in K,$$

and $\bar{\delta} = \{\bar{\delta}_i\}_{i \geq 0}$ is a higher derivation of K such that $\bar{\delta}_i|_A = \delta_i$ for every $i \geq 0$. Set $K_0 = \{\xi \in K \mid \Phi(\xi) = \xi\}$, and call it the *invariant subfield* of $\bar{\delta}$. Then it is clear that $K_0 \cap A = A_0$. The following results are found in [54, Chapter I].

Lemma. *With the above notations and assumptions, the following assertions hold.*

(1) A_0 *is factorially closed in* A. *Namely, if an element* $a \in A_0$ *is factored in* A *as* $a = bc$, *then* $b, c \in A_0$.

(2) *Let* A^* *be the (multiplicative) group of units of* A. *Then* $A^* \subset A_0$, *whence* $A^* = A_0^*$.

(3) A_0 *is integrally closed in* A.

(4) $K_0 \cap A = A_0$, *and* $K_0 = Q(A_0)$.

(5) K_0 *is algebraically closed in* K.

(6) *If* A *is normal (resp. factorial[2]) and* A_0 *is noetherian, then so is* A_0.

Proof. (1) The lfihd δ corresponds to a G_a-action $\varphi(a) = \sum_{i=0}^{\infty} \delta_i(t)$ on A, and $a \in A_0$ if and only if $\varphi(a) = a$. Let $\deg \varphi(a)$ be the t-degree of a polynomial $\varphi(a)$ in $A[t]$. Since $\varphi(a) = \varphi(b)\varphi(c)$, we have $\deg \varphi(b) + \deg \varphi(c) = \deg \varphi(a) = 0$, whence $\deg \varphi(b) = \deg \varphi(c) = 0$ because A is an integral domain. Hence $b, c \in A_0$.

(2) Let $u \in A^*$. Then $uv = 1$ for some $v \in A^*$. Then (1) implies that $u, v \in A_0$ because $1 \in A_0$. So, $A^* \subseteq A_0^*$. The inclusion $A_0^* \subseteq A^*$ is clear.

(3) Suppose that $a \in A$ satisfies an equation

$$a^n + c_1 a^{n-1} + \cdots + c_{n-1} a + c_n = 0, \quad c_1, \ldots, c_n \in A_0,$$

which yields an equation in $A[t]$

$$\varphi(a)^n + c_1 \varphi(a)^{n-1} + \cdots + c_{n-1} \varphi(a) + c_n = 0.$$

Suppose that $d := \deg \varphi(a) > 0$. Then the left side of the above equation has the t-degree $nd > 0$, which is a contradiction. Hence $\deg \varphi(a) = 0$, i.e., $a \in A_0$.

(4) Let $\xi = b/a$ be an element of K_0 with $a, b \in A$. Since it implies that $a\varphi(b) = b\varphi(a)$, we have $ab_n = ba_n$, where $b_n = \delta_n(b)$ and $a_n = \delta_n(a)$ are the coefficients of the top terms of t-polynomials of $\varphi(b)$ and $\varphi(a)$ respectively. Here note that $\deg \varphi(b) = \deg \varphi(a)$. By the iterativity condition (6.3), both b_n and a_n are elements of A_0. Hence $\xi = b_n/a_n \in Q(A_0) = K_0$.

(5) Suppose that $\xi \in K$ satisfies an algebraic equation over K_0

$$\xi^n + z_1 \xi^{n-1} + \cdots + z_{n-1} \xi + z_n = 0, \quad z_1, \ldots, z_n \in K_0.$$

Writing $\xi = b/a$ with $a, b \in A$ and applying Φ to the above equation, we have

$$\left(\frac{\varphi(b)}{\varphi(a)} \right)^n + z_1 \left(\frac{\varphi(b)}{\varphi(a)} \right)^{n-1} + \cdots + z_{n-1} \left(\frac{\varphi(b)}{\varphi(a)} \right) + z_n = 0.$$

---

[2]Namely, A is a unique factorization domain (UFD).

For a general element $\lambda \in k$, $\frac{\varphi(b)}{\varphi(a)}\big|_{t=\lambda}$ is an element of $K = Q(A)$. Since k is an infinite field and since an algebraic equation has finitely many roots, there exist an infinite subset S of k and elements c, d of A such that $c\varphi(b)|_{t=\lambda} = d\varphi(a)|_{t=\lambda}$ for every $\lambda \in S$. This implies that $c\varphi(b) = d\varphi(a)$. As in the proof of (4), take the coefficients a_n, b_n of top-degree terms of $\varphi(a), \varphi(b)$ respectively. Then we have $cb_n = da_n$. Hence we have

$$\xi = \frac{b}{a} = \left(\frac{\varphi(b)}{\varphi(a)}\right)\bigg|_{t=0} = \frac{d}{c} = \frac{b_n}{a_n} \in K_0.$$

(6) The assertion follows from (1) and (3). $\qquad\square$

6.2.2

Let δ be a lfihd on A. Let $a \in A$. The δ-*length* of a is $\deg_t \varphi(a)$. We denote it by $\text{length}_\delta(a)$ or $\text{length}(a)$ if there are no fear of confusion. For $a \in A \setminus (0)$, $a \in A_0$ if and only if $\text{length}_\delta(a) = 0$. We define $\text{length}_\delta(0) = -\infty$. We also define the δ-*exponent* (or simply *exponent*) of a by $\max\{r \mid \delta_{p^r}(a) \neq 0\}$.

6.2.2.1

Lemma. *The following assertions hold.*

(1) *For an element $a \in A \setminus (0)$, we have $\text{length}\,\delta_i(a) < \text{length}(a)$ for every $i > 0$.*

(2) *Let x be an element of $A \setminus A_0$ of minimal δ-length. Then $\delta_i(x) \in A_0$ for every $i > 0$, which is equivalent to the condition that $\delta_{p^s}(x) \in A_0$ for every $s \geq 0$. Hence we have*

$$\varphi(x) = x + \delta_1(x)t + \delta_p(x)t^p + \cdots + \delta_{p^r}(x)t^{p^r},$$

where r is the δ-exponent of x.

(3) *Let $a \in A \setminus A_0$, let $n = \text{length}(a)$ and let $\delta_n(a)$ be the coefficient of the term t^n in $\varphi(a)$. Then $\delta_n(a) \in A_0$ and n is divisible by p^r.*

Proof. (1) Let $n = \text{length}(a)$. Then $\delta_n(a) \neq 0$ and $\delta_j(a) = 0$ for every $j > n$. Then $\delta_n\delta_i(a) = 0$ by the iterativity condition (6.3) because $n+i > n$. Hence $\text{length}\,\delta_i(a) < \text{length}(a) = n$.

(2) By the choice of x as an element of $A \setminus A_0$ with minimal δ-length, it is immediate that $\delta_i(x) \in A_0$ because $\text{length}\,\delta_i(x) < \text{length}(x)$, whence $\text{length}\,\delta_i(x) = 0$. The condition that $\delta_i(x) \in A_0$ for every $i > 0$ is equivalent

to the condition that $\delta_{p^s}(x) \in A_0$ for every $s \geq 0$. Note that $\varphi_1(x) = x + \delta_1(x)t$. As an induction hypothesis, assume that

$$\varphi_1 \varphi_p \cdots \varphi_{p^i}(x) = x + \delta_1(x)t + \cdots + \delta_{p^i}(x)t^{p^i}.$$

Then we have

$$\varphi_1 \cdots \varphi_{p^i} \varphi_{p^{i+1}}(x) = \varphi_{p^{i+1}}(x + \delta_1(x)t + \cdots + \delta_{p^i}(x)t^{p^i})$$

$$= \varphi_{p^{i+1}}(x) + \delta_1(x)t + \cdots + \delta_{p^i}(x)t^{p^i}$$

$$= x + \delta_1(x)t + \cdots + \delta_{p^i}(x)t^{p^i} + \delta_{p^{i+1}}(x)t^{p^{i+1}}.$$

Thus we obtain the formula for $\varphi(x)$ by induction on the δ-exponent of x.

(3) By the condition (6.3), $\delta_n(a) \in A_0$. Let $n = n_0 + n_1 p + \cdots + n_s p^s$ be the p-adic expansion of n. Then we have $s \geq r$. In fact, if $s < r$ then $n < p^r$, which contradicts the choice of the element x. Furthermore, we have

$$\delta_n(a) = \frac{(\delta_1)^{n_0}(\delta_p)^{n_1} \cdots (\delta_{p^s})^{n_s}}{(n_0)!(n_1)! \cdots (n_s)!}(a).$$

Let u be an integer such that $n_0 = \cdots = n_{u-1} = 0$ and $n_u \neq 0$. If $u \geq r$ then n is divisible by p^r. Suppose that $u < r$. Let

$$b = \frac{(\delta_{p^u})^{n_u-1}(\delta_{p^{u+1}})^{n_{u+1}} \cdots (\delta_{p^s})^{n_s}}{(n_u - 1)!(n_{u+1})! \cdots (n_r)!}(a) \quad \text{if } n_u > 1$$

and

$$b = \frac{(\delta_{p^{u+1}})^{n_{u+1}} \cdots (\delta_{p^s})^{p^s}}{(n_{u+1})! \cdots (n_s)!}(a) \quad \text{if } n_u = 1.$$

Then $b \in A \setminus A_0$ and $\text{length}(b) = p^u < p^r$ because $\delta_i(b) = \delta_i \delta_{n'}(a) = 0$ by (6.3) if $i > p^u$, where $n' = (n_u - 1)p^u + n_{u+1}p^{u+1} + \cdots + n_r p^r$. This contradicts the choice of the element x. Hence $u \geq r$. $\qquad \square$

6.2.2.2

Theorem. *Let the notations and assumptions be the same as in Lemma 6.2.2.1. Let $a_0 = \delta_{p^r}(x)$. Then $A[a_0^{-1}] = A_0[a_0^{-1}][x]$, where $A[a_0^{-1}]$ and $A_0[a_0^{-1}]$ are the localizations of A and A_0 with respect to the multiplicative set $\{a_0^s \mid s = 0, 1, 2, \ldots\}$.*

Proof. The lfihd δ is extended to a lfihd of $A[a_0^{-1}]$ by setting $\delta_i(a/a_0^s) = \delta_i(a)/a_0^s$ for every $i \geq 0$. Then $A_0[a_0^{-1}]$ is the ring of invariants of the extended lfihd. Then we have

$$\varphi\left(\frac{x}{a_0}\right) = \frac{x}{a_0} + \frac{\delta_1(x)}{a_0}t + \frac{\delta_p(x)}{a_0}t^p + \cdots + t^{p^r}$$

because $\delta_{p^r}(x)/a_0 = 1$. By replacing A by $A[a_0^{-1}]$, we may assume that $\delta_{p^r}(x) = 1$. We show that $A = A_0[x]$. Namely, for every $a \in A$, we prove by induction on $n := \text{length}(a)$ that $a \in A_0[x]$. If $n = p^r$ then $\text{length}(a - \delta_{p^r}(a)x) < p^r$, whence $a - \delta_{p^r}(a)x \in A_0$. Suppose that $n > p^r$. By Lemma 6.2.2.1, (3), n is divisible by p^r. So, write $n = mp^r$. Then $\text{length}(a - \delta_{p^n}(a)x^m) < n$. By induction, $a - \delta_{p^n}(a)x^m \in A_0[x]$, whence $a \in A_0[x]$. □

A different proof of the theorem is available in [54, Lemma 1.5]. In the case of positive characteristic, the element x as in Lemma 6.2.2.1 is called a *local slice* and Theorem 6.2.2.2 is called the *local slice theorem* (see [17] for the case of characteristic zero). A *slice* is a local slice x with the coefficient $\delta_{p^r}(x)$ of the highest t-term equal to 1. If A has a slice x, Theorem 6.2.2.2 is improved to the effect that $A = A_0[x]$. Even if A is an affine k-domain, the ring of invariants A_0 under an effective G_a-action (or equivalently, a nonzero lfihd) is not necessarily an affine-domain if $\dim A \geq 4$ as shown by many counterexamples to the fourteenth problem of Hilbert [17, Chap. 7]. If A has a slice, then A_0 is affine domain.

6.2.2.3

As a corollary of Theorem 6.2.2.2, we prove the following result (see [52]).

Theorem. *Let $X = \text{Spec } A$ be an affine surface such that the following conditions are satisfied.*

(1) *The coordinate ring A is a factorial domain.*
(2) *$A^* = k^*$.*
(3) *X has a non-trivial G_a-action.*

Then X is isomorphic to the affine plane $\mathbb{A}^2 = \text{Spec } k[x,y]$ such that y is invariant under the G_a-action and x satisfies

$$\varphi(x) = x + \delta_1(x)t + \delta_p(x)t^p + \cdots + \delta_{p^r}(x)t^{p^r}, \quad \delta_{p^i}(x) \in k[y] \ (0 \leq i \leq r),$$

where $\delta = \{\delta_i\}_{i \geq 0}$ is the lfihed on A associated with the G_a-action.

Proof. Let δ be the lfihd on A associated with the G_a-action and let $A_0 = \text{Ker } \delta$. By a theorem of Zariski (see [67, Theorem 4, p.51] or [17, Zariski's finiteness theorem, p.180]), A_0 is finitely generated over k, i.e., the coordinate ring of an affine curve. By Lemma 6.2.1, A_0 is factorial, and $A_0^* \subseteq A^* = k*$ by the condition (2). Hence $A_0^* = k^*$. Let $C = \text{Spec } A_0$ and let $q : X \to C$ be the morphism induced by the inclusion $A_0 \hookrightarrow A$. Then C

is a smooth, factorial curve. For distinct two points $P, Q \in C$, the divisor $P - Q \sim 0$ because $\mathrm{Pic}(C) = (0)$ as A_0 is factorial. Hence C is a rational curve with $A_0^* = k^*$. This implies that A_0 is a polynomial ring $k[y]$.

Take an element $x \in A$ with the minimal length p^r. Let $Z = \mathrm{Spec}\, k[x, y]$. Then q is factored as

$$q : X \xrightarrow{\rho} Z \xrightarrow{\pi} C .$$

By Theorem 6.2.2.2, ρ is a birational morphism. If $y - \beta$ divides $x - \alpha$ for $\alpha, \beta \in k$, we replace x by $(x - \alpha)/(y - \beta)$. Note that $(x - \alpha)/(y - \beta)$ has the same length as x. So, after these replacements, we may assume that $y - \beta$ does not divide $x - \alpha$ for all $\alpha, \beta \in k$. Let P be a point of C. Then the point P is defined by $y - \beta$ for some $\beta \in k$. Since $y - \beta$ is a prime element of A_0, it is also a prime element of A by Lemma 6.2.1. This implies that $A/(y - \beta)$ is an affine domain of dimension one. We have a canonical inclusion $k[x] \hookrightarrow A/(y - \beta)$. In fact, if the canonical homomorphism $k[x, y]/(y - \beta) \to A/(y - \beta)$ is not injective, there exists a polynomial $f(x) \in k[x]$ such that $f(x)$ is divisible by $y - \beta$ in A. Hence $x - \alpha$ is divisible by $y - \beta$ for some $\alpha \in k$. But this does not occur by the assumption. Let $F_P := q^{-1}(P)$ be the fiber of q over P. Then the injection $k[x] \hookrightarrow A/(y - \beta)$ implies that the induced morphism $\rho|_{F_P} : F_P \to \mathbb{A}^1 = \mathrm{Spec}\, k[x]$ is a quasi-finite morphism, and $F_P = \mathrm{Spec}\, A/(y - \beta) \neq \emptyset$ by the condition (2). Hence no curves on X are contracted to points by the morphism $\rho : X \to \mathbb{A}^2$. By Zariski Main Theorem, this implies that ρ is an open immersion and $\mathbb{A}^2 \setminus \rho(X)$ has codimension ≥ 2. Since $\rho(X)$ is an affine open set of $\mathbb{A}^2$, we can conclude that ρ is an isomorphism. This entails the statement of Theorem. $\qquad\square$

6.2.3

Let δ be a truncated ihd of p-exponent n of an affine k-domain A. Namely, $\delta = \{\delta_i \mid 0 \leq i < p^n\}$ is associated to an α_{p^n}-action on $\mathrm{Spec}\, A$. The associated coaction

$$\varphi : A \to k[t] \otimes_k A, \quad t^{p^n} = 0$$

is extended to the quotient field $K = Q(A)$ as

$$\Phi : K \to k[t] \otimes_k K, \quad \Phi\left(\frac{b}{a}\right) = \frac{\varphi(b)}{\varphi(a)}, \quad a, b \in A$$

because $\varphi(a)^{-1} \in K[t]$. Let A_0 (resp. K_0) be the ring (resp. field) of invariants in A (resp. K), which are defined in the same way as for a lfihd. Then we have the following result.

Lemma. *With the same notations and assumptions as above, the following assertions hold.*

(1) *The k-subalgebra $A^{(p^n)}$ is k-isomorphic to $A \otimes_k (k, \varphi^n)$, where $\varphi^n :$ $\alpha \mapsto \alpha^{p^n}$ by the abuse of notations (cf. I.2.2). If A is regular (Cohen-Macaulay, normal, resp,) then so is $A^{(p^n)}$. If A is regular, A is faithfully flat over $A^{(p^n)}$.*

(2) $A^{(p^n)}$ *is a subalgebra of A_0 and A_0 is an affine domain over k. Furthermore, A is a finite A_0-module, $A_0 = K_0 \cap A$ and $K_0 = Q(A_0)$.*

(3) *If A is normal, so is A_0.*

Proof. (1) The first assertion is clear. We prove the rest of assertions for $n = 1$. Let $\mathfrak{m}$ be a maximal ideal of A and let $\mathfrak{m}^{(p)} = \mathfrak{m} \cap A^{(p)}$. Then $\mathfrak{m}^{(p)} = \mathfrak{m} \otimes_k (k, \varphi)$ if $A^{(p)}$ is identified with $A \otimes_k (k, \varphi)$. Further, the local ring $A^{(p)}_{\mathfrak{m}^{(p)}}$ is identified with $A_{\mathfrak{m}} \otimes_k (k, \varphi)$. If $(a_1, \ldots, a_s)$ is a regular sequence of $A_{\mathfrak{m}}$, then $(a \otimes 1, \ldots, a_s \otimes 1)$ is a regular sequence for $A^{(p)}_{\mathfrak{m}^{(p)}}$. So, $A^{(p)}_{\mathfrak{m}^{(p)}}$ satisfies the condition (R_s) if so does $A_{\mathfrak{m}}$. If $(a_1, \ldots, a_d)$ is a regular system of parameters of $A_{\mathfrak{m}}$, then $(a \otimes 1, \ldots, a_d \otimes 1)$ is the one for $A_{\mathfrak{m}} \otimes_k (k, \varphi)$. The assertions then follow from these observations. See [EGA, Chap. IV].

(2) Lemma 6.1.2 holds as it is for $r, s < n$. Then it is clear that $A^{(p^n)} \subseteq A_0$. Since A is finitely generated over k and purely inseparable over $A^{(p^n)}$, A is integral over A_0. Hence A is a finite A_0-module. By definition of A_0 and K_0, it is clear that $A_0 = K_0 \cap A$. Note that, for all $a \in A$, the coefficient of the highest degree t-term in $\varphi(a)$ is in A_0. By making use of this fact, we can prove $K_0 = Q(A_0)$ by the same argument as in Lemma 6.2.1.

(3) Let ξ be an element of K_0 which is integral over A_0. Since A is normal by assumption, $\xi \in A$. Hence $\xi \in K_0 \cap A = A_0$. So, A_0 is normal.

$\square$

Let $X = \operatorname{Spec} A$ with an α_{p^n}-action. The affine variety $Y = \operatorname{Spec} A_0$ is called the *algebraic quotient* and is denoted by X/α_{p^n}. If X is a projective variety with an α_{p^n}-action, there exists an affine open covering $\mathcal{U} = \{U_\lambda\}_{\lambda \in \Lambda}$ such that $U_\lambda = \operatorname{Spec} A_\lambda$ is stable under the α_{p^n}-action. In fact, since the α_{p_n}-action does not move X set-theoretically, we can easily obtain an α_{p^n}-stable affine open set as the complement of a very ample divisor. Let $\delta^{(\lambda)}$ be the truncated ihd of A_λ obtained from the α_{p^n}-coaction $\varphi_\lambda : A_\lambda \to k[t] \otimes_k A_\lambda$, where $t^{p^n} = 0$. Then $\delta^{(\lambda)}|_{A_{\lambda\mu}} = \delta^{(\mu)}|_{A_{\lambda\mu}}$, where

$U_\lambda \cap U_\mu = \operatorname{Spec} A_{\lambda\mu}$. Let $A_0^{(\lambda)} = \operatorname{Ker} \delta^{(\lambda)}$. Then the $V_\lambda := \operatorname{Spec} A_0^{(\lambda)}$ patch together to give a projective variety $Y = \cup_{\lambda \in \Lambda} V_\lambda$ and a finite, purely inseparable morphism $q : X \to Y$. We call Y simply the *quotient* and the morphism q the *quotient morphism*. We denote Y also by X/α_{p^n}.

The following two propositions show that α_p (or μ_p, resp.) quotients of the affine plane can have rather complex singular points.

6.2.3.1

Proposition. *Let $p = 2m + 1$ be a prime number > 2 and let $\operatorname{Spec} B$ be an affine normal hypersurface in $\mathbb{A}^3 = \operatorname{Spec} k[x, y, t]$ defined by an equation $y^2 = x^p + f(t)$ with $f(t) \in k[t]$. Let $t = \tau^p$ and $f(t) = g(\tau)^p$. Set*

$$u = \frac{y}{(x + g(\tau))^m}, \quad x = u^2 - g(\tau), \quad y = u^p .$$

Set a derivation δ on $k[u, \tau]$ by

$$\delta = 2u\frac{\partial}{\partial \tau} + g'(\tau)\frac{\partial}{\partial u}.$$

Then the following assertions hold.

(1) *We have the inclusion relations*

$$k[y, t] \subset B = k[x, y, t]/(y^2 = x^p + f(t)) \subset k[u, \tau], \quad B = \operatorname{Ker} \delta .$$

(2) *If $p = 3$, δ defines an α_p-action on $A := k[u, \tau]$ if and only if $g''(\tau) = 0$, and δ defines a μ_p-action if and only if $g''(\tau) = -1$.*

(3) *Suppose that $p = 3$ and $\delta^3 = 0$. Then $Y := \operatorname{Spec} B$ is the normal quotient surface X/α_3 with $X = \operatorname{Spec} k[u, \tau]$. If $\delta^3 = 0$, Y is smooth if and only if $Y \cong \mathbb{A}^2$. If $\delta^3 = 0$ and Y is singular, then every singular point of Y is a Gorenstein singular point of type $y^2 + x^3 + t^{3\ell+1} = 0$ for $\ell > 0$ [3].*

---

[3]The resolution graph is rather complicated to obtain. For example, if $\ell = 1$, Lemma 2.8.1 applied to a triple covering $x^3 = -(y^2 + t^4)$ by the argument in the proof of Proposition 6.2.3.2 below gives the following graph which is different from the graph in the case of characteristic zero:

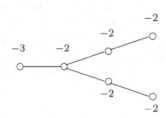

(4) *Suppose that $p = 3$ and $\delta^3 = \delta$. Then $Y = \operatorname{Spec} B$, which is the quotient surface X/μ_3, is isomorphic to a hypersurface $y^2 = x^3 + t^2$ whose singular point is a rational double point $(x, y, t) = (0, 0, 0)$ of type A_2.*

Proof. (1) It is clear that we have the stated inclusions. To show that $B = \operatorname{Ker} \delta$, it suffices to notice that $Q(B)$ is a purely inseparable extension of degree p and B is normal by the assumption, and to use Lemma 1.2.2.

(2) We can write $f(t) = f_0(t^3) + f_1(t^3)t + f_2(t^3)t^2$ with $f_i(t^3) \in k[t^3]$ for $i = 0, 1, 2$. Accordingly, $g(\tau) = g_0(\tau^3) + g_1(\tau^3)\tau + g_2(\tau^3)\tau^2$ with $g_i(\tau^3) \in k[\tau^3]$ for $i = 0, 1, 2$. By a straightforward computation, we have

$$\delta^3(u) = -g'(\tau)g''(\tau)$$
$$\delta^3(\tau) = -ug''(\tau) .$$

Hence we have $\delta^3 = -g''(\tau)\delta$. Then the assertion follows from Lemma 1.5.3.

(3) Note that $g''(\tau) = 0$ if and only if $g_2(\tau^3) = 0$. Since $f_2(t^3) = g_2(\tau^3)^3$, the assumption implies $f_2(t^3) = 0$. By the Jacobian criterion, Y is smooth if and only if the equation $f_1(t^3) = 0$ has no roots. Hence $f_1(t^3) \in k^*$. We may assume that $f_1(t^3) = 1$. By replacing x by $x + g_0(\tau^3)$, the equation of Y becomes $t = y^2 - x^3$. Thus B is a polynomial ring $k[x, y]$. Hence $Y \cong \mathbb{A}^2$. The converse is clear. Suppose that Y is singular. Since the equation is $y^2 = x^3 + f_0(t^3) + f_1(t^3)t$, we may assume that $f_0(t^3) = 0$ after replacing x by $x + g_0(\tau^3)$. A singular point of Y is then given by $(x, y, t) = (0, 0, c)$ for $c \in k$ such that $f_1(c^3) = 0$. Locally near this point, by replacing $t - c$ by t, the equation is changed to

$$y^2 = x^3 + f_0(t^3 + c^3) + f_1(t^3 + c^3)(t + c)$$
$$= x^3 + \big(f_0(t^3 + c^3) + cf_1(t^3 + c^3)\big) + f_1(t^3 + c^3)t ,$$

where $f_1(t^3 + c^3) = t^{3\ell}h(t^3)$ with $h(0) \neq 0$, which we can rewrite

$$y^2 = x^3 + t^{3\ell+1}h(t^3).$$

Since $3\ell + 1 \not\equiv 0 \pmod 3$, we can find a series $m(t^3) \in k[[t^3]]$ such that $h(t^3) = m(t^3)^{3\ell+1}$. Then, by replacing $tm(t^3)$ by t, the equation is, locally at the singular point $(0, 0, c)$, reduced to $y^2 = x^3 + t^{3\ell+1}$.

(4) The condition $g''(\tau) = -1$ implies that $g_2(\tau^3) = 1$, hence $f_2(t^3) = g_2(\tau^3)^3 = 1$. Then the equation is changed as follows:

$$y^2 = x^3 + f_0(t^3) + f_1(t^3)t + t^2$$
$$= x^3 + f_0(t^3) + (t^2 + f_1(t^3)t + f_1(t^3)^2) - f_1(t^3)^2$$
$$= \{x^3 + f_0(t^3) - f_1(t^3)^2\} + (t - f_1(t^3))^2 .$$

Hence the equation is changed to $y^2 = x^3 + t^2$. The rest is proved by the standard argument.   □

Remark. In the case $p > 3$ the computation becomes very complex. But it is a good exercise to find an argument to avoid a straightforward computation and state the result corresponding to the case $p = 3$.

6.2.3.2

Proposition. *Let* $\delta = x^{p^r} \frac{\partial}{\partial x} - y^{p^r} \frac{\partial}{\partial y}$ *with* $r > 0$ *be a derivation on a polynomial ring* $k[x, y]$. *Then the following assertions hold.*

(1) $\delta^p = 0$. *Hence* δ *defines an* α_p-*action on* $\mathbb{A}^2 = \operatorname{Spec} k[x, y]$.

(2) $B := \operatorname{Ker} \delta = k[x^p, y^p, x^{p^r} y + y^{p^r} x] \cong k[X, Y, Z]/(Z^p = X^{p^r} Y + Y^{p^r} X)$.

(3) *Let* $X = \operatorname{Spec} B$. *Then* X *has a unique singular point* $(X, Y, Z) = (0, 0, 0)$, *whose minimal resolution has the following dual graph with* $(p^r + 1)$ *branches* B_i $(1 \le i \le p^r + 1)$ *of* (-2)-*curves of length* $p - 1$.

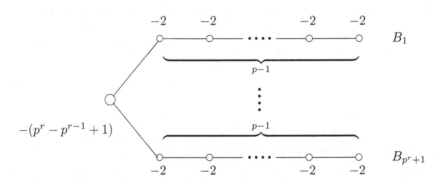

(4) *The hypersurface* $Z^p = X^{p^r} Y + Y^{p^r} X$ *is a rational surface.*

Proof. (1) Since $\delta^2(x) = \delta^2(y) = 0$, it follows that $\delta^p(x) = \delta^p(y) = 0$. Hence $\delta^p = 0$.

(2) Clearly, $x^{p^r} y + y^{p^r} x \in \operatorname{Ker} \delta$. Since $k(x^p, y^p, x^{p^r} y + y^{p^r} x)$ is a purely inseparable extension of degree p of $k(x^p, y^p)$, it follows that $Q(B) = k(x^p, y^p, x^{p^r} y + y^{p^r} x)$ by Lemma 1.2.1.

On the other hand, we have a surjection

$$\theta : \ k[X,Y,Z]/(Z^p - (X^{p^r}Y + Y^{p^r}X)) \longrightarrow k[x^p, y^p, x^{p^r}y + y^{p^r}x]$$

given by $X \mapsto x^p$, $Y \mapsto y^p$ and $Z \mapsto x^{p^r}y + y^{p^r}x$. It is readily verified that the hypersurface $Z^p = X^{p^r}Y + Y^{p^r}X$ is a normal irreducible hypersurface since the unique singular point $(X, Y, Z) = (0,0,0)$ is isolated. Hence $k[X,Y,Z]/(Z^p - (X^{p^r}Y + Y^{p^r}X))$ is an integral domain. If θ is not injective, $\text{Ker}\,\theta$ is a prime ideal of height > 0, which is a contradiction. Hence θ induces an isomorphism between the hypersurface $Z^p = X^{p^r}Y + Y^{p^r}X$ and $X = \text{Spec}\,B$.

(3) Consider the weighted embedded resolution X' of the hypersurface X at the point $(X, Y, Z) = (0,0,0)$. Namely, let $Z = X^{p^{r-1}}v$ and $Y = Xu$. Then we have

$$X^{p^r}v^p = X^{p^r+1}(u + u^{p^r}),$$

whence $v^p = X(u + u^{p^r})$. By blowing up the origin $(X, u) = (0, 0)$, we obtain a surface X'' with following picture, which we compare with the one in 2.8:

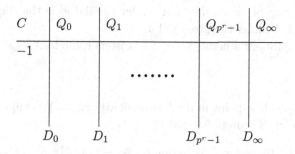

Here C is the exceptional curve, and the curves D_i $(0 \le i \le p^r - 1)$ correspond to the roots of $u + u^{p^r} = 0$. Further, we have to consider the curve D_∞ corresponding to $u = \infty$. The ramification data are $(a, b_0, b_1, \ldots, b_\infty) = (1, 1, \ldots, 1)$, whence $d(a, b_i) = p - 1$ and

$$\widetilde{C}^2 = \frac{1}{p}(-1 - (p-1)(p^r + 1)) = -(p^r - p^{r-1} + 1) \ .$$

We thus obtain the resolution graph of the minimal resolution of singularity by Lemma 2.8.1.

(4) With the notations in the proof of (3), the hypersurface in question is birational to a hypersurface $v^p = X(u + u^{p^r})$. Since $X = v^p/(u + u^{p^r})$, the second hypersurface is rational. $\qquad\square$

6.2.4

The subgroup scheme α_{p^n} is identified with $\mathrm{Ker}\,(F_{G_a})^n$, where F_{G_a} is the Frobenius morphism of G_a. If $1 \leq r < n$, then α_{p^r}, identified with $\mathrm{Ker}\,(F_{G_a})^r$, is naturally considered as a subgroup scheme of α_{p^n}. Further, the quotient $\alpha_{p^n}/\alpha_{p^r}$ is isomorphic to $\alpha_{p^{n-r}}$. In fact, if we write $\alpha_{p^n} = \mathrm{Spec}\,k[t]/(t^{p^n})$, the subgroup scheme α_{p^r} is the spectrum of the truncation of $k[t]/(t^{p^n})$ by (t^{p^r}), that is to say, $(k[t]/(t^{p^n}))/((t^{p^r})/(t^{p^n})) \cong k[t]/(t^{p^r})$, and the quotient $\alpha_{p^n}/\alpha_{p^r}$ is the spectrum of the ring of invariants of $k[t]/(t^{p^n})$ by α_{p^r}. Hence the ring of invariants is isomorphic to $k[t^{p^r}]/(t^{p^n}) \cong k[t]/(t^{p^{n-r}})$.

Let $\delta = \{\delta\}_{0 \leq i < p^n}$ be the (truncated) ihd associated with an α_{p^n}-action on $X = \mathrm{Spec}\,A$. Then the restricted α_{p^r}-action on X is associated with the ihd $\delta^{(r)} := \{\delta_i\}_{0 \leq i < p^r}$ which is the *truncation* of δ at the *level* r. Let A' be the ring of invariants of A under the α_{p^r}-action (equivalently, by $\delta^{(r)}$) and let $X' = \mathrm{Spec}\,A'$. Then $\delta' = \{\delta_i\}_{p^r \leq i < p^n}$ acts on A' as a (truncated) ihd of length $n - r$. Namely, the quotient group scheme $\alpha_{p^n}/\alpha_{p^r}$ acts on X' via the (truncated) ihd δ'. The following assertions hold.

(1) The ring of invariants of A' under the ihd δ' is the ring of invariants A_0 of A under the given ihd δ.
(2) The quotient scheme $X'/\alpha_{p^{n-r}}$ is isomorphic to X/α_{p^n}.

6.2.4.1

There is a subtle point in the foregoing argument. We will explain it by an example (cf. Theorem 6.2.2.3).

Remark. Consider a G_a-action on $\mathbb{A}^2 = \mathrm{Spec}\,k[x, y]$ defined by

$$\varphi(y) = y$$
$$\varphi(x) = x + \delta_1(x)t + \delta_p(x)t^p + \cdots + \delta_{p^n}(x)t^{p^n}, \quad \delta_{p^i}(x) \in k[y]\ (0 \leq i \leq n).$$

Set $a_{p^i} = \delta_{p^i}(x)$ for the simplicity of the notations. Let $\alpha_{p^i} = \mathrm{Ker}\,F_{G_a}^i$. Then α_p-action on $A = k[x, y]$ is given by a k-derivation δ_1 such that $\delta(y) = 0$ and $\delta_1(x) = a_0$. Hence the ring of invariants A_1 is $k[x^p, y]$. Then $\alpha_p = \alpha_{p^2}/\alpha_p$ acts on $\mathrm{Spec}\,k[x^p, y]$ by the associated k-derivation δ_p which acts on A_1 by $\delta_p(y) = 0$ and $\delta_p(x^p) = a_p$. Here is a subtle point. By Lemma 6.1.2, $\delta_p(x^p) = \delta_1(x)^p = a_0^p$ in $A = k[x, y]$. But $\delta_p(x^p) = a_p$ in A_1. This strange phenomenon occurs because x is not an element of A_1. Hence the α_p-action on $\mathrm{Spec}\,k[x^p, y]$ is not the same as the α_p-action induced by the restriction of the G_a-action onto $\mathrm{Spec}\,k[x^p, y]$. The ring of invariants of A_1

under the $\alpha_p = \alpha_{p^2}/\alpha_p$-action is $A_p = k[x^{p^2}, y]$. The $\alpha_p = \alpha_{p^3}/\alpha_{p^2}$-action on $\operatorname{Spec} A_p$ is given by a k-derivation δ_{p^2} on A_p such that $\delta_{p^2}(y) = 0$ and $\delta_{p^2}(x^{p^2}) = a_{p^2}$. The same situation occurs afterwards.

6.3   Miscellaneous problems

In the case of characteristic zero, there are abundant results on locally nilpotent derivations acting on affine domains (see [17]). It seems, at least to the authors, that the validity of most of them have not been checked for the ground field of characteristic $p > 0$. We will consider some of them.

6.3.1   *When is $f\delta$ an lfihd if so is δ ?*

Lemma. *Let A be an affine domain over k and let $\delta = \{\delta_i\}_{i \geq 0}$ be an lfihd. Let f be a nonzero element of A such that $\delta_i(f) \in fA$ for every $i > 0$. Then $f \in A_0$, where A_0 is the ring of δ-invariants of A.*

Proof. Suppose that $f \notin A_0$. Then there exists $i > 0$ such that $\delta_i(f) \neq 0$. Write $\delta_i(f) = fa$ with $a \in A$. By Lemma 6.2.2.1, we have

$$\operatorname{length}(\delta_i(f)) < \operatorname{length}(f),$$

while we have

$$\operatorname{length}(f) \leq \operatorname{length}(f) + \operatorname{length}(a) = \operatorname{length}(fa) = \operatorname{length}(\delta_i(f)).$$

This is a contradiction. Hence $f \in A_0$. $\qquad\square$

Corollary. *Let A and δ be the same as in Lemma 6.3.1. Let f be an element of A such that $f\delta = \{f\delta_i\}_{i \geq 0}$ is an lfihd. Then $f \in A_0$.*

Proof. Note that $(f\delta_i)(f) \in fA$ for every $i > 0$. Now apply the above lemma for the lfihd $f\delta$ and the element f. $\qquad\square$

6.3.2 $\mathbb{A}^1$-*fibrations and G_a-actions*

Let $X = \operatorname{Spec} A$ be an affine variety. An $\mathbb{A}^1$-fibration on X is a dominant morphism $q : X \to Y$ such that $q^{-1}(U) \cong U \times \mathbb{A}^1$ for a non-empty open set U of Y. Suppose that X has an effective G_a-action. By Theorem 6.2.2.2, the morphism $q : X \to Y$ with $Y = \operatorname{Spec} A_0$ is an $\mathbb{A}^1$-fibration provided A_0 is an affine domain, hence the coordinate ring of an affine variety Y. The converse holds if Y is an affine variety.

Theorem. *Let $q : X \to Y$ be an $\mathbb{A}^1$-fibration between affine varieties X and Y. Then there exists a G_a-action on X such that q is generically (over Y) the quotient morphism. If the coordinate ring of Y is factorially closed in the coordinate ring of X then q is the quotient morphism.*

Proof. Let A and B be the coordinate rings of X and Y, respectively. Then there exist elements $b \in B$ and $u \in A$ such that $A[b^{-1}] = B[b^{-1}][u]$. Let $b_0, \ldots, b_r$ be elements of B such that $(b_0, \ldots, b_r) \neq (0, \ldots, 0)$. Define a B-algebra homomorphism $\varphi : B[u] \to k[t] \otimes_k B[u]$ by
$$\varphi(u) = 1 \otimes u + t \otimes b_0 + \cdots + t^{p^r} \otimes b_r .$$
Then it is straightforward to verify the commutativity of the following diagram

$$
\begin{array}{ccc}
B[u] & \xrightarrow{\ \varphi\ } & k[t] \otimes_k B[u] \\
{\scriptstyle\varphi}\downarrow & & \downarrow{\scriptstyle\Delta \otimes \mathrm{id}_{B[u]}} \\
k[t] \otimes_k B[u] & \xrightarrow{\ \mathrm{id}_{k[t]} \otimes \varphi\ } & k[t] \otimes_k k[t] \otimes_k B[u] ,
\end{array}
$$

where $\Delta(t) = t \otimes 1 + 1 \otimes t$. Hence a lfihd $\delta = \{\delta_i\}_{i \geq 0}$ on $B[u]$ is defined for $f(u) \in B[u]$ by
$$\varphi(f(u)) = f(u) + \delta_1(f(u))t + \delta_2(f(u))t^2 + \cdots + \delta_i(f(u))t^i + \cdots$$
$$= f(\varphi(u)) = f(u + b_0 t + \cdots + b_r t^{p^r}) .$$
Now write $A = k[a_1, \ldots, a_n]$ as a finitely generated k-domain. Since $a_i \in B[b^{-1}][u]$, there exist $f_i(u) \in B[u]$ such that $a_i = b^{-m} f_i(u)$ with a non-negative integer m for every $1 \leq i \leq n$. Define a new lfihd $\widetilde{\delta} = \{\widetilde{\delta}_i\}_{i \geq 0}$ on $B[u]$ by $\widetilde{\delta}_i = b^{im}\delta_i$. Namely, we consider a B-algebra homomorphism $\widetilde{\varphi} : B[u] \to B[u][t]$ by replacing t in a polynomial $\varphi(f(u))$ by $b^m t$. Then $\widetilde{\varphi}$ extends naturally to $B[b^{-1}][u]$ as a $B[b^{-1}]$-algebra homomorphism. Its restriction onto A satisfies
$$\widetilde{\varphi}(a_i) = \widetilde{\varphi}(b^{-m} f_i(u)) = b^{-m} f_i(u + b_0 b^m t + \cdots + b^{mp^r} b_r t^{p^r}) \in A[t] .$$
Hence $\widetilde{\delta}$ restricted onto A is an lfihd. Since the ring of invariants of $B[b^{-1}][u]$ is $B[b^{-1}]$, if B is factorially closed in A, it follows that the ring of invariants of A is $A \cap B[b^{-1}] = B$. Thus the quotient morphism $X \to X/G_a$ coincides with the given q. $\qquad\square$

6.3.3 $\ G_a$-actions on $\mathbb{A}^3$

In characteristic zero, the quotient of the affine 3-space $\mathbb{A}^3$ under an effective G_a-action is isomorphic to the affine plane $\mathbb{A}^2$. What will be the corresponding result in the case of positive characteristic?

6.3.3.1

Proposition. *Let $A = k[x, y, z]$ be a polynomial ring in three variables and let δ be a non-trivial lfihd on A. Suppose that δ has a slice, say u. Then A_0 is a polynomial ring in two variables.*

Proof. By Theorem 6.2.2.2 and a remark following it, we have $A = A_0[u] \cong k[x, y, z]$. Then, by the cancellation theorem for $\mathbb{A}^2$ (see [57, Chapter 3, Theorem 2.3.1], which holds also in the case of positive characteristic), A_0 is isomorphic to a polynomial ring in two variables. $\qquad\square$

6.3.3.2

Let $\delta = \{\delta_i\}_{i \geq 0}$ and $\Delta = \{\Delta_i\}_{i \geq 0}$ be two nonzero lfihds on an affine k-domain A. We say that δ and Δ are *commutative* if $\delta_i \Delta_j = \Delta_j \delta_i$ for all $i, j \geq 0$.

Lemma. *Let δ and Δ be the same as above. Suppose $\operatorname{Ker} \delta \subseteq \operatorname{Ker} \Delta$. Then $\operatorname{Ker} \delta = \operatorname{Ker} \Delta$.*

Proof. Suppose that $\operatorname{Ker} \delta \subsetneqq \operatorname{Ker} \Delta$. Let $A_0 = \operatorname{Ker} \delta$. By Theorem 6.2.2.2, there exist $a_0 \in A_0$ and $x \in A$ such that $A_0[a_0^{-1}] = A_0[a_0^{-1}][x]$. Let $a \in \operatorname{Ker} \Delta \setminus A_0$. Write $a = f(x) \in A_0[a_0^{-1}][x]$. Since Δ is nonzero, $\operatorname{length}_\Delta(x) > 0$. Namely, $\varphi_\Delta(x) := \sum_{i \geq 0} \Delta_i(x) t^i$ has positive t-degree. Meanwhile, since $a \in \operatorname{Ker} \Delta$, $\varphi_\Delta(a) = a$ and $\varphi_\Delta(f(x)) = f(\varphi_\Delta(x))$ has positive t-degree. This is a contradiction. $\qquad\square$

We define δ and Δ to be *independent* if $\operatorname{Ker} \delta \neq \operatorname{Ker} \Delta$.

Theorem. *Let δ and Δ be two independent commutative lfihds on $A := k[x, y, z]$. Then $\operatorname{Ker} \delta$ and $\operatorname{Ker} \Delta$ are isomorphic to a polynomial ring in two variables.*

Proof. Let $A_0 = \operatorname{Ker} \delta$. We show that A_0 is isomorphic to a polynomial ring in two variables. For $\operatorname{Ker} \Delta$, we have only to exchange the roles of δ and Δ in the subsequent arguments. Since δ and Δ are commutative, it is easy to show that Δ induces a lfihd of A_0, which we denote by $\Delta|_{A_0}$. Further, since δ and Δ are independent, Δ restricted on A_0 is nonzero. On the other hand, A_0 is a factorial affine domain of dimension 2 by Zariski's finiteness theorem [17, p.180] and Lemma 6.2.1. Since $A_0^* \subseteq A^* = k^*$, it follows by Theorem 6.2.2.3 that A_0 is isomorphic to a polynomial ring in two variables. $\qquad\square$

6.3.3.3

The following result is more restrictive, but shows clearly what has to be shown in a possible proof corresponding to the one in the case of characteristic zero.

Theorem. *Let* $A := k[x, y, z]$ *and let* δ *be a nonzero lfihd of* A. *Let* $q : X \to Y := X/G_a$ *be the quotient morphism, where* $Y = \operatorname{Spec} A_0$. *Suppose that the following two conditions are satisfied.*

(1) Y *is smooth.*
(2) *There exists a (linear) hyperplane* H *of* $X \cong \mathbb{A}^3$ *such that* $q|_H : H \to Y$ *is generically separable. Namely,* H *meets a general fiber* F *of the morphism* q *transversally in as many points as* $\deg q|_H$.

Then Y *is isomorphic to* $\mathbb{A}^2$.

Proof. By the condition (2), the log Kodaira dimension $\overline{\kappa}(Y)$ is not greater than the one $\overline{\kappa}(H)$ for H, which is isomorphic to $\mathbb{A}^2$. See [57] for the definition of log Kodaira dimension, and [40] or [57, Chapter 1, 1.17] for a relevant result in the case of positive characteristic. Since $\overline{\kappa}(H) = -\infty$, it follows that $\overline{\kappa}(Y) = -\infty$. By [57, Chapter 3, Theorem 1.3.2] which is valid also in the case of positive characteristic, Y has an $\mathbb{A}^1$-fibration. By Theorem 6.3.2, Y has a non-trivial G_a-action. Since A_0 is factorial and $A_0^* = k^*$, Theorem 6.2.2.3 shows that $Y \cong \mathbb{A}^2$. □

6.3.4　*A problem of Abhyankar-Moh*

One of renowned theorems in affine algebraic geometry in characteristic zero is a theorem due to Abhyankae-Moh-Suzuki (see [54, Chapter II]), which is stated as follows.

Assume that the field k *has characteristic zero. Let* $C_0 = \{f = 0\}$ *be a curve in the affine plane* $\mathbb{A}^2 = \operatorname{Spec} k[x, y]$ *such that* C_0 *is isomorphic to the affine line. Then the curve* $C_\lambda = \{f = \lambda\}$ *is also isomorphic to* $\mathbb{A}^1$, *and* f *is a generator of* $k[x, y]$. *Namely,* $k[x, y] = k[f, g]$ *for some* $g \in k[x, y]$.

In positive characteristic, we can ask the following problem.

Problem. Let $f(x, y) \in k[x, y]$ be a polynomial such that the curve $f = 0$ is isomorphic to $\mathbb{A}^1$. Is the curve $f = \lambda$ isomorphic to $\mathbb{A}^1$ for every $\lambda \in k$?

It is not possible to ask if f is a generator of $k[x, y]$ because $k[x, y] \otimes_{k[f]}$ $k(f)$ is in general a nontrivial $k(f)$-form of $\mathbb{A}^1$. There is an illuminative example due to M. Nagata, which we will consider in 6.3.4.2, and there are several references on this problem [20, 58, 61].

6.3.4.1

Let $f(x, y)$ be a polynomial as in the above problem. Let $\Lambda = \{C_\lambda \mid \lambda \in k\}$ be the linear pencil on $\mathbb{A}^2 = \operatorname{Spec} k[x, y]$ defined by f, where C_λ is the curve defined by $f = \lambda$. Embed $\mathbb{A}^2$ into $\mathbb{P}^2$ in the standard way as $\mathbb{A}^2 = \mathbb{P}^2 \setminus \ell_\infty$, where ℓ_∞ is the line at infinity. Let $\overline{\Lambda}$ be the linear pencil on $\mathbb{P}^2$ generated by the closures $\overline{C}_\lambda$ of C_λ for general $\lambda \in k$. Then an extra member $\overline{C}_\infty = d\ell_\infty$ is adjoined, where $d = \deg f(x, y)$. Let $\sigma : V \to \mathbb{P}^2$ be a minimal sequence of blowing-ups which eliminates the base points of $\overline{\Lambda}$, and let $\rho : V \to \mathbb{P}^1$ be the morphism induced by the proper transform of $\overline{\Lambda}$. Further, let S be the last (-1)-curve in the blowing-up process σ. In each step of the process σ, an exceptional curve, say E, is produced. If E belongs to a some member of the proper transform of $\overline{\Lambda}$, it is in the member corresponding to $d\ell_\infty$. In fact, if E is produced by the blowing-up $\sigma_Q : V'' \to V'$ of a point Q on a surface V' which appears midway between V and $\mathbb{P}^2$, the proper transform of $\overline{\Lambda}$ on V' is generated by a general member, say G, and the member corresponding to $d\ell_\infty$. Then E is not contained in the member $\sigma'_Q(G)$ because it is a general member of the proper transform of $\overline{\Lambda}$ on V'' and hence must be irreducible. Hence it is contained in the member corresponding to $d\ell_\infty$. Further, the center of the first blowing-up of σ lies on the line ℓ_∞, and the consecutive centers of the blowing-ups are uniquely determined as the base points of the pencils produced by $\overline{\Lambda}$. Hence all the centers of blowing-ups afterwards lie on the inverse images of ℓ_∞, and the affine plane $\mathbb{A}^2$ embedded into $\mathbb{P}^2$ remains as an open set in the surface V.

Lemma. *With the above notations and assumptions, the following assertions hold.*

(1) *The generic fiber $\overline{C}_\eta := V \times_{\mathbb{P}^1} \operatorname{Spec} k(\mathbb{P}^1)$ as well as general fibers of the morphism ρ has one place at the intersection point with S. Hence $\rho|_S : S \to \mathbb{P}^1$ is a purely inseparable morphism of degree $q := p^N$, say. The intersection point of $\overline{C}_\eta$ with S might be a (geometrically) singular point.*

(2) *Let $T \to S$ be a purely inseparable covering such that T is a complete smooth rational curve. Let W be the normalization of $(V \times_{\mathbb{P}^1} S) \times_S T$*

and let $\tau : W \to T$ be the composite of the normalization morphism $W \to V \times_{\mathbb{P}^1} T$ and the base change $\rho_T : V \times_{\mathbb{P}^1} T \to T$. We choose T to satisfy the following conditions.

(a) *The generic fiber $\overline{F}_\zeta := W \times_T \operatorname{Spec} k(T)$ of τ is a complete smooth curve. Namely, a possible singular point of $\overline{C}_\eta$ is resolved by the field extension $k(\mathbb{P}^1) \hookrightarrow k(T)$.*

(b) *The closure T' in W of the point of $\overline{F}_\zeta$ which lies over the point $\overline{C}_\eta \backslash C_\eta$ is a cross-section of the morphism τ, where C_η is the generic member of the pencil Λ on $\mathbb{A}^2$.*

Let $\theta : W \to V$ be the normalization morphism followed by the first projection. Then θ is a purely inseparable morphism of degree $p^M = [k(T) : k(\mathbb{P}^1)]$, where $k(\mathbb{P}^1)$ is identified with the rational field $k(f)$ generated by $f(x, y)$.

(3) *For every $\lambda \in k$, the curve $\overline{C}_\lambda$ is an irreducible reduced curve, and coincides with the fiber $\rho^*(\lambda)$ on the open set $\mathbb{A}^2$ naturally embedded in V.*

(4) *The following conditions are equivalent.*

   (i) *The surface W is smooth.*

   (ii) *Let z be an inhomogeneous coordinate of T such that the point $z = \infty$ lies over the point $\lambda = \infty$ of $\mathbb{P}^1$. Then the hypersurface $z^{p^M} = f(x, y)$ in the affine 3-space $\operatorname{Spec} k[x, y, z]$ is isomorphic to $\mathbb{A}^2 = \operatorname{Spec} k[z, w]$ for some element w.*

   (iii) *C_λ is isomorphic to $\mathbb{A}^1$ for every $\lambda \in k$.*

Proof. (1) Note that the curve C_0 defined by $f = 0$ has one place at infinity by the assumption. By the generic irreducibility theorem of Ganong [54], the generic fiber $\overline{C}_\eta$ as well as general fibers of ρ has only one place at the intersection point with S. Hence $\rho|_S$ is generically one-to-one. So, $\rho|_S$ is purely inseparable.

(2) The statement is almost proved in itself. Since the point of $\overline{F}_\zeta$ lying over the point $\overline{C}_\eta \setminus C_\eta$ is a $k(T)$-rational point, there is a birational mapping $\alpha : T \to T'$ to a subvariety $T' \subset W$ such that $\tau_{T'} \cdot \alpha = \operatorname{id}$. Then α is an isomorphism, and hence T' is a cross-section. Since $k(W) = k(V) \otimes_{k(\mathbb{P}^1)} k(T)$, we have $[k(W) : k(V)] = [k(T) : k(\mathbb{P}^1)] = p^M$, and θ is a purely inseparable morphism of degree p^M.

(3) Let R be the second last exceptional curve of the process σ. Then $(S \cdot R) = 1$, and R belongs to the fiber at infinity $\rho^*(\infty)$ of ρ. Hence it is of the form $\rho^*(\infty) = qR + \cdots$. For $P \in \mathbb{P}^1$, let ν_P be the number of irreducible components of $\rho^*(P)$. Likewise, for $Q \in T$, let ν_Q be the number

of irreducible components of $\tau^*(Q)$. Since θ is a purely inseparable, finite morphism, we have $\nu_Q = \nu_P$ if $P \in \mathbb{P}^1$ and Q is the unique point of T lying over P. Furthermore, the projection formula $\theta_*(\theta^*(D)) = (\deg \theta)D$ for $D \in C\ell(V)$ implies that $\theta^* \otimes \mathbb{Q} : C\ell(V) \otimes \mathbb{Q} \to C\ell(W) \otimes \mathbb{Q}$ is an isomorphism. Let $U = \mathbb{P}^2 \setminus \ell_\infty$ and $\widetilde{U} = W \setminus (T' \cup \tau^*(\infty))$. The above construction implies that $\theta|_{\widetilde{U}} : \widetilde{U} \to U$ is a finite, purely inseparable morphism. Hence $C\ell(U) \otimes \mathbb{Q} \cong C\ell(\widetilde{U}) \otimes \mathbb{Q}$. Since $U \cong \mathbb{A}^2$, we have $C\ell(\widetilde{U}) \otimes \mathbb{Q} = 0$. Hence we have

$$\operatorname{rank} C\ell(V) = \operatorname{rank} C\ell(W) = 2 + \sum_{P \in \mathbb{P}^1} (\nu_P - 1) = 1 + \nu_{P_\infty}.$$

This implies that $\nu_P = 1$ for every $P \neq \infty$. Hence $\overline{C}_\lambda$ is irreducible and coincides with $\rho^*(P)$ if $\lambda \neq \infty$. Suppose that C_λ is not reduced. Then $f - \lambda = g^\mu$ with $g \in k[x, y]$ and $\mu \geq 2$. Since $\rho^*(\lambda) = \mu G$ with the closure G of the curve $g = 0$ in V, we have

$$(\rho^*(\lambda) \cdot S) = q = \mu(G \cdot S).$$

Hence μ is a power of p, i.e., $\mu = p^m$ with $m \geq 1$. Then $f = g^{p^m} + \lambda = (g + \lambda^{p^{-m}})^{p^m}$, which is a contradiction because f is reduced.

(4) (i) implies (ii). Let $\overline{F}_0$ be the fiber of τ such that $\theta(\overline{F}_0) = \overline{C}_0$. Since $C_0 \cong \mathbb{A}^1$, it follows that $\theta^{-1}(C_0) = \overline{F}_0 \setminus (\overline{F}_0 \cap T') \cong \mathbb{A}^1$. Since W is smooth and T' is a cross-section of τ, we know that $\overline{F}_0 \cong \mathbb{P}^1$. By the arithmetic genus formula and by the fact that all closed fibers are linearly equivalent to $\overline{F}_0$, it follows that all irreducible fibers are isomorphic to $\mathbb{P}^1$, and the arithmetic genus of the generic fiber $\overline{F}_\zeta$ is zero. Since $\overline{F}_\zeta$ has a $k(T)$-rational point, $\overline{F}_\zeta \cong \mathbb{P}^1_{k(T)}$. This implies that τ is a $\mathbb{P}^1$-fibration with a unique singular fiber $\tau^{-1}(\infty)$. Hence $\widetilde{U} := W \setminus (\tau^{-1}(\infty) \cup T') \cong \mathbb{A}^2$. Then with the choice of an inhomogeneous coordinate z of T' as indicated in the statement, $\widetilde{U}$ is given as a hypersurface $z^{p^M} = f(x, y)$ in the affine 3-space $\operatorname{Spec} k[x, y, z]$.

(ii) implies (iii). Since the canonical surjection $\mathbb{A}^2 \to \mathbb{A}^1 = \operatorname{Spec} k[z]$ induced by τ restricted onto $\widetilde{U}$ is an $\mathbb{A}^1$-bundle, the curve C_λ defined by $f(x, y) = \lambda$ is isomorphic to the curve $\tau^{-1}(\lambda^{p^{-M}})$ minus its intersection point with T', which is isomorphic to $\mathbb{A}^1$.

(iii) implies (i). Since almost all fibers of τ becomes isomorphic to $\mathbb{P}^1$ under the hypothesis (iii), the morphism $\tau : W \to T$ is a $\mathbb{P}^1$-fibration. Suppose that W has a singular point Q lying over a fiber $\overline{F}$. If $Q \notin T'$, Q has cyclic quotient singularity, and its minimal resolution graph Γ is a linear chain, i.e., a linear chain consisting of smooth rational curves with self-intersection ≤ -2. Suppose that Q is a unique singular point of W lying on the fiber $\overline{F}$. Then the proper transform $\overline{F}'$ plus Γ is the reduced support of a singular fiber of a $\mathbb{P}^1$-fibration. Since the multiplicity of $\overline{F}'$ is equal to one and $(\overline{F}')^2 = -1$, there must exist another irreducible component with self-intersection -1 which must be a component of Γ. This is a contradiction. The cases where there are more singular points lying on $\overline{F}$ with possibly one on T' can be treated in the same fashion. $\qquad\square$

6.3.4.2

The following is a result based on an example of M. Nagata.

Proposition. *Assume that the curve $C = \{f(x,y) = 0\}$ has the following parametrization:*

$$x = t^{pn} + t, \quad y = t^{p^2}, \quad \gcd(p,n) = 1.$$

Then the following assertions hold.

(1) *The curve $C_\lambda = \{f(x,y) = \lambda\}$ is isomorphic to $\mathbb{A}^1$ for every $\lambda \in k$.*
(2) *The hypersurface $z^{p^2} = f(x,y)$ is isomorphic to $\mathbb{A}^2 = \operatorname{Spec} k[z,w]$ with z as a generator of a polynomial ring.*
(3) *If $p = 2$ and $n = 3$, the generic fiber C_η is a form of $\mathbb{A}^1$ with arithmetic genus 1, and the fiber $\rho^*(\infty)$ of the fibration $\rho : V \to \mathbb{P}^1$ is a degenerate fiber 2Γ, where Γ has type II^* (see 3.1.5). Hence the curve S is the locus of moving singularities.*

Proof. (1) We have $f(x,y) = (x^p - y^n)^p - y$. Hence $f(x,y) = 0$ is a smooth curve. By the parametrization, we have an isomorphism between $k[x,y]/(f(x,y))$ and $k[t]$ given by $x = t^{pn} + t, y = t^{p^2}$ and $t = x - (x^p - y^n)^n$. For every $\lambda \in k$, $f(x,y) = \lambda$ is rewritten in the form

$$\left((x - \lambda^{p^{-2}})^p - y^n\right)^p - y = 0.$$

Hence it is isomorphic to the curve $f(x,y) = 0$ by the substitution $(x - \lambda^{p^{-2}}, y) \mapsto (x,y)$.

(2) We can write the equation $z^{p^2} = f(x,y)$ as $y = (x^p - y^n - z^p)^p$. Set $u = x^p - y^n - z^p$. Then $y = u^p$ and $u = x^p - u^{pn} - z^p = (x - u^n - z)^p$. Set $v = x - u^n - z$. Then we have $u = v^p$ and

$$x = v + v^{pn} + z, \quad y = v^{p^2}.$$

Since $v = x - (x^p - y^n - z^p)^n - z$, the hypersurface is isomorphic to $\mathbb{A}^2 = \operatorname{Spec} k[v,z]$, and the polynomial ring $k[v,z]$ contains z as a generator.

(3) Since the linear pencil $\overline{\Lambda}$ is generated by the curve $\overline{C}_0$ and $6\ell_\infty$, by tracking the proper transforms of these two members under the embedded blowing-ups, we can see how the elimination of base points of $\overline{\Lambda}$ is achieved. Let (X_0, X_1, X_2) be a system of homogeneous coordinates such that $x = X_1/X_0$ and $y = X_2/X_0$. The line at infinity ℓ_∞ is defined by $X_0 = 0$, and $\ell_\infty \cap \overline{C}_0$ is given by $(0, 1, 0)$ with respect to this coordinate system. Set $u = X_0/X_1$ and $v = X_2/X_1$. Then the curve $\overline{C}_0$ is defined by $u^2 + v^6 + u^5 v = 0$ and $6\ell_\infty$ by $u^6 = 0$. Then a natural sequence of the embedded blowing-ups of base points leads to the proper transform of $6\ell_\infty$ given as 2Γ when the base points are eliminated, where Γ has type II^* in 3.1.5. Hence every fiber of the fibration $\rho : V \to \mathbb{P}^1$ has arithmetic genus 1. This implies that the generic fiber C_η is a form of $\mathbb{A}^1$ of arithmetic genus 1, and the fiber $\overline{C}_\lambda$ has a cuspidal singularity at the intersection point $S \cap \overline{C}_\lambda$ with $(S \cdot \overline{C}_\lambda) = 2$. $\square$

Chapter 7

Unified p-group scheme

7.1 Group scheme G_λ and pseudo-derivation

Both finite group schemes α_p and $\mathbb{Z}/p\mathbb{Z}$ appear as kernels of the Frobenius endomorphism F and $F - \mathrm{id}$ of the additive group scheme, and one is infinitesimal and the other is reduced. Our objective is to deal with these two group schemes simultaneously as fibers over an affine scheme.

Let R be a noetherian integral domain, e.g., the base field k, a discrete valuation ring $(\mathcal{O}, \mathfrak{m})$ or an affine domain which is an integral domain finitely generated over the field k and let $S = \operatorname{Spec} R$. Let λ be a nonzero element of R. Let $G_{\lambda,S}$ be the kernel of the endomorphism $F - \lambda^{p-1}\mathrm{id}$ of the additive group scheme $G_{a,S}$. Hence $G_{\lambda,S} = \operatorname{Spec} R[t]/(t^p - \lambda^{p-1}t)$, where we denote the residue class of t by ξ and sometimes confuse ξ with t. With this confusion, we have the equality $t^p = \lambda^{p-1}t$. We also write $G_{\lambda,S}$ as G_λ if the existence of S is clear from the context. If $\lambda \notin \mathfrak{p}$ for a prime ideal $\mathfrak{p}$ of R, then $(G_\lambda)_{\mathfrak{p}} \cong \mathbb{Z}/p\mathbb{Z}$ because $t/\lambda \in R_{\mathfrak{p}}[t]$ and $(t/\lambda)^p = t/\lambda$. We can then identify $(G_\lambda)_{\mathfrak{p}}$ with a discrete group $\{\lambda, 2\lambda, \ldots, (p-1)\lambda\}$. On the other hand, $G_\lambda \otimes_R R/\mathfrak{p} \cong \alpha_p$, which is a local infinitesimal group scheme.

According to the terminology in 1.5, the multiplication m, the inverse i and the unit e of G_λ are given respectively by R-algebra homomorphisms $\Delta(\xi) = \xi \otimes 1 + 1 \otimes \xi$, $\iota(\xi) = -\xi$ and $\varepsilon(\xi) = 0$.

7.1.1 G_λ-action on an affine scheme

Let $X = \operatorname{Spec} A$ be an affine scheme over S. *We assume that the ideal λA is a prime ideal.* We call such an R-algebra A a G_λ-*algebra* over R if X has a G_λ-action. A G_λ-action on X is associated with an R-algebra homomorphism

$$\varphi : A \to A \otimes_R R[\xi], \quad \xi^p = \lambda\xi$$

329

such that

$$(\varphi \otimes \mathrm{id}_{R[\xi]})\varphi = (\mathrm{id}_A \otimes \Delta)\varphi \tag{7.1}$$

$$(\mathrm{id}_A \otimes \varepsilon) = \mathrm{id}_A. \tag{7.2}$$

Write

$$\varphi(a) = \delta_0(a) + \delta_1(a)\xi + \delta_2(a)\xi^2 + \cdots + \delta_{p-1}(a)\xi^{p-1},$$

where δ_i is an R-module endomorphism of A. The equality (7.1) implies that

$$\sum_{i,j=0}^{p-1} \delta_i\delta_j(a)\xi^i \otimes \xi^j = \sum_{\ell=0}^{p-1} \sum_{i+j=\ell} \delta_\ell(a)(\xi \otimes 1 + 1 \otimes \xi)^{i+j}$$

$$= \sum_{i,j=0}^{p-1} \binom{i+j}{j} \delta_{i+j}(a)\xi^i \otimes \xi^j, \tag{7.3}$$

whence we have

$$\delta_i\delta_j = \begin{cases} \binom{i+j}{j}\delta_{i+j}, & 0 \le i+j < p \\ 0 & p \le i+j \end{cases}. \tag{7.4}$$

This in turn implies that

$$\delta_i = \frac{1}{i!}\delta^i \quad (0 \le i < p) \quad \text{and} \quad \delta^p = 0 \tag{7.5}$$

for $\delta = \delta_1$. In fact, for $0 < i < p$, we have

$$\delta_i\delta_{p-i} = \frac{1}{i!(p-i)!}\delta^p$$

whence $\delta^p = 0$. By the equality (7.2) we have $\delta_0 = \mathrm{id}_A$. Since φ is an R-algebra homomorphism, we have

$$\varphi(ab) = \varphi(a)\varphi(b) \quad \text{and} \quad \varphi(\alpha a) = \alpha\varphi(a), \quad \alpha \in R.$$

Expanding $\varphi(ab)$ and $\varphi(a)\varphi(b)$ as the sums of terms of ξ^i with coefficients in A and comparing the coefficients of the ξ^i-terms on both sides, we obtain

$$\delta(ab) = a\delta(b) + \delta(a)b + \sum_{i=1}^{p-1} \frac{\lambda^{p-1}}{i!(p-i)!}\delta^i(a)\delta^{p-i}(b), \tag{7.6}$$

where the relation $\xi^p = \lambda^{p-1}\xi$ is taken into account. Such an R-linear endomorphism δ satisfying the formula (7.6) is called an *pseudo-derivation* of A. Since $\varphi(a^p) = \varphi(a)^p$, it follows that $\delta(a^p) = \lambda^{p-1}\delta(a)^p$ for $a \in A$.

Let $\sigma : A \to A$ be an automorphism defined by $\sigma(a) = \varphi(a)|_{\xi=\lambda}$. Then $\sigma^i(a) = \varphi(a)|_{\xi=i\lambda}$ for $0 < i < p$ and $\sigma^p = \mathrm{id}\,_A(=1)$. If $\lambda \notin \mathfrak{p}$ then σ is a generator of the p-group $(G_\lambda)_\mathfrak{p}$. Define the trace T by

$$T = 1 + \sigma + \sigma^2 + \cdots + \sigma^{p-1}.$$

Lemma. *The R-algebra endomorphisms T and $\sigma - 1$ of A are given by the formulas in terms of λ and δ.*

$$T = (\lambda\delta)^{p-1} \tag{7.7}$$

$$\sigma - 1 = (\lambda\delta) + \frac{1}{2!}(\lambda\delta)^2 + \cdots + \frac{1}{(p-1)!}(\lambda\delta)^{p-1}. \tag{7.8}$$

Proof. The second formula follows from the definition. To prove the first formula, we use

$$\sigma^i = 1 + (i\lambda\delta) + \frac{1}{2!}(i\lambda\delta)^2 + \cdots + \frac{1}{(p-1)!}(i\lambda\delta)^{p-1}$$

for $0 \le i < p$ and the following congruence relations

$$\tfrac{1}{i!}(1^i + 2^i + \cdots + (p-1)^i) \equiv 0 \pmod{p} \quad \text{for} \quad 1 \le i < p - 1$$

$$\tfrac{1}{(p-1)!}(1^{p-1} + 2^{p-1} + \cdots + (p-1)^{p-1}) \equiv 1 \pmod{p}.$$

$\square$

7.1.2 *Invariant subalgebra under $G_\lambda \otimes_R R[\lambda^{-1}]$*

If G_λ acts on $X = \operatorname{Spec} A$ with an affine domain A, the G_λ-invariant subalgebra B is given by

$$B = \{a \in A \mid \varphi(a) = a\}.$$

We consider the following localizations:

$$R' = R[\lambda^{-1}], \quad A' = A_{R'} = A \otimes_R R', \quad \delta' = \lambda\delta, \quad \xi' = \xi/\lambda.$$

Since $\xi'^p = \xi'$, φ induces an R-algebra homomorphism $\varphi' : A' \to A' \otimes_R R[\xi']$ and hence $X' = \operatorname{Spec} A'$ has a $\mathbb{Z}/p\mathbb{Z}$-action. We set $B' = \operatorname{Ker} \delta'$ and

$$\sigma' = 1 + \delta' + \frac{1}{2!}\delta'^2 + \cdots + \frac{1}{(p-1)!}\delta'^{p-1}.$$

Clearly $\sigma' = \sigma$. We also set $T' = 1 + \sigma' + \sigma'^2 + \cdots + \sigma'^{p-1}$. We refer the following result to [59].

Lemma. *With the notations and assumptions, we assume that the $\mathbb{Z}/p\mathbb{Z}$-action on X' is nontrivial. Then we have the following assertions.*

(1) $B' = \mathrm{Ker}\,(\sigma' - 1) = \mathrm{Im}\,(T')$.
(2) *There exists an element z' of A' such that $\delta'(z') = 1$. With this element z', it holds that $z'^{P} - z' = b' \in B'$ and $A' = B'[z']$. Such an element z' is determined by A' up to an element of B'. Namely, if z'' satisfies $\delta'(z'') = 1$, then $z' - z'' \in B$.*
(3) *Conversely, if $A' = B'[z']$ with $z'^{P} - z' = b' \in B'$, then $\delta'(z') \in \{1, 2, \ldots, p - 1\}$. We may assume that $\delta'(z') = 1$ by replacing z' by z'/i if $\delta'(z') = i$.*

7.1.3   Invariant subalgebra in terms of δ, $\sigma - 1$ and T

By making use of Lemma 7.1.2, we try to describe $B = A^{G_\lambda}$ in terms of δ, σ and T.

Lemma. *The following assertions hold true.*

(1) $B = \mathrm{Ker}\,\delta$.
(2) $B = \mathrm{Ker}\,(\sigma - 1) = \left(\cup_{s \geq 1} \lambda^{-s} \mathrm{Im}\,(\delta^{p-1})\right) \cap A$.
(3) *There exists an element $z \in A$ such that $\delta(z) = \lambda^r$ with $r \geq 0$.*

Proof.   (1) $\varphi(a) = a$ if and only if

$$\delta(a)\xi + \frac{1}{2!}\delta^2(a)\xi^2 + \cdots + \frac{1}{(p-1)!}\delta^{p-1}(a)\xi^{p-1} = 0, \qquad (7.9)$$

for $\xi = \lambda, 2\lambda, \ldots, (p-1)\lambda$. Substituting $i\lambda$ for ξ in the above equation, we have

$$\delta(a)(i\lambda) + \frac{1}{2!}\delta^2(a)(i\lambda)^2 + \cdots + \frac{1}{(p-1)!}\delta^{p-1}(a)(i\lambda)^{p-1} = 0$$

for $i = 1, \ldots, p-1$. These $(p-1)$ equations are written in the matrix form as a system of linear equations

$$\begin{pmatrix} 1 & 2 & \cdots & p-1 \\ 1^2 & 2^2 & \cdots & (p-1)^2 \\ \vdots & \vdots & \cdots & \vdots \\ 1^{p-1} & 2^{p-1} & \cdots & (p-1)^{p-1} \end{pmatrix} \begin{pmatrix} \delta(a)\lambda \\ \frac{1}{2!}\delta^2(a)\lambda^2 \\ \vdots \\ \frac{1}{(p-1)!}\delta^{p-1}(a)\lambda^{p-1} \end{pmatrix} = \begin{pmatrix} 0 \\ 0 \\ \vdots \\ 0 \end{pmatrix}.$$

The coefficient matrix has a nonzero determinant as the Vandermonde determinant. Hence $\frac{1}{i!}\delta^i(a)\lambda^i = 0$ for $1 \leq i \leq p-1$. In particular, $\delta(a) = 0$ if $\varphi(a) = a$. The converse holds clearly.

(2) By (7.8), $B \subseteq \mathrm{Ker}\,(\sigma - 1)$. Suppose that $\sigma(a) = a$. By (7.9), we have

$$i(\lambda\delta(a)) + i^2 \frac{1}{2!}(\lambda\delta(a))^2 + \cdots + i^{p-1} \frac{1}{(p-1)!}(\lambda\delta(a))^{p-1} = 0$$

for $1 \le i \le p - 1$. In the matrix form, we have the same system of linear equations as in (1) above. Hence we have $\delta(a) = 0$. So, $\mathrm{Ker}\,(\sigma - 1) \subseteq B$.

Set $M := \left(\cup_{s \ge 1} \lambda^{-s} \mathrm{Im}\,(\delta^{p-1})\right) \cap A$. By (7.5), $\delta^p = 0$. Hence $M \subseteq B$. Suppose that $\delta(a) = 0$. In A', $\delta'(a) = 0$ as well. Hence $a = (\delta')^{p-1}(a')$ with $a' \in A'$ by Lemma 7.1.2 since $T' = (\delta')^{p-1} = (\lambda\delta)^{p-1}$. This implies that $\lambda^r a = \delta^{p-1}(a_1)$ with $a_1 \in A$ and $r \in \mathbb{Z}$. Hence $a \in M$. So, $B \subseteq M$.

(3) By Lemma 7.1.2, there exists an element $z' = z/\lambda^r$ with $z \in A$ and $r \ge 0$ such that $\delta'(z') = 1$ and $z \notin \lambda A$. This implies that $\delta(z) = \lambda^r$. $\qquad\square$

7.2 Structure of A with G_λ-action

With the same notations and assumptions as in section 7.1, we consider how the R-algebra A is expressed over the subalgebra B. For this purpose, we introduce the notion of λ-*depth* of the pseudo-derivation δ as the smallest integer r such that $\delta(z) = \lambda^r$ with $r \ge 0$ and some $z \in A$. Such an element z exists by Lemma 7.1.3. If z is an element of A yielding the λ-depth r, then $z \notin \lambda A$. In fact, if $z \in \lambda A$, write $z = \lambda z_1$. Then $\delta(z_1) = \lambda^{r-1}$ with $r - 1 \ge 0$, which contradicts the definition of r.

7.2.1 *Case where λ-depth is zero*

Lemma. *With the previous notations and assumptions, assume that λ-depth r of δ is zero. Then $A = B[z]$ with $z^p - \lambda^{p-1}z = c$ with $c \in B \setminus \lambda B$. The G_λ-action is given by $\sigma(z) = z + \lambda$ and $\sigma|_B = \mathrm{id}_B$. Furthermore, the following assertions hold.*

(1) *B is finitely generated over R.*
(2) *A is faithfully flat over B, and B is smooth (resp. normal) if A is smooth (resp. normal).*

Proof. If $r = 0$, we can take $z \in A$ such that $\delta(z) = 1$. Then $\delta(z^p) = \lambda^{p-1}\delta(z)^p = \lambda^{p-1}$. Hence $\delta(z^p - \lambda^{p-1}z) = 0$ and $z^p - \lambda^{p-1}z = c \in B$. If $c \in \lambda B$, write $c = \lambda c_1$. Then $z^p = \lambda^{p-1}z + \lambda c_1 \in \lambda A$. Since λA is a prime ideal of A by assumption, $z \in \lambda A$, which contradicts the minimality of r.

We show that $A = B[z]$. The proof is by induction on the δ-*length* $\ell(a)$ of elements $a \in A$, which is by definition the smallest integer n such that $\delta^n(a) \neq 0$ and $\delta^{n+1}(a) = 0$. We denote by $\ell(a)$ the δ-length of $a \in A$. Since $\delta^p = 0$, we have $\ell(a) < p$. If $\ell(a) = 0$ then $a \in B$. Suppose that n is the maximum of $\ell(a)$ when a moves in A and that all elements $a' \in A$ with $\ell(a') < n$ belong to $B[z]$. Suppose that $\ell(a) = n$. Then $\delta^{n+1}(a) = 0$ and hence $\delta^n(a) \in B$. Since $\varphi(z^n) = \varphi(z)^n = (z + \xi)^n$, whence its expansion in ξ-power terms ends at the ξ^n, and since

$$\varphi(z^n) = z^n + \delta(z^n)\xi + \frac{1}{2!}\delta^2(z^n)\xi^2 + \cdots + \frac{1}{n!}\delta^n(z^n)\xi^n,$$

it follows that $\delta^n(z^n) = n!$. Let $a_1 = a - \frac{1}{n!}\delta^n(a)z^n$. Then we have

$$\delta^n(a_1) = \delta^n(a) - \frac{1}{n!}\delta^n(a)n! = 0.$$

Hence $\ell(a_1) < n$ and $a_1 \in B[z]$. Then $a \in B[z]$. This argument shows that $B = A[z]$.

It is then clear that A is a finite B-module. Since R is noetherian, by the well-known argument, B is then finitely generated over R. Furthermore, A is a free B-module, hence faithfully B-flat. In particular, by the flat descent, B is smooth (resp. normal) if A is smooth (resp. normal).   □

7.2.2   *Case where λ-depth is positive*

With the same settings as in subsection 7.1.3, we consider the case where the λ-depth of δ is positive.

Lemma. *Assume that $r > 0$ for the λ-depth r of δ. Let $z \in A$ be an element such that $\delta(z) = \lambda^r$. Let $w = z/\lambda^r$ and let $\widetilde{A} = A[w]$. Then the following assertions hold.*

(1) z *satisfies an equation* $z^p - \lambda^{(r+1)(p-1)}z = c$ *with* $c \in B$ *and* $c \notin \lambda B$. *Further,* $\sigma(z) = z + \lambda^{r+1}$ *and* $c = N(z) := \prod_{i=0}^{p-1}\sigma^i(z)$, *i.e.,* c *is the norm of* z.
(2) A *is a finite B-module. For every $a \in A$, the equation* $f_a(t) = \prod_{i=0}^{p-1}(t - \sigma^i(a))$ *is a monic polynomial in $B[t]$ and satisfies $f(a) = 0$.*

Proof.   (1) Note that $\delta(z) = \lambda^r$ and $\delta(z^p) = \lambda^{p-1}\delta(z)^p = \lambda^{pr+p-1}$. Hence $\delta(z^p - \lambda^{(r+1)(p-1)}z) = 0$. So, $z^p - \lambda^{(r+1)(p-1)}z = c \in B \setminus \lambda B$, where $c \notin \lambda B$ because $z \notin \lambda A$ and λA is a prime ideal. Dividing the obtained equation of z by λ^{pr}, we obtain an equation $w^p - \lambda^{p-1}w = c/\lambda^{pr}$. Note that $\sigma^i(w) = w + i\lambda$ for $0 \le i < p$. It is then easy to see that $f(t) := \prod_{i=0}^{p-1}(t - \sigma^i(w)) = t^p - \lambda^{p-1}t - N(w)$. Hence $c/\lambda^{pr} = N(w)$. This entails $N(z) = \lambda^{pr}N(w) = c$.

(2) Straightforward. □

An R-algebra A which satisfies the condition that λA is a prime ideal and has a G_λ-action is said to be of *simple type* if $A = B[z]/(z^p - \lambda^{(r+1)(p-1)}z - N(z))$ with an element z specified in Lemma 7.2.2. It is not known if every R-algebra with G_λ-action is of simple type. Generically, A is of simple type since $A \otimes_R R[\lambda^{-1}]$ is a simple extension of the above type over $B[\lambda^{-1}]$. If A is of simple type, then A is a free B-module of rank p. Hence B is finitely generated over R, and B is smooth (resp. normal) if A is smooth (resp. normal). Let $Y = \operatorname{Spec} B$ and let $q : X \to Y$ be the morphism induced by the inclusion $B \hookrightarrow A$; q is called the *quotient morphism* by G_λ. The locus $V(\lambda B)$ is the branch locus of q. In fact, q is étale outside $V(\lambda B)$ and purely inseparable over $V(\lambda B)$.

7.3 Pseudo-derivation as a lift of p-closed derivation

We consider the case where $R = k[[\lambda]]$ and $A = R[[x, y]]$ with two independent variables x, y. Our objective is to lift an α_p-action given on $k[[x, y]]$ to an R-trivial G_λ-action on A. Note that an α_p-action is given by a k-derivation $\bar\delta$ on $k[[x, y]]$ such that $\bar\delta^p = 0$. *We limit ourselves to the case $p = 2$ which will ease the subsequent computation.*

7.3.1 *Characterization of a p-closed derivation in the case $p = 2$*

The following characterization is possible only in the case $p = 2$. If $p > 2$, we need much involved computations.

Lemma. *Let $\bar\delta$ be a k-derivation on $k[[x, y]]$, and write it as*

$$\bar\delta = f\frac{\partial}{\partial x} + g\frac{\partial}{\partial y}, \quad f, g \in k[[x, y]].$$

We assume that $f, g \in k[x, y]$ and $\bar\delta$ is reduced, i.e., $\gcd(f, g) = 1$. Then $\bar\delta^2 = 0$ if and only if $f, g \in k[[x^2, y^2]]$.

Proof. We assume that $fg \ne 0$. Otherwise, the assumption that $\bar\delta$ is reduced implies that $\bar\delta = \partial/\partial x$ or $\bar\delta = \partial/\partial y$. The assertion holds clearly in this case. Write $f = f_0 + f_1x + f_2y + f_3xy$ and $g = g_0 + g_1x + g_2y + g_3xy$ with $f_i, g_j \in k[x^2, y^2]$. Then we have

$$\overline{\delta}^2(x) = \overline{\delta}(f) = f(f_1 + f_3y) + g(f_2 + f_3x)$$
$$\overline{\delta}^2(y) = \overline{\delta}(g) = f(g_1 + g_3y) + g(g_2 + g_3x) .$$

Suppose that $\overline{\delta}^2 = 0$. Since $\overline{\delta}^2(x) = \overline{\delta}^2(y) = 0$, we have $f(f_1 + f_3y) = g(f_2 + f_3x)$ and $f(g_1 + g_3y) = g(g_2 + g_3y)$. Since $\gcd(f, g) = 1$, the first relation implies that f divides $f_2 + f_3x$. Since $\deg(f_2 + f_3x) < \deg f$, this is impossible. Hence $f_2 = f_3 = 0$ and $f_1 + f_3y = 0$ which implies $f_1 = 0$. Thus $f = f_0 \in k[x^2, y^2]$. Similarly, $g = g_0 \in k[x^2, y^2]$. The converse is clear. $\square$

7.3.2 *A recursive method of lifting a p-closed derivation*

We present here a construction of a pseudo-derivation on A from a p-closed derivation $\overline{\delta}$ on $k[[x, y]]$.

Lemma. *Let $\overline{\delta}$ be a p-closed derivation on $k[[x, y]]$ of the form*

$$\overline{\delta} = (ax^2 + by^2)\frac{\partial}{\partial x} + (cx^2 + dy^2)\frac{\partial}{\partial y}, \quad a, b, c, d \in k, \quad ad + bc \neq 0.$$

Define $\delta(x)$ and $\delta(y)$ in a recursive way by

$$\delta(x) = a(x^2 + \lambda x\delta(x)) + b(y^2 + \lambda y\delta(y))$$
$$\delta(y) = c(x^2 + \lambda x\delta(x)) + d(y^2 + \lambda y\delta(y)).$$

Then the following assertions hold.

(1)

$$\delta(x) = \frac{ax^2 + by^2 + \lambda(ad + bc)x^2y}{1 + \lambda(ax + dy) + \lambda^2(ad + bc)xy}$$
$$\delta(y) = \frac{cx^2 + dy^2 + \lambda(ad + bc)xy^2}{1 + \lambda(ax + dy) + \lambda^2(ad + bc)xy} ,$$

where $\delta(x)$ and $\delta(y)$ are determined as elements of $k[[\lambda, x, y]]$.

(2) *δ is a pseudo-derivation of $A := k[[\lambda, x, y]]$ such that $\sigma := \mathrm{id} + \lambda\delta$ is an $k[[\lambda]]$-automorphism of A such that*

$$\sigma(x) = \frac{x + \lambda(dx + by)y}{1 + \lambda(ax + by) + \lambda^2(ad + bc)xy}$$
$$\sigma(y) = \frac{y + \lambda(cx + ay)x}{1 + \lambda(ax + dy) + \lambda^2(ad + bc)xy} .$$

(3) $\bar{\delta} = \delta \pmod{\lambda A}$.

Proof. We can extend δ to a mapping from $k[\lambda, x, y]$ to $k[[\lambda, x, y]]$ by setting $\sigma(fg) = \sigma(f)\sigma(g)$, $\sigma(f+g) = \sigma(f)+\sigma(g)$ and $\sigma|_{k[[\lambda]]} = \mathrm{id}$. Since it is clear that σ is continuous with respect to the maximal ideal (λ, x, y)-adic topology, it extends to a $k[[\lambda]]$-algebra homomorphism. If we can show that $\sigma^2 = \mathrm{id}$, then $\sigma^{-1} = \mathrm{id} + \lambda\delta$. Then it follows that σ is an automorphism and defines a G_λ-action on A. Hence it suffices to show that $\sigma^2 = \mathrm{id}$. This is done by a straightforward computation. $\square$

PART III

Rational double points

Chapter 1

Basics on rational double points

1.1 Artin theory of rational singularities

There is a very beautiful theory of rational surface singularities by M. Artin [2] and [3]. We first recall the outline of this theory which is characteristic free of the ground field. Let k be an algebraically closed field which we fix as the ground field. Let V be a smooth algebraic surface and let $X = \sum_{i=1}^{n} X_i$ be a reduced divisor on V such that every X_i is a complete curve. A birational proper morphism $\pi : V \to W$ from V to a normal algebraic surface W is called a *contraction* of X if $\pi(X)$ is a point P of W and π induces an isomorphism between $V \setminus X$ and $W \setminus \{P\}$. Such a contraction, if it exists, is unique up to isomorphism of W by Zariski Main Theorem. By Mumford [63], the intersection matrix

$$((X_i \cdot X_j))_{1 \le i,j \le n}$$

is negative-definite. In fact, by replacing W by an affine open neighborhood of P and then by embedding W into a projective surface, we may assume that W is projective. Then V is a smooth complete algebraic surface, hence a projective surface. Let $\overline{H}$ be a hypersurface section of W. Then $\overline{H}$ is ample. Let $H = \pi^*(\overline{H})$. Then $(H^2) > 0$ and $(H \cdot Z) = 0$ for every divisor Z on V such that $\mathrm{Supp}(Z) = X$. By Hodge index theorem, we have $(Z^2) < 0$. Hence $((X_i \cdot X_j))$ is negative definite.

Assume that V is smooth and projective. We say that a divisor $Z = \sum_{i=1}^{n} r_i X_i$ is *positive* if it is effective, i.e., every $r_i \ge 0$, and $r_i > 0$ for some i, and we confuse the support $\cup_{i=1}^{r} X_i$ of X with the reduced divisor X itself. We say that a divisor Z has *support in* X if $\mathrm{Supp}(Z) := \cup_{r_i \ne 0} X_i \subseteq X$. For such a positive divisor, we consider a subscheme $Z = (\mathrm{Supp}(Z), \mathcal{O}_Z)$ of V, where $\mathcal{O}_Z$ is defined by an exact sequence

$$0 \longrightarrow \mathcal{O}_V(-Z) \longrightarrow \mathcal{O}_V \longrightarrow \mathcal{O}_Z \longrightarrow 0.$$

The Euler-Poincaré characteristic $\chi(Z)$ is defined by $\chi(Z) = \chi(\mathcal{O}_Z) = h^0(\mathcal{O}_Z) - h^1(\mathcal{O}_Z)$ and the *arithmetic genus* $p(Z)$ [1]

$$p(Z) = \frac{1}{2}(Z \cdot Z + K_V) + 1.$$

Then, by the Riemann-Roch theorem, we have $\chi(Z) = 1 - p(Z)$. Since $h^0(\mathcal{O}_Z) \geq 1$, it follows that $p(Z) \leq h^1(\mathcal{O}_Z)$, where the equality holds if and only if $H^0(\mathcal{O}_Z) \cong k$.

1.1.1

The following result is due to M. Artin [2, Theorem 1.7].

Theorem. *Let Z be a positive divisor with support in X and let $s = \#\{i \mid r_i > 0\}$. Then the following conditions are equivalent.*

(1) $H^1(\mathcal{O}_Z) = 0$.
(2) $\operatorname{Pic}(Z/k) = H^1(\mathcal{O}_Z^*) \cong \mathbb{Z}^s$.
(3) *For every positive divisor $Y \leq Z$, i.e., $Z - Y \geq 0$, we have $p(Y) \leq 0$.*

1.1.2

Let W be a normal projective surface with a unique singular point P and let $\pi : V \to W$ be a resolution of singularity. Let $X = \sum_{i=1}^{n} X_i$ be the exceptional locus of π. Then there exists a spectral sequence

$$E_2^{p,q} = H^p(W, R^q \pi_* \mathcal{O}_V) \Longrightarrow H^{p+q}(V, \mathcal{O}_V).$$

Since $R^q \pi_* \mathcal{O}_V = 0$ for $q > 1$, $\pi_* \mathcal{O}_V = \mathcal{O}_W$ and $R^1 \pi_* \mathcal{O}_V$ is supported on the point P, we have the following long exact sequence of lower degree terms

$$0 \to H^1(\mathcal{O}_W) \to H^1(\mathcal{O}_V) \to H^0(R^1 \pi_* \mathcal{O}_V) \to H^2(\mathcal{O}_W) \to H^2(\mathcal{O}_V) \to 0.$$

Hence we have $\chi(\mathcal{O}_W) - \chi(\mathcal{O}_V) = h^0(R^1 \pi_* \mathcal{O}_V) \geq 0$, where $h^0(R^1 \pi_* \mathcal{O}_V) \geq p(Z)$ for any positive divisor Z with support in X. In fact, since $R^1 \pi_* \mathcal{O}_V$ is supported by the point $P \in W$, we have $R^1 \pi_* \mathcal{O}_V \cong \varprojlim_Z H^1(\mathcal{O}_Z)$ by [24, Th. 11.1]. If $Y \geq Z$ with positive divisors Y, Z supported by X, we have $h^1(\mathcal{O}_Y) \geq h^1(\mathcal{O}_Z) \geq p(Z)$. Hence $h^0(R^1 \pi_* \mathcal{O}_V) \geq p(Z)$. In particular, $\chi(\mathcal{O}_W) \geq \chi(\mathcal{O}_V)$ (see [2, Lemma 2.2]).

---

[1] We used to denote it by $p_a(Z)$. Here we use $p(Z)$ for a not necessarily reduced curve Z by retaining Artin's original notations.

Note that $\chi(\mathcal{O}_V)$ is independent of the choice of a resolution of singularity $\pi : V \to W$. In fact, by factorization of a birational morphism $f : V_1 \to V_2$ of smooth projective surfaces into a composite of local blowing-ups, it suffices to show that $\chi(\mathcal{O}_{V_1}) = \chi(\mathcal{O}_{V_2})$ if $f : V_1 \to V_2$ is the blowing-up of a single point P. Then, by the above argument, $\chi(\mathcal{O}_{V_2}) - \chi(\mathcal{O}_{V_1}) = h^0(R^1 f_* \mathcal{O}_{V_1})$, where we have by [EGA, Chapter III, (4.2.1)],

$$R^1 f_* \mathcal{O}_{V_1} = \varprojlim_{r \to \infty} H^1(X, \mathcal{O}_{Z_r}),$$

where $Z_r = (E, \mathcal{O}_{rE})$ with the exceptional curve $E = f^{-1}(P)$. We have an exact sequence of sheaves on E

$$0 \longrightarrow \mathcal{O}_{\mathbb{P}^1}(r) \longrightarrow \mathcal{O}_{Z_{r+1}} \longrightarrow \mathcal{O}_{Z_r} \longrightarrow 0,$$

where $r > 0$. Since $H^1(\mathcal{O}_{Z_1}) = H^1(\mathbb{P}^1, \mathcal{O}_{\mathbb{P}^1}(r)) = 0$ for $r > 0$, it follows by induction on r that $H^1(X, \mathcal{O}_{Z_r}) = 0$ for every $r > 0$. Hence $R^1 f_* \mathcal{O}_{V_1} = 0$ and $\chi(\mathcal{O}_{V_2}) = \chi(\mathcal{O}_{V_1})$.

1.1.3 *Contractibility*

We quote the following result of M. Artin [2, Theorem 2.3], where it is assumed that V is only smooth along X.

Theorem. *Let V be a smooth projective surface and let $X = \sum_{i=1}^{n} X_i$ be a reduced divisor. Then the following conditions are equivalent.*

(1) *X is contractible, i.e., there exists a contraction $\pi : V \to W$ such that $\pi(X) = P$, and $\chi(\mathcal{O}_V) = \chi(\mathcal{O}_W)$ holds.*
(2) *The intersection matrix $((X_i \cdot X_j))_{1 \le i,j \le n}$ is negative-definite, and $p(Z) \le 0$ for every positive divisor Z with support in X.*

Now let $X = \sum_{i=1}^{n} X_i$ be a divisor such that the intersection matrix $((X_i \cdot X_j))_{1 \le i,j \le n}$ is negative-definite. Then there exists a nonzero divisor $Z = \sum_{i=1}^{n} r_i X_i$ such that $(Z \cdot X_i) \le 0$ for every i. Then Z is a positive divisor. In fact, write $Z = A - B$ with effective divisors A, B such that A and B have no common components. Then $(A \cdot B) \ge 0$ and $(B^2) \le 0$, where $(B^2) = 0$ if and only if $B = 0$. We have

$$0 \ge (Z \cdot B) = (A \cdot B) - (B^2) \ge 0,$$

whence $(B^2) = 0$ and hence $B = 0$. This implies $Z = A$ and Z is a positive divisor. If X is connected, then $\mathrm{Supp}(Z) = X$. In fact, if $r_i = 0$ for some i, choose X_i, say X_0, such that $X_0 \not\subseteq \operatorname{Supp}(Z)$ but $X_0 \cap \operatorname{Supp}(Z) \neq \emptyset$. Then $(Z \cdot X_0) > 0$, which is a contradiction to $(Z \cdot X_0) \leq 0$. [2]

Let $\mathcal{Z}(X)$ be the set of positive divisors Z with support on X such that $(Z \cdot X_i) \leq 0$ for every i. We define an order $Z \leq Z'$ if $r_i \leq r_i'$ for every i, where $Z = \sum_{i=1}^{n} r_i X_i$ and $Z' = \sum_{i=1}^{n} r_i' X_i$. Then, by M. Artin [3], the set $\mathcal{Z}(X)$ contains the smallest element. In fact, it suffices to show that if $Z_0 = \sum_{i=1}^{n} s_i X_i$ such that $s_i = \min(r_i, r_i')$ for every i then $(Z_0 \cdot X_i) \leq 0$ for every i. Suppose that $r_i \leq r_i'$ for i. Then we have

$$(Z_0 \cdot X_i) = r_i(X_i^2) + \sum_{j \neq i} s_j(X_j \cdot X_i)$$

$$\leq r_i(X_i^2) + \sum_{j \neq i} r_j(X_j \cdot X_i) = (Z \cdot X_i) \leq 0.$$

Hence $Z_0 \in \mathcal{Z}(X)$. This shows the existence of the smallest positive divisor Z in $\mathcal{Z}(X)$, which we call the *fundamental cycle* with support on X. Since Z is the smallest element of $\mathcal{Z}(X)$, if $Y \not\geq Z$, then $Y \notin \mathcal{Z}(X)$ and hence $(Y \cdot X_i) > 0$ for some X_i. By an easy computation using the definition of $p(Y + X_i)$ shows that

$$p(Y + X_i) = p(Y) + p(X_i) + (Y \cdot X_i) - 1 \geq p(Y)$$

because $p(X_i)$ is larger than or equal to the geometric genus of X_i. If $0 < Y < Z$ then $Y \not\geq Z$, and $(Y \cdot X_i) > 0$. Write $Y = \sum_{i=1}^{n} s_i X_i$ and $Z = \sum_{i=1}^{n} r_i X_i$. Then $s_i < r_i$. In fact, if $s_i = r_i$ then

$$(Y \cdot X_i) = s_i(X_i^2) + \sum_{j \neq i} s_j(X_j \cdot X_i)$$

$$\leq r_i(X_i^2) + \sum_{j \neq i} r_j(X_j \cdot X_i) = (Z \cdot X_i) \leq 0,$$

which is a contradiction. Hence $Y + X_i \leq Z$. Starting with some irreducible component $Y = X_1$ and iterating the above argument, we can show by induction on $\sum_{i=1}^{n} r_i$ that $p(Z) \geq 0$. This is a remarkable property of the fundamental cycle of X.

1.1.4 *Rational singularities*

Let P be a singular point of a normal algebraic surface W and let $\pi : V \to W$ be a resolution of singularity at P. By definition, P has

---

[2]In [3, Prop.2], it is shown that if there exists a positive divisor $Z = \sum_{i=1}^{n} r_i X_i$ such that $(Z \cdot X_i) \leq 0$ for every i then $((X_i \cdot X_j))_{1 \leq i,j \leq n}$ is negative semi-definite. If further $(Z^2) < 0$ then $((X_i \cdot X_j))_{1 \leq i,j \leq n}$ is negative-definite.

rational singularity if $R^1\pi_*\mathcal{O}_V = 0$. Since $R^1\pi_*\mathcal{O}_V$ is a coherent $\mathcal{O}_W$-sheaf supported by the point P, this is equivalent to the condition that $\chi(\mathcal{O}_V) = \chi(\mathcal{O}_W)$ by 1.1.2. This condition is independent of the choice of the resolution $\pi : V \to W$. Hence we may assume that π is *minimal* in the sense that the exceptional locus $X = \pi^{-1}(P)$ contains no (-1)-curve, i.e., an irreducible curve E such that $E \cong \mathbb{P}^1$ and $(E^2) = -1$.

For a geometric local ring $(A, \mathfrak{m})$ of dimension d with $A/\mathfrak{m} = k$, it is well-known (see [66]) that $\dim_k A/\mathfrak{m}^{n+1}$ is written by a $\mathbb{Q}$-polynomial $\lambda(A, n)$ in n if $n \gg 0$, where

$$\dim_k A/\mathfrak{m}^{n+1} = \lambda(A, n) = \frac{\mu(A)}{d!} n^d + \text{ terms of degree} < d.$$

We call $\mu(A)$ the *multiplicity* of A. If $d = 2$, then it is easy to show that

$$\dim_k \mathfrak{m}^n/\mathfrak{m}^{n+1} = \mu(A)n + \text{ a constant}$$

if $n \gg 0$. If $A = \mathcal{O}_{W,P}$ then $\mu(A)$ is called the multiplicity of a singular point P of W.

Theorem. *With the above notations, the following assertions hold.*

(1) *If P has rational singularity then $p(Z) = 0$ for the fundamental cycle Z on the exceptional locus X.*

(2) *Conversely, if $p(Z) = 0$ for the fundamental cycle, then P has rational singularity.*

(3) *Assume that P is a rational singular point of W. Then the following assertions hold for every integer $n > 0$.*

   (i) *For $A = \mathcal{O}_{W,P}$, let $W_{P,n} = \operatorname{Spec} A/\mathfrak{m}^n$ be the nth infinitesimal neighborhood of P. Then $\pi^*(W_{P,n}) \cong nZ$ as schemes, where Z is the fundamental cycle.*

   (ii) $H^0(nZ, \mathcal{O}_{nZ}) \cong A/\mathfrak{m}^n$.

   (iii) $\dim_k \mathfrak{m}^n/\mathfrak{m}^{n+1} = -(Z^2)n + 1$. *In particular, the multiplicity of P is equal to $-(Z^2)$.*

Proof. (1) $p(Z) \geq 0$ for the fundamental cycle Z by a remark at the end of 1.1.3 and $p(Z) \leq 0$ by $\chi(\mathcal{O}_V) = \chi(\mathcal{O}_W)$ and Theorem 1.1.3. Hence $p(Z) = 0$.

(2) See [3, Theorem 3].

(3) See [3, Theorem 4]. $\qquad\qquad\qquad\qquad\qquad\qquad\qquad\qquad\qquad\qquad$ □

If P is a rational singularity of W with multiplicity 2, we call it a *rational double point*. By the assertion (3), (iii) of the above theorem, we have then

$(Z^2) = -2$ and $\dim_k \mathfrak{m}/\mathfrak{m}^2 = 3$. Hence $\widehat{\mathcal{O}}_{W,P} \cong k[[x,y,z]]/(f)$. Namely, W is locally analytically described as a hypersurface $f = 0$. In this sense, a rational double point is called a *hypersurface singularity*.

We add here the following result.

Corollary. *Let W be a projective normal surface with a unique singular point P and let $\pi : V \to W$ be a minimal resolution of singularity. Let X be the exceptional locus and let Z be the fundamental cycle on X. Then the following assertions are equivalent.*

(1) *(W, P) is a rational singularity.*
(2) *$p(Z') \le 0$ for every positive divisor Z' with support on X.*
(3) *$H^1(Y, \mathcal{O}_Y) = 0$ for every positive divisor Y with support in X.*

Proof. (1) $\Rightarrow$ (2). Since $\chi(\mathcal{O}_W) = \chi(\mathcal{O}_V)$, Theorem 1.1.3 implies that $p(Z') \le 0$.

(2) $\Rightarrow$ (1). The condition (2) implies that $p(Z) = 0$ since $p(Z) \ge 0$ by a remark at the end of 1.1.3. Hence P has rational singularity by Theorem 1.1.4.

(2) $\Rightarrow$ (3). Note that $Y \le rZ$ for some integer $r > 0$. Then we have an exact sequence

$$0 \longrightarrow \mathcal{O}_V(-Y)/\mathcal{O}_V(-rZ) \longrightarrow \mathcal{O}_{rZ} \longrightarrow \mathcal{O}_Y \longrightarrow 0.$$

This yields a surjection $H^1(rZ, \mathcal{O}_{rZ}) \twoheadrightarrow H^1(Y, \mathcal{O}_Y)$. Since

$$R^1 f_* \mathcal{O}_V = \varprojlim_{r \to \infty} H^1(rZ, \mathcal{O}_{rZ}) \tag{1.1}$$

and since we have a surjection $H^1(sZ, \mathcal{O}_{sZ}) \twoheadrightarrow H^1(rZ, \mathcal{O}_{rZ})$ if $s \ge r$, the assumption that P has rational singularity implies that $H^1(rZ, \mathcal{O}_{rZ}) = 0$ for every $r > 0$. Hence $H^1(Y, \mathcal{O}_Y) = 0$.

(3) $\Rightarrow$ (1). This is straightforward by the relation (1.1). $\qquad\square$

If $p(Z) = 0$, we have

$$p(rZ) = \frac{1}{2}(r^2 - r)(Z^2) - r + 1.$$

Hence, if $(Z^2) < 0$ and $r \ge 2$ then $p(rZ) < 0$.

1.1.5   *Du Val list of rational double singularities*

Let P be a rational double point of W and let $\pi : V \to W$ be a minimal resolution of singularity at P. Let $X = \sum_{i=1}^n X_i$ be the exceptional locus and let Z be the fundamental cycle. Then we know that $p(Z) = 0$ and

$(Z^2) = -2$. Hence $(K \cdot Z) = 0$, where $K = K_V$ is the canonical divisor. Write $Z = \sum_{i=1}^{n} r_i X_i$ with all $r_i > 0$. Since the intersection matrix $((X_i \cdot X_j))_{1 \leq i,j \leq n}$ is negative-definite, we have $(X_i^2) < 0$. If $(K \cdot X_i) < 0$ for some i, then X_i is a (-1)-curve because $p(X_i) = 0$. This contradicts the minimality of π. Hence $(K \cdot X_i) \geq 0$ for every i. Note that $p(X_i) \leq 0$ by Corollary 1.1.4. This implies that every X_i is a (-2)-curve, i.e., $X_i \cong \mathbb{P}^1$ and $(X_i^2) = -2$.

An irreducible component X_i is a *branching component* of X if X_i meets more than two other irreducible components of X. By making use of the negative-definiteness of $((X_i \cdot X_j))_{1 \leq i,j \leq n}$, we can readily show the following assertions.

Lemma. (1) *No three components of X pass through the same point, i.e., X is a divisor with normal crossings.*

(2) *X does not contain a loop.*

(3) *Four or more components do not meet an irreducible component other than these four components.*

(4) *There is at most one branching component. If there exists a branching component X_0, there are three linear chains L_1, L_2, L_3 of irreducible components of X.*

(5) *Assume that three linear chains L_1, L_2, L_3 have respectively $n_1 - 1, n_2 - 1, n_3 - 1$ irreducible components. Assume further that $n_1 \leq n_2 \leq n_3$. Then $\{n_1, n_2, n_3\}$ is a Platonic triplet, i.e., it is one of the following triplets $\{2, 2, m\}$ $(m \geq 2)$, $\{2, 3, 3\}$, $\{2, 3, 4\}$, $\{2, 3, 5\}$.*

Proof. (1) Suppose that three irreducible components, X_1, X_2, X_3 say, meets in a point. Then $(X_1 + X_2 + X_3)^2 = 0$, which contradicts the negative definiteness of the intersection matrix.

(2) Suppose that $X_1, X_2, \ldots, X_s$ is a loop in X if arranged in this order. Then $(X_1 + X_2 + \cdots + X_s)^2 = 0$, which is also a contradiction.

(3) Suppose that four components X_1, X_2, X_3, X_4 meet the component X_0. Then $(2X_0 + X_1 + X_2 + X_3 + X_4)^2 = 0$ which is a contradiction.

(4) Suppose that $X_1, X_2, \ldots, X_s$ is a linear chain with X_1 meeting extra X_0, X_0' and X_s meeting extra X_{s+1}, X_{s+1}'. Then $\{X_0 + X_0' + 2(X_1 + \cdots + X_s) + X_{s+1} + X_{s+1}'\}^2 = 0$, which is a contradiction.

(5) Suppose that the linear chain L_i is $X_{n_i-1}^{(i)} + X_{n_i-2}^{(i)} + \cdots + X_1^{(i)}$ for $i = 1, 2, 3$. Let ℓ be the least common multiple of n_1, n_2, n_3. Let

$$D = \ell X_0 + \sum_{j=1}^{n_1-1} \frac{\ell j}{n_1} X_j^{(1)} + \sum_{j=1}^{n_2-1} \frac{\ell j}{n_2} X_j^{(2)} + \sum_{j=1}^{n_3-1} \frac{\ell j}{n_3} X_j^{(3)}.$$

Then an easy computation shows that

$$(D^2) = \ell^2 \left\{ 1 - \frac{1}{n_1} - \frac{1}{n_2} - \frac{1}{n_3} \right\} < 0.$$

Hence the inequality $\frac{1}{n_1} + \frac{1}{n_2} + \frac{1}{n_3} > 1$ holds, and $\{n_1, n_2, n_3\}$ is a Platonic triplet. $\qquad\square$

By the above lemma, all possible dual graphs of rational double points are exhausted by the following configurations A_n ($n \geq 1$), D_n ($n \geq 4$), E_6, E_7, E_8. We call each of these graphs a *Du Val graph*.

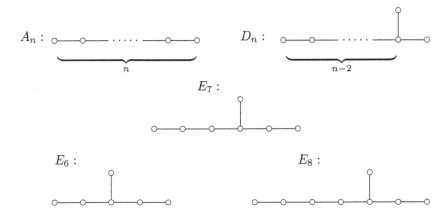

1.1.6 *Du Val graphs and analytic isomorphisms*

We remarked after the proof of Theorem 1.1.4 that the complete local ring $\widehat{\mathcal{O}}_{W,P}$ has a hypersurface singularity $k[[x, y, z]]/(f)$, for which the origin $(0, 0, 0)$ is a singular point. We say that two germs (W, P) and (W', P') are locally *analytically isomorphic* if $\widehat{\mathcal{O}}_{W,P} \cong \widehat{\mathcal{O}}_{W',P'}$. In the case of characteristic zero, it is known, e.g., see Durfee [13], that each rational double point is locally analytically isomorphic to a hypersurface singular point $(0, 0, 0)$ defined by the following weighted-homogeneous, defining equation (called a *normal form* of singularity).

name	equation	weights of x, y, z		
A_n	$x^{n+1} + y^2 + z^2$	$1/(n+1)$	$1/2$	$1/2$
D_n	$x^{n-1} + xy^2 + z^2$	$1/(n-1)$	$(n-2)/2(n-1)$	$1/2$
E_6	$x^4 + y^3 + z^2$	$1/4$	$1/3$	$1/2$
E_7	$x^3 y + y^3 + z^2$	$2/9$	$1/3$	$1/2$
E_8	$x^5 + y^3 + z^2$	$1/5$	$1/3$	$1/2$

In positive characteristics, the bijective correspondence between normal forms and Du Val graphs of rational double points holds only if $p > 5$. If $p \leq 5$, the correspondence is such a relation that finitely many normal forms (up to certain classification invariants) to one Du Val graph (see Greuel-Kröning [21]).

1.1.7 *Classification of rational double points by Artin and Lipman*

Lipman [49] classified the E_8 singularities in all characteristics, and then Artin [5] supplemented the Lipman's list for other singularities. We present here the list by Artin. The verification of the list was omitted by Artin because it is rather tedious. The authors tried to present a rather transparent and sophisticated argument for the verification without reaching a success. It should be considered to be a good problem for young researchers. The arguments in the subsequent sections 1.4 and 1.6 about determinacy and algorithm, though they mainly consist of long and sometimes tedious computations, show that the normal forms by Artin exhaust all possibilities of rational double points at least in characteristic 2, and in other characteristic as well by similar computations. In fact, a rational double point is defined analytically as a hypersurface singularity by a formal power series equation $f(x, y, z) = 0$ and we run the algorithm in 1.6 for $f(x, y, z)$. Then we reach either one of the normal forms of Artin or a non-rational double point.

Hence we state

Theorem. *The correspondence between the Du Val graphs and normal forms (up to contact equivalence) is given by the following tables.*

In characteristic $p = 5$.

name	normal form for $f \in k[[x, y, z]]$	
A_n, D_n, E_6, E_7	*classical forms*	
E_8	E_8^0	$z^2 + x^3 + y^5$
	E_8^1	$z^2 + x^3 + y^5 + xy^4$

In characteristic $p = 3$.

name	normal form for $f \in k[[x, y, z]]$	
A_n, D_n	*classical forms*	
E_6	E_6^0	$z^2 + x^3 + y^4$
	E_6^1	$z^2 + x^3 + y^4 + x^2 y^2$
E_7	E_7^0	$z^2 + x^3 + xy^3$
	E_7^1	$z^2 + x^3 + xy^3 + x^2 y^2$
E_8	E_8^0	$z^2 + x^3 + y^5$
	E_8^1	$z^2 + x^3 + y^5 + x^2 y^3$
	E_8^2	$z^2 + x^3 + y^5 + x^2 y^2$

In characteristic $p = 2$.

name	normal form for $f \in k[[x, y, z]]$		
A_n		$z^{n+1} + xy$	
D_{2m}	D_{2m}^0	$z^2 + x^2 y + xy^m$	$m \geq 2$
	D_{2m}^r	$z^2 + x^2 y + xy^m + xy^{m-r}z$	$m \geq 2, \ 1 \leq r \leq m-1$
D_{2m+1}	D_{2m+1}^0	$z^2 + x^2 y + y^m z$	$m \geq 2$
	D_{2m+1}^r	$z^2 + x^2 y + y^m z + xy^{m-r}z$	$m \geq 2, \ 1 \leq r \leq m-1$
E_6	E_6^0	$z^2 + x^3 + y^2 z$	
	E_6^1	$z^2 + x^3 + y^2 z + xyz$	
E_7	E_7^0	$z^2 + x^3 + xy^3$	
	E_7^1	$z^2 + x^3 + xy^3 + x^2 yz$	
	E_7^2	$z^2 + x^3 + xy^3 + y^3 z$	
	E_7^3	$z^2 + x^3 + xy^3 + xyz$	
E_8	E_8^0	$z^2 + x^3 + y^5$	
	E_8^1	$z^2 + x^3 + y^3 + xy^3 z$	
	E_8^2	$z^2 + x^3 + y^5 + xy^2 z$	
	E_8^3	$z^2 + x^3 + y^5 + y^3 z$	
	E_8^4	$z^2 + x^3 + y^5 + xyz$	

A singularity is called *taut* if there is a bijective correspondence between normal forms and Du Val graphs, *pseudo-taut* if a correspondence is such that finitely many normal forms are related to one Du Val graph.

Therefore, rational double points in characteristic $p \neq 2$ are taut except the following cases:

$$E_8 \text{ for } p = 5, \qquad E_6, E_7, E_8 \text{ for } p = 3.$$

The exceptional cases, E_8 in $p = 5$, E_6, E_7, E_8 in $p = 3$, are all pseudo-taut singularities.

However, rational double points in characteristic 2 are all pseudo-taut except for the case of type A_n.

1.2 Lipman's local study of rational singularity

The base field k is an algebraically closed field of characteristic $p \geq 0$. Let $(R, \mathfrak{m})$ be a two-dimensional normal local ring. If R is the local ring $\mathcal{O}_{W,P}$ of a projective normal surface W defined over k at a singular point P, we say that $(R, \mathfrak{m})$ is a *geometric local ring*. In this section, we follow several results in Lipman [49] and present a way to describe the divisor class group $C\ell(R)$.

1.2.1 *The divisor class group of R*

Let $f : X \to \operatorname{Spec} R$ be a desingularization such that $X \setminus f^{-1}(\{\mathfrak{m}\}) \cong \operatorname{Spec} R \setminus \{\mathfrak{m}\}$. We assume that such a desingularization exists. Write $f^{-1}(\{\mathfrak{m}\})_{\mathrm{red}} = E_1 + E_2 + \cdots + E_n$, where the E_i are irreducible components. Then the intersection matrix $((E_i \cdot E_j))$ is negative definite. Let $\operatorname{Pic}(X)$ be the Picard group which is isomorphic to $H^1(X, \mathcal{O}_X^*)$. Let $\operatorname{Pic}(U)$ be the Picard group of $U = \operatorname{Spec} R \setminus \{\mathfrak{m}\}$, which is equal to the divisor class group $C\ell(R)$ of the ring R because R is normal.

Let $\rho : \operatorname{Pic}(X) \to \operatorname{Pic}(U)$ be the restriction map $D \mapsto D|_U$. Let $\mathbb{E} = \oplus_{i=1}^n \mathbb{Z} E_i$. Then there is a canonical homomorphism $\mathbb{E} \to \operatorname{Pic}(X)$ defined by $D \in \mathbb{E} \mapsto \mathcal{O}_X(D) \in \operatorname{Pic}(X)$, which is injective. In fact, if $D = \sum_{i=1}^n s_i E_i \sim 0$ on X then $(D^2) = 0$ and hence $s_i = 0$ for all i because $((E_i \cdot E_j))$ is negative definite.

Let $D \in \operatorname{Ker} \rho$. Then $\rho(D) = (\varphi)$ for $\varphi \in \Gamma(U, \mathcal{O}_U) = R$. Then $\operatorname{Supp}(D - (\varphi)_X) \subseteq \cup_{i=1}^n E_i$. Thus $D - (\varphi)_X \in \mathbb{E}$. Hence we have

Lemma. *There exists an exact sequence*

$$0 \to \mathbb{E} \to \operatorname{Pic}(X) \xrightarrow{\rho} \operatorname{Pic}(U) \to 0.$$

Proof. It is enough to show ρ is surjective. Suppose $D' \in \operatorname{Pic}(U)$, which is considered as a divisor on $X \setminus f^{-1}(\{\mathfrak{m}\})$. Let $D = \overline{D'}$, where $\overline{D'} = \sum_j m_j \overline{D}_j$ if $D' = \sum_j m_j D'_j$. Then $\rho(D) = D'$. $\square$

1.2.2   The group H

Let $\mathbb{E}^* = \mathrm{Hom}(\mathbb{E}, \mathbb{Z}) = \oplus_{i=1}^n \mathbb{Z}e_i$ with the dual basis $\{e_i\}$ of $\{E_i\}$, i.e., $e_i(E_j) = \delta_{ij}$. Then there exists a group homomorphism $\theta : \mathrm{Pic}(X) \to \mathbb{E}^*$ defined by

$$\theta(\Delta)(E_i) = \frac{1}{d_i}(\Delta \cdot E_i), \quad \Delta \in \mathrm{Pic}(X),$$

where

$$d_i = \gcd\{\deg_{E_i} \mathcal{L} \mid \mathcal{L} \in \mathrm{Pic}(E_i)\} \geq 1.$$

For the definition of $\deg_{E_i}(\mathcal{L})$, we refer to [49]. If E_i is a smooth curve defined over k, we have $d_i = 1$. However, if E_i is a curve defined over a non-closed field, it might be larger than 1. Let $\mathrm{Pic}^0(X) = \{D \in \mathrm{Pic}(X) \mid (D \cdot E_i) = 0 \text{ for all } i\}$. Then $\mathrm{Pic}^0(X) \cap \mathbb{E} = (0)$. In fact, if $(D \cdot E_i) = 0$ for $D = \sum_{i=1}^n \alpha_i E_i$ and for all i, then $\alpha_i = 0$ for all i because $((E_i \cdot E_j))$ is negative-definite.

Let

$$H = \mathrm{Coker}\,(\mathbb{E} \xrightarrow{\theta} \mathbb{E}^*) = \bigoplus_{i=1}^n \mathbb{Z}e_i \bigg/ \left\langle \sum_{j=1}^n \frac{1}{d_j}(E_i \cdot E_j)e_j \right\rangle.$$

In fact, if we write $\theta(E_i) = \sum_{j=1}^n \alpha_j^{(i)} e_j$, then $\theta(E_i)(E_j) = \frac{1}{d_j}(E_i \cdot E_j) = \alpha_j^{(i)}$. Then H is a finite abelian group of order equal to

$$|H| = \frac{1}{d_1 d_2 \cdots d_n} \det((E_i \cdot E_j)).$$

Let $G = \mathrm{Coker}\,(\mathrm{Pic}(X) \xrightarrow{\theta} \mathbb{E}^*)$. Then we have the following result (see [49, p.225 and §15].

Lemma. (1) *We have an exact sequence of abelian groups*

$$0 \to \mathrm{Pic}^0(X) \to \mathrm{Pic}(U) \to H \to G \to 0.$$

(2) *The above exact sequence is independent of the choice of desingularization $f : X \to \mathrm{Spec}\,(R)$. Hence it is determined only by R.*

1.2.3   Case of geometric local rings

We assume that the local ring R is equal to $\mathcal{O}_{W,P}$ for a singular point P of a normal projective surface W defined over the field k. If the characteristic of k is zero, by the description of the local fundamental group of W at

P (see Mumford [63]), the group H is the local first homology group of W at P, i.e., the local fundamental group modulo its commutator group. We compute here the group H for each class in the Du Val list of rational double points. The given computation is effective for all characteristics.

Lemma. *The group H is given as below for each class of the Du Val list.*

class	H		
A_n	$\mathbb{Z}/(n+1)$		
$D_n \ (n \geq 4)$	$\mathbb{Z}/2 \times \mathbb{Z}/2$	$n \equiv 0$	$(\mathrm{mod}\ 2)$
	$\mathbb{Z}/4$	$n \equiv 1$	$(\mathrm{mod}\ 2)$
E_6	$\mathbb{Z}/3$		
E_7	$\mathbb{Z}/2$		
E_8	(0)		

Proof. We consider only the class D_n with $n \geq 4$. The configuration of the exceptional curves is given by

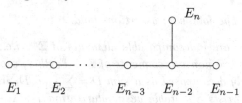

Corresponding to the irreducible components E_i, we have the generators e_i of the group H for $i = 1, \ldots, n$. The relations among the e_i's are given as follows, where we put $x = e_1, y = e_{n-1}$ and $z = e_n$,

$$e_1 = x, e_2 = 2x, \ldots, e_{n-2} = (n-2)x$$
$$e_{n-2} = 2y = 2z$$
$$(n-3)x - 2(n-2)x + y + z = 0.$$

From the last relation, we have $z = (n-1)x - y$. Plugging this into the second last relation, we have $(n-2)x = 2y = 2(n-1)x - 2y$. Hence we obtain $2y = (n-2)x$ and $4y = 2(n-1)x$. This implies $2x = 0$. Suppose $n \equiv 0 \ (\mathrm{mod}\ 2)$. Then $2y = 0$ and $z = x + y$, whence $H \cong \mathbb{Z}/2 \times \mathbb{Z}/2$. Suppose $n \equiv 1 \ (\mathrm{mod}\ 2)$. Then $2x = 0, x = 2y$ and $z = -y$, whence $H \cong \mathbb{Z}/4$.

The other classes can be treated in similar ways.   □

1.2.4　*Vanishing of the group H and factoriality of R*

Note that the intersection matrix $((E_i \cdot E_j))$ is negative definite. Let $D = \sum_{i=1}^{n} s_i E_i$ be an element of $\mathbb{E}$. Then $D \geq 0$, i.e., $s_i \geq 0$ for all i, if $(D \cdot E_i) \leq 0$ for all i. In fact, write $D = A - B$ with effective divisors A, B such that A, B are supported by $\cup_{i=1}^{n} E_i$ and A, B have no common irreducible components. Then we have

$$0 \geq (D \cdot B) = (A \cdot B) - (B^2), \quad (A \cdot B) \geq 0.$$

Hence $(B^2) = (A \cdot B) - (D \cdot B) \geq 0$. Since $(B^2) \leq 0$, it follows that $(B^2) = 0$ and hence $B = 0$, i.e., $D = A \geq 0$. The converse does not necessarily hold.

Let $\mathbb{E}^+ = \{D \in \mathbb{E} \mid (D \cdot E_i) \leq 0 \text{ for every } i\}$. Suppose that $H = (0)$. This implies that $\theta : \mathbb{E} \to \mathbb{E}^*$ is an isomorphism. Hence, by the definition of θ, there exist divisors $D_1, \ldots, D_n \in \mathbb{E}$ such that

$$(D_i \cdot E_j) = -d_j \delta_{ij}, \quad 1 \leq i, j \leq n.$$

Then $D_i \in \mathbb{E}^+$ by the above remark. The following result holds (see [49, §19].

Lemma.　*The following assertions hold.*

(1) *The D_i are indecomposable elements of $\mathbb{E}^+$, i.e., any D_i cannot be a sum of two nonzero elements of $\mathbb{E}^+$. Further, each member D of $\mathbb{E}^+$ is uniquely decomposed as a sum $D = \sum_{i=1}^{n} a_i D_i$ in $\mathbb{E}^+$ with $a_i \geq 0$.*
(2) *There exists a suitable desingularization $f : X \to \operatorname{Spec}(R)$ such that the converse of the above assertion implies the vanishing of H. Namely, if the unique factorization holds in $\mathbb{E}^+$ with respect to indecomposable elements $D_1, \ldots, D_n$ of $\mathbb{E}^+$, then $H = (0)$.*

The following is a main result (see [49, Lemma 14.3 and Proposition 17.1] for the first assertion and [*ibid.*; Theorem 20.1]).

Theorem.　*Let $(R, \mathfrak{m})$ be a two-dimensional normal local ring. Suppose that R has rational singularity. Then the following assertions hold.*

(1) *The divisor class group $C\ell(R) \cong \operatorname{Pic}(U)$ is a finite group. Suppose further that for every height one prime ideal $\mathfrak{p}$ of R, the integral closure of $R/\mathfrak{p}$ in the quotient field is a local ring (unibranch condition of $R/\mathfrak{p}$). Then $\operatorname{Pic}(U) \cong H$.*
(2) *Let $\widehat{R}$ (resp. R^*) be the completion (resp. the henselization) of R. Then the following conditions are equivalent.*

(a) $H = (0)$.

(b) $\widehat{R}$ *is factorial.*

(c) R^* *is factorial.*

1.3 Right and contact equivalence, Milnor number and Tjurina number

Our next objective is to study the resolution $\pi : V \to W$ locally over the singular point P. Namely, let $R = \widehat{\mathcal{O}}_{W,P}$ be the completion of $\mathcal{O}_{W,P}$ and let $Y = V \times_W \operatorname{Spec} R$. Then $\pi_R : Y \to \operatorname{Spec} R$ is a proper morphism whose closed fiber over the point $\widehat{\mathfrak{m}}_{W,P}$ is the exceptional locus X of π. If P is a rational double point, R is written in a hypersurface form $k[[x, y, z]]/(f)$. We look for normal forms in the next section by limiting ourselves to hypersurface singularities. It is natural to introduce the following notion of equivalence.

1.3.1

Definition. Let $k[[x]] = k[[x_1, \ldots, x_n]]$ be the formal power series in n-variables. Let $k[[x]]^*$ or $k[[x_1, \ldots, x_n]]^*$ denote the unit groups. Let $f, g \in k[[x]]$.

(1) f is *right equivalent* to g (denoted as $f \overset{r}{\sim} g$) if there exists $\varphi \in \operatorname{Aut}(k[[x]])$ such that $f = \varphi(g)$, i.e., f is obtained from g by an analytic change of coordinates of $k[[x]]$.

(2) f is *contact equivalent* to g (denoted as $f \overset{c}{\sim} g$) if there exist $\varphi \in \operatorname{Aut}(k[[x]])$ and $u \in k[[x]]^*$ such that $f = u \cdot \varphi(g)$, i.e., $k[[x]]/(f) \cong k[[x]]/(g)$.

1.3.2

For $f \in k[[x]]$, the *Jacobian ideal* $j(f)$ is the ideal of $k[[x]]$ generated by partial derivatives $\frac{\partial f}{\partial x_1}, \ldots, \frac{\partial f}{\partial x_n}$. The *Tjurina ideal* $Tj(f)$ of f is the ideal of $k[[x]]$ generated by f and its partial derivatives $\frac{\partial f}{\partial x_1}, \ldots, \frac{\partial f}{\partial x_n}$. If $k[[x]]/j(f)$ is a finite k-algebra, i.e., $\mathfrak{m}^k \subseteq j(f)$, we call the k-dimension of $k[[x]]/j(f)$ the *Milnor number* of f (denoted by $\mu(f)$). If $k[[x]]/j(f)$ is not finite, we set $\mu(f) = \infty$. If $k[[x]]/Tj(f)$ is a finite k-algebra, its k-dimension is called the *Tjurina number* of f (denoted by $\tau(f)$). If $k[[x]]/Tj(f)$ is not finite, we set $\tau(f) = \infty$. If φ is a k-automorphism of $k[[x]]$ and $u \in k[[x]]^*$, it follows that $j(\varphi(f)) = \varphi(j(f))$ and $Tj(u\varphi(f)) = \varphi(Tj(f))$. Hence if $f \overset{r}{\sim} g$ then $\mu(f) = \mu(g)$, and if $f \overset{c}{\sim} g$ then $\tau(f) = \tau(g)$. See [10] for details.

The following results are known. The first two results are immediate by the definition.

(1) For $f \in k[[x]]$, we have $\tau(f) \le \mu(f)$ because $j(f) \subseteq Tj(f)$.
(2) An element $f \in k[[x]]$ is *weighted homogeneous* of weight N with respect to the positive weights $\mathrm{wt}(x_i) = w_i$ $(1 \le i \le n)$ if f consists of monomials $cx_1^{a_1} \cdots x_n^{a_n}$ such that $\sum_{i=1}^{n} a_i w_i = N$. If the characteristic is zero, we have the relation

$$Nf = \sum_{i=1}^{n} w_i x_i \frac{\partial f}{\partial x_i}.$$

Hence $j(f) = Tj(f)$, which implies $\mu(f) = \tau(f)$. This is not the case if the characteristic is positive.
(3) Suppose that the characteristic is zero. In view of the defining equations of rational double points which are weighted homogeneous (see 1.1.3), we have the following table.

type	A_n	D_n	E_6	E_7	E_8
$\mu(f)$	n	n	6	7	8

(4) In the case of positive characteristic, the Tjurina number of each Du Val type is given as following tables. For type D_{2m}^r and D_{2m+1}^r in characteristic 2, m and r satisfy $m \ge 2$ and $0 \le r \le m - 1$.

char.	$p > 0$		$p \ge 3$	$p = 2$	
type	$A_n(p \nmid n+1)$	$A_n(p \mid n+1)$	$D_n(n \ge 4)$	D_{2m}^r	D_{2m+1}^r
τ	n	$n+1$	n	$4m - r$	$4m - r$

char.	$p \ge 5$	$p = 3$		$p = 2$		$p \ge 5$	$p = 3$		$p = 2$			
type	E_6	E_6^0	E_6^1	E_6^0	E_6^1	E_7	E_7^0	E_7^1	E_7^0	E_7^1	E_7^2	E_7^3
τ	6	9	7	8	6	7	9	7	14	12	10	8

char.	$p \ge 7$	$p = 5$		$p = 3$			$p = 2$				
type	E_8	E_8^0	E_8^1	E_8^0	E_8^1	E_8^2	E_8^0	E_8^1	E_8^2	E_8^3	E_8^4
τ	8	10	8	12	10	8	16	14	12	10	8

1.4 Determinacy

Let $J^{(k)} = k[[x]]/\mathfrak{m}^{k+1} = k[[x_1, \ldots, x_n]]/\mathfrak{m}^{k+1}$ be the space of k-jets. For a nonzero power series $f \in k[[x]]$, we write $f^{(k)} \in J^{(k)}$ the power series of f up to and including the terms of order k. We call $f^{(k)}$ the k-jets of f.

Definition. $f \in k[[x]]$ is called *contact k-determined* (resp. *right k-determined*) if each $g \in k[[x]]$ with $f^{(k)} = g^{(k)}$ is contact equivalent (resp. right equivalent) to f. The minimal such k is called the *contact determinacy* (resp. *right determinacy*) of f.

1.4.1

The following result is found in Boubakri-Greuel-Markwig [10].

Theorem. *Let $f \in \mathfrak{m}^2 \setminus \{0\}$ and $k \in \mathbb{N}$ and let $j(f)$ denote the Jacobian ideal of f.*

(1) If $\mathfrak{m}^{k+2} \subset \mathfrak{m}^2 \cdot j(f)$, then f is right $(2k - \mathrm{ord}\,(f) + 2)$-determined.
(2) If $\mathfrak{m}^{k+2} \subset \mathfrak{m} \cdot \langle f \rangle + \mathfrak{m}^2 \cdot j(f)$, then f is contact $(2k - \mathrm{ord}\,(f) + 2)$- determined.

1.4.2 *Characteristic $\neq 2$*

Theorem 1.4.1 gives only overestimates of right and contact determinacies of f which are not optimal in general.

Proposition. *For rational double points in characteristic different from 2, the contact determinacy is nothing but the maximal degree of monomials occuring in the normal form in 1.1.6.*

For characteristic 0, it is straightforward by the following lemma, which is originally due to V.I. Arnold (see Durfee [13, Lemma 8.2] and its references), and remark after that.

Lemma. *Assume that the characteristic is zero. If $\mathfrak{m}^{k+1} \subset \mathfrak{m}^2 \cdot j(f)$ then f is right k-determined.*

Remark. If a hypersurface singularity is right k-determined, then it is also contact k-determined.

1.4.3 Characteristic 2

The following result is due to Greuel-Kröning [21, Remark 2.7].

Proposition. *For characteristic 2, the contact determinacy is given as follows.*

A_n	D_{2m}^r $(r \geq 0)$	D_{2m+1}^r $(r \geq 0)$	E_6^r	$E_7^{0,1}$	$E_7^{2,3}$	E_8^r
$n+1$	$\max\{2r, m+1\}$	$\max\{2r+1, m+1\}$	3	5	4	5

Here $E_7^{0,1}$ *(resp. $E_7^{2,3}$) stands for both cases E_7^0 and E_7^1 (resp. E_7^2 and E_7^3).*

We explain below how to obtain this list by calculation only in the cases E_7^r and D_{2m}^r.

1.4.3.1 Calculation

The most important thing in conducting calculation in the special low characteristic is how to treat units which can be usually ignored in many cases of the characteristic zero. We are going to exhibit calculations in some cases for determining the contact determinacies so that the readers can feel how peculiar the world is in the characteristic two. See Yanagisawa [94] for more details. The following result is called the *Unit Lemma*.

Lemma.

(1) Let the characteristic be a prime p. Let

$$f(x,y,z) = z^{p^\ell k} + x^{s_1} y^{t_1} z^h + x^{s_2} y^{t_2} + x^{s_3} y^{t_3}$$

be local equation where h, k, ℓ are positive integers satisfying $(h,p) = (k,p) = 1$. Then $f(x,y,z)$ is contact equivalent to

$$g(x,y,z) = U_0 z^{p^\ell k} + U_1 x^{s_1} y^{t_1} z^h + U_2 x^{s_2} y^{t_2} + U_3 x^{s_3} y^{t_3}$$

for any $U_i \in k[[x,y,z]]^*$ $(i = 0,1,2,3)$ if and only if the condition

$$s_2 t_3 \not\equiv s_3 t_2 \pmod{p}$$

is satisfied.

(2) Assume the characteristic is 2 and let

$$f(x,y,z) = z^2 + x^{s_1} y^{t_1} z + x^{s_2} y^{t_2} + x^{s_3} y^{t_3}$$

be a local equation. Then $f(x, y, z)$ is contact equivalent to
$$g(x, y, z) = U_0 z^2 + U_1 x^{s_1} y^{t_1} z + U_2 x^{s_2} y^{t_2} + U_3 x^{s_3} y^{t_3}$$
for any $U_i \in k[[x, y, z]]^*$ $(i = 0, 1, 2, 3)$ if and only if the condition
$$s_2 t_3 \not\equiv s_3 t_2 \pmod{2}$$
is satisfied.

(3) Assume the characteristic is 2. Let
$$f(x, y, z) = z^2 + x^{s_1} y^{t_1} + x^{s_2} y^{t_2}$$
be a local equation. Then $f(x, y, z)$ is contact equivalent to
$$g(x, y, z) = U_0 z^2 + U_1 x^{s_1} y^{t_1} + U_2 x^{s_2} y^{t_2}$$
for any $U_i \in k[[x, y, z]]^*$ $(i = 0, 1, 2)$ if and only if the condition
$$s_1 t_2 \not\equiv s_2 t_1 \pmod{2}$$
is satisfied.

Proof.

(1) We may assume $U_0 = 1$. Moreover, since $p^\ell k - h$ is prime to p, we can transform $z \mapsto U_1^{1/(p^\ell k - h)} z$ to obtain
$$\begin{aligned} g(x, y, z) &\overset{c}{\sim} U_1^{p^\ell k/(p^\ell k - h)} z^{p^\ell k} + U_1^{p^\ell k/(p^\ell k - h)} x^{s_1} y^{t_1} z^h \\ &\quad + U_2 x^{s_2} y^{t_2} + U_3 x^{s_3} y^{t_3} \\ &\overset{c}{\sim} z^{p^\ell k} + x^{s_1} y^{t_1} z^h + U_2' x^{s_2} y^{t_2} + U_3' x^{s_3} y^{t_3}, \end{aligned}$$
where U_2', U_3' are some units. Now we make changes $x \mapsto \alpha x$, $y \mapsto \beta y$, $z \mapsto \gamma z$, so that we have equalities of coefficients:
$$\gamma^{p^\ell k} = \alpha^{s_1} \beta^{t_1} \gamma^h = U_2' \alpha^{s_2} \beta^{t_2} = U_3' \alpha^{s_3} \beta^{t_3}.$$
By the first equation, we can determine $\gamma = (\alpha^{s_1} \beta^{t_1})^{1/(p^\ell k - h)}$. Thus it suffices to solve
$$\begin{cases} \alpha^{s_1 \cdot \frac{p^\ell k}{p^\ell k - h} - s_2} = U_2' \beta^{t_2 - t_1 \cdot \frac{p^\ell k}{p^\ell k - h}}, \\ \alpha^{s_1 \cdot \frac{p^\ell k}{p^\ell k - h} - s_3} = U_3' \beta^{t_3 - t_1 \cdot \frac{p^\ell k}{p^\ell k - h}}. \end{cases}$$
Now these equations have a solution, that is, α and β can be expressed by U_2' and U_3', if and only if the determinant of the matrix
$$\begin{pmatrix} \dfrac{s_1 p^\ell k}{p^\ell k - h} - s_2 & \dfrac{t_1 p^\ell k}{p^\ell k - h} - t_2 \\ \dfrac{s_1 p^\ell k}{p^\ell k - h} - s_3 & \dfrac{t_1 p^\ell k}{p^\ell k - h} - t_3 \end{pmatrix}$$
has a numerator which is relatively prime to p. Thus we have
$$s_2 t_3 - s_3 t_2 \not\equiv 0 \pmod{p}$$
by a direct calculation.

(2) We can argue similarly as in the case (1).

(3) This is the special case of (2).

$\square$

1.4.3.2 Case E_7^r

Here we verify the determinacy for E_7^r to be given as in the table in Proposition 1.4.3. The calculation consists of several steps below. Let $f_r(x, y, z)$ be a defining equation of E_7^r,

$$f_r(x, y, z) = z^2 + x^3 + xy^3 + m_r(x, y, z),$$

$$\text{where } m_r(x, y, z) = \begin{cases} 0 & r = 0 \\ x^2yz & r = 1 \\ y^3z & r = 2 \\ xyz & r = 3 \,. \end{cases}$$

(1) USE OF THEOREM 1.4.1.

By Theorem of 1.4.1, we know E_7^r is contact 8-, 8-, 6-, 8-determined for $r = 0, 1, 2, 3$, respectively. In the case $r = 0$, for example, we have

$$\mathfrak{m} \cdot (f_0) + \mathfrak{m}^2 \cdot j(f_0)$$
$$= (x, y, z) \cdot (z^2 + x^3 + xy^3) + (x^2, y^2, z^2, xy, yz, zx) \cdot (x^2 + y^3, xy^2).$$

Hence one can check easily that this ideal contains $\mathfrak{m}^6$, thus $k = 4$ and the determinacy is at most $2 \cdot 4 - 2 + 2 = 8$. Therefore $f_0 \overset{c}{\sim} g$ for all $g \in k[[x, y, z]]$ with $f_0^{(8)} = g^{(8)}$.

(2) DESCENDING ℓ-DETERMINACY TO $(\ell - 1)$-DETERMINACY. We abbreviate this claim by Det $_\ell$, for short. This is indeed a claim that Det $_\ell$ can be reduced to Det $_{\ell-1}$ under the assumption that it is ℓ-determined with $\ell > \deg f_r(x, y, z)$. For this, one needs to show that

$$f_r(x, y, z) + \sum_{a+b+c=\ell} d_{abc}x^ay^bz^c \overset{c}{\sim} f_r(x, y, z).$$

(3) A VARIANT OF THE UNIT LEMMA.

We frequently use the following lemma, which is an easy variant of Lemma 1.4.3.1.

Lemma. *For $\alpha, \beta, \gamma, \delta \in k[[x, y, z]]^*$ and the singularity f_r of type E_7^r with $r = 0, 1, 2$ or 3, we have*

$$\alpha z^2 + \beta x^3 + \gamma xy^3 + \delta m_r(x, y, z) \overset{c}{\sim} f_r(x, y, z).$$

(4) Det $_8$ AND Det $_7$ FOR E_7^r FOR $r = 0, 1, 3$.
Since

$$f_r(x, y, z) + \sum_{a+b+c=\ell} d_{abc} x^a y^b z^c$$

$$= \left(1 + \sum_{c \geq 2} d_{abc} x^a y^b z^{c-2} \right) z^2 + \left(1 + \sum_{a \geq 3} d_{abc} x^{a-3} y^b z^c \right) x^3$$

$$+ \left(1 + \sum_{a \geq 1, b \geq 3} d_{abc} x^{a-1} y^{b-3} z^c \right) xy^3 + m_r(x, y, z)$$

$$+ A y^\ell + B y^{\ell-1} z$$

$$\overset{c}{\sim} z^2 + x^3 + xy^3 + m_r(x, y, z) + u y^\ell + v y^{\ell-1} z$$

$$= z^2 + x^3 + (x + u y^{\ell-3} + v y^{\ell-4} z) y^3 + m_r(x, y, z)$$

for $A, B \in k$ and $u, v \in k[[x]]^*$ by Lemma 1.4.3.1, we make a change of
variables $X = x + u y^{\ell-3} + v y^{\ell-4} z$. Then we have

$$x^3 = X^3 + (u y^{\ell-3} + v y^{\ell-4} z) X^2 + O(2(\ell - 3) + 1)$$

$$m_r(x, y, z) = m_r(X, y, z) + \begin{cases} 0 & r = 0 \\ O(2(\ell - 3) + 2) & r = 1 \\ u y^{\ell-2} z + v y^{\ell-3} z^2 & r = 3, \end{cases}$$

where $O(2(\ell - 3) + 1)$ is a sum of terms in y, z, X of degree $\geq 2(\ell - 3) + 1$.
Note that the large O term is negligible because $2(\ell - 3) + 1 \geq \ell + 1$ since
it is assumed to be ℓ-determined.

Thus the equation $z^2 + x^3 + (x + u y^{\ell-3} + v y^{\ell-4} z) y^3 + m_r(x, y, z)$ can
be written using the Unit Lemma as

$$(r = 0, 1) \quad z^2 + X^3 + (1 + u X y^{\ell-6} + v X y^{\ell-7} z) X y^3 + m_r(X, y, z)$$
$$(r = 3) \quad (1 + v y^{\ell-3}) z^2 + X^3 + (1 + u X y^{\ell-6} + v X y^{\ell-7} z) X y^3$$
$$+ X y z + u y^{\ell-2} z$$

$$\overset{c}{\sim} f_r(x, y, z) + \begin{cases} 0 & r = 0 \\ 0 & r = 1 \\ u y^{\ell-2} z & r = 3 . \end{cases}$$

For $r = 3$ case, the last equation can be written as

$$z^2 + X^3 + Xy^3 + Xyz + uy^{\ell-2}z$$
$$= z^2 + \tilde{X}^3 + u\tilde{X}^2 y^{\ell-5}z + u^2\tilde{X}y^{2(\ell-5)}z^2 + u^3 y^{3(\ell-5)}z^3 + \tilde{X}y^3$$
$$+ \tilde{X}yz + uy^{\ell-4}z^2$$
$$= (1 + uy^{\ell-4} + u^2\tilde{X}y^{2(\ell-5)} + u^3 y^{3(\ell-5)}z)z^2 + \tilde{X}^3 + \tilde{X}y^3$$
$$+ (1 + u\tilde{X}y^{\ell-6})\tilde{X}yz$$
$$\overset{c}{\sim} f_3(x, y, z)$$

by change of variable $X = \tilde{X} + uy^{\ell-5}z$. Therefore, all singularities E_7^r with $r = 0, 1, 2, 3$ are at most 6-determined up to here.

(5) E_7^r IS AT MOST 5-DETERMINED FOR $0 \leq r \leq 3$.

By the same argument as in (4), we have

$$f_r(x, y, z) + \sum_{a+b+c=6} d_{abc}x^a y^b z^c$$
$$\overset{c}{\sim} z^2 + X^3 + Xy^3 + uX^2 y^3 + vX^2 y^2 z + m_r(x, y, z)$$
$$= z^2 + X^3 + (1 + uX)Xy^3 + vX^2 y^2 z$$

$$+ m_r(X, y, z) + \begin{cases} 0 & r = 0 \\ O(8) & r = 1 \\ 0 & r = 2 \\ uy^4 z + vy^3 z^2 & r = 3 \end{cases}$$

$$= \begin{cases} z^2 + X^3 + (1 + uX)Xy^3 + vX^2 y^2 z & r = 0 \\ z^2 + X^3 + (1 + uX)Xy^3 + (1 + vy)X^2 yz & r = 1 \\ z^2 + X^3 + (1 + uX)Xy^3 + vX^2 y^2 z + y^3 z & r = 2 \\ (1 + vy^3)z^2 + X^3 + (1 + uX)Xy^3 + vX^2 y^2 z + Xyz + uy^4 z & r = 3 \end{cases}$$

$$\overset{c}{\sim} \begin{cases} z^2 + X^3 + Xy^3 + vX^2 y^2 z & r = 0 \\ z^2 + X^3 + Xy^3 + X^2 yz & r = 1 \\ z^2 + X^3 + Xy^3 + vX^2 y^2 z + y^3 z & r = 2 \\ z^2 + X^3 + Xy^3 + vX^2 y^2 z + Xyz + uy^4 z & r = 3 \end{cases}$$

for $u, v \in k$.

Thus we are done for $r = 1$. For $r = 0$, we need a change of variables $Y = y + vXz$ to make the following calculation

$$z^2 + X^3 + Xy^3 + vX^2y^2z = z^2 + X^3 + Xy^2(y + vXz)$$
$$= z^2 + X^3 + X(Y + vXz)^2Y = z^2 + (1 + vYz)X^3 + XY^3$$

and are done by the Unit Lemma.

For $r = 2$, a change of variables $Y = y + vXz$ gives

$$z^2 + X^3 + Xy^3 + vX^2y^2z + y^3z = z^2 + X^3 + Xy^2(y + vXz) + y^3z$$
$$= z^2 + X^3 + X(Y + vXz)^2Y + (Y + vXz)^3z$$
$$= (1 + v^2X^3Y + vXY^2 + v^2X^2Yz + v^3X^3z^2)z^2 + X^3 + XY^3 + Y^3z$$

and we are done by the Unit Lemma.

For $r = 3$, we have the following calculation

$$z^2 + X^3 + Xy^3 + vX^2y^2z + Xyz + uy^4z$$
$$= z^2 + X^3 + Xy^3 + (1 + vXy)Xyz + uy^4z$$
$$\overset{c}{\sim} z^2 + X^3 + Xy^3 + Xyz + u'y^4z = z^2 + X^3 + (X + u'yz)y^3 + Xyz$$
$$= (1 + u'y^2 + u'^2y^2\widetilde{X} + u'^3y^3z)z^2 + \widetilde{X}^3 + \widetilde{X}y^3 + (1 + u'\widetilde{X})\widetilde{X}yz$$

and so we are done by the Unit Lemma, where u' is another unit and $\widetilde{X} = X + u'yz$.

(6) E_7^0 AND E_7^1 ARE PRECISELY 5-DETERMINED.

For E_7^0, consider the singularities defined by

$$g_0(x, y, z) := z^2 + x^3 + xy^3 + y^4z + z^5.$$

Then $g_0(x, y, z)^{(4)} = f_0(x, y, z)$ but $g_0(x, y, z)$ is not contact equivalent to f_0 because $\tau(f_0) = 14$ and $\tau(g_0) = 12$. Thus E_7^0 is not 4-determined by 1.3.2. We only compute $\tau(f_0) = 14$. By the definition, the Tjurina ideal is

$$Tj(f_0) = \left(z^2 + x^3 + xy^3, x^2 + y^3, xy^2\right).$$

Hence we have $z^2 = 0, x^2 = y^3, xy^2 = 0$ and $y^5 = y^2x^2 = x(xy^2) = 0$ modulo the ideal $Tj(f_0)$. This implies

$$k[[x, y, z]]/(Tj(f_0)) = \left((k + ky + ky^2 + ky^3 + ky^4) + (kx + kxy)\right)$$
$$+ \left((k + ky + ky^2 + ky^3 + ky^4) + (kx + kxy)\right)z,$$

whence $\tau(f_0) = \dim_k k[[x, y, z]]/Tj(f_0) = 14$.

For E_7^1 case, consider the singularity

$$g_1(x, y, z) := z^2 + x^3 + xy^3 + x^2yz + y^2z^3.$$

Then $\tau(g_1) = 16$ and $\tau(f_1) = 12$ with $g_1^{(4)} = f_1$. Hence E_7^1 is not 4-determined.

(7) E_7^2 AND E_7^3 ARE PRECISELY 4-DETERMINED.

For the 4-determinedness, we will show the claim Det_5. We have the following equation after incorporating the terms divisible by monomials appearing in $f_r(x, y, z)$ into single terms with unital coefficients and using the Unit Lemma.

$$f_r(x, y, z) + \sum_{a+b+c=5} d_{abc}x^a y^b z^c$$

$$\overset{c}{\sim} f_r(x, y, z) + ux^2 y^2 z + vy^4 z + wy^5$$

$$= \begin{cases} z^2 + x^3 + xy^3 + (1+vy)y^3 z + ux^2 y^2 z + wy^5 & r = 2 \\ z^2 + x^3 + xy^3 + (1+uxy)xyz + +vy^4 z + wy^5 & r = 3 \end{cases}$$

$$\overset{c}{\sim} \begin{cases} z^2 + x^3 + xy^3 + y^3 z + u'x^2 y^2 z + w'y^5 & r = 2 \\ z^2 + x^3 + xy^3 + xyz + v'y^4 z + w'y^5 & r = 3, \end{cases}$$

where $u, v, w, u', v', w' \in k$.

For the E_7^2 case, we kill the $x^2 y^2 z$ term by a change of variable $X = x + u'y^2 z$. Then we have

$$z^2 + x^3 + xy^3 + y^3 z + u'x^2 y^2 z + w'y^5$$
$$= z^2 + x^2(x + u'y^2 z) + xy^3 + y^3 z + w'y^5$$
$$= (1 + u'^2 X y^4)z^2 + X^3 + Xy^3 + y^3 z + w'y^5 + u'y^5 z$$
$$\overset{c}{\sim} z^2 + X^3 + Xy^3 + y^3 z + w''y^5,$$

where $w'' \in k[[x, y, z]]$. We note that the coefficient w' of the y^5 term is unchanged as seen from the first and second lines. By a further change of variable $\widetilde{X} = X + w''y^2$ and by setting $w'' = c_0 + c_1\widetilde{X} + c_2 y + c_3 z + O(2)$ with $c_0, c_1, c_2, c_3 \in k$, the above calculation continues to

$$= z^2 + X^3 + (X + w''y^2)y^3 + y^3 z$$
$$= z^2 + \widetilde{X}^3 + w''\widetilde{X}^2 y^2 + w''^2 \widetilde{X}y^4 + \widetilde{X}y^3 + y^3 z + O(6)$$
$$= (z + c_0^{1/2}\widetilde{X}y)^2 + (1 + c_1 y^2)\widetilde{X}^3 + (1 + c_2\widetilde{X} + w''^2 y)\widetilde{X}y^3$$
$$\qquad\qquad + c_3\widetilde{X}^2 y^2 z + y^3 z + O(6)$$
$$\overset{c}{\sim} z^2 + \widetilde{X}^3 + \widetilde{X}y^3 + c_3\widetilde{X}^2 y^2 z + y^3 z.$$

We can kill the $c_3 \widetilde{X}^2 y^2 z$ by the first change of variable without producing a y^5 term with nonzero coefficient. Hence we have

$$f_r(x, y, z) + \sum_{a+b+c=5} d_{abc} x^a y^b z^c$$

$$\overset{c}{\sim} z^2 + \widetilde{X}^3 + \widetilde{X} y^3 + y^3 z.$$

Comparison with the singularity $g = z^2 + x^3$ tells us that E_7^2 is not 3-determined because $f_2^{(3)} = g^{(3)}$ but $\tau(g) = \infty \neq 10 = \tau(f_2)$.

Finally we treat the remaining E_7^3 case. In view of the calculation at the beginning with $r = 2$, we can start from the equation $z^2 + x^3 + xy^3 + xyz + vy^4 z + wy^5$ with $v, w \in k[[x, y, z]]$. After changing variables $X = x + vyz + wy^2$, we have

$$z^2 + x^3 + xy^3 + xyz + vy^4 z + wy^5$$

$$= (1 + v^2 Xy^2 + vy^2)z^2 + X^3 + (1 + w^2 y)Xy^3$$

$$+ (1 + vX)Xyz + wX^2 y^2 + wy^3 z + O(6)$$

$$\overset{c}{\sim} z^2 + X^3 + Xy^3 + Xyz + w' X^2 y^2 + w' y^3 z$$

$$= z^2 + X^3 + (X + w'z)y^3 + Xyz + w' X^2 y^2.$$

A further change of variables $\widetilde{X} = X + w'z$ allows us to continue the above equations to

$$\overset{c}{\sim} (1 + w'^2 X + w'^3 z + w'y + w'^3 y^2)z^2 + \widetilde{X}^3 + \widetilde{X} y^3$$

$$+ \widetilde{X} yz + w' \widetilde{X}^2 z + w' \widetilde{X}^2 y^2$$

$$\overset{c}{\sim} z^2 + \widetilde{X}^3 + \widetilde{X} y^3 + \widetilde{X} yz + \xi \widetilde{X}^2 z + \eta \widetilde{X}^2 y^2,$$

where $w', \xi, \eta \in k[[x, y, z]]$. Now, putting $Y = y + \eta \widetilde{X}$, we can continue the above calculation to

$$\overset{c}{\sim} z^2 + (1 + \eta^2 Y)\widetilde{X}^3 + \widetilde{X} Y^3 + \widetilde{X} Yz + \eta \widetilde{X}^2 z + \xi \widetilde{X}^2 z$$

$$\overset{c}{\sim} z^2 + \widetilde{X}^3 + \widetilde{X} Y^3 + \widetilde{X} Yz + (\eta' + \xi') \widetilde{X}^2 z.$$

If $\eta' + \xi' = 0$ then the last equation is nothing but f_3. Thus we assume that $A := \eta' + \xi' \neq 0$. Then

$$z^2 + \widetilde{X}^3 + \widetilde{X} Y^3 + \widetilde{X} Yz + A \widetilde{X}^2 z = z^2 + \widetilde{X}^3 + \widetilde{X} Y^3 + \widetilde{X}(Y + A\widetilde{X})z$$

$$= z^2 + (1 + A^2 \widetilde{Y} + A^3 \widetilde{X})\widetilde{X}^3 + \widetilde{X} \widetilde{Y}^3 + \widetilde{X} \widetilde{Y} z + A \widetilde{X}^2 \widetilde{Y}^2$$

$$\overset{c}{\sim} z^2 + \widetilde{X}^3 + \widetilde{X} \widetilde{Y}^3 + \widetilde{X} \widetilde{Y} z + B \widetilde{X}^2 \widetilde{Y}^2$$

with $\widetilde{Y} = Y + A\widetilde{X}$. Solve the following Artin-Schreier equation in α in the Henselian ring $R := k[[\widetilde{X}, \widetilde{Y}, z]] = k[[x, y, z]]$ with $B \in R$

$$\alpha^2 + \alpha + B = 0,$$

and make a change of variables $Z = z + \alpha\widetilde{X}\widetilde{Y}$ with a solution $\alpha \in R$. Then we have

$$z^2 + \widetilde{X}^3 + \widetilde{X}\widetilde{Y}^3 + \widetilde{X}\widetilde{Y}z + B\widetilde{X}^2\widetilde{Y}^2 \overset{c}{\sim} Z^2 + \widetilde{X}^3 + \widetilde{X}\widetilde{Y}^3 + \widetilde{X}\widetilde{Y}Z.$$

Comparison with the singularity $g = z^2 + x^3 + xyz$ tells us that E_7^3 is not 3-determined because $f_3^{(3)} = g^{(3)}$ but $\tau(g) = \infty \neq 8 = \tau(f_3)$.

1.4.3.3   Case D_{2m}^r

Let $f_{m,r}(x, y, z)$ be a defining equation of D_{2m}^r with $m \geq 2, 0 \leq r \leq m - 1$,

$$f_{m,r}(x, y, z) = z^2 + x^2 y + xy^m + m_r(x, y, z),$$

$$m_r(x, y, z) = \begin{cases} 0 & r = 0 \\ xy^{m-r}z & 1 \leq r \leq m - 1 \ . \end{cases}$$

(1) USE OF THEOREM 1.4.1.

By Theorem 1.4.1, we know D_{2m}^0 is contact $2m$-determined and D_{2m}^r is contact $(2m + 2r - 2)$-determined for any r. In fact, one can show that if $r = 0$ we have

$$\mathfrak{m} \cdot (f_{m,0}) + \mathfrak{m}^2 \cdot j(f_{m,0})$$
$$= (x, y, z) \cdot (z^2 + x^2 y + xy^m) + (x^2, y^2, z^2, xy, yz, zx) \cdot (y^m, x^2 + mxy^{m-1})$$
$$\supseteq \mathfrak{m}^{m+2}$$

and that if $r > 0$ we have

$$\mathfrak{m} \cdot (f_{m,r}) + \mathfrak{m}^2 \cdot j(f_{m,r})$$
$$= (xz^2, yz^2, z^3)$$
$$+ (y^{m+2} + y^{m-r+2}z, y^{m+1}z, x^4, x^3 y, x^2 y^2, x^3 z, x^2 yz, xy^{m-r+2}, xy^{m-r+1}z)$$
$$\supseteq \mathfrak{m}^{m+r+1}.$$

Thus the condition in Theorem 1.4.1 holds with $k = m$ for $r = 0$ and with $k = m + r - 1$ for $r > 0$. Hence $f_{m,r}$ is at least $2m$-determined if $r = 0$ and $(2m + 2r - 2)$-determined if $r > 0$.

(2) UNIT LEMMA.

As a variant of Lemma 1.4.3.1, we have the following Unit Lemma.

Lemma. *The following assertions hold.*

(1) For $\alpha, \beta, \gamma \in k[[x, y, z]]^*$ and the singularity $f_{m,0}$ of type D_{2m}^0, we have

$$\alpha z^2 + \beta x^2 y + \gamma x y^m \overset{c}{\sim} f_{m,0}(x, y, z).$$

(2) For $\alpha, \beta, \gamma, \delta \in k[[x, y, z]]^*$ and the singularity $f_{m,r}$ of type D_{2m}^r with $1 \le r \le m - 1$, we have

$$\alpha z^2 + \beta x^2 y + \gamma x y^m + \delta x y^{m-r} z \overset{c}{\sim} f_{m,r}(x, y, z).$$

(3) CLAIM: Det_ℓ FOR D_{2m}^0 FOR $m + 2 \le \ell \le 2m$.

We claim that the determinacy for D_{2m}^0 can be reduced to $\ell - 1$ under the assumption that it is ℓ-determined with $m + 2 \le \ell \le 2m$. Therefore one needs to show

$$f_{m,0}(x, y, z) + \sum_{a+b+c=\ell} d_{abc} x^a y^b z^c \overset{c}{\sim} f_{m,0}(x, y, z).$$

Since

$$f_{m,0}(x, y, z) + \sum_{a+b+c=\ell} d_{abc} x^a y^b z^c = z^2 + x^2 y + x y^m + \sum_{a+b+c=\ell} d_{abc} x^a y^b z^c$$

$$= \left(1 + \sum_{c \ge 2} d_{abc} x^a y^b z^{c-2}\right) z^2 + \left(1 + \sum_{a \ge 2, b \ge 1} d_{abc} x^{a-2} y^{b-1} z^c\right) x^2 y$$

$$+ \left(1 + \sum_{a \ge 1, b \ge m} d_{abc} x^{a-1} y^{b-m} z^c\right) x y^m$$

$$+ A x^\ell + B x^{\ell-1} z + C y^\ell + D y^{\ell-1} z$$

$$\overset{c}{\sim} z^2 + x^2 y + x y^m + u_1 x^\ell + u_2 x^{\ell-1} z + v_1 y^\ell + v_2 y^{\ell-1} z$$

for $A, B, C, D \in k$ and $u_1, u_2, v_1, v_2 \in k[[x, y, z]]$ by the Unit Lemma, we make a change of variables $Y = y + u_1 x^{\ell-2} + u_2 x^{\ell-3} z$. Then we have $x y^m = x Y^m + O(\ell + 1)$ and we thus proceed

$$z^2 + x^2 y + x y^m + u_1 x^\ell + u_2 x^{\ell-1} z + v_1 y^\ell + v_2 y^{\ell-1} z$$

$$\overset{c}{\sim} z^2 + x^2 (Y + u_1 x^{\ell-2} + u_2 x^{\ell-3} z)^m + x Y^m + v_1 (Y + u_1 x^{\ell-2} + u_2 x^{\ell-3} z)^\ell$$

$$+ v_2 (Y + u_1 x^{\ell-2} + u_2 x^{\ell-3} z)^{\ell-1} z$$

$$= z^2 + x^2 Y + x Y^m + v_1 Y^\ell + v_2 Y^{\ell-1} z + O(\ell + 1)$$

$$\overset{c}{\sim} z^2 + x^2 Y + x Y^m + v_1 Y^\ell + v_2 Y^{\ell-1} z.$$

We again make a change of variables $X = x + v_2 Y^{\ell-m-1} z$, we have

$$z^2 + x^2Y + xY^m \mid v_1Y^\ell + v_2Y^{\ell-1}z$$
$$= (1 + v_2^2Y^{2(\ell-1-m)})z^2 + X^2Y + XY^m + v_1Y^\ell \quad (X := x + v_2Y^{\ell-1-m}z)$$
$$\overset{c}{\sim} z^2 + X^2Y + XY^m + v_1Y^\ell.$$

(i) When ℓ is even, we have

$$z^2 + X^2Y + XY^m + v_1Y^\ell$$
$$= (z + v_1^{\frac{1}{2}}Y^{\frac{1}{2}})^2 + X^2Y + XY^m$$
$$\overset{c}{\sim} Z^2 + X^2Y + XY^m.$$

(ii) When ℓ is odd, let α be $v_1^{\frac{1}{2}}$ for $\ell \le 2m - 3$, or a solution of Artin-Schreier equation $t^2 + t + v_1 = 0$ for $\ell = 2m - 1$, then by changing variable $\tilde{X} := X + \alpha Y^{\frac{\ell-1}{2}}$, we have

$$z^2 + x^2Y + xY^m + v_1Y^\ell$$
$$\overset{c}{\sim} z^2 + \tilde{X}^2Y + \tilde{X}Y^m + O(\ell + 1).$$

Therefore the claim Det_ℓ is verified. for $m + 2 \le \ell \le 2m$ and D_{2m}^0 is $m + 1 = \max\{2r, m + 1\}$-determined.

(4) CLAIM: Det_ℓ FOR D_{2m}^r.

We claim that the determinacy can be reduced to $\ell - 1$ under the assumption that it is ℓ-determined with $\max\{m+2, 2r+1\} \le \ell \le 2m+2r-2$.

Let us start from

$$f_{m,r}(x, y, z) + \sum_{a+b+c=\ell} d_{abc}x^ay^bz^c,$$

where $f_{m,r}(x, y, z) = z^2 + x^2y + xy^m + xy^{m-r}z$ and this can be transformed to

$$z^2 + x^2y + xy^m + xy^{m-r}z + \sum_{a+b+c=\ell} d_{abc}x^ay^bz^c$$

$$= \left(1 + \sum_{c\ge 2} d_{abc}x^ay^bz^{c-2}\right)z^2 + \left(1 + \sum_{a\ge 2,b\ge 1} d_{abc}x^{a-2}y^{b-1}z^c\right)x^2y$$

$$+ \left(1 + \sum_{a\ge 1,b\ge m} d_{abc}x^{a-1}y^{b-m}z^c\right)xy^m + xy^{m-r}z$$

$$+ Ax^\ell + Bx^{\ell-1}z + Cy^\ell + Dy^{\ell-1}z$$

$$\overset{c}{\sim} z^2 + x^2y + xy^m + xy^{m-r}z + u_1x^\ell + u_2x^{\ell-1}z + v_1y^\ell + v_2y^{\ell-1}z$$

for $A, B, C, D \in k$ and $u_1, u_2, v_1, v_2 \in k[[x, y, z]]$ by the Unit Lemma. Change the variable $Y = y + u_1 x^{\ell-2} + u_2 x^{\ell-3} z$, then we have

$$xY^m = x(y + u_1 x^{\ell-2} + u_2 x^{\ell-3} z)^m$$

$$= xy^m + \begin{cases} mxy^{m-1}(u_1 x^{\ell-2} + u_2 x^{\ell-3} z) + O(m + 2\ell - 5) & (m \geq 3) \\ x(u_1 x^{\ell-2} + u_2 x^{\ell-3} z)^2 & (m = 2) \end{cases}$$

$$= xy^m + O(\ell + 1)$$

$$\overset{c}{\sim} xy^m.$$

On the other hand, in case of $r < m - 1$, we have

$$xY^{m-r} z = x(y + u_1 x^{\ell-2} + u_2 x^{\ell-3} z)^{m-r} z$$

$$= xy^{m-r} z + O(\ell + 1)$$

$$\overset{c}{\sim} xy^{m-r} z,$$

and in case of $r = m - 1$,

$$xY^{m-r} z = xYz = x(y + u_1 x^{\ell-2} + u_2 x^{\ell-3} z)z$$

$$= xyz + u_1 x^{\ell-1} z + u_2 x^{\ell-2} z^2.$$

We have

$$z^2 + x^2 y + xy^m + xy^{m-r} z + u_1 x^\ell + u_2 x^{\ell-1} z + v_1 y^\ell + v_2 y^{\ell-1} z$$

$$\overset{c}{\sim} \begin{cases} z^2 + x^2 Y + xY^m + xY^{m-r} z + v_1 Y^\ell + v_2 Y^{\ell-1} z & (r > m - 1) \\ (1 + u_2 x^{\ell-2})z^2 + x^2 Y + xY^m + xYz + u_1 x^{\ell-1} z + v_1 Y^\ell + v_2 Y^{\ell-1} z \\ & (r = m - 1) \end{cases}$$

$$\overset{c}{\sim} \begin{cases} z^2 + x^2 Y + xY^m + xY^{m-r} z + v_1 Y^\ell + v_2 Y^{\ell-1} z & (r > m - 1) \\ z^2 + x^2 Y + xY^m + xYz + u_1 x^{\ell-1} z + v_1 Y^\ell + v_2 Y^{\ell-1} z \\ & (r = m - 1) \end{cases}$$

and for the case $r = m - 1$, a change of variables $\tilde{Y} = Y + u_1 x^{\ell-3} z$ gives

$$x\tilde{Y}^m = x(Y + u_1 x^{\ell-3} z)^m$$

$$= xY^m + \begin{cases} mu_1 x^{\ell-2} Y^{m-1} z + O(m + \ell - 1) & (m \geq 3) \\ u_1^2 x^{2(\ell-3)+1} z^2 & (m = 2) \end{cases}$$

$$x\tilde{Y}z = xYz + u_1 x^{\ell-2} z^2,$$

thus we have

$$z^2 + x^2 Y + xY^m + xYz + u_1 x^{\ell-1} z + v_1 Y^\ell + v_2 Y^{\ell-1} z$$

$$= (1 + u_1 x^{\ell-2})z^2 + x^2 \tilde{Y} + x\tilde{Y}^m + x\tilde{Y}z + v_1 \tilde{Y}^\ell + v_2 \tilde{Y}^{\ell-1} z + O(\ell + 1)$$

$$\overset{c}{\sim} z^2 + x^2 \tilde{Y} + x\tilde{Y}^m + x\tilde{Y}z + v_1 \tilde{Y}^\ell + v_2 \tilde{Y}^{\ell-1} z.$$

Now we have

$$f_{m,r}(x,y,z) + \sum_{a+b+c=\ell} d_{abc}x^a y^b z^c$$

$$= z^2 + x^2 y + xy^m + xy^{m-r}z + \sum_{a+b+c=\ell} d_{abc}x^a y^b z^c$$

$$\overset{c}{\sim} z^2 + x^2 y + xy^m + xy^{m-r}z + v_1 y^\ell + v_2 y^{\ell-1}z.$$

We show the last equation is contact equivalent to the defining equation $f_{m,r}(x,y,z)$. First of all, by a change of variables x by $X = x + v_2 y^{\ell-m-1}z$, we have

$$z^2 + x^2 y + xy^m + xy^{m-r}z + v_1 y^\ell + v_2 y^{\ell-1}z$$

$$= z^2 + (X + v_2 y^{\ell-m-1}z)^2 y + Xy^m + (X + v_2 y^{\ell-m-1}z)y^{m-r}z + v_1 y^\ell$$

$$= (1 + v_2^2 y^{2(\ell-m-1)+1} + v_2 y^{\ell-r-1})z^2 + X^2 y + Xy^m + Xy^{m-r}z + v_1 y^\ell$$

$$\overset{c}{\sim} z^2 + X^2 y + Xy^m + Xy^{m-r}z + v_1 y^\ell.$$

Since $(v_1 - v_{10})y^\ell = O(\ell+1)$ for the constant term v_{10} of $v_1 = v_{10} + O(1)$, the hypothesis on Det_ℓ allows us to kill the terms in $v_1 - v_{10}$ and assume that $v_1 \in k$ in the last equation.

Now we divide into the four cases depending on ℓ.

(I) Case ℓ is even.

Make a change of variables z by $Z = z + v_1^{\frac{1}{2}} y^{\frac{\ell}{2}}$. Since $m - r + \frac{\ell}{2} - m = \frac{\ell-2r}{2} \geq \frac{1}{2}$ by the assumption $\ell \geq \max\{m+2, 2r+1\}$, we are done by a subsequent calculation

$$z^2 + X^2 y + Xy^m + Xy^{m-r}z + v_1 y^\ell$$

$$= Z^2 + X^2 y + Xy^m + Xy^{m-r}Z + v_1 Xy^{m-r+\frac{\ell}{2}}$$

$$= Z^2 + X^2 y + (1 + v_1 y^{\frac{\ell}{2}-r})Xy^m + Xy^{m-r}Z$$

$$\overset{c}{\sim} Z^2 + X^2 y + Xy^m + Xy^{m-r}Z.$$

(II) Case ℓ is odd and $\ell \geq 2m$.

Make a change of variables $X = x + v_1 y^{\ell-m}$. Since $\ell \geq 2m$ we have

$$z^2 + x^2 y + xy^m + xy^{m-r}z + v_1 y^\ell$$

$$= z^2 + X^2 y + Xy^m + Xy^{m-r}z + v_1^2 y^{2(\ell-m)+1} + v_1 y^{\ell-m+m-r}z$$

$$= z^2 + X^2 y + Xy^m + Xy^{m-r}z + v_1 y^{\ell-m+m-r}z + O(\ell+1)$$

$$\overset{c}{\sim} z^2 + X^2 y + Xy^m + Xy^{m-r}z + v_1 y^{\ell-m+m-r}z$$

$$\overset{c}{\sim} z^2 + X^2 y + Xy^m + Xy^{m-r}z + v_1 y^{\ell-r}z.$$

By transferring the variable X to $\tilde{X} = X + v_1 y^{\ell-r-m} z$, we are done by

$$z^2 + X^2 y + Xy^m + Xy^{m-r}z + v_1 y^{\ell-r}z$$

$$= z^2 + \tilde{X}^2 y + \tilde{X}y^m + \tilde{X}y^{m-r}z + v_1^2 y^{2(\ell-r-m)+1}z^2 + v_1 y^{\ell-r-m+m-r}z^2$$

$$= (1 + v_1^2 y^{2(\ell-r-m)+1} + v_1 y^{\ell-r-m+m-r})z^2 + \tilde{X}^2 y + \tilde{X}y^m + \tilde{X}y^{m-r}z$$

$$\overset{c}{\sim} z^2 + \tilde{X}^2 y + \tilde{X}y^m + \tilde{X}y^{m-r}z.$$

(III) CASE ℓ IS ODD AND $\ell = 2m - 1$.

Let us α be a root of an equation $\xi^2 + \xi + v_1 = 0$ of Artin-Schreier type and transfer the variable x to $X = x + \alpha y^{m-1}$. Then we have

$$z^2 + x^2 y + xy^m + xy^{m-r}z + v_1 y^{2m-1}$$

$$= z^2 + X^2 y + \alpha^2 y^{2(m-1)+1} + Xy^m + \alpha y^{2m-1}$$

$$\qquad\qquad + Xy^{m-r}z + \alpha y^{2m-r-1}z + v_1 y^{2m-1}$$

$$= z^2 + X^2 y + Xy^m + Xy^{m-r}z + \alpha y^{2m-r-1}z.$$

Transferring the variable X to $\tilde{X} = X + \alpha y^{m-r-1}z$ again, we have

$$z^2 + X^2 y + Xy^m + Xy^{m-r}z + \alpha y^{2m-r-1}$$

$$= z^2 + \tilde{X}^2 y + \tilde{X}y^m + \tilde{X}y^{m-r}z + \alpha^2 y^{2(m-r-1)+1}z^2 + \alpha y^{m-r-1+m-r}z^2$$

$$= (1 + \alpha^2 y^{2(m-r)-1} + \alpha y^{2(m-r)-1})z^2 + \tilde{X}^2 y + \tilde{X}y^m + \tilde{X}y^{m-r}z$$

$$\overset{c}{\sim} z^2 + \tilde{X}^2 y + \tilde{X}y^m + \tilde{X}y^{m-r}z.$$

(IV) CASE ℓ IS ODD AND $\ell < 2m - 1$, I.E., ℓ ODD AND $\ell \le 2m - 3$.

Transferring the variable x to $X = x + v_1^{\frac{1}{2}} y^{\frac{\ell-1}{2}}$, since $\frac{\ell-1}{2} + m \ge \ell + 1$ as $\ell \le 2m - 3$, we have

$$z^2 + x^2 y + xy^m + xy^{m-r}z + v_1 y^\ell$$

$$= z^2 + X^2 y + Xy^m + v_1^{\frac{1}{2}} y^{\frac{\ell-1}{2}+m} + Xy^{m-r}z + v_1^{\frac{1}{2}} y^{\frac{\ell-1}{2}+m-r}z$$

$$= z^2 + X^2 y + Xy^m + Xy^{m-r}z + v_1^{\frac{1}{2}} y^{\frac{\ell-1}{2}+m-r}z + O(\ell+1)$$

$$\overset{c}{\sim} z^2 + X^2 y + Xy^m + Xy^{m-r}z + v_1^{\frac{1}{2}} y^{\frac{\ell-1}{2}+m-r}z.$$

Again, by transferring the variable X to $\tilde{X} + v_1^{\frac{1}{2}} y^{\frac{\ell-1}{2}-r}z$, we are done as

$$z^2 + X^2 y + Xy^m + Xy^{m-r}z + v_1^{\frac{1}{2}} y^{\frac{\ell-1}{2}+m-r}z$$

$$= z^2 + \tilde{X}^2 y + v_1 y^{\ell-2r}z^2 + \tilde{X}y^m + \tilde{X}y^{m-r}z + v_1^{\frac{1}{2}} y^{\frac{\ell-1}{2}-r+m-r}z^2$$

$$= (1 + v_1 y^{\ell-2r} + v_1^{\frac{1}{2}} y^{\frac{\ell-1}{2}-r+m-r})z^2 + \tilde{X}^2 y + \tilde{X}y^m + \tilde{X}y^{m-r}z$$

$$\overset{c}{\sim} z^2 + \tilde{X}^2 y + \tilde{X}y^m + \tilde{X}y^{m-r}z.$$

(5) D_{2m}^r IS PRECISELY $\max\{2r, m+1\}$-DETERMINED.

For D_{2m}^0, it is enough to consider the non-normal singularity defined by $g_0 = z^2 + x^2 y$ because $g_0^{(m)} = f_{m,0}^{(m)}$. Since $\tau(g_0) = \infty$ and $\tau(f_{m,0}) = 4m$, we have $f_{m,0} \overset{c}{\not\sim} g_0$. Let us consider D_{2m}^r with $r > 0$. Suppose $m + 1 \geq 2r$. The singularity $g_1 = z^2 + x^2 y + xy^{m-r}z$ is the same as the mth jet of $f_{m,r}$, i.e., $g_1^{(m)} = f_{m,r}^{(m)}$, but g_1 is non-normal. Suppose $m + 1 < 2r$. The singularity $g_2 = z^2 + x^2 y + xy^m + xy^{m-r}z + y^{2r}$ which is contact equivalent to g_1 and hence g_2 is not contact equivalent to $f_{m,r}$.

1.5   Simple singularities

Simple singularity is a generalization of rational double point in arbitrary dimension.

For characteristic $\neq 2$, a simple singularity is an n-dimensional hypersurface singularity which comes from one of the following curve singularities by successive suspension.

A_n $(n \geq 1)$	$x^2 + y^{n+1}$ for $p = 0$ or $p \geq 3$
D_n $(n \geq 4)$	$x^2 y + y^{n-1}$ for $p = 0$ or $p \geq 3$
E_6	$x^3 + y^4$ for $p = 0$ or $p \geq 5$
E_7	$x^3 + xy^3$ for $p = 0$ or $p \geq 5$
E_8	$x^3 + y^5$ for $p = 0$ or $p \geq 7$
E_6^0	$x^3 + y^4$ for $p = 3$
E_6^1	$x^3 + y^4 + x^2 y^2$ for $p = 3$
E_7^0	$x^3 + xy^3$ for $p = 3$
E_7^1	$x^3 + xy^3 + x^2 y^2$ for $p = 3$
E_8^0	$x^3 + y^5$ for $p = 3$
E_8^1	$x^3 + y^5 + x^2 y^3$ for $p = 3$
E_8^2	$x^3 + y^5 + x^2 y^2$ for $p = 3$
E_8^0	$x^3 + y^5$ for $p = 5$
E_8^1	$x^3 + y^5 + xy^4$ for $p = 5$

Here suspension of an n-dimensional hypersurface singularity $f = f(x, y, x_2, \ldots, x_n)$ is an $(n+1)$-dimensional hypersurface singularity of the form

$$x_{n+1}^2 + f(x, y, x_2, \ldots, x_n).$$

For characteristic 2, we need to divide into two cases. For even dimension $2n$ $(n \geq 1)$, simple singularity is one which is obtained from a rational double point by successive double suspensions. For odd dimension $2n - 1$ $(n \geq 1)$, simple singularity is one which is obtained from one of the following curve singularities by successive double suspensions.

A_{2m-1} $(m \geq 1)$	$x^2 + xy^m$
A_{2m}^0 $(m \geq 1)$	$x^2 + y^{2m+1}$
A_{2m}^r $(m \geq 1, 1 \leq r \leq m-1)$	$x^2 + y^{2m+1} + xy^{2m-r}$
D_{2m} $(m \geq 2)$	$x^2 y + xy^m$
D_{2m+1}^0 $(m \geq 2)$	$x^2 y + y^{2m}$
D_{2m+1}^r $(m \geq 2, 1 \leq r \leq m-1)$	$x^2 y + y^{2m} + xy^{2m-r}$
E_6^0	$x^3 + y^4$
E_6^1	$x^3 + y^4 + xy^3$
E_7	$x^3 + xy^3$
E_8	$x^3 + y^5$

Here double suspension in characteristic 2 of an n-dimensional hypersurface singularity $f(x_0, x_1, \ldots, x_n)$ is an $(n+2)$-dimensional hypersurface singularity of the form

$$x_{n+1}x_{n+2} + f(x_0, x_1, \ldots, x_n).$$

For more details on one-dimensional simple singularities in characteristic 2, see Kiyek-Steinke [45].

1.5.1 Determinacies for simple singularities

Proposition. *(1) Contact determinacies of simple singularities in arbitrary dimension and characteristic $\neq 2$ are the same as those for the surface case.*

(2) Contact determinacies of simple singularities in even dimension and in characteristic 2 are the same as those for the surface case.

(3) Contact determinacies of simple singularities in odd dimension and in characteristic 2 are the same as those for the curve case as indicated below.

A_{2m-1}	A_{2m}^0, A_{2m}^r	D_{2m}	D_{2m+1}^0, D_{2m+1}^r	E_6^0, E_6^1	E_7	E_8
$2m-1$ $(m \geq 2)$	$2m+1$	$2m-2 (m \geq 3)$	$2m$	4	4	5
2 $(m=1)$		3 $(m=2)$				

Proof. (1) A simple singularity in arbitrary dimension and characteristic $\neq 2$ is given by successive suspensions from a simple curve singularity. Since suspension does not affect determinacy, the assertion follows.

(2) and (3). In characteristic 2, arbitrary simple singularity is given by successive double suspensions from a simple surface singularity if dimension is even, or from a simple curve singularity if dimension is odd. Double suspension does not affect the determinacy, either. See Ito-Saito [36] for the proof in characteristic 2. □

1.6 Algorithm in characteristic 2 – detecting the type of rational double point

Let $f \in \mathfrak{m}^2 \subset k[[x, y, z]]$ be a hypersurface singularity in characteristic 2. We here give an algorithm due to Greuel and Kröning [21] to determine the type of $f = 0$ provided it gives a rational double point.

We below frequently use Proposition 1.4.3 for the determinacy of rational double points in $p = 2$ and the unit lemma 1.4.3.1.

1.6.1 *Look at the quadratic part*

By the theory of quadratic forms in characteristic 2, the 2-jet of f is contact equivalent to one of the following three types.

(1) $f^{(2)} \overset{c}{\sim} xy + z^2$,
(2) $f^{(2)} \overset{c}{\sim} xy$,
(3) $f^{(2)} \overset{c}{\sim} z^2$.

1.6.2 *Quadratic Part - Case (1)*

If $f^{(2)}$ has type (1), f is a rational double point of type A_1. This assertion follows from the determinacy table 1.4.3 because A_1 singularity is 2-determined.

1.6.3 *Quadratic Part - Case (2)*

Proposition. *Assume that $f^{(2)}$ has type (2). If a nonzero monomial of the form cz^ℓ with $c \in k^*$, we let k be the smallest of such integers ℓ. Then the following assertions hold.*

(1) *If k exists, then $k \geq 3$ and $f \overset{c}{\sim} xy + z^k$. Hence f is a rational double point of type A_{k-1}.*
(2) *If k does not exist, the singularity of f is not rational. Furthermore, $f \overset{c}{\sim} xy$. This singularity is called of type $1A$ (or A_∞) in 1.6.9.*

Proof. (1) If k exists, then $k \geq 3$ clearly. Hence we can proceed

$$f = xy + \sum_{a+b+c=k} \alpha_{abc} x^a y^b z^c + O(k+1)$$

$$= x(y + \sum_{a \geq 1} \alpha_{abc} x^{a-1} y^b z^c) + \sum_{b+c=k} \alpha_{0bc} y^b z^c + O(k+1)$$

$$= xY + \sum_{b+c=k} \alpha_{0bc} Y^b z^c + O(k+1) \qquad (Y := y + \sum_{a \geq 1} \alpha_{abc} x^{a-1} y^b z^c)$$

$$= (x + \sum_{b \geq 1} \alpha_{0bc} Y^{b-1} z^c) Y + \alpha_{00k} z^k + O(k+1)$$

$$= XY + \alpha_{00k} z^k + O(k+1) \qquad (X := x + \sum_{b \geq 1} \alpha_{0bc} Y^{b-1} z^c)$$

$$\overset{c}{\sim} XY + z^k.$$

(2) If k does not exist, the above calculation lead to $f \overset{c}{\sim} xy$. The classification of equations of rational double points denies rationality of double point singularity f.

$\square$

1.6.4 Quadratic Part - Case (3)

Since $f^{(2)} \overset{c}{\sim} z^2$, one can write f as

$$f \overset{c}{\sim} z^2 + \varphi(x,y)z + \psi(x,y)$$

with $\varphi(x,y) = O(2), \psi(x,y) = O(3)$. We divide it into four subcases according to the degree 3 part of $\psi(x,y)$,

(i) $\psi^{(3)} = 0$,
(ii) $\psi^{(3)} \overset{c}{\sim} x^2 y + xy^2$,
(iii) $\psi^{(3)} \overset{c}{\sim} x^2 y$,
(iv) $\psi^{(3)} \overset{c}{\sim} x^3$.

In fact, the 3-jet $\psi^{(3)}$ of ψ defines a set of three lines on $\mathbb{A}^2$ passing through the origin. The case (i) corresponds to $\psi^{(3)}(x,y) = 0$, the case (ii) to the case where $\psi^{(3)}(x,y) = 0$ consists of three distinct lines, the case (iii) to the case where $\psi^{(3)}(x,y) = 0$ consists of one double line and another line, and finally the case (iv) to the case where $\psi^{(3)}(x,y) = 0$ is a triple line.

1.6.5　*Case (3)-(i)*

For this case, f is not rational double point. In fact, by looking at $\varphi^{(2)}(x, y)$ and arguing as in 1.6.4 for $\psi^{(3)}(x, y, z)$, we know that it is contact equivalent to one of

$$f \overset{c}{\sim} z^2 + O(4),$$
$$f \overset{c}{\sim} z^2 + x^2 z + O(4),$$
$$f \overset{c}{\sim} z^2 + xyz + O(4).$$

These are non-rational double points of type $2A, 2B, 2C$ in 1.6.9, respectively.

1.6.6　*Case (3)-(ii)*

Since f can be written as

$$z^2 + x^2 y + xy^2 + \varphi(x, y)z + O(4), \quad \varphi(x, y) = ax^2 + bxy + cy^2, \quad a, b, c \in k$$

up to contact equivalence, it follows that

$$
\begin{aligned}
f \overset{c}{\sim} {}& z^2 + x^2 y + xy^2 + (ax^2 + bxy + cy^2)z \\
= {}& z^2 + x^2(y + az) + xy^2 + (bxy + cy^2)z \\
= {}& (1 + a^2 x + abx + a^2 cz)z^2 + x^2 Y + xY^2 + (bxY + cY^2)z, \\
& \hspace{6.5cm} (Y := y + az) \\
\overset{c}{\sim} {}& z^2 + x^2 Y + xY^2 + bxYz + cY^2 z \\
= {}& z^2 + x^2 Y + (x + cz)Y^2 + bxYz \\
= {}& (1 + bcY + c^2 Y)z^2 + X^2 Y + XY^2 + bXYz, \quad (X := x + cz) \\
\overset{c}{\sim} {}& z^2 + X^2 Y + XY^2 + bXYz.
\end{aligned}
$$

This is a rational double point of type

$$
\begin{cases}
D_4^0 & \text{if } b = 0 \\
D_4^1 & \text{if } b \neq 0.
\end{cases}
$$

1.6.7　*Case (3)-(iii)*

By the similar change of variables as in the previous case (3)-(ii), we can write

$$f \overset{c}{\sim} z^2 + x^2 y + axyz + by^2 z + O(4),$$

because

$$z^2 + x^2y + (axy + by^2 + cx^2)z + O(4)$$
$$= z^2 + x^2(y + cz) + axyz + by^2z + O(4)$$
$$= (1 + acx + bc^2z)z^2 + x^2Y + axYz + bY^2z + O(4),$$

where $Y := y + cz$.

This case is divided into four subcases.

1.6.7.1 Case (3)-(iii-1) $ab \neq 0$

For this case, f is contact equivalent to

$$f \overset{c}{\sim} z^2 + x^2y + xyz + y^2z + O(4)$$

which is of type D_5^1.

1.6.7.2 Case (3)-(iii-2) $a = 0, b \neq 0$

For this case, f is contact equivalent to

$$f \overset{c}{\sim} z^2 + x^2y + y^2z + O(4)$$

which is of type D_5^0.

1.6.7.3 Case (3)-(iii-3) $a \neq 0, b = 0$

$$f \overset{c}{\sim} z^2 + x^2y + xyz + O(4),$$

this can be written as

$$Z^2 + X^2Y + XYZ + \epsilon Y^k + O(k+1)$$

up to contact equivalence.

Proof. For $k \geq 4$,

$$z^2 + x^2y + xyz + \sum_{a+b+c=k} \alpha_{abc} x^a y^b z^c + O(k+1)$$

$$= (1 + \sum_{c \geq 2} \alpha_{abc} x^a y^b z^{c-2})z^2 + (1 + \sum_{a \geq 2, b \geq 1} \alpha_{abc} x^{a-2} y^{b-1} z^c)x^2y$$

$$+ (1 + \sum_{a \geq 1, b \geq 1, c \geq 1} \alpha_{abc} x^{a-1} y^{b-1} z^{c-1})xyz$$

$$+ \alpha x^{k-1}z + \beta x^k + \gamma xy^{k-1} + \delta y^{k-1}z + \epsilon y^k + O(k+1)$$

$$\overset{c}{\sim} z^2 + x^2 Y + xYz + \gamma x Y^{k-1} + \delta Y^{k-1} z + \epsilon Y^k + O(k+1)$$
$$\text{where } Y := y + \beta x^{k-2} + (\beta + \alpha) x^{k-3} z$$
$$= z^2 + x^2 Y + xY(z + \gamma Y^{k-2}) + \delta Y^{k-1} z + \epsilon Y^k + O(k+1)$$
$$\overset{c}{\sim} Z^2 + x^2 Y + xYZ + \delta Y^{k-1} Z + \epsilon Y^k + O(k+1) \quad \text{where } Z := z + \gamma Y^{k-2}$$
$$= Z^2 + x^2 Y + (x + \delta Y^{k-2}) YZ + \epsilon Y^k + O(k+a).$$

$\square$

Therefore, if there exists $\epsilon \neq 0$ for some $k \geq 4$ where k is smallest integer with $\gamma \neq 0$, then f is of type D^r_{k+2} with $r = \frac{k-1}{2}$ for odd k, or $r = \frac{k}{2}$ for even k. If there exists no such k, then it is of type D^1_∞ in 1.6.9.

Note that these D^r_{k+2} can be rewritten as

$$
\begin{cases}
D^r_{2m+1} \text{ with } \begin{cases} r = \frac{k-1}{2} \\ m = \frac{k+1}{2} \end{cases} & \text{for odd } k \\[2em]
D^r_{2m} \text{ with } \begin{cases} r = \frac{k}{2} \\ m = \frac{k+2}{2} \end{cases} & \text{for even } k
\end{cases}
$$

and that they satisfy $m - r = 1$.

1.6.7.4　 *Case (3)-(iii-4)* $a = b = 0$

For the case $a = 0$ in 1.6.7,

$$f \overset{c}{\sim} z^2 + x^2 y + O(4).$$

Proposition.　*Let k be an integer with $k \geq 4$, then f is contact equivalent to*

$$f \overset{c}{\sim} z^2 + x^2 y + \alpha xy^{k-2} z + \beta y^{k-1} z + \gamma xy^{k-1} + O(k+1).$$

Proof.　If we start from the equation

$$f = z^2 + x^2 y + \sum_{a+b+c=k} \alpha_{abc} x^a y^b z^c + O(k+1)$$

we have

$$f \overset{c}{\sim} z^2 + x^2 y + A x^{k-1} z + B x^k + C y^k$$
$$+ \alpha xy^{k-2} z + \beta y^{k-1} z + \gamma xy^{k-1} + O(k+1)$$

by the same treatment of units as in the previous case $a \neq 0$.

Since
$$x^2y + Ax^{k-1}z + Bx^k = x^2(y + Ax^{k-3}z + Bx^{k-2})$$
change of coordinate $Y := y + Ax^{k-3}z + Bx^{k-2}$ gives
$$f \overset{c}{\sim} z^2 + x^2y + Cy^k + \alpha xy^{k-2}z + \beta y^{k-1}z + \gamma xy^{k-1} + O(k+1).$$

Change of coordinate $Z := z + C^{\frac{1}{2}}y^{\frac{k}{2}}$ for even k, change of coordinate $X := x + C^{\frac{1}{2}}y^{\frac{k-1}{2}}$ for odd k, gives the conclusion. $\qquad\square$

Now suppose one of α, β or γ be nonzero for some k (here take k smallest among such ones), or otherwise it is not a rational double point $\overset{c}{\sim} z^2 + x^2y$ which is of type D_∞^0 in 1.6.9.

Then one of the following cases occurs.

(1) When $\gamma \neq 0$, it is of type
$$\begin{cases} D_{2k-2}^1 \text{ for } \alpha \neq 0 \\ D_{2k-2}^0 \text{ for } \alpha = 0. \end{cases}$$

(2) When $\gamma = 0$ and $\beta \neq 0$, it is of type
$$\begin{cases} D_{2k-1}^1 \text{ for } \alpha \neq 0 \\ D_{2k-1}^0 \text{ for } \alpha = 0. \end{cases}$$

(3) Suppose that $\gamma = \beta = 0$, then $\alpha \neq 0$, that is,
$$f \overset{c}{\sim} z^2 + x^2y + xy^{k-2}z + O(k+1).$$

Then f can be written as
$$f \overset{c}{\sim} z^2 + x^2y + xy^{k-2}z + \delta y^{\ell-1}z + \epsilon xy^{\ell-1} + O(\ell+1),$$
for some $\ell > k$ with one of δ and ϵ being nonzero. It is of type
$$\begin{cases} D_{2\ell-2}^{\ell-k+1} \text{ for } \epsilon \neq 0 \\ D_{2\ell-1}^{\ell-k+1} \text{ for } \epsilon = 0, \delta \neq 0. \end{cases}$$

Furthermore, if there does not exist such $\ell > k$, then it is of type D_∞^{k-2} in 1.6.9.

Proof. (1) Since $\gamma \neq 0$,
$$\overset{c}{\sim} z^2 + x^2y + xy^{k-1} + \alpha xy^{k-2}z + \beta y^{k-1}z + O(k+1)$$
$$= z^2 + x^2y + (x + \beta z)y^{k-1} + \alpha xy^{k-2}z + O(k+1)$$
$$= (1 + \beta^2 y + \alpha\beta y^{k-2})z^2 + X^2y + Xy^{k-1} + \alpha Xy^{k-2}z + O(k+1)$$
$$\overset{c}{\sim} z^2 + X^2y + Xy^{k-1} + \alpha Xy^{k-2}z + O(k+1) \text{ where } X := x + \beta z.$$

Thus we get a conclusion from the table 1.1.7 and determinacy theorem 1.4.3.

(2) Since $\gamma = 0$ and $\beta \neq 0$,

$$f \overset{c}{\sim} z^2 + x^2 y + y^{k-1} z + \alpha x y^{k-2} z + O(k+1).$$

Thus we get a conclusion.

(3) Let us write

$$f = z^2 + x^2 y + x y^{k-2} z + \sum_{a+b+c=\ell} \alpha_{abc} x^a y^b z^c + O(\ell+1)$$

with some $\ell > k$.

Then by the same argument as before using unit lemma 1.4.3.1, we have

$$f \overset{c}{\sim} z^2 + x^2 y + x y^{k-2} z + \delta y^{\ell-1} z + \epsilon x y^{\ell-1} + A x^\ell + B x^{\ell-1} z + C y^\ell$$
$$= z^2 + x^2 (y + A x^{\ell-2} + B x^{\ell-3} z) + x y^{k-2} z$$
$$\quad + \delta y^{\ell-1} z + \epsilon x y^{\ell-1} + C y^\ell + O(\ell+1)$$
$$\overset{c}{\sim} z^2 + x^2 y + x y^{k-2} z + \delta y^{\ell-1} z + \epsilon x y^{\ell-1} + C y^\ell + O(\ell+1)$$

and $C y^\ell$ can be put into z or x according to ℓ being even or odd as before. The type of singularity is clear by the table 1.1.7 and determinacy theorem 1.4.3.   □

1.6.8   *Case (3)-(iv)*

In this case, we can write f as

$$f = z^2 + x^3 + \varphi(x, y) z + O(4) \quad \text{with } \varphi(x, y) = c x^2 + a x y + b y^2$$

for some $a, b, c \in k$.

Then by almost same argument as in Case (3)-(ii) 1.6.6, we have

$$f = z^2 + x^3 + c x^2 z + a x y z + b y^2 z + O(4)$$
$$= (1 + c^2 X + a c y) z^2 + X^3 + a X y z + b y^2 z + O(4), \text{ where } X = x + c z$$
$$\overset{c}{\sim} z^2 + X^3 + a X y z + b y^2 z + O(4).$$

This case is divided into three subcases.

1.6.8.1   *Case (3)-(iv-1) $b \neq 0$*

Since f is contact equivalent to

$$z^2 + x^3 + y^2 z + a x y z + O(5),$$

it is of type

$$\begin{cases} E_6^0 & \text{for } a = 0 \\ E_6^1 & \text{for } a \neq 0. \end{cases}$$

1.6.8.2 Case (3)-(iv-2) $b = 0, a \neq 0$

Since
$$f \overset{c}{\sim} z^2 + x^3 + xyz + O(4),$$
we can write it as
$$f \overset{c}{\sim} z^2 + x^3 + xyz + \sum_{a+b+c=4} x^a y^b z^c + O(5)$$

$$= (1 + \sum_{c \geq 2} \alpha_{abc} x^a y^b z^{c-2}) z^2 + (1 + \sum_{a \geq 3} \alpha_{abc} x^{a-3} y^b z^c) x^3$$

$$+ (1 + \sum_{a \geq 1, b \geq 1, c \geq 1} \alpha_{abc} x^{a-1} y^{b-1} z^{c-1}) xyz + a y^3 z + b x^2 y^2 + c x y^3 + d y^4$$

$$\overset{c}{\sim} z^2 + x^3 + (x + a y^2) yz + b x^2 y^2 + c x y^3 + d y^4 + O(5)$$

$$\overset{c}{\sim} z^2 + X^3 + Xyz + (a+b) X^2 y^2 + c X y^3 + d y^4 + O(5),$$

where $X := x + a y^2$.

Now, take s and t as
$$\begin{cases} t^2 = d \\ s^2 + s + (a+b) = 0, \end{cases}$$
and change of coordinate $Z := z + sXy + ty^2$, then we have
$$f \overset{c}{\sim} Z^2 + X^3 + XyZ + (c + d^{\frac{1}{2}}) X y^3 + O(5).$$

Thus we can rewrite in this case as
$$f \overset{c}{\sim} Z^2 + X^3 + XyZ + c X y^3 + O(5),$$
which is of type E_8^4 when $c \neq 0$. For the case $c = 0$, it is contact equivalent to
$$f \overset{c}{\sim} z^2 + x^3 + xyz + d y^5 + O(6).$$

Proof.
$$f \overset{c}{\sim} z^2 + x^3 + xyz + \sum_{a+b+c=5} \alpha_{abc} x^a y^b z^c + O(6)$$

$$= (1 + \sum_{c \geq 2} \alpha_{abc} x^a y^b z^{c-2}) z^2 + (1 + \sum_{a \geq 3} \alpha_{abc} x^{a-3} y^b z^c) x^3$$

$$+ (1 + \sum_{a \geq 1, b \geq 1, c \geq 1} \alpha_{abc} x^{a-1} y^{b-1} z^{c-1}) xyz$$

$$+ a y^4 z + b x^2 y^3 + c x y^4 + d y^5 + O(6)$$

$$\overset{c}{\sim} z^2 + X^3 + Xyz + (a+b) X^2 y^3 + c X y^4 + d y^5 + O(6),$$

where $X := x + a y^3$

$$\overset{c}{\sim} Z^2 + X^3 + XyZ + d y^5 + O(6), \quad \text{where } Z := z + (a+b) X y^2 + c y^3.$$

$\square$

Then it is of type

$$\begin{cases} E_8^4 & \text{for } d \neq 0 \\ \text{not rational double point (of type 3F in 1.6.9)} & \text{for } d = 0 \,. \end{cases}$$

1.6.8.3   Case (3)-(iv-3) $a = b = 0$

In the case of $a = b = 0$ in 1.6.8, f can start from

$$f \overset{c}{\sim} z^2 + x^3 + \alpha xy^2 z + \beta y^3 z + \gamma xy^3 + O(5).$$

Proof.   Since $a = b = 0$ in 1.6.8, we have

$$f \overset{c}{\sim} z^2 + x^3 + O(4)$$

$$= z^2 + x^3 + \sum_{a+b+c=4} \alpha_{abc} x^a y^b z^c + O(5)$$

$$= (1 + \sum_{c \geq 2} \alpha_{abc} x^a y^b z^{c-2}) z^2 + (1 + \sum_{a \geq 3} \alpha_{abc} x^{a-3} y^b z^c) x^3$$

$$\quad + \alpha x^2 yz + \beta xy^2 z + \gamma y^3 z + \delta x^2 y^2 + \epsilon xy^3 + \varphi y^4 + O(5)$$

$$\overset{c}{\sim} z^2 + x^3 + \alpha x^2 yz + \beta xy^2 z + \gamma y^3 z + \delta x^2 y^2 + \epsilon xy^3 + \varphi y^4 + O(5)$$

$$\overset{c}{\sim} Z^2 + x^3 + \alpha x^2 yZ + \beta xy^2 Z + \gamma y^3 Z + \epsilon xy^3 + O(5),$$

$$\quad \text{with } Z := z + \delta^{\frac{1}{2}} xy + \varphi^{\frac{1}{2}} y^2$$

$$\overset{c}{\sim} Z^2 + X^3 + \beta Xy^2 Z + \gamma y^3 Z + \epsilon Xy^3 + O(5),$$

$$\quad \text{with } X := x + \alpha yZ.$$

$\square$

Now we divide into further five cases according to the combination of α, β and γ as follows.

1.6.8.4   Case (3)-(iv-3-1) $\beta\gamma \neq 0$

Since $\gamma \neq 0$ and $\beta \neq 0$, we have

$$f \overset{c}{\sim} z^2 + x^3 + \alpha xy^2 z + y^3 z + xy^3 + O(5)$$

$$= (1 + \alpha^2 xY + \alpha Y^2 + \alpha^2 Yz + \alpha^3 z^2) z^2 + x^3 + xY^3 + Y^3 z + O(5),$$

$$\quad \text{with } Y := y + \alpha z$$

$$\overset{c}{\sim} z^2 + x^3 + xy^3 + y^3 z,$$

which is of type E_7^2.

1.6.8.5 *Case (3)-(iv-3-2) $\beta = 0, \gamma \neq 0$*

Since $\gamma \neq 0$ and $\beta = 0$, we have

$$f \overset{c}{\sim} z^2 + x^3 + \alpha xy^2 z + xy^3 + O(5)$$
$$= (1 + \alpha^2 xY)z^2 + x^3 + xY^3 + O(5), \text{ with } Y := y + \alpha z$$
$$\overset{c}{\sim} z^2 + x^3 + xY^3 + O(5).$$

Next we look into the degree 5 terms. Using the same argument before, we have

$$f = z^2 + x^3 + xy^3 + \sum_{a+b+c=5} \alpha_{abc} x^a y^b z^c + O(6)$$
$$\overset{c}{\sim} z^2 + x^3 + xy^3 + \delta x^2 y^2 z + \epsilon y^4 z + \varphi y^5 + O(6)$$
$$= (z + \varphi^{\frac{1}{2}} Xy)^2 + X^3 + (1 + \varphi^2 y)Xy^3 + \epsilon y^4 z, \text{ with } X := x + \delta y^2 z + \varphi y^2$$
$$\overset{c}{\sim} Z^2 + X^3 + Xy^3 + \epsilon y^4 Z,$$

therefore it is of type E_7^0 for $\epsilon = 0$, or type E_7^1 for $\epsilon \neq 0$.

1.6.8.6 *Case (3)-(iv-3-3) $\beta \neq 0, \gamma = 0$*

Start from $f \overset{c}{\sim} z^2 + x^3 + \alpha xy^2 z + y^3 z + O(5)$. After coordinate changes $Y := y + \alpha x$ and $X := x + \alpha^2 Y z$, we have

$$f \overset{c}{\sim} z^2 + X^3 + Y^3 z + O(5).$$

Next we need to look into the degree 5 terms. We can transform as before, we have

$$f = z^2 + x^3 + y^3 z + \sum_{a+b+c=5} \alpha_{abc} x^a y^b z^c + O(6)$$
$$\overset{c}{\sim} z^2 + x^3 + y^3 z + Ax^2 y^2 z + Bx^2 y^3 + Cxy^4 + Dy^5 + O(6)$$
$$= Z^2 + (1 + B^2 x)x^3 + y^3 Z + Ax^2 y^2 Z + Cxy^4 + Dy^5 + O(6),$$
$$\text{with } Z := z + Bx^2$$
$$\overset{c}{\sim} Z^2 + x^3 + y^3 Z + Ax^2 y^2 Z + Cxy^4 + Dy^5 + O(6)$$
$$\overset{c}{\sim} Z^2 + x^3 + Y^3 Z + CxY^4 + DY^5 + O(6), \text{ with } Y := y + Ax^2.$$

If $D \neq 0$, then we may assume $D = 1$ and change coordinates $X := x + \alpha^2 y^2$ and $\widetilde{Z} := Z + \alpha Xy$ with $\alpha^4 + \alpha + C = 0$, we have

$$f = Z^2 + x^3 + Y^3 Z + CxY^4 + Y^5 + O(6)$$
$$\overset{c}{\sim} \widetilde{Z}^2 + X^3 + Y^5 + Y^3 \widetilde{Z},$$

thus it is of type E_8^3.

Suppose $D = 0$, then

$$f \overset{c}{\sim} z^2 + x^3 + y^3 z + Cxy^4 + O(6),$$

which is not a rational double point. (Of type 3E in 1.6.9.)

1.6.8.7 Case (3)-(iv-3-4) $\beta = \gamma = 0, \alpha \neq 0$

We can start from $f \overset{c}{\sim} z^2 + x^3 + xy^2 z + O(5)$. By the same argument as before, we have

$$f \overset{c}{\sim} z^2 + x^3 + xy^2 z + Ay^4 z + Bx^2 y^3 + Cxy^4 + Dy^5 + O(6)$$
$$\overset{c}{\sim} z^2 + X^3 + Xy^2 z + Ay^4 z + CXy^4 + Dy^5 + O(6), \text{ with } X := x + By^3$$
$$\overset{c}{\sim} Z^2 + \tilde{X}^3 + \tilde{X}y^2 Z + Dy^5 + (A^2 + C)\tilde{X}y^4 + O(6),$$

with $X := X + Ay^2 + A^{\frac{1}{2}}y^3, Z := z + A^{\frac{1}{2}}\tilde{X}y$.

When $D \neq 0$, we may assume $D = 1$ and we have

$$f \overset{c}{\sim} Z^2 + \tilde{X}^3 + \tilde{X}y^2 Z + y^5 + (A^2 + C)\tilde{X}y^4 + O(6)$$
$$= Z^2 + (1 + (A^2 + C)\tilde{X}Y)\tilde{X}^3 + \tilde{X}Y^2 Z + Y^5 + O(6),$$

$$\text{with } Y := y + (A^2 + C)\tilde{X}$$

$$\overset{c}{\sim} Z^2 + \tilde{X}^3 + Y^5 + \tilde{X}Y^2 Z + O(6),$$

which is of type E_8^2.

Suppose $D = 0$, we have

$$f = Z^2 + \tilde{X}^3 + \tilde{X}y^2 Z + (A^2 + C)\tilde{X}y^4 + O(6)$$

which is not a rational double point. (Of type 3D in 1.6.9.)

1.6.8.8 Case (3)-(iv-3-5) $\beta = \gamma = \alpha = 0$

Let us start from $f = z^2 + x^3 + O(5)$. Same argument allows us to start from

$$f \overset{c}{\sim} z^2 + x^3 + Ax^2 y^2 z + Bxy^3 z + Cy^4 z + Dx^2 y^3 + Exy^4 + Fy^5 + O(6).$$

By changing the coordinate $X := x + Ay^2 z + Dy^3 + E^{\frac{1}{2}}y^2$, this is contact equivalent to

$$(z + E^{\frac{1}{4}}Xy)^2 + X^3 + BXy^3 z + Cy^4 z + Fy^5 + O(6)$$
$$\overset{c}{\sim} Z^2 + X^3 + BXy^3 Z + Cy^4 Z + Fy^5 + O(6), \text{ with } Z := z + E^{\frac{1}{4}}Xy.$$

When $F \neq 0$, we may assume $F = 1$ and we have

$$f \overset{c}{\sim} z^2 + x^3 + y^5 + Bxy^3z + Cy^4z + O(6)$$

$$\overset{c}{\sim} z^2 + x^3 + Y^5 + BxY^3z + O(6), \text{ with } Y := y + Cz,$$

which is of type E_8^0 for $B = 0$, or E_8^1 for $B \neq 0$.

Next, suppose $F = 0$, we have $f = z^2 + x^3 + Bxy^3z + Cy^4z + O(6)$, which is not rational double point. (More precisely, this is of type

$$\begin{cases} 3C \text{ for } B \neq 0 \\ 3A \text{ for } B = C = 0 \\ 3B \text{ for } B = 0, C \neq 0 \end{cases}$$

with the notation in 1.6.9.)

1.6.9 *Non-rational double points*

These non-rational double points are listed into the following table which is the same as the one in [21], whereas the case 3B is missing in [*ibid.*].

ours	G-K [21]	equation
1A	A_∞	$f \overset{c}{\sim} xy$
1B	D_∞^0	$f \overset{c}{\sim} z^2 + x^2y$
1C	D_∞^n	$f \overset{c}{\sim} z^2 + x^2y + xy^nz$
2A		$f \overset{c}{\sim} z^2 + O(4)$
2B		$f \overset{c}{\sim} z^2 + x^2z + O(4)$
2C		$f \overset{c}{\sim} z^2 + xyz$
3A		$f \overset{c}{\sim} z^2 + x^3 + O(6)$
3B	missed	$f \overset{c}{\sim} z^2 + x^3 + xy^4 + O(6)$
3C		$f \overset{c}{\sim} z^2 + x^3 + xy^3z + dy^4z + O(6)$
3D		$f \overset{c}{\sim} z^2 + x^3 + xy^2z + dxy^4 + O(6)$
3E		$f \overset{c}{\sim} z^2 + x^3 + y^3z + dxy^4 + O(6)$
3F		$f \overset{c}{\sim} z^2 + x^3 + xyz + O(6)$

Greuel-Kröning [21] denotes D_∞^0 by $D_\infty(0)$, D_∞^n by $D_\infty(n)$, and refers the remaining cases to step numbers where the corresponding equations appear.

1.6.10 *Flowchart*

The above algorithm to determine the type of rational (or non-rational) double points in characteristic 2 is summarized into the following flowchart.

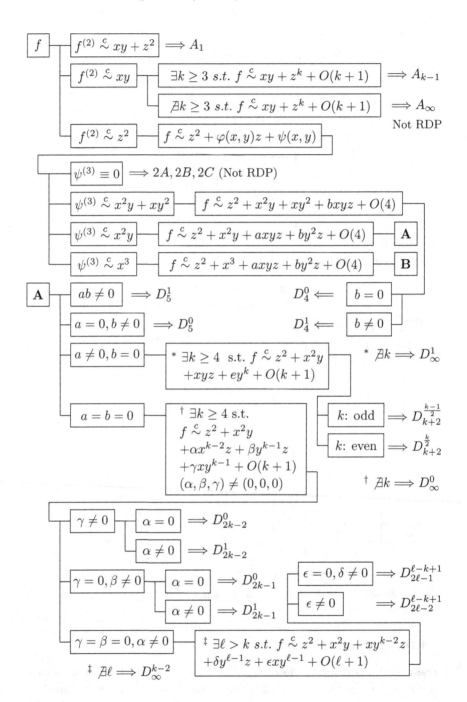

B

$b \neq 0$

$a = 0 \implies E_6^0$

$a \neq 0 \implies E_6^1$

$b = 0, a \neq 0$

$f \overset{c}{\sim} z^2 + x^3 + xyz + cxy^3 + O(5)$

$c \neq 0 \implies E_7^3$

$c = 0$

$f \overset{c}{\sim} z^2 + x^3 + xyz + dy^5 + O(6)$

$d \neq 0 \implies E_8^4$

$d = 0 \implies 3F$ (Not RDP)

$b = a = 0$

$f \overset{c}{\sim} z^2 + x^3 + \alpha xy^2 z + \beta y^3 z + \gamma xy^3 z + O(5)$

$\beta\gamma \neq 0 \implies E_7^2$

$\beta = 0, \gamma \neq 0$

$f \overset{c}{\sim} z^2 + x^3 + xy^3 + \epsilon y^4 z + O(6)$

$\epsilon = 0 \implies E_7^0$

$\epsilon \neq 0 \implies E_7^1$

$D \neq 0 \implies E_8^3$

$D = 0 \implies 3E$
Not RDP

$\beta \neq 0, \gamma = 0$

$f \overset{c}{\sim} z^2 + x^3 + y^3 z$
$+ Cxy^4 + Dy^5 + O(6)$

$\beta = \gamma = 0, \alpha \neq 0$

$f \overset{c}{\sim} z^2 + x^3 + xy^2 z$
$Ay^4 z + Cxy^4 + Dy^5 + O(6)$

$D \neq 0 \implies E_8^2$

$D = 0 \implies 3D$
Not RDP

$F \neq 0, B = 0 \implies E_8^0$

$FB \neq 0 \implies E_8^1$

$F = 0 \implies$
$3C (B \neq 0)$
$3A$
$\quad (B = C = 0)$
$3B$
$\quad (B = 0,$
$\qquad C \neq 0)$

$\gamma = \beta = \alpha = 0$

$f \overset{c}{\sim} z^2 + x^3 + Bxy^3 z$
$+ Cy^4 z + Fy^5 + O(6)$

Chapter 2

Deformation of rational double points

2.1 Versal deformation

Let k be an algebraically closed field of characteristic $p \geq 0$ and $k[[x, y, z_1, z_2, \cdots, z_{n-1}]]$ be the formal power series ring over k with maximal ideal $\mathfrak{m}$. For a nonzero power series $f \in k[[x, y, z_1, z_2, \cdots, z_{n-1}]]$, it is called a *hypersurface singularity* if $f \in \mathfrak{m}^2$. A singularity $f \in k[[x, y, z_1, z_2, \cdots, z_{n-1}]]$ is called an *isolated singularity* if there exists $l > 0$ such that the Tjurina ideal

$$\mathrm{Tj}\,(f) := \left\langle f, \frac{\partial f}{\partial x}, \frac{\partial f}{\partial y}, \frac{\partial f}{\partial z_1}, \cdots, \frac{\partial f}{\partial z_{n-1}} \right\rangle$$

contains $\mathfrak{m}^l$. Then the *Tjurina number*

$$\tau := \tau(f) = \dim_k k[[x, y, z_1, z_2, \cdots, z_{n-1}]]/\mathrm{Tj}\,(f)$$

is finite.

We define an algebraic representative of the versal deformation of the singular point $P \in V(f) = \mathrm{Spec}\,(k[[x, y, z_1, z_2, \cdots, z_{n-1}]]/(f))$ for $f \in k[x, y, z_1, z_2, \cdots, z_{n-1}]$ as follows:

Choose $g_1, g_2, \cdots, g_\tau \in k[x, y, z_1, z_2, \cdots, z_{n-1}]$ which represent a k-basis of $k[[x, y, z_1, z_2, \cdots, z_{n-1}]]/Tj(f)$, and define

$$F(x, y, z_1, z_2, \cdots, z_{n-1})$$

$$:= f(x, y, z_1, z_2, \cdots, z_{n-1}) + \sum_{i=1}^\tau t_i g_i(x, y, z_1, z_2, \cdots, z_{n-1})$$

$$\in k[x, y, z_1, z_2, \cdots, z_{n-1}][t_1, t_2, \cdots, t_\tau].$$

Set $\mathrm{Def}\,(f) := \mathrm{Spec}\,k[t_1, t_2, \cdots, t_\tau] \cong \mathbb{A}^\tau$. We call it the *versal deformation space* of f and $F(x, y, z_1, z_2, \cdots, z_{n-1})$ an *algebraic representative* of the

389

versal deformation of f. Hereafter, we treat only the case where f is finitely determined. Namely we assume that $f \in k[x, y, z_1, z_2, \cdots, z_{n-1}]$. We define the natural map $\varphi : V(F) \to \mathrm{Def}\,(f)$, where $V(F)$ is the closed subset of $\mathrm{Spec}\,k[x, y, z_1, z_2, \cdots, z_{n-1}, t_1, t_2, \cdots, t_r]$ defined by $F = 0$.

A reference for singularities and deformations is a book by Greuel-Lossen-Shustin [22].

2.2 Equisingular loci

Let $f \in k[[x, y, z_1, z_2, \cdots, z_{n-1}]]$ be an isolated hypersurface singularity. We define the *equisingular locus* $\mathrm{ES}(f)$ of f in an intuitive way by

$$ES(f) := \{P \in \mathrm{Def}\,(f) \mid \varphi^{-1}(P) \text{ has the same singularity as } f = 0 \text{ has.}\}$$

as a subset of $\mathrm{Def}\,(f)$.

It is known that the equisingular locus is trivial in characteristic zero, i.e., $\mathrm{ES}(f) = \{$ point of origin $\}$ for f which defines a rational double point (see [74] remark after Corollary 1.14). The readers are advised to make calculations with standard form of f.

Unless in characteristic zero, there exists nontrivial equisingular loci for many rational double points in positive characteristic, especially, in low characteristic.

We exhibit below the results on all equisingular loci in arbitrary characteristic. But we will give a proof for only the cases E_7^r and D_{2m}^r in characteristic 2, but the proof need tedious calculations though they are important and similar calculations. More precisely, see Hirokado-Ito-Saito [30] for characteristic $p \geq 3$ and Ito-Saito [36] for characteristic 2.

2.2.1 Type A_n in positive characteristic

Theorem. *Consider the rational double point $z^{n+1} + xy$ of type A_n ($n \geq 1$) in characteristic $p > 0$. The equisingular locus is trivial if $p \nmid (n+1)$. If $p \mid (n+1)$, write $n+1 = p^e m$ with $(p, m) = 1$. Then the equisingular locus is*

$$xy + (z^{p^e} + \alpha)^m$$

with one-dimensional parameter α inside the $(n+1)$-dimensional versal deformation space

$$xy + z^{n+1} + a_n y^n + a_{n-1} y^{n-1} + \cdots + a_0.$$

Remark. The Tjurina number of A_n for any characteristic $p > 0$ is $n+1$ if $p \mid (n+1)$. Otherwise, it is n which is the same as in characteristic 0 (1.3.2).

2.2.2   *Characteristic $p > 5$*

Theorem. *Except for a rational double point of type A_n with $p \mid (n+1)$, all other rational double points have trivial equisingular loci if the characteristic is greater than five.*

Remark. We give versal deformation spaces of type other than the exceptional types in characteristic 2, 3 and 5 to be treated in 2.2.3, 2.2.4 and 2.2.5 below.

$D_k(p > 2):$ $z^2 + x^2 y + y^{k-1} + a_{k-2}y^{k-2} + a_{k-3}y^{k-3} + \cdots + a_0 + bx$

$E_6(p > 3):$ $z^2 + x^3 + y^5 + a_1 xy^2 + a_2 xy + a_3 x + a_4 y^2 + a_5 y + a_6$

$E_7(p > 3):$ $z^2 + x^3 + xy^3 + a_1 y^4 + a_2 y^3 + a_3 y^2 + a_4 xy + a_5 y + a_6 x + a_7$

$E_8(p > 5):$ $z^2 + x^3 + y^5 + a_1 xy^3 + a_2 y^3 + a_3 xy^2 + a_4 y^2$

$$+ a_5 xy + a_6 y + a_7 x + a_8.$$

In each case the versal deformation space is parametrized by the a_i's and b.

2.2.3   *Characteristic 5*

Theorem. *For rational double points of type other than A_k in characteristic 5, only the singularity of type E_8^0 has nontrivial equisingular locus*

$$z^2 + x^3 + y^5 + a_{10}$$

inside the 10-dimensional versal deformation space

$$z^2 + x^3 + y^5 + a_1 xy^4 + a_2 xy^3 + a_3 xy^2 + a_4 xy + a_5 x + a_6 y^4 + a_7 y^3 + a_8 y^2 + a_9 y + a_{10}.$$

2.2.4   Characteristic 3

Theorem. *For rational double points of type other than A_k in characteristic 3, only the singularities of type $E_6^0, E_6^1, E_7^0, E_8^0, E_8^1$ have nontrivial equisingular locus as given below.*

$$
\begin{aligned}
E_6^0 : &\quad z^2 + x^3 + y^4 + a_8 y + a_9 \\
E_6^1 : &\quad z^2 + x^3 + y^4 + x^2 y^2 + a_4 y^3 \\
E_7^0 : &\quad z^2 + x^3 + xy^3 + a_6 x \\
E_7^1 : &\quad \text{trivial} \\
E_8^0 : &\quad z^2 + x^3 + y^5 + a_9 y^3 + a_{12} \\
E_8^1 : &\quad z^2 + x^3 + y^5 + x^2 y^3 + a_7 y^3 \\
E_8^2 : &\quad \text{trivial}
\end{aligned}
$$

inside the versal deformation spaces

$$
\begin{aligned}
E_6^0 : \quad & z^2 + x^3 + y^4 \\
& + (a_1 y^2 + a_2 y + a_3)x^2 + (a_4 y^2 + a_5 y + a_6)x + a_7 y^2 + a_8 y + a_9 \\
E_6^1 : \quad & z^2 + x^3 + y^4 + x^2 y^2 \\
& + a_1 x^2 + a_2 xy + a_3 x + a_4 y^3 + a_5 y^2 + a_6 y + a_7 \\
E_7^0 : \quad & z^2 + x^3 + xy^3 \\
& + (a_1 y^2 + a_2 y + a_3)x^2 + (a_4 y^2 + a_5 y + a_6)x + a_7 y^2 + a_8 y + a_9 \\
E_7^1 : \quad & z^2 + x^3 + xy^3 + x^2 y^2 \\
& + a_1 x^2 + a_2 xy + a_3 x + a_4 y^3 + a_5 y^2 + a_6 y + a_7 \\
E_8^0 : \quad & z^2 + x^3 + y^5 \\
& + (a_1 y^3 + a_2 y^2 + a_3 y + a_4)x^2 + (a_5 y^3 + a_6 y^2 + a_7 y + a_8)x \\
& + a_9 y^3 + a_{10} y^2 + a_{11} y + a_{12} \\
E_8^1 : \quad & z^2 + x^3 + y^5 + x^2 y^3 \\
& + (a_1 y^2 + a_2 y + a_3)x^2 + (a_4 y^2 + a_5 y + a_6)x \\
& + a_7 y^3 + a_8 y^2 + a_9 y + a_{10} \\
E_8^2 : \quad & z^2 + x^3 + y^5 + x^2 y^2 \\
& + a_1 x^2 + a_2 xy + a_3 x + a_4 y^4 + a_5 y^3 + a_6 y^2 + a_7 y + a_8.
\end{aligned}
$$

2.2.5 Characteristic 2

Theorem. *For all rational double points in characteristic 2, and even-dimensional simple singularities in characteristic 2, the following is an exhaustive list of non-trivial equisingular families.*

For type A_n, the same as given in 2.2.1.

For type D_n, equisingular loci are given by the following equations inside the versal deformation spaces.

If the condition is too complicated to write down, we indicate hints of conditions among versal parameters which are necessary to determine the loci.

$$D_{2n}^0(n \geq 2): \quad z^2 + x^2 y + x(y+\beta)^n + \sum_{i=0}^{n-1} a_i y^i$$

$D_{2n}^r(n \geq 2): \quad a_{2r}, \ldots, a_{n+r-1}$ *and* β *(2.2.6.5 final step) are free parameters, $c_i = 0$ for every i, other coefficients are either 0 or functions of free parameters.*

$$D_{2n+1}^0(n \geq 2): \quad z^2 + x^2 y + y^n z + \sum_{i=0}^{n-1} a_i y^i$$

$D_{2n+1}^r(n \geq 2): \quad a_{2r+1}, \ldots, a_{n+r-1}$ *and* β *are free parameters, $b_i = 0$ for every i, other coefficients are either 0 or functions of free parameters.*

The versal deformation spaces are given below.

$$D_{2n}^0(n \geq 2): \quad z^2 + x^2 y + xy^n$$

$$+ \sum_{i=0}^{n-1} a_i y^i + \sum_{i=0}^{n-1} b_i xy^i + \sum_{i=0}^{n-1} c_i y^i z + \sum_{i=0}^{n-1} d_i xy^i z$$

$$D_{2n}^r(n \geq 2): \quad z^2 + x^2 y + xy^n + xy^{n-r} z$$

$$+ \sum_{i=0}^{n+r-1} a_i y^i + \sum_{i=0}^{n-r-1} b_i xy^i + \sum_{i=0}^{n-r-1} c_i y^i z + \sum_{i=0}^{n-r-1} d_i xy^i z$$

$D^0_{2n+1}(n \geq 2):\quad z^2 + x^2y + y^n z$

$$+ \sum_{i=0}^{n-1} a_i y^i + \sum_{i=0}^{n-1} b_i xy^i + \sum_{i=0}^{n-1} c_i y^i z + \sum_{i=0}^{n-1} d_i xy^i z$$

$D^r_{2n+1}(n \geq 2):\quad z^2 + x^2y + y^n z + xy^{n-r}z$

$$+ \sum_{i=0}^{n+r-1} a_i y^i + \sum_{i=0}^{n-r-1} b_i xy^i + \sum_{i=0}^{n-r-1} c_i y^i z + \sum_{i=0}^{n-r-1} d_i xy^i z.$$

For type E, equisingular loci are given by the following equations inside the versal deformation spaces with or without conditions among parameters.

$E^0_6:\quad z^2 + x^3 + y^2 z + a_4 z$

$E^1_6:\quad$ *trivial*

$E^0_7:\quad z^2 + x^3 + xy^3 + a_8 y^4 + a_9 y^3 + a_{10} y^2 + a_{13} x + a_{14}, \quad a_{13} + a_9^2 = 0$

$E^1_7:\quad z^2 + x^3 + xy^3 + y^4 z + a_7 y^3 + a_8 y^2 + a_{11} x + a_{12}, \quad a_{11} + a_7^2 = 0$

$E^2_7:\quad z^2 + x^3 + xy^3 + y^3 z + a_6 y^2 + a_9 x + a_{10} \quad$ *with* $a_9 = a_{10}, a_9 \neq 1$

$E^3_7:\quad z^2 + x^3 + xy^3 + xyz + a_3 y^4, \quad a_3 \neq 1$

$E^0_8:\quad z^2 + x^3 + y^5 + a_{12} y^2 + a_{14} y + a_{15} x + a_{16}$

$E^1_8:\quad z^2 + x^3 + y^5 + xy^3 z + a_1 y^3 z + a_2 xy^2 z + a_3 y^2 z + a_4 xyz$

$\qquad\qquad + a_5 yz + a_6 xz + a_7 z + a_{10} y^2 + a_{12} y + a_{13} x + a_{16},$

$\qquad\qquad a_3 = a_1 a_2, a_4 = a_2 a^2, a_5 = a_1 a_2^2, a_6 = a_2^3, a_7 = a_1 a_2^3,$

$\qquad\qquad a_{12} = a_2^4, a_{13} = a_1^2$

$E^2_8:\quad z^2 + x^3 + y^5 + xy^2 z + a_2 y^2 z + a_5 xz + a_6 z + a_{10} y + a_{11} x,$

$\qquad\qquad a_6 = a_2 a_5, a_{11} = a_2^2, a_{10} = a_5^2$

$E^3_8:\quad z^2 + x^3 + y^5 + y^3 z + a_6 y^2, \quad a_6 \neq 1$

$E^4_8:\quad$ *trivial.*

The versal deformation spaces are given below.

$E_6^0:$ $z^2 + x^3 + y^2z + (a_1xy + a_2y + a_3x + a_4)z + a_5xy + a_6y + a_7x + a_8$

$E_6^1:$ $z^2 + x^3 + y^2z + xyz + (a_1y + a_2)z + a_3y^2 + a_4y + a_5x + a_6$

$E_7^0:$ $z^2 + x^3 + xy^3 + (a_1y^4 + a_2y^3 + a_3y^2 + a_4xy + a_5y + a_6x + a_7)z$
$\qquad + a_8y^4 + a_9y^3 + a_{10}y^2 + a_{11}xy + a_{12}y + a_{13}x + a_{14}$

$E_7^1:$ $z^2 + x^3 + xy^3 + y^4z + (a_1y^3 + a_2y^2 + a_3xy + a_4y + a_5x + a_6)z$
$\qquad + a_7y^3 + a_8y^2 + a_9xy + a_{10}y + a_{11}x + a_{12}$

$E_7^2:$ $z^2 + x^3 + xy^3 + y^3z + (a_1y^2 + a_2xy + a_3y + a_4x + a_5)z$
$\qquad + a_6y^2 + a_7xy + a_8y + a_9x + a_{10}$

$E_7^3:$ $z^2 + x^3 + xy^3 + xyz + (a_1y + a_2)z$
$\qquad + a_3y^4 + a_4y^3 + a_5y^2 + a_6y + a_7x + a_8$

$E_8^0:$ $z^2 + x^3 + y^5$
$\qquad + (a_1xy^3 + a_2y^3 + a_3xy^2 + a_4y^2 + a_5xy + a_6y + a_7x + a_8)z$
$\qquad + a_9xy^3 + a_{10}y^3 + a_{11}xy^2 + a_{12}y^2 + a_{13}xy + a_{14}y + a_{15}x + a_{16}$

$E_8^1:$ $z^2 + x^3 + y^5 + xy^3z$
$\qquad + (a_1y^3 + a_2xy^2 + a_3y^2 + a_4xy + a_5y + a_6x + a_7)z$
$\qquad + a_8y^3 + a_9xy^2 + a_{10}y^2 + a_{11}xy + a_{12}y + a_{13}x + a_{14}$

$E_8^2:$ $z^2 + x^3 + y^5 + xy^2z + (a_1y^3 + a_2y^2 + a_3xy + a_4y + a_5x + a_6)z$
$\qquad + a_7y^3 + a_8y^2 + a_9xy + a_{10}y + a_{11}x + a_{12}$

$E_8^3:$ $z^2 + x^3 + y^5 + y^3z + (a_1xy + a_2y + a_3x + a_4)z$
$\qquad + a_5xy^2 + a_6y^2 + a_7xy + a_8y + a_9x + a_{10}$

$E_8^4:$ $z^2 + x^3 + y^5 + xyz + (a_1y + a_2)z$
$\qquad + a_3y^4 + a_4y^3 + a_5y^2 + a_6y + a_7x + a_8.$

In the higher-dimensional case for simple singularities, the equisingular locus is given by

$$z_3z_4 + \cdots + z_{n-1}z_n + \text{(equation in the surface case)}.$$

Remark. In Theorem 2.2.5, the defining equation of E_7^1 is

$$z^2 + x^3 + xy^3 + y^4z$$

instead of the equation in 1.1.7

$$z^2 + x^3 + xy^3 + x^2yz$$

by a technical reason for calculation.

These two equations are contact equivalent, for

$$z^2 + x^3 + xy^3 + x^2yz = (1 + y^2X)z^2 + X^3 + Xy^3 + y^4z, \quad X := x + yz.$$

Hence, the dimension of equisingular locus is unchanged.

2.2.6 Proof of Theorem 2.2.5

We provide calculation for simple singularities of types E_7^r and D_{2n}^r. Equisingular loci of other types can be obtained similarly. Here we use the algorithm 1.6 frequently.

2.2.6.1 Case E_7^0

Normal form of E_7^0 is $\boxed{z^2 + x^3 + xy^3 = 0}$.

Step 0. Deformation space

$$f(x, y, z) := z^2 + x^3 + xy^3$$

$$\frac{\partial f}{\partial x} = x^2 + y^3, \quad \frac{\partial f}{\partial y} = xy^2, \quad \frac{\partial f}{\partial z} = 0$$

$$T^1(X) := k[[x, y, z]] / \left(\frac{\partial f}{\partial x}, \frac{\partial f}{\partial y}, \frac{\partial f}{\partial z}, f \right) = k[[x, y, z]]/(x^2 + y^3, xy^2, z^2)$$

$$= \langle 1, x, y, z, xz, yz, xy, xyz, y^2, y^3, y^4, y^2z, y^3z, y^4z \rangle_k,$$

where the last term signifies the k vector space generated by elements inside brackets. Now set

$$\tilde{f}(x, y, z) := z^2 + x^3 + xy^3 + f_1 y^4 z + e_1 y^4 + e_2 y^3 z + d_1 y^3 + d_2 y^2 z + d_3 xyz$$
$$+ c_1 y^2 + c_2 xz + c_3 yz + c_4 xy + b_1 x + b_2 y + b_3 z + a$$

and

$$T^1(\mathscr{X}) := k[a, b_1, b_2, b_3, c_1, c_2, c_3, c_4, d_1, d_2, d_3, e_1, e_2, f_1][[x, y, z]]/(\tilde{f}(x, y, z)).$$

Take a point on the versal deformation space, which we denote simply by the corresponding parameters $(a, b_1, \ldots, e_2, f_1)$, and look at the fiber of this point. Suppose this fiber has a singular point $(x, y, z) = (\alpha, \beta, \gamma)$. In order to analyze the singularity, we translate the singular point to the origin $(x, y, z) = (0, 0, 0)$. We call this fiber the *central fiber*.

Step 1. Translation: $x \mapsto x + \alpha,\ y \mapsto y + \beta,\ z \mapsto z + \gamma$

$$
\begin{aligned}
F := & \tilde{f}(x + \alpha, y + \beta, z + \gamma) \\
= & (z + \gamma)^2 + (x + \alpha)^3 + (x + \alpha)(y + \beta)^3 \\
& + f_1(y + \beta)^4(z + \gamma) + e_1(y + \beta)^4 + e_2(y + \beta)^3(z + \gamma) \\
& + d_1(y + \beta)^3 + d_2(y + \beta)^2(z + \gamma) + d_3(x + \alpha)(y + \beta)(z + \gamma) \\
& + c_1(y + \beta)^2 + c_2(x + \alpha)(z + \gamma) + c_3(y + \beta)(z + \gamma) + c_4(x + \alpha)(y + \beta) \\
& + b_1(x + \alpha) + b_2(y + \beta) + b_3(z + \gamma) + a \\
= & f_1 y^4 z + (e_1 + f_1\gamma)y^4 + xy^3 + e_2 y^3 z \\
& + x^3 + (\alpha + e_2\gamma + d_1)y^3 + \beta xy^2 + (e_2\beta + d_2)y^2 z + d_3 xyz \\
& + \alpha x^2 + (\alpha\beta + e_2\beta\gamma + d_1\beta + d_2\gamma + c_1)y^2 + z^2 \\
& + (e_2\beta^2 + d_3\alpha + c_3)yz + (d_3\beta + c_2)xz + (\beta^2 + d_3\gamma + c_4)xy \\
& + (\alpha^2 + \beta^3 + d_3\beta\gamma + c_2\gamma + c_4\beta + b_1)x \\
& + (\alpha\beta^2 + e_2\beta^2\gamma + d_1\beta^2 + d_3\alpha\gamma + c_3\gamma + c_4\alpha + b_2)y \\
& + (f_1\beta^4 + e_2\beta^3 + d_2\beta^2 + d_3\alpha\beta + c_2\alpha + c_3\beta + b_3)z \\
& + \alpha^3 + \alpha\beta^3 + \gamma^2 + f_1\beta^4\gamma + e_1\beta^4 + e_2\beta^3\gamma + d_1\beta^3 + d_2\beta^2\gamma + d_3\alpha\beta\gamma \\
& + c_1\beta^2 + c_2\alpha\gamma + c_3\beta\gamma + c_4\alpha\beta + b_1\alpha + b_2\beta + b_3\gamma + a.
\end{aligned}
$$

By the Jacobian criterion, we have

$$
\begin{aligned}
\alpha^3 + \alpha\beta^3 + \gamma^2 &+ f_1\beta^4\gamma + e_1\beta^4 + e_2\beta^3\gamma + d_1\beta^3 + d_2\beta^2\gamma + d_3\alpha\beta\gamma \\
&+ c_1\beta^2 + c_2\alpha\gamma + c_3\beta\gamma + c_4\alpha\beta + b_1\alpha + b_2\beta + b_3\gamma + a = 0,
\end{aligned}
\tag{2.1}
$$

$$
\alpha^2 + \beta^3 + d_3\beta\gamma + c_2\gamma + c_4\beta + b_1 = 0,
\tag{2.2}
$$

$$
\alpha\beta^2 + e_2\beta^2\gamma + d_1\beta^2 + d_3\alpha\gamma + c_3\gamma + c_4\alpha + b_2 = 0,
\tag{2.3}
$$

$$
f_1\beta^4 + e_2\beta^3 + d_2\beta^2 + d_3\alpha\beta + c_2\alpha + c_3\beta + b_3 = 0.
\tag{2.4}
$$

These (vanishing) conditions ensure that the coefficient of constant and linear terms are 0. Hence f starts with the quadratic terms. We then apply the algorithm 1.6 to f.

Step 2. Determine the degree 2 terms

By the above explanations, F is written as

$$F = f_1 y^4 z + (e_1 + f_1 \gamma) y^4 + x y^3 + e_2 y^3 z$$
$$+ x^3 + (\alpha + e_2 \gamma + d_1) y^3 + \beta x y^2 + (e_2 \beta + d_2) y^2 z + d_3 x y z$$
$$+ \alpha x^2 + (\alpha \beta + e_2 \beta \gamma + d_1 \beta + d_2 \gamma + c_1) y^2 + z^2$$
$$+ (e_2 \beta^2 + d_3 \alpha + c_3) y z + (d_3 \beta + c_2) x z + (\beta^2 + d_3 \gamma + c_4) x y$$
$$= \{ z + \alpha^{\frac{1}{2}} x + (\alpha \beta + e_2 \beta \gamma + d_1 \beta + d_2 \gamma + c_1)^{\frac{1}{2}} y \}^2$$
$$+ (e_2 \beta^2 + d_3 \alpha + c_3) y z + (d_3 \beta + c_2) x z + (\beta^2 + d_3 \gamma + c_4) x y$$
$$+ x^3 + (\alpha + e_2 \gamma + d_1) y^3 + \beta x y^2 + (e_2 \beta + d_2) y^2 z + d_3 x y z$$
$$+ (e_1 + f_1 \gamma) y^4 + x y^3 + e_2 y^3 z + f_1 y^4 z.$$

Since X has a singularity of type E_7^0, by the flowchart, the coefficients of yz, xz, xy terms are all 0, i.e.,

$$e_2 \beta^2 + d_3 \alpha + c_3 = 0, \tag{2.5}$$

$$d_3 \beta + c_2 = 0, \tag{2.6}$$

$$\beta^2 + d_3 \gamma + c_4 = 0. \tag{2.7}$$

Step 3. Determine next the degree 3 terms

Under the conditions above, we have

$$F = \{ z + \alpha^{\frac{1}{2}} x + (\alpha \beta + e_2 \beta \gamma + d_1 \beta + d_2 \gamma + c_1)^{\frac{1}{2}} y \}^2$$
$$+ x^3 + (\alpha + e_2 \gamma + d_1) y^3 + \beta x y^2 + (e_2 \beta + d_2) y^2 z + d_3 x y z$$
$$+ (e_1 + f_1 \gamma) y^4 + x y^3 + e_2 y^3 z + f_1 y^4 z$$

$$= z'^2 + x^3 + (\alpha + e_2\gamma + d_1)y^3 + \beta xy^2$$
$$+ (e_2\beta + d_2)y^2\{z' + \alpha^{\frac{1}{2}}x + (\alpha\beta + e_2\beta\gamma + d_1\beta + d_2\gamma + c_1)^{\frac{1}{2}}y\}$$
$$+ d_3xy\{z' + \alpha^{\frac{1}{2}}x + (\alpha\beta + e_2\beta\gamma + d_1\beta + d_2\gamma + c_1)^{\frac{1}{2}}y\} + (e_1 + f_1\gamma)y^4$$
$$+ xy^3 + e_2y^3\{z' + \alpha^{\frac{1}{2}}x + (\alpha\beta + e_2\beta\gamma + d_1\beta + d_2\gamma + c_1)^{\frac{1}{2}}y\}$$
$$+ f_1y^4\{z' + \alpha^{\frac{1}{2}}x + (\alpha\beta + e_2\beta\gamma + d_1\beta + d_2\gamma + c_1)^{\frac{1}{2}}y\}$$
$$\quad (z' := z + \alpha^{\frac{1}{2}}x + (\alpha\beta + e_2\beta\gamma + d_1\beta + d_2\gamma + c_1)^{\frac{1}{2}}y)$$
$$= z'^2 + x^3 + d_3\alpha^{\frac{1}{2}}x^2y$$
$$+ \{\beta + \alpha^{\frac{1}{2}}(e_2\beta + d_2) + d_3(\alpha\beta + e_2\beta\gamma + d_1\beta + d_2\gamma + c_1)^{\frac{1}{2}}\}xy^2$$
$$+ \{\alpha + e_2\gamma + d_1 + (e_2\beta + d_2)(\alpha\beta + e_2\beta\gamma + d_1\beta + d_2\gamma + c_1)^{\frac{1}{2}}\}y^3$$
$$+ (e_2\beta + d_2)y^2z' + d_3xyz'$$
$$+ \{e_1 + f_1\gamma + e_2(\alpha\beta + e_2\beta\gamma + d_1\beta + d_2\gamma + c_1)^{\frac{1}{2}}\}y^4 + (1 + e_2\alpha^{\frac{1}{2}})xy^3$$
$$+ e_2y^3z' + f_1(\alpha\beta + e_2\beta\gamma + d_1\beta + d_2\gamma + c_1)^{\frac{1}{2}}y^5$$
$$+ f_1\alpha^{\frac{1}{2}}xy^4 + f_1y^4z'.$$

Since X has a singularity of type E_7^0, we must have $\psi^{(3)} \overset{c}{\sim} x^3$ (see the flowchart)

$$x^3 + d_3\alpha^{\frac{1}{2}}x^2y$$
$$+ \{\beta + \alpha^{\frac{1}{2}}(e_2\beta + d_2) + d_3(\alpha\beta + e_2\beta\gamma + d_1\beta + d_2\gamma + c_1)^{\frac{1}{2}}\}xy^2$$
$$+ \{\alpha + e_2\gamma + d_1 + (e_2\beta + d_2)(\alpha\beta + e_2\beta\gamma + d_1\beta + d_2\gamma + c_1)^{\frac{1}{2}}\}y^3$$
$$= (x + \{\alpha + e_2\gamma + d_1 + (e_2\beta + d_2)(\alpha\beta + e_2\beta\gamma + d_1\beta + d_2\gamma + c_1)^{\frac{1}{2}}\}^{\frac{1}{3}}y)^3$$
$$(=: x'^3).$$

By comparison of coefficients of corresponding terms of the equality, we have

$$d_3\alpha^{\frac{1}{2}} = \{\alpha + e_2\gamma + d_1 + (e_2\beta + d_2)(\alpha\beta + e_2\beta\gamma + d_1\beta + d_2\gamma + c_1)^{\frac{1}{2}}\}^{\frac{1}{3}},$$
$$\tag{2.8}$$

$$\beta + \alpha^{\frac{1}{2}}(e_2\beta + d_2) + d_3(\alpha\beta + e_2\beta\gamma + d_1\beta + d_2\gamma + c_1)^{\frac{1}{2}}$$
$$= \{\alpha + e_2\gamma + d_1 + (e_2\beta + d_2)(\alpha\beta + e_2\beta\gamma + d_1\beta + d_2\gamma + c_1)^{\frac{1}{2}}\}^{\frac{2}{3}}. \tag{2.9}$$

With the conditions (2.8) and (2.9) take into account, we have

$$
\begin{aligned}
F = {}& z'^2 + x'^3 + (e_2\beta + d_2)y^2 z' \\
&+ d_3(x' + \{\alpha + e_2\gamma + d_1 \\
&\qquad\qquad + (e_2\beta + d_2)(\alpha\beta + e_2\beta\gamma + d_1\beta + d_2\gamma + c_1)^{\frac{1}{2}}\}^{\frac{1}{3}}y)yz' \\
&+ \{e_1 + f_1\gamma + e_2(\alpha\beta + e_2\beta\gamma + d_1\beta + d_2\gamma + c_1)^{\frac{1}{2}}\}y^4 \\
&+ (1 + e_2\alpha^{\frac{1}{2}})(x' + \{\alpha + e_2\gamma + d_1 \\
&\qquad\qquad + (e_2\beta + d_2)(\alpha\beta + e_2\beta\gamma + d_1\beta + d_2\gamma + c_1)^{\frac{1}{2}}\}^{\frac{1}{3}}y)y^3 \\
&+ e_2 y^3 z' + f_1(\alpha\beta + e_2\beta\gamma + d_1\beta + d_2\gamma + c_1)^{\frac{1}{2}}y^5 \\
&+ f_1\alpha^{\frac{1}{2}}(x' + \{\alpha + e_2\gamma + d_1 \\
&\qquad\qquad + (e_2\beta + d_2)(\alpha\beta + e_2\beta\gamma + d_1\beta + d_2\gamma + c_1)^{\frac{1}{2}}\}^{\frac{1}{3}}y)y^4 \\
&+ f_1 y^4 z' \\
= {}& z'^2 + x'^3 + d_3 x' yz \\
&+ (e_2\beta + d_2 + d_3\{\alpha + e_2\gamma + d_1 \\
&\qquad\qquad + (e_2\beta + d_2)(\alpha\beta + e_2\beta\gamma + d_1\beta + d_2\gamma + c_1)^{\frac{1}{2}}\}^{\frac{1}{3}})y^2 z' \\
&+ \{e_1 + f_1\gamma + e_2(\alpha\beta + e_2\beta\gamma + d_1\beta + d_2\gamma + c_1)^{\frac{1}{2}} \\
&+ (1 + e_2\alpha^{\frac{1}{2}})\{\alpha + e_2\gamma + d_1 \\
&\qquad\qquad + (e_2\beta + d_2)(\alpha\beta + e_2\beta\gamma + d_1\beta + d_2\gamma + c_1)^{\frac{1}{2}}\}^{\frac{1}{3}}\}y^4 \\
&+ (1 + e_2\alpha^{\frac{1}{2}})x'y^3 + e_2 y^3 z' \\
&+ f_1(\alpha\beta + e_2\beta\gamma + d_1\beta + d_2\gamma + c_1)^{\frac{1}{2}}y^5 + f_1\alpha^{\frac{1}{2}}x'y^4 \\
&+ f_1\alpha^{\frac{1}{2}}\{\alpha + e_2\gamma + d_1 \\
&\qquad\qquad + (e_2\beta + d_2)(\alpha\beta + e_2\beta\gamma + d_1\beta + d_2\gamma + c_1)^{\frac{1}{2}}\}^{\frac{1}{3}}y^5 \\
&+ f_1 y^4 z'.
\end{aligned}
$$

Since X has a singularity of type E_7^0, we have

$$e_2\beta + d_2 = 0, \tag{2.10}$$
$$d_3 = 0. \tag{2.11}$$

Step 4. Determine the degree 4 terms

From (2.6) and (2.11), we have $c_2 = 0$. Moreover, from (2.8)-(2.11), we have $\beta = \alpha + e_2\gamma + d_1 = 0$, which implies $c_3 = c_4 = d_2 = 0$ from (2.5) and (2.7). Also, from (2.3) and (2.4), we see that $b_2 = b_3 = 0$ since $\beta = 0$. Hence

$$F = z'^2 + x'^3 + \{e_1 + f_1\gamma + e_2c_1^{\frac{1}{2}}\}y^4 + (1 + e_2\alpha^{\frac{1}{2}})x'y^3 + e_2y^3z'$$
$$+ f_1c_1^{\frac{1}{2}}y^5 + f_1\alpha^{\frac{1}{2}}x'y^4 + f_1y^4z'.$$

Now setting $z'' := z' + \{e_1 + f_1\gamma + e_2c_1^{\frac{1}{2}}\}y^2$, we have

$$F = z''^2 + x'^3 + (1 + e_2\alpha^{\frac{1}{2}})x'y^3 + e_2y^3(z'' + \{e_1 + f_1\gamma + e_2c_1^{\frac{1}{2}}\}y^2)$$
$$+ f_1c_1^{\frac{1}{2}}y^5 + f_1\alpha^{\frac{1}{2}}x'y^4 + f_1y^4(z'' + \{e_1 + f_1\gamma + e_2c_1^{\frac{1}{2}}\}y^2)$$
$$= z''^2 + x'^3 + (1 + e_2\alpha^{\frac{1}{2}})x'y^3 + e_2y^3z''$$
$$+ \{e_2\{e_1 + f_1\gamma + e_2c_1^{\frac{1}{2}}\} + f_1c_1^{\frac{1}{2}}\}y^5 + f_1\alpha^{\frac{1}{2}}x'y^4 + f_1y^4z''$$
$$+ f_1\{e_1 + f_1\gamma + e_2c_1^{\frac{1}{2}}\}y^6.$$

Since X has a singularity of type E_7^0, we have

$$e_2 = 0. \tag{2.12}$$

Step 5. Determine the degrees 5 terms

From (2.8)-(2.12), we have

$$F = z''^2 + x'^3 + x'y^3 + f_1c_1^{\frac{1}{2}}y^5 + f_1\alpha^{\frac{1}{2}}x'y^4 + f_1y^4z'' + O(6)$$
$$= z''^2 + x'^3 + U_1x'y^3 + f_1c_1^{\frac{1}{2}}y^5 + f_1y^4z'' + O(6),$$

where $U_1 := 1 + f_1\alpha^{\frac{1}{2}}y$. Since the index of determinacy of E_7^0 is 5, we have

$$f_1 = 0. \tag{2.13}$$

Step 6. Equisingular locus

From (2.13), we have $F = z''^2 + x'^3 + U_1x'y^3 + O(6)$, which gives the singularity of type E_7^0. Since we see also that $a = \gamma^2$, $b_1 = \alpha^2$ and $d_1 = \alpha$ from (2.1), (2.2) and (2.8), the equation of equisingular locus of E_7^0 is

$$z^2 + x^3 + xy^3 + e_1y^4 + d_1y^3 + c_1y^2 + b_1x + a = 0$$

with $b_1 = d_1^2$, $\dim \mathrm{ES}(E_7^0) = 4$.

2.2.6.2   *Case* E_7^1

We restate Remark 2.2.5 in the form of lemma.

Lemma. *The normal form of singularity of type* E_7^1

$$z^2 + x^3 + xy^3 + x^2yz = 0$$

is contact equivalent to the equation

$$\boxed{z^2 + x^3 + xy^3 + y^4z = 0} \, .$$

Hence, we use this equation to start with.

Step 0. Deformation space

$$f(x, y, z) := z^2 + x^3 + xy^3 + y^4z.$$

$$\frac{\partial f}{\partial x} = x^2 + y^3, \quad \frac{\partial f}{\partial y} = xy^2, \quad \frac{\partial f}{\partial z} = y^4.$$

$$T^1(X) = k[[x, y, z]] / \left(\frac{\partial f}{\partial x}, \frac{\partial f}{\partial y}, \frac{\partial f}{\partial z}, f \right) = k[[x, y, z]] / (x^2 + y^3, xy^2, y^4, z^2)$$

$$= \langle 1, x, y, z, xz, yz, xy, xyz, y^2, y^3, y^2z, y^3z \rangle_k.$$

Now set

$$\tilde{f}(x, y, z) := z^2 + x^3 + xy^3 + x^2yz + e_1y^3z + d_1y^3 + d_2y^2z + d_3xyz$$
$$+ c_1y^2 + c_2xz + c_3yz + c_4xy + b_1x + b_2y + b_3z + a$$

and

$$T^1(\mathscr{X}) = k[a, b_1, b_2, b_3, c_1, c_2, c_3, c_4, d_1, d_2, d_3, e_1][[x, y, z]] / (\tilde{f}(x, y, z)).$$

Step 1. Translation: $x \mapsto x + \alpha, \ y \mapsto y + \beta, \ z \mapsto z + \gamma$

$$(z + \gamma)^2 + (x + \alpha)^3 + (x + \alpha)(y + \beta)^3 + (y + \beta)^4(z + \gamma)$$
$$+ e_1(y + \beta)^3(z + \gamma) + d_1(y + \beta)^3 + d_2(y + \beta)^2(z + \gamma)$$
$$+ d_3(x + \alpha)(y + \beta)(z + \gamma) + c_1(y + \beta)^2 + c_2(x + \alpha)(z + \gamma)$$
$$+ c_3(y + \beta)(z + \gamma) + c_4(x + \alpha)(y + \beta) + b_1(x + \alpha) + b_2(y + \beta)$$
$$+ b_3(z + \gamma) + a$$

$$= z^2 + \gamma^2 + x^3 + \alpha x^2 + \alpha^2 x + \alpha^3$$
$$+ xy^3 + \beta xy^2 + \beta^2 xy + \beta^3 x + \alpha y^3 + \alpha \beta y^2 + \alpha \beta^2 y + \alpha \beta^3 + y^4 z + \gamma y^4$$
$$+ \beta^4 z + \beta^4 \gamma + e_1(y^3 z + \beta y^2 z + \beta^2 yz + \beta^3 z + \gamma y^3 + \beta \gamma y^2 + \beta^2 \gamma y + \beta^3 \gamma)$$
$$+ d_1(y^3 + \beta y^2 + \beta^2 y + \beta^3) + d_2(y^2 z + \gamma y^2 + \beta^2 z + \beta^2 \gamma)$$
$$+ d_3(xyz + \alpha yz + \beta xz + \gamma xy + \beta \gamma x + \alpha \gamma y + \alpha \beta z + \alpha \beta \gamma)$$
$$+ c_1(y^2 + \beta^2) + c_2(xz + \gamma x + \alpha z + \alpha \gamma) + c_3(yz + \gamma y + \beta z + \beta \gamma)$$
$$+ c_4(xy + \beta x + \alpha y + \alpha \beta) + b_1(x + \alpha) + b_2(y + \beta) + b_3(z + \gamma) + a$$
$$= z^2 + x^3 + xy^3 + y^4 z + e_1 y^3 z + \gamma y^4 + (\alpha + e_1 \gamma + d_1)y^3 + \beta xy^2$$
$$+ (e_1 \beta + d_2)y^2 z + d_3 xyz + \alpha x^2 + (\alpha \beta + e_1 \beta \gamma + d_1 \beta + d_2 \gamma + c_1)y^2$$
$$+ (e_1 \beta^2 + d_3 \alpha + c_3)yz + (d_3 \beta + c_2)xz + (\beta^2 + d_3 \gamma + c_4)xy$$
$$+ (\alpha^2 + \beta^3 + d_3 \beta \gamma + c_2 \gamma + c_4 \beta + b_1)x$$
$$+ (\alpha \beta^2 + e_1 \beta^2 \gamma + d_1 \beta^2 + d_3 \alpha \gamma + c_3 \gamma + c_4 \alpha + b_2)y$$
$$+ (\beta^4 + e_1 \beta^3 + d_2 \beta^2 + d_3 \alpha \beta + c_2 \alpha + c_3 \beta + b_3)z$$
$$+ \beta^4 \gamma + \alpha^3 + \alpha \beta^3 + \gamma^2 + e_1 \beta^3 \gamma + d_1 \beta^3 + d_2 \beta^2 \gamma + d_3 \alpha \beta \gamma$$
$$+ c_1 \beta^2 + c_2 \alpha \gamma + c_3 \beta \gamma + c_4 \alpha \beta + b_1 \alpha + b_2 \beta + b_3 \gamma + a.$$

From the condition that the central fiber has the singularity, we have

$$\begin{cases} \beta^4 \gamma + \alpha^3 + \alpha \beta^3 + \gamma^2 + e_1 \beta^3 \gamma + d_1 \beta^3 + d_2 \beta^2 \gamma + d_3 \alpha \beta \gamma \\ \quad + c_1 \beta^2 + c_2 \alpha \gamma + c_3 \beta \gamma + c_4 \alpha \beta + b_1 \alpha + b_2 \beta + b_3 \gamma + a = 0, \quad (2.14) \\ \qquad\qquad\qquad\qquad \alpha^2 + \beta^3 + d_3 \beta \gamma + c_2 \gamma + c_4 \beta + b_1 = 0, \quad (2.15) \\ \qquad \alpha \beta^2 + e_1 \beta^2 \gamma + d_1 \beta^2 + d_3 \alpha \gamma + c_3 \gamma + c_4 \alpha + b_2 = 0, \quad (2.16) \\ \qquad\qquad \beta^4 + e_1 \beta^3 + d_2 \beta^2 + d_3 \alpha \beta + c_2 \alpha + c_3 \beta + b_3 = 0. \quad (2.17) \end{cases}$$

Step 2. Determine the degree 2 terms

Under the conditions above, we have

$$F := \tilde{f}(x + \alpha, y + \beta, z + \gamma)$$
$$= z^2 + x^3 + xy^3 + y^4 z + e_1 y^3 z$$
$$\quad + \gamma y^4 + (\alpha + e_1 \gamma + d_1)y^3 + \beta xy^2 + (e_1 \beta + d_2)y^2 z + d_3 xyz$$
$$\quad + \alpha x^2 + (\alpha \beta + e_1 \beta \gamma + d_1 \beta + d_2 \gamma + c_1)y^2$$
$$\quad + (e_1 \beta^2 + d_3 \alpha + c_3)yz + (d_3 \beta + c_2)xz + (\beta^2 + d_3 \gamma + c_4)xy$$

$$= \{z + \alpha^{\frac{1}{2}}x + (\alpha\beta + e_1\beta\gamma + d_1\beta + d_2\gamma + c_1)^{\frac{1}{2}}y\}^2$$
$$+ (e_1\beta^2 + d_3\alpha + c_3)yz + (d_3\beta + c_2)xz + (\beta^2 + d_3\gamma + c_4)xy$$
$$+ x^3 + (\alpha + e_1\gamma + d_1)y^3 + \beta xy^2 + (e_1\beta + d_2)y^2z + d_3xyz$$
$$+ xy^3 + y^4z + e_1y^3z + \gamma y^4.$$

Since X has a singularity of type E_7^1,

$$\begin{cases} e_1\beta^2 + d_3\alpha + c_3 = 0, & (2.18) \\ d_3\beta + c_2 = 0, & (2.19) \\ \beta^2 + d_3\gamma + c_4 = 0. & (2.20) \end{cases}$$

Step 3. Determine the degree 3 terms

Under the conditions above, we have

$$F = \{z + \alpha^{\frac{1}{2}}x + (\alpha\beta + e_1\beta\gamma + d_1\beta + d_2\gamma + c_1)^{\frac{1}{2}}y\}^2 + x^3$$
$$+ (\alpha + e_1\gamma + d_1)y^3 + \beta xy^2 + (e_1\beta + d_2)y^2z + d_3xyz + xy^3 + y^4z$$
$$+ e_1y^3z + \gamma y^4$$
$$= z'^2 + x^3 + (\alpha + e_1\gamma + d_1)y^3 + \beta xy^2$$
$$+ (e_1\beta + d_2)y^2\{z' + \alpha^{\frac{1}{2}}x + (\alpha\beta + e_1\beta\gamma + d_1\beta + d_2\gamma + c_1)^{\frac{1}{2}}y\}$$
$$+ d_3xy\{z' + \alpha^{\frac{1}{2}}x + (\alpha\beta + e_1\beta\gamma + d_1\beta + d_2\gamma + c_1)^{\frac{1}{2}}y\}$$
$$+ xy^3 + y^4\{z' + \alpha^{\frac{1}{2}}x + (\alpha\beta + e_1\beta\gamma + d_1\beta + d_2\gamma + c_1)^{\frac{1}{2}}y\}$$
$$+ e_1y^3\{z' + \alpha^{\frac{1}{2}}x + (\alpha\beta + e_1\beta\gamma + d_1\beta + d_2\gamma + c_1)^{\frac{1}{2}}y\} + \gamma y^4$$
$$(z' := z + \alpha^{\frac{1}{2}}x + (\alpha\beta + e_1\beta\gamma + d_1\beta + d_2\gamma + c_1)^{\frac{1}{2}}y)$$
$$= z'^2 + x^3 + d_3\alpha^{\frac{1}{2}}x^2y$$
$$+ \{\beta + \alpha^{\frac{1}{2}}(e_1\beta + d_2) + d_3A\}xy^2 + \{\alpha + e_1\gamma + d_1 + (e_1\beta + d_2)A\}y^3$$
$$+ (e_1\beta + d_2)y^2z' + d_3xyz'$$
$$+ (e_1A + \gamma)y^4 + (1 + e_1\alpha^{\frac{1}{2}})xy^3 + e_1y^3z' + y^4z' + \alpha^{\frac{1}{2}}xy^4 + Ay^5.$$
$$(A := (\alpha\beta + e_1\beta\gamma + d_1\beta + d_2\gamma + c_1)^{\frac{1}{2}})$$

Since X has a singularity of type E_7^1, we can write

$$x^3 + d_3\alpha^{\frac{1}{2}}x^2y + \{\beta + \alpha^{\frac{1}{2}}(e_1\beta + d_2) + d_3A\}xy^2$$
$$+ \{\alpha + e_1\gamma + d_1 + (e_1\beta + d_2)A\}y^3$$
$$= (x + \{\alpha + e_1\gamma + d_1 + (e_1\beta + d_2)A\}^{\frac{1}{3}}y)^3 (=: x'^3).$$

Thus we have

$$\begin{cases} d_3\alpha^{\frac{1}{2}} = \{\alpha + e_1\gamma + d_1 + (e_1\beta + d_2)A\}^{\frac{1}{3}}, & (2.21) \\ \beta + \alpha^{\frac{1}{2}}(e_1\beta + d_2) + d_3A = \{\alpha + e_1\gamma + d_1 + (e_1\beta + d_2)A\}^{\frac{2}{3}}. & (2.22) \end{cases}$$

Under the conditions (2.21) and (2.22), we have

$$\begin{aligned} F &= z'^2 + x'^3 + (e_1\beta + d_2)y^2z' \\ &\quad + d_3(x' + \{\alpha + e_1\gamma + d_1 + (e_1\beta + d_2)A\}^{\frac{1}{3}}y)yz' + (e_1A + \gamma)y^4 \\ &\quad + (1 + e_1\alpha^{\frac{1}{2}})(x' + \{\alpha + e_1\gamma + d_1 + (e_1\beta + d_2)A\}^{\frac{1}{3}}y)y^3 + e_1y^3z' \\ &\quad + y^4z' + \alpha^{\frac{1}{2}}(x' + \{\alpha + e_1\gamma + d_1 + (e_1\beta + d_2)A\}^{\frac{1}{3}}y)y^4 + Ay^5 \\ &= z'^2 + x'^3 + (e_1\beta + d_2 + d_3\{\alpha + e_1\gamma + d_1 + (e_1\beta + d_2)A\}^{\frac{1}{3}})y^2z' \\ &\quad + d_3x'yz' + \{e_1A + \gamma + (1 + e_1\alpha^{\frac{1}{2}})\{\alpha + e_1\gamma + d_1 + (e_1\beta + d_2)A\}^{\frac{1}{3}}\}y^4 \\ &\quad + (1 + e_1\alpha^{\frac{1}{2}})x'y^3 + e_1y^3z' + y^4z' + \alpha^{\frac{1}{2}}x'y^4 \\ &\quad + \{\alpha^{\frac{1}{2}}\{\alpha + e_1\gamma + d_1 + (e_1\beta + d_2)A\}^{\frac{1}{3}} + A\}y^5. \end{aligned}$$

Since X has a singularity of type E_7^1, we have

$$\begin{cases} e_1\beta + d_2 + d_3\{\alpha + e_1\gamma + d_1 + (e_1\beta + d_2)A\}^{\frac{1}{3}} = 0, & (2.23) \\ d_3 = 0. & (2.24) \end{cases}$$

Step 4. Determine the degrees 4 terms

Under the conditions above, we have

$$\begin{aligned} F &= \{z' + \{e_1A + \gamma + (1 + e_1\alpha^{\frac{1}{2}})\{\alpha + e_1\gamma + d_1 + (e_1\beta + d_2)A\}^{\frac{1}{3}}\}^{\frac{1}{2}}y^2\}^2 \\ &\quad + x'^3 + (1 + e_1\alpha^{\frac{1}{2}})x'y^3 + e_1y^3z' + y^4z' + \alpha^{\frac{1}{2}}x'y^4 \\ &\quad + \{\alpha^{\frac{1}{2}}\{\alpha + e_1\gamma + d_1 + (e_1\beta + d_2)A\}^{\frac{1}{3}} + A\}y^5 \\ &= z''^2 + x'^3 + (1 + e_1\alpha^{\frac{1}{2}})x'y^3 \\ &\quad + e_1y^3\{z' + \{e_1A + \gamma + (1 + e_1\alpha^{\frac{1}{2}})\{\alpha + e_1\gamma + d_1 \\ &\qquad\qquad\qquad\qquad\qquad\qquad + (e_1\beta + d_2)A\}^{\frac{1}{3}}\}^{\frac{1}{2}}y^2\} \\ &\quad + y^4\{z' + \{e_1A + \gamma + (1 + e_1\alpha^{\frac{1}{2}})\{\alpha + e_1\gamma + d_1 \\ &\qquad\qquad\qquad\qquad\qquad\qquad + (e_1\beta + d_2)A\}^{\frac{1}{3}}\}^{\frac{1}{2}}y^2\} \\ &\quad + \alpha^{\frac{1}{2}}x'y^4 + \{\alpha^{\frac{1}{2}}\{\alpha + e_1\gamma + d_1 + (e_1\beta + d_2)A\}^{\frac{1}{3}} + A\}y^5 \\ &\quad (z'' := z' + \\ &\qquad \{e_1A + \gamma + (1 + e_1\alpha^{\frac{1}{2}})\{\alpha + e_1\gamma + d_1 + (e_1\beta + d_2)A\}^{\frac{1}{3}}\}^{\frac{1}{2}}y^2) \end{aligned}$$

$$= z''^2 + x'^3 + (1 + e_1\alpha^{\frac{1}{2}})x'y^3$$
$$+ e_1 y^3\{z'' + \{e_1 A + \gamma + (1 + e_1\alpha^{\frac{1}{2}})\{\alpha + e_1\gamma + d_1$$
$$+ (e_1\beta + d_2)A\}^{\frac{1}{3}}\}^{\frac{1}{2}}y^2\}$$
$$+ y^4\{z'' + \{e_1 A + \gamma + (1 + e_1\alpha^{\frac{1}{2}})\{\alpha + e_1\gamma + d_1$$
$$+ (e_1\beta + d_2)A\}^{\frac{1}{3}}\}^{\frac{1}{2}}y^2\}$$
$$+ \alpha^{\frac{1}{2}}x'y^4 + \{\alpha^{\frac{1}{2}}\{\alpha + e_1\gamma + d_1 + (e_1\beta + d_2)A\}^{\frac{1}{3}} + A\}y^5$$
$$= z''^2 + x'^3 + (1 + e_1\alpha^{\frac{1}{2}})x'y^3 + y^4 z' + e_1 y^3 z' + \alpha^{\frac{1}{2}}x'y^4$$
$$+ \{\alpha^{\frac{1}{2}}\{\alpha + e_1\gamma + d_1 + (e_1\beta + d_2)A\}^{\frac{1}{3}} + A$$
$$+ e_1\{e_1 A + \gamma + (1 + e_1\alpha^{\frac{1}{2}})\{\alpha + e_1\gamma + d_1 + (e_1\beta + d_2)A\}^{\frac{1}{3}}\}^{\frac{1}{2}}\}y^5$$
$$+ \{e_1 A + \gamma + (1 + e_1\alpha^{\frac{1}{2}})\{\alpha + e_1\gamma + d_1 + (e_1\beta + d_2)A\}^{\frac{1}{3}}\}^{\frac{1}{2}}y^6 z''.$$

Since X has a singularity of type E_7^1, we have

$$e_1 = 0. \tag{2.25}$$

Step 5. Equisingular locus

From (2.23)-(2.25), we have $d_2 = d_3 = e_1 = 0$. Moreover, from (2.21) and (2.22), we have $\beta = 0$ and $d_1 = \alpha$. Also, from (2.14)-(2.20), we have $c_2 = c_3 = c_4 = b_2 = b_3 = 0$, $b_1 = \alpha^2$, and $a = \gamma^2$. Hence the equation of equisingular locus of E_7^1 is
$z^2 + x^3 + xy^3 + y^4 z + d_1 y^3 + c_1 y^2 + b_1 x + a = 0$ with $b_1 = d_1^2$, $\dim \mathrm{ES}(E_7^1) = 3$.

2.2.6.3 *Case E_7^2*

Normal form of E_7^2 is $\boxed{z^2 + x^3 + xy^3 + y^3 z = 0}$.

Step 0. Deformation space
$$f(x, y, z) := z^2 + x^3 + xy^3 + y^3 z.$$
$$\frac{\partial f}{\partial x} = x^2 + y^3, \quad \frac{\partial f}{\partial y} = xy^2 + y^2 z, \quad \frac{\partial f}{\partial z} = y^3.$$
$$T^1(X) = k[[x, y, z]]/\left(\frac{\partial f}{\partial x}, \frac{\partial f}{\partial y}, \frac{\partial f}{\partial z}, f\right) = k[[x, y, z]]/(x^2 + y^3, xy^2 + y^2 z, z^2)$$
$$= \langle 1, x, y, z, xz, yz, xy, y^2, y^2 z, xyz \rangle_k.$$

Now set
$$\tilde{f}(x, y, z) := z^2 + x^3 + xy^3 + y^3 z$$
$$+ d_1 y^2 z + d_2 xyz + c_1 y^2 + c_2 xz + c_3 yz + c_4 xy$$
$$+ b_1 x + b_2 y + b_3 z + a$$

and

$$T^1(\mathscr{X}) = k[a, b_1, b_2, b_3, c_1, c_2, c_3, c_4, d_1, d_2][[x, y, z]]/(\tilde{f}(x, y, z)).$$

Step 1. Translation: $x \mapsto x + \alpha$, $y \mapsto y + \beta$, $z \mapsto z + \gamma$

$$
\begin{aligned}
&(z + \gamma)^2 + (x + \alpha)^3 + (x + \alpha)(y + \beta)^3 + (y + \beta)^3(z + \gamma) \\
&\quad + d_1(y + \beta)^2(z + \gamma) + d_2(x + \alpha)(y + \beta)(z + \gamma) \\
&\quad + c_1(y + \beta)^2 + c_2(x + \alpha)(z + \gamma) + c_3(y + \beta)(z + \gamma) + c_4(x + \alpha)(y + \beta) \\
&\quad + b_1(x + \alpha) + b_2(y + \beta) + b_3(z + \gamma) + a \\
&= z^2 + \gamma^2 + x^3 + \alpha x^2 + \alpha^2 x + \alpha^3 + xy^3 + \beta xy^2 + \beta^2 xy + \beta^3 x + \alpha y^3 \\
&\quad + \alpha\beta y^2 + \alpha\beta^2 y + \alpha\beta^3 + d_1(y^2 z + \gamma y^2 + \beta^2 z + \beta^2 \gamma) \\
&\quad + d_2(xyz + \alpha yz + \beta xz + \gamma xy + \beta\gamma x + \alpha\gamma y + \alpha\beta z + \alpha\beta\gamma) + c_1(y^2 + \beta^2) \\
&\quad + c_2(xz + \gamma x + \alpha z + \alpha\gamma) + c_3(yz + \gamma y + \beta z + \beta\gamma) \\
&\quad + c_4(xy + \beta x + \alpha y + \alpha\beta) + b_1(x + \alpha) + b_2(y + \beta) + b_3(z + \gamma) + a \\
&= xy^3 + y^3 z + x^3 + (\alpha + \gamma)y^3 + \beta xy^2 + (\beta + d_1)y^2 z + d_2 xyz \\
&\quad + \alpha x^2 + (\alpha\beta + \beta\gamma + d_1\gamma + c_1)y^2 + z^2 + (\beta^2 + d_2\alpha + c_3)yz + (d_2\beta + c_2)xz \\
&\quad + (\beta^2 + d_2\gamma + c_4)xy + (\alpha^2 + \beta^3 + d_2\beta\gamma + c_2\gamma + c_4\beta + b_1)x \\
&\quad + (\alpha\beta^2 + \beta^2\gamma + d_2\alpha\gamma + c_3\gamma + c_4\alpha + b_2)y \\
&\quad + (\beta^3 + d_1\beta^2 + d_2\alpha\beta + c_2\alpha + c_3\beta + b_3)z + \alpha^3 + \alpha\beta^3 + \gamma^2 + \beta^3\gamma + d_1\beta^2\gamma \\
&\quad + d_2\alpha\beta\gamma + c_1\beta^2 + c_2\alpha\gamma + c_3\beta\gamma + c_4\alpha\beta + b_1\alpha + b_2\beta + b_3\gamma + a.
\end{aligned}
$$

From the condition that the central fiber has the singularity, we have

$$
\left\{
\begin{aligned}
&\alpha^3 + \alpha\beta^3 + \gamma^2 + \beta^3\gamma + d_1\beta^2\gamma + d_2\alpha\beta\gamma \\
&\quad + c_1\beta^2 + c_2\alpha\gamma + c_3\beta\gamma + c_4\alpha\beta + b_1\alpha + b_2\beta + b_3\gamma + a = 0, \quad (2.26) \\
&\alpha^2 + \beta^3 + d_2\beta\gamma + c_2\gamma + c_4\beta + b_1 = 0, \quad (2.27) \\
&\alpha\beta^2 + \beta^2\gamma + d_2\alpha\gamma + c_3\gamma + c_4\alpha + b_2 = 0, \quad (2.28) \\
&\beta^3 + d_1\beta^2 + d_2\alpha\beta + c_2\alpha + c_3\beta + b_3 = 0. \quad (2.29)
\end{aligned}
\right.
$$

Step 2. Determine the degree 2 terms

Under the conditions above, we have

$$F := \tilde{f}(x+\alpha, y+\beta, z+\gamma)$$
$$= xy^3 + y^3 z + x^3 + (\alpha+\gamma)y^3 + \beta xy^2 + (\beta+d_1)y^2 z + d_2 xyz$$
$$+ \alpha x^2 + (\alpha\beta + \beta\gamma + d_1\gamma + c_1)y^2 + z^2$$
$$+ (\beta^2 + d_2\alpha + c_3)yz + (d_2\beta + c_2)xz + (\beta^2 + d_2\gamma + c_4)xy$$
$$= \{z + \alpha^{\frac{1}{2}}x + (\alpha\beta + \beta\gamma + d_1\gamma + c_1)^{\frac{1}{2}}y\}^2$$
$$+ (\beta^2 + d_2\alpha + c_3)yz + (d_2\beta + c_2)xz + (\beta^2 + d_2\gamma + c_4)xy$$
$$+ x^3 + (\alpha+\gamma)y^3 + \beta xy^2 + (\beta+d_1)y^2 z + d_2 xyz + xy^3 + y^3 z.$$

Since X has a singularity of type E_7^2,

$$\begin{cases} \beta^2 + d_2\alpha + c_3 = 0, & (2.30) \\ d_2\beta + c_2 = 0, & (2.31) \\ \beta^2 + d_2\gamma + c_4 = 0. & (2.32) \end{cases}$$

Step 3. Determine the degree 3 terms

Under the conditions above, we have

$$F = \{z + \alpha^{\frac{1}{2}}x + (\alpha\beta + \beta\gamma + d_1\gamma + c_1)^{\frac{1}{2}}y\}^2$$
$$+ x^3 + (\alpha+\gamma)y^3 + \beta xy^2 + (\beta+d_1)y^2 z + d_2 xyz + xy^3 + y^3 z$$
$$= z'^2 + x^3 + (\alpha+\gamma)y^3 + \beta xy^2$$
$$+ (\beta+d_1)y^2\{z' + \alpha^{\frac{1}{2}}x + (\alpha\beta + \beta\gamma + d_1\gamma + c_1)^{\frac{1}{2}}y\}$$
$$+ d_2 xy\{z' + \alpha^{\frac{1}{2}}x + (\alpha\beta + \beta\gamma + d_1\gamma + c_1)^{\frac{1}{2}}y\}$$
$$+ xy^3 + y^3\{z' + \alpha^{\frac{1}{2}}x + (\alpha\beta + \beta\gamma + d_1\gamma + c_1)^{\frac{1}{2}}y\}$$
$$(z' := z + \alpha^{\frac{1}{2}}x + (\alpha\beta + \beta\gamma + d_1\gamma + c_1)^{\frac{1}{2}}y)$$
$$= z'^2 + x^3 + d_2\alpha^{\frac{1}{2}}x^2 y$$
$$+ \{\beta + \alpha^{\frac{1}{2}}(\beta+d_1) + d_2(\alpha\beta + \beta\gamma + d_1\gamma + c_1)^{\frac{1}{2}}\}xy^2$$
$$+ \{\alpha+\gamma + (\beta+d_1)(\alpha\beta + \beta\gamma + d_1\gamma + c_1)^{\frac{1}{2}}\}y^3 + (\beta+d_1)y^2 z'$$
$$+ d_2 xyz' + (\alpha\beta + \beta\gamma + d_1\gamma + c_1)^{\frac{1}{2}}y^4 + (1 + \alpha^{\frac{1}{2}})xy^3 + y^3 z'.$$

Since X has a singularity of type E_7^2, we can write

$$x^3 + d_2\alpha^{\frac{1}{2}}x^2 y + \{\beta + \alpha^{\frac{1}{2}}(\beta+d_1) + d_2(\alpha\beta + \beta\gamma + d_1\gamma + c_1)^{\frac{1}{2}}\}xy^2$$
$$+ \{\alpha+\gamma + (\beta+d_1)(\alpha\beta + \beta\gamma + d_1\gamma + c_1)^{\frac{1}{2}}\}y^3$$
$$= (x + \{\alpha+\gamma + (\beta+d_1)(\alpha\beta + \beta\gamma + d_1\gamma + c_1)^{\frac{1}{2}}\}^{\frac{1}{3}}y)^3 (=: x'^3).$$

Thus we have

$$\begin{cases} d_2 \alpha^{\frac{1}{2}} = \{\alpha + \gamma + (\beta + d_1)(\alpha\beta + \beta\gamma + d_1\gamma + c_1)^{\frac{1}{2}}\}^{\frac{1}{3}}, & (2.33) \\ \beta + \alpha^{\frac{1}{2}}(e_2\beta + d_1) + d_2(\alpha\beta + \beta\gamma + d_1\gamma + c_1)^{\frac{1}{2}} \\ \qquad = \{\alpha + \gamma + (\beta + d_1)(\alpha\beta + \beta\gamma + d_1\gamma + c_1)^{\frac{1}{2}}\}^{\frac{2}{3}}. & (2.34) \end{cases}$$

Under the conditions (2.33) and (2.34), we have

$$F = z'^2 + x'^3 + (\beta + d_1)y^2 z'$$
$$+ d_2(x' + \{\alpha + \gamma + (\beta + d_1)(\alpha\beta + \beta\gamma + d_1\gamma + c_1)^{\frac{1}{2}}\}^{\frac{1}{3}} y) yz'$$
$$+ (\alpha\beta + \beta\gamma + d_1\gamma + c_1)^{\frac{1}{2}} y^4$$
$$+ (1 + \alpha^{\frac{1}{2}})(x' + \{\alpha + \gamma$$
$$\qquad\qquad + (\beta + d_1)(\alpha\beta + \beta\gamma + d_1\gamma + c_1)^{\frac{1}{2}}\}^{\frac{1}{3}} y) y^3 + y^3 z'$$
$$= z'^2 + x'^3 + d_2 x' yz$$
$$+ (\beta + d_1 + d_2\{\alpha + \gamma + (\beta + d_1)(\alpha\beta + \beta\gamma + d_1\gamma + c_1)^{\frac{1}{2}}\}^{\frac{1}{3}})y^2 z'$$
$$+ ((\alpha\beta + \beta\gamma + d_1\gamma + c_1)^{\frac{1}{2}}$$
$$+ (1 + \alpha^{\frac{1}{2}})\{\alpha + \gamma + (\beta + d_1)(\alpha\beta + \beta\gamma + d_1\gamma + c_1)^{\frac{1}{2}}\}^{\frac{1}{3}})y^4$$
$$+ (1 + \alpha^{\frac{1}{2}})x' y^3 + y^3 z'.$$

Since X has a singularity of type E_7^2, we have

$$\begin{cases} \beta + d_1 = 0, & (2.35) \\ d_2 = 0. & (2.36) \end{cases}$$

Step 4. Equisingular locus

From (2.31) and (2.36), we have $c_2 = 0$. Moreover, from (2.33)-(2.36), we have $\beta = d_1 = 0$, which implies $c_3 = c_4 = 0$ and $\alpha = \gamma$ from (2.30), (2.32) and (2.33). Also, from (2.28) and (2.29), we see that $b_2 = b_3 = 0$ since $\beta = 0$. Hence

$$F = z'^2 + x'^3 + \{c_1^{\frac{1}{2}} + (1 + \alpha^{\frac{1}{2}})(\alpha + \gamma)^{\frac{1}{3}}\}y^4 + (1 + \alpha^{\frac{1}{2}})x' y^3 + y^3 z'.$$

Now setting $z'' := z' + \{c_1^{\frac{1}{2}} + (1 + \alpha^{\frac{1}{2}})(\alpha + \gamma)^{\frac{1}{3}}\}y^2$, we have

$$F = z''^2 + x'^3 + (1 + \alpha^{\frac{1}{2}})x' y^3 + y^3(z'' + \{c_1^{\frac{1}{2}} + (1 + \alpha^{\frac{1}{2}})(\alpha + \gamma)^{\frac{1}{3}}\}y^2)$$
$$= z''^2 + x'^3 + (1 + \alpha^{\frac{1}{2}})x' y^3 + y^3 z'' + \{c_1^{\frac{1}{2}} + (1 + \alpha^{\frac{1}{2}})(\alpha + \gamma)^{\frac{1}{3}}\}y^5.$$

If $1 + \alpha^{\frac{1}{2}} \neq 0$, this gives the equation of E_7^2. Since we see also that $a = \alpha^2$ and $b_1 = \alpha^2$ from (2.26) and (2.27), the equation of equisingular locus of E_7^2 is

$$z^2 + x^3 + xy^3 + y^3 z + c_1 y^2 + b_1 x + a = 0$$

with $a = b_1$, $\dim \mathrm{ES}(E_7^2) = 2$.

2.2.6.4   *Case E_7^3*

Normal form of E_7^3 is $\boxed{z^2 + x^3 + xy^3 + xyz = 0}$.

Step 0. Deformation space

$$f(x, y, z) := z^2 + x^3 + xy^3 + xyz.$$

$$\frac{\partial f}{\partial x} = x^2 + y^3 + yz, \quad \frac{\partial f}{\partial y} = xy^2 + xz, \quad \frac{\partial f}{\partial z} = xy.$$

$$T^1(X) = k[[x, y, z]] / \left(\frac{\partial f}{\partial x}, \frac{\partial f}{\partial y}, \frac{\partial f}{\partial z}, f \right)$$

$$= k[[x, y, z]] / (x^2 + y^3 + yz, xy^2 + xz, xy, z^2)$$

$$= \langle 1, x, y, z, yz, y^2, y^3, y^4 \rangle_k.$$

Now set

$$\tilde{f}(x, y, z) := z^2 + x^3 + xy^3 + xyz + ey^4 + dy^3 + c_1 y^2 + c_2 yz + b_1 x + b_2 y + b_3 z + a$$

and

$$T^1(\mathscr{X}) = k[a, b_1, b_2, b_3, c_1, c_2, d, e][[x, y, z]] / (\tilde{f}(x, y, z)).$$

Step 1. Translation: $x \mapsto x + \alpha,\ y \mapsto y + \beta,\ z \mapsto z + \gamma$

$$(z + \gamma)^2 + (x + \alpha)^3 + (x + \alpha)(y + \beta)^3 + (x + \alpha)(y + \beta)(z + \gamma) + e(y + \beta)^4$$
$$+ d(y + \beta)^3 + c_1(y + \beta)^2 + c_2(y + \beta)(z + \gamma) + b_1(x + \alpha) + b_2(y + \beta)$$
$$+ b_3(z + \gamma) + a$$
$$= z^2 + \gamma^2 + x^3 + \alpha x^2 + \alpha^2 x + \alpha^3 + xy^3 + \beta xy^2 + \beta^2 xy + \beta^3 x + \alpha y^3$$
$$+ \alpha\beta y^2 + \alpha\beta^2 y + \alpha\beta^3 + xyz + \alpha yz + \beta xz + \gamma xy + \beta\gamma x + \alpha\gamma y + \alpha\beta z$$
$$+ \alpha\beta\gamma + e(y^4 + \beta^4) + d(y^3 + \beta y^2 + \beta^2 y + \beta^3) + c_1(y^2 + \beta^2)$$
$$+ c_2(yz + \gamma y + \beta z + \beta\gamma) + b_1(x + \alpha) + b_2(y + \beta) + b_3(z + \gamma) + a$$
$$= xy^3 + ey^4 + x^3 + (\alpha + d)y^3 + \beta xy^2 + xyz$$
$$+ \alpha x^2 + (\alpha\beta + d\beta + c_1)y^2 + z^2 + (\alpha + c_2)yz + \beta xz + (\beta^2 + \gamma)xy$$
$$+ (\alpha^2 + \beta^3 + \beta\gamma + b_1)x + (\alpha\beta^2 + \alpha\gamma + d\beta^2 + c_2\gamma + b_2)y$$
$$+ (\alpha\beta + c_2\beta + b_3)z + \alpha^3 + \alpha\beta^3 + \gamma^2 + e\beta^4 + d\beta^3 + \alpha\beta\gamma + c_1\beta^2 + c_2\beta\gamma$$
$$+ b_1\alpha + b_2\beta + b_3\gamma + a.$$

From the condition that the central fiber has the singularity, we have

$$
\begin{cases}
\alpha^3 + \alpha\beta^3 + \gamma^2 + e\beta^4 + d\beta^3 + \alpha\beta\gamma + c_1\beta^2 + c_2\beta\gamma \\
\qquad\qquad\qquad +b_1\alpha + b_2\beta + b_3\gamma + a = 0, & (2.37) \\
\qquad\qquad\qquad \alpha^2 + \beta^3 + \beta\gamma + b_1 = 0, & (2.38) \\
\qquad\qquad \alpha\beta^2 + \alpha\gamma + d\beta^2 + c_2\gamma + b_2 = 0, & (2.39) \\
\qquad\qquad\qquad \alpha\beta + c_2\beta + b_3 = 0. & (2.40)
\end{cases}
$$

Step 2. Determine the degree 2 terms

Under the conditions above, we have

$$
\begin{aligned}
F := &\tilde{f}(x + \alpha, y + \beta, z + \gamma) \\
= &\, xy^3 + ey^4 + x^3 + (\alpha + d)y^3 + \beta xy^2 + xyz \\
&+ \alpha x^2 + (\alpha\beta + d\beta + c_1)y^2 + z^2 + (\alpha + c_2)yz + \beta xz + (\beta^2 + \gamma)xy \\
= &\, \{z + \alpha^{\frac{1}{2}}x + (\alpha\beta + d\beta + c_1)^{\frac{1}{2}}y\}^2 + (\alpha + c_2)yz + \beta xz + (\beta^2 + \gamma)xy \\
&+ x^3 + (\alpha + d)y^3 + \beta xy^2 + xyz + ey^4 + xy^3.
\end{aligned}
$$

Since X has a singularity of type E_7^3,

$$
\begin{cases}
\alpha + c_2 = 0, & (2.41) \\
\beta = 0, & (2.42) \\
\beta^2 + \gamma = 0. & (2.43)
\end{cases}
$$

Step 3. Determine the degree 3 terms

Under the conditions above, we have

$$
\begin{aligned}
F = &\,(z + \alpha^{\frac{1}{2}}x + c_1^{\frac{1}{2}}y)^2 + x^3 + (\alpha + d)y^3 + \beta xy^2 + xyz + ey^4 + xy^3 \\
= &\, z'^2 + x^3 + (\alpha + d)y^3 + xy(z' + \alpha^{\frac{1}{2}}x + c_1^{\frac{1}{2}}y) + ey^4 + xy^3 \\
&\qquad\qquad (z' := z + \alpha^{\frac{1}{2}}x + c_1^{\frac{1}{2}}y) \\
= &\, z'^2 + x^3 + \alpha^{\frac{1}{2}}x^2y + c_1^{\frac{1}{2}}xy^2 + (\alpha + d)y^3 + xyz' + ey^4 + xy^3
\end{aligned}
$$

Since X has a singularity of type E_7^3, we can write

$$
x^3 + \alpha^{\frac{1}{2}}x^2y + c_1^{\frac{1}{2}}xy^2 + (\alpha + d)y^3 = (x + (\alpha + d)^{\frac{1}{3}}y)^3 (=: x'^3).
$$

Thus we have

$$
\begin{cases}
\alpha^{\frac{1}{2}} = (\alpha + d)^{\frac{1}{3}}, & (2.44) \\
c_1^{\frac{1}{2}} = (\alpha + d)^{\frac{2}{3}}. & (2.45)
\end{cases}
$$

Under the conditions (2.44) and (2.45), we have

$$F = z'^2 + x'^3 + (x' + (\alpha + d)^{\frac{1}{3}}y)yz' + ey^4 + (x' + (\alpha + d)^{\frac{1}{3}}y)y^3$$
$$= z'^2 + x'^3 + (\alpha + d)^{\frac{1}{3}}y^2z' + x'yz' + (e + (\alpha + d)^{\frac{1}{3}})y^4 + x'y^3.$$

Since X has a singularity of type E_7^3, we have

$$(\alpha + d)^{\frac{1}{3}} = 0, \tag{2.46}$$

which implies $\alpha = d = c_1 = c_2 = 0$ from (2.41), (2.44) and (2.45).

Step 4. Equisingular locus

From (2.46), we have

$$F = z'^2 + x'^3 + x'y^3 + x'yz + ey^4.$$

Now setting $z'' := z' + e^{\frac{1}{2}}y^2$, we have

$$F = z''^2 + x'^3 + x'y^3 + x'y(z'' + e^{\frac{1}{2}}y^2) = z''^2 + x'^3 + (1 + e^{\frac{1}{2}})x'y^3 + x'yz''.$$

If $1 + e^{\frac{1}{2}} \neq 0$, this gives the equation of E_7^3. Since we see also that $a = b_1 = b_2 = b_3 = 0$ from (2.37)-(2.40), the equation of equisingular locus of E_7^3 is

$$z^2 + x^3 + xy^3 + xyz + ey^4 = 0, \quad \dim \mathrm{ES}(E_7^3) = 1.$$

2.2.6.5 Case D_{2n}^r

Normal form of the singularity D_{2n}^r is $\boxed{z^2 + x^2y + xy^n + xy^{n-r}z = 0}$.

We see that

$$T^1(X) = k[[x, y, z]]/\left(\frac{\partial f}{\partial x}, \frac{\partial f}{\partial y}, \frac{\partial f}{\partial z}, f\right)$$
$$= \langle 1, y, y^2, \dots, y^{n+r-1}, x, xy, xy^2, \dots, xy^{n-r-1},$$
$$z, yz, y^2z, \dots, y^{n-r-1}z, xz, xyz, xy^2z, \dots, xy^{n-r-1}z\rangle_k.$$

Now set

$$\tilde{f}(x, y, z) := z^2 + x^2y + xy^n + xy^{n-r}z$$
$$+ \overset{n+r-1}{\underset{i=0}{\sum}} a_iy^i + \overset{n-r-1}{\underset{i=0}{\sum}} (b_ixy^i + c_iy^iz + d_ixy^iz)$$

and

$$T^1(\mathcal{X}) = k[a_0, \dots, a_{n+r-1}, b_0, \dots, b_{n-r-1},$$
$$c_0, \dots, c_{n-r-1}, d_0, \dots, d_{n-r-1}][[x, y, z]]/(\tilde{f}(x, y, z)).$$

Step 1. We set all the terms of degree 0 and 1 are zero. Here we must add the terms x^2, xy^i $(i \leq n-1)$ and $y^j z$ $(j \leq n-r)$. Thus we consider

$$F := z^2 + x^2 y + xy^n + xy^{n-r} z + ex^2$$

$$+ \sum_{i=2}^{n+r-1} a_i y^i + \sum_{i=1}^{n-1} b_i xy^i + \sum_{i=1}^{n-r} c_i y^i z + \sum_{i=0}^{n-r-1} d_i xy^i z$$

$$= (z + e^{\frac{1}{2}} x + a_2^{\frac{1}{2}} y)^2 + b_1 xy + c_1 yz + d_0 xz + O(3).$$

Step 2. Determine the degree 2 terms

Since X has a singularity of type D_{2n}^r, we have

$$b_1 = c_1 = d_0 = 0.$$

Under the condition above, we have

$$F = z_2{}^2 + x^2 y + a_3 y^3 + b_2 xy^2 + c_2 y^2 z + d_1 xyz + O(4),$$

where $z_2 := z + e^{\frac{1}{2}} x + a_2^{\frac{1}{2}} y$.

Step 3. Determine the degree 3 terms

We can describe

$$F = z_2{}^2 + x^2 y + a_3 y^3 + b_2 xy^2 + c_2 y^2 z + d_1 xyz + O(4)$$

$$= z_2{}^2 + x^2 y + a_3 y^3 + b_2 xy^2 + c_2 y^2 (z_2 + e^{\frac{1}{2}} x + a_2^{\frac{1}{2}} y)$$

$$+ d_1 xy(z_2 + e^{\frac{1}{2}} x + a_2^{\frac{1}{2}} y) + O(4)$$

$$= z_2{}^2 + (1 + d_1 e^{\frac{1}{2}}) x^2 y + (a_3 + a_2^{\frac{1}{2}} c_2) y^3$$

$$+ (b_2 + c_2 e^{\frac{1}{2}} + a_2^{\frac{1}{2}} d_1) xy^2 + c_2 y^2 z_2 + d_1 xyz_2 + O(4).$$

Since X has a singularity of type D_{2n}^r, we can describe the degree 3 term $Ax^2 y + Bxy^2 + Cy^3$ as $X^2 Y$ after suitable coordinate changes. This implies that

$$b_2 + c_2 e^{\frac{1}{2}} + a_2^{\frac{1}{2}} d_1 = 0,$$

in which case we have

$$F = z_2{}^2 + (1 + d_1 e^{\frac{1}{2}})(x + \left(\frac{a_3 + a_2^{\frac{1}{2}} c_2}{1 + d_1 e^{\frac{1}{2}}}\right)^{\frac{1}{2}} y)^2 y + c_2 y^2 z_2 + d_1 xyz_2 + O(4)$$

$$= z_2{}^2 + (1 + d_1 e^{\frac{1}{2}}) x_3^2 y + c_2 y^2 z_2$$

$$+ d_1 (1 + d_1 e^{\frac{1}{2}})(x_3 + \left(\frac{a_3 + a_2^{\frac{1}{2}} c_2}{1 + d_1 e^{\frac{1}{2}}}\right)^{\frac{1}{2}} y) yz_2 + O(4)$$

$$= z_2{}^2 + (1 + d_1 e^{\frac{1}{2}}) x_3^2 y + (c_2 + d_1 \left(\frac{a_3 + a_2^{\frac{1}{2}} c_2}{1 + d_1 e^{\frac{1}{2}}}\right)^{\frac{1}{2}}) y^2 z_2 + d_1 x_3 yz_2 + O(4),$$

where $x_3 := x + \left(\dfrac{a_3 + a_2^{\frac{1}{2}} c_2}{1 + d_1 e^{\frac{1}{2}}}\right)^{\frac{1}{2}} y$. Since X has a singularity of type D_{2n}^r, the coefficients of the terms $y^2 z_2$ and $x_3 yz$ must be also zero. Thus $c_2 + d_1 \left(\dfrac{a_3 + a_2^{\frac{1}{2}} c_2}{1 + d_1 e^{\frac{1}{2}}}\right)^{\frac{1}{2}} = d_1 = 0$, which implies

$$b_2 = c_2 = d_1 = 0.$$

Step $k_{\leq n-r+1}$. Determine the degree k ($3 \leq k \leq n - r + 1$) term

We can describe

$$F = z^2 + x^2 y + a_k y^k + b_{k-1} xy^{k-1} + c_{k-1} y^{k-1} z + d_{k-2} xy^{k-2} z + O(k+1).$$

Case (i): k is even ($4 \leq k \leq n - r + 1$)

We can describe

$$
\begin{aligned}
F &= z_{k-2}^2 + x_{k-1}^2 y + a_k y^k + b_{k-1} xy^{k-1} + c_{k-1} y^{k-1} z + d_{k-2} xy^{k-2} z \\
&\qquad\qquad + O(k+1) \\
&= z_{k-2}^2 + x_{k-1}^2 y + a_k y^k + b_{k-1}(x_{k-1} + a_3^{\frac{1}{2}} y + \cdots + a_{k-1}^{\frac{1}{2}} y^{\frac{k-2}{2}}) y^{k-1} \\
&\quad + c_{k-1} y^{k-1} (z_{k-2} + e^{\frac{1}{2}}(x_{k-1} + a_3^{\frac{1}{2}} y + \cdots + a_{k-1}^{\frac{1}{2}} y^{\frac{k-2}{2}}) \\
&\qquad\qquad\qquad + a_2^{\frac{1}{2}} y + \cdots + a_{k-2}^{\frac{1}{2}} y^{\frac{k-2}{2}}) \\
&\quad + d_{k-2}(x_{k-1} + a_3^{\frac{1}{2}} y + \cdots + a_{k-1}^{\frac{1}{2}} y^{\frac{k-2}{2}}) y^{k-2} \\
&\qquad \times (z_{k-2} + e^{\frac{1}{2}}(x_{k-1} + a_3^{\frac{1}{2}} y \\
&\qquad\qquad + \cdots + a_{k-1}^{\frac{1}{2}} y^{\frac{k-2}{2}}) + a_2^{\frac{1}{2}} y + \cdots + a_{k-2}^{\frac{1}{2}} y^{\frac{k-2}{2}}) + O(k+1) \\
&= z_{k-2}^2 + x_{k-1}^2 y \\
&\quad + (a_k + a_3^{\frac{1}{2}} b_{k-1} + a_2^{\frac{1}{2}} c_{k-1} + a_3^{\frac{1}{2}} c_{k-1} e^{\frac{1}{2}} + a_2^{\frac{1}{2}} a_3^{\frac{1}{2}} d_{k-2} \\
&\qquad\qquad\qquad + a_3 d_{k-2} e^{\frac{1}{2}}) y^k \\
&\quad + (b_{k-1} + c_{k-1} e^{\frac{1}{2}} + a_2^{\frac{1}{2}} d_{k-2} + a_3^{\frac{1}{2}} d_{k-2} e^{\frac{1}{2}}) x_{k-1} y^{k-1} \\
&\quad + (c_{k-1} + a_3^{\frac{1}{2}} d_{k-2}) y^{k-1} z_{k-2} + d_{k-2} x_{k-1} y^{k-2} z_{k-2} + O(k+1).
\end{aligned}
$$

Since X has a singularity of type D_{2n}^r, we have

$$c_{k-1} + a_3^{\frac{1}{2}} d_{k-2} = 0,$$

in which case

$$F = (z_{k-2} + (a_k + a_3^{\frac{1}{2}} b_{k-1} + a_2^{\frac{1}{2}} c_{k-1} + a_3^{\frac{1}{2}} c_{k-1} e^{\frac{1}{2}} + a_2^{\frac{1}{2}} a_3^{\frac{1}{2}} d_{k-2}$$
$$+ a_3 d_{k-2} e^{\frac{1}{2}})^{\frac{1}{2}} y^{k/2})^2$$
$$+ x_{k-1}^2 y + (b_{k-1} + c_{k-1} e^{\frac{1}{2}} + a_2^{\frac{1}{2}} d_{k-2} + a_3^{\frac{1}{2}} d_{k-2} e^{\frac{1}{2}}) x_{k-1} y^{k-1}$$
$$+ d_{k-2} x_{k-1} y^{k-2} z_{k-2} + O(k+1)$$
$$= z_k^2 + x_{k-1}^2 y + (b_{k-1} + c_{k-1} e^{\frac{1}{2}} + a_2^{\frac{1}{2}} d_{k-2} + a_3^{\frac{1}{2}} d_{k-2} e^{\frac{1}{2}}) x_{k-1} y^{k-1}$$
$$+ d_{k-2} x_{k-1} y^{k-2}$$
$$\times (z_k + (a_k + a_3^{\frac{1}{2}} b_{k-1} + a_2^{\frac{1}{2}} c_{k-1} + a_3^{\frac{1}{2}} c_{k-1} e^{\frac{1}{2}}$$
$$+ a_2^{\frac{1}{2}} a_3^{\frac{1}{2}} d_{k-2} + a_3 d_{k-2} e^{\frac{1}{2}})^{\frac{1}{2}} y^{k/2}) + O(k+1),$$

where $z_k := z_{k-2} + (a_k + a_3^{\frac{1}{2}} b_{k-1} + a_2^{\frac{1}{2}} c_{k-1} + a_3^{\frac{1}{2}} c_{k-1} e^{\frac{1}{2}} + a_2^{\frac{1}{2}} a_3^{\frac{1}{2}} d_{k-2} + a_3 d_{k-2} e^{\frac{1}{2}})^{\frac{1}{2}} y^{k/2}$. Since X has a singularity of type D_{2n}^0, the coefficients of the terms $x_{k-1} y^{k-2}$ and $x_{k-1} y^{k-2} z_k$ must be also zero. Thus $b_{k-1} + c_{k-1} e^{\frac{1}{2}} + a_2^{\frac{1}{2}} d_{k-2} + a_3^{\frac{1}{2}} d_{k-2} e^{\frac{1}{2}} = d_{k-2} = 0$, which implies

$$b_{k-1} = c_{k-1} = d_{k-2} = 0.$$

Case (ii) : k is odd $(5 \leqq k \leqq n - r + 1)$

We can describe

$$F = z_{k-1}^2 + x_{k-2}^2 y + a_k y^k + b_{k-1} xy^{k-1}$$
$$+ c_{k-1} y^{k-1} z + d_{k-2} xy^{k-2} z + O(k+1)$$
$$= z_{k-1}^2 + x_{k-2}^2 y + a_k y^k + b_{k-1}(x_{k-2} + a_3^{\frac{1}{2}} y + \cdots + a_{k-2}^{\frac{1}{2}} y^{\frac{k-3}{2}}) y^{k-1}$$
$$+ c_{k-1} y^{k-1} (z_{k-1} + e^{\frac{1}{2}} (x_{k-1} + a_3^{\frac{1}{2}} y + \cdots + a_{k-2}^{\frac{1}{2}} y^{\frac{k-3}{2}})$$
$$+ a_2^{\frac{1}{2}} y + \cdots + a_{k-2}^{\frac{1}{2}} y^{\frac{k-1}{2}})$$
$$+ d_{k-2}(x_{k-2} + a_3^{\frac{1}{2}} y + \cdots + a_{k-2}^{\frac{1}{2}} y^{\frac{k-3}{2}}) y^{k-2}$$
$$\times (z_{k-1} + e^{\frac{1}{2}} (x_{k-2} + a_3^{\frac{1}{2}} y + \cdots + a_{k-2}^{\frac{1}{2}} y^{\frac{k-3}{2}})$$
$$+ a_2^{\frac{1}{2}} y + \cdots + a_{k-2}^{\frac{1}{2}} y^{\frac{k-1}{2}}) + O(k+1)$$
$$= z_{k-1}^2 + x_{k-2}^2 y + (a_k + a_3^{\frac{1}{2}} b_{k-1} + a_2^{\frac{1}{2}} c_{k-1} + a_3^{\frac{1}{2}} c_{k-1} e^{\frac{1}{2}}$$
$$+ a_2^{\frac{1}{2}} a_3^{\frac{1}{2}} d_{k-2} + a_3 d_{k-2} e^{\frac{1}{2}}) y^k$$
$$+ (b_{k-1} + c_{k-1} e^{\frac{1}{2}} + a_2^{\frac{1}{2}} d_{k-2} + a_3^{\frac{1}{2}} d_{k-2} e^{\frac{1}{2}}) x_{k-2} y^{k-1}$$
$$+ (c_{k-1} + a_3^{\frac{1}{2}} d_{k-2}) y^{k-1} z_{k-1} + d_{k-2} x_{k-2} y^{k-2} z_{k-1} + O(k+1).$$

Since X has a singularity of type D_{2n}^0, we can describe the term $Ax_{k-2}^2 y + Bx_{k-2}y^{k-1} + Cy^k$ as X^2Y after suitable coordinate changes. This implies that

$$b_{k-1} + c_{k-1}e^{\frac{1}{2}} + a_2^{\frac{1}{2}}d_{k-2} + a_3^{\frac{1}{2}}d_{k-2}e^{\frac{1}{2}} = 0,$$

in which case we have

$$F = z_{k-1}^2 + (x_{k-2} + (a_k + a_3^{\frac{1}{2}}b_{k-1} + a_2^{\frac{1}{2}}c_{k-1} + a_3^{\frac{1}{2}}c_{k-1}e^{\frac{1}{2}}$$
$$+ a_2^{\frac{1}{2}}a_3^{\frac{1}{2}}d_{k-2} + a_3 d_{k-2}e^{\frac{1}{2}})^{\frac{1}{2}}y^{\frac{k-1}{2}})^2 y$$
$$+ (c_{k-1} + a_3^{\frac{1}{2}}d_{k-2})y^{k-1}z_{k-1} + d_{k-2}x_{k-2}y^{k-2}z_{k-1} + O(k+1)$$
$$= z_{k-1}^2 + x_k^2 y + (c_{k-1} + a_3^{\frac{1}{2}}d_{k-2})y^{k-1}z_{k-1}$$
$$+ d_{k-2}(x_k + (a_k + a_3^{\frac{1}{2}}b_{k-1} + a_2^{\frac{1}{2}}c_{k-1} + a_3^{\frac{1}{2}}c_{k-1}e^{\frac{1}{2}}$$
$$+ a_2^{\frac{1}{2}}a_3^{\frac{1}{2}}d_{k-2} + a_3 d_{k-2}e^{\frac{1}{2}})^{\frac{1}{2}}y^{\frac{k-1}{2}})y^{k-2}z_{k-1} + O(k+1),$$

where $x_k := x_{k-2} + (a_k + a_3^{\frac{1}{2}}b_{k-1} + a_2^{\frac{1}{2}}c_{k-1} + a_3^{\frac{1}{2}}c_{k-1}e^{\frac{1}{2}} + a_2^{\frac{1}{2}}a_3^{\frac{1}{2}}d_{k-2} + a_3 d_{k-2}e^{\frac{1}{2}})^{\frac{1}{2}}y^{\frac{k-1}{2}}$. Since X has a singularity of type D_{2n}^0, the coefficients of the terms $y^{k-1}z_{k-1}$ and $x_k y^{k-2}z_{k-1}$ must be also zero. Thus $c_{k-1} + a_3^{\frac{1}{2}}d_{k-2} = d_{k-2} = 0$, which implies

$$b_{k-1} = c_{k-1} = d_{k-2} = 0.$$

Step $n - r + 2$. Determine the degree $n - r + 2$ term

Case (i): $n - r$ is even

We can describe

$$F = z_{n-r}^2 + x_{n-r+1}^2 y + xy^{n-r}z$$
$$+ a_{n-r+2}y^{n-r+2} + b_{n-r+1}xy^{n-r+1} + O(n-r+3).$$

We can see that $b_{n-r+1} = 0$ by similar argument in the previous steps. So we focus on the term $xy^{n-r}z$.

$$F = z_{n-r}^2 + x_{n-r+1}^2 y + xy^{n-r}z + a_{n-r+2}y^{n-r+2} + \text{(other terms)}$$
$$= z_{n-r}^2 + x_{n-r+1}^2 y$$
$$+ (x_{n-r+1} + a_3^{\frac{1}{2}}y + \cdots + a_{n-r+1}^{\frac{1}{2}}y^{\frac{n-r}{2}})y^{n-r}$$
$$\times (z_{n-r} + e^{\frac{1}{2}}(x_{n-r+1} + a_3^{\frac{1}{2}}y + \cdots + a_{n-r+1}^{\frac{1}{2}}y^{\frac{n-r}{2}})$$
$$+ a_2^{\frac{1}{2}}y + \cdots + a_{n-r}^{\frac{1}{2}}y^{\frac{n-r}{2}})$$
$$+ a_{n-r+2}y^{n-r+2} + \text{(other terms)}$$

$$= (z_{n-r} + a_{n-r+2}^{\frac{1}{2}} y^{\frac{n-r+2}{2}})^2 + x_{n-r+1}^2 y + x_{n-r+1} y^{n-r} z_{n-r}$$

$$+ \sum_{i=1}^{\frac{n-r}{2}} (a_{2i}^{\frac{1}{2}} x_{n-r+1} y^{n-r+i} + a_{2i+1}^{\frac{1}{2}} y^{n-r+i} z_{n-r}) + e^{\frac{1}{2}} x_{n-r+1}^2 y^{n-r}$$

$$+ (\text{the term } y^i\text{'s with } i \geq n-r+2) + (\text{other terms})$$

$$= z_{n-r+2}^2 + x_{n-r+1}^2 y + x_{n-r+1} y^{n-r} (z_{n-r+2} + a_{n-r+2}^{\frac{1}{2}} y^{\frac{n-r+2}{2}})$$

$$+ \sum_{i=1}^{\frac{n-r}{2}} (a_{2i}^{\frac{1}{2}} x_{n-r+1} y^{n-r+i} + a_{2i+1}^{\frac{1}{2}} y^{n-r+i} (z_{n-r+2} + a_{n-r+2}^{\frac{1}{2}} y^{\frac{n-r+2}{2}}))$$

$$+ e^{\frac{1}{2}} x_{n-r+1}^2 y^{n-r} + (\text{the term } y^i\text{'s with } i \geq n-r+2) + (\text{other terms})$$

$$= z_{n-r+2}^2 + x_{n-r+1}^2 y + x_{n-r+1} y^{n-r} z_{n-r+2}$$

$$+ \sum_{i=1}^{\frac{n-r+2}{2}} a_{2i}^{\frac{1}{2}} x_{n-r+1} y^{n-r+i} + \sum_{i=1}^{\frac{n-r}{2}} a_{2i+1}^{\frac{1}{2}} y^{n-r+i} z_{n-r+2} + e^{\frac{1}{2}} x_{n-r+1}^2 y^{n-r}$$

$$+ (\text{the term } y^i\text{'s with } i \geq n-r+2) + (\text{other terms}).$$

Case (ii): $n-r$ is odd

We can describe

$$F = z_{n-r+1}^2 + x_{n-r}^2 y + xy^{n-r} z$$
$$+ a_{n-r+2} y^{n-r+2} + b_{n-r+1} xy^{n-r+1} + O(n-r+3).$$

We can see that $b_{n-r+1} = 0$ by similar argument in the previous steps. So we focus on the term $xy^{n-r} z$.

$$F = z_{n-r+1}^2 + x_{n-r}^2 y + xy^{n-r} z + a_{n-r+2} y^{n-r+2} + (\text{other terms})$$
$$= z_{n-r+1}^2 + x_{n-r}^2 y$$
$$+ (x_{n-r} + a_3^{\frac{1}{2}} y + \cdots + a_{n-r}^{\frac{1}{2}} y^{\frac{n-r-1}{2}}) y^{n-r}$$
$$\times (z_{n-r+1} + e^{\frac{1}{2}} (x_{n-r} + a_3^{\frac{1}{2}} y + \cdots + a_{n-r}^{\frac{1}{2}} y^{\frac{n-r-1}{2}})$$
$$+ a_2^{\frac{1}{2}} y + \cdots + a_{n-r+1}^{\frac{1}{2}} y^{\frac{n-r+1}{2}})$$
$$+ a_{n-r+2} y^{n-r+2} + (\text{other terms})$$
$$= z_{n-r+1}^2 + (x_{n-r} + a_{n-r+2}^{\frac{1}{2}} y^{\frac{n-r+1}{2}})^2 y + x_{n-r} y^{n-r} z_{n-r+1}$$
$$+ \sum_{i=1}^{\frac{n-r+1}{2}} a_{2i}^{\frac{1}{2}} x_{n-r} y^{n-r+i} + \sum_{i=1}^{\frac{n-r-1}{2}} a_{2i+1}^{\frac{1}{2}} y^{n-r+i} z_{n-r+1} + e^{\frac{1}{2}} x_{n-r}^2 y^{n-r}$$
$$+ (\text{the term } y^i\text{'s with } i \geq n-r+2) + (\text{other terms})$$

$$= z_{n-r+1}^2 + x_{n-r+2}^2 y + (x_{n-r+2} + a_{n-r+2}^{\frac{1}{2}} y^{\frac{n-r+1}{2}}) y^{n-r} z_{n-r+1}$$

$$+ \sum_{i=1}^{\frac{n-r+1}{2}} a_{2i}^{\frac{1}{2}} (x_{n-r+2} + a_{n-r+2}^{\frac{1}{2}} y^{\frac{n-r+1}{2}}) y^{n-r+i} + \sum_{i=1}^{\frac{n-r-1}{2}} a_{2i+1}^{\frac{1}{2}} y^{n-r+i} z_{n-r+1}$$

$$+ e^{\frac{1}{2}} (x_{n-r+2} + a_{n-r+2}^{\frac{1}{2}} y^{\frac{n-r+1}{2}})^2 y^{n-r}$$

$$+ \text{(the term } y^i\text{'s with } i \geq n-r+2) + \text{(other terms)}$$

$$= z_{n-r+1}^2 + x_{n-r+2}^2 y + x_{n-r+2} y^{n-r} z_{n-r+1}$$

$$+ \sum_{i=1}^{\frac{n-r+1}{2}} \left(a_{2i}^{\frac{1}{2}} x_{n-r} y^{n-r+i} + a_{2i+1}^{\frac{1}{2}} y^{n-r+i} z_{n-r+1} \right) + e^{\frac{1}{2}} x_{n-r}^2 y^{n-r}$$

$$+ \text{(the term } y^i\text{'s with } i \geq n-r+2) + \text{(other terms)}.$$

Step $k_{\leq n}$. Determine the degree k $(n-r+3 \leq k \leq n)$ term

We can calculate as the same as in the previous steps. We have

$$F = z'^2 + x'^2 y' + (1 + a_{2r}^{\frac{2}{2}}) x' y'^n + x' y'^{n-r} z'$$

$$+ \left\{ \begin{array}{l} \displaystyle\sum_{i=1}^{(n+r-1)/2} a_{2i}^{\frac{1}{2}} x' y'^{n-r+i} + \sum_{i=1}^{(n+r-3)/2} a_{2i+1}^{\frac{1}{2}} y'^{n-r+i} z' \quad (n+r-1 : \text{even}) \\[3ex] \displaystyle\sum_{i=1}^{(n+r-2)/2} a_{2i}^{\frac{1}{2}} x' y'^{n-r+i} + \sum_{i=1}^{(n+r-2)/2} a_{2i+1}^{\frac{1}{2}} y'^{n-r+i} z' \quad (n+r-1 : \text{odd}) \end{array} \right\}$$

$$+ e^{\frac{1}{2}} x'^2 y'^{n-r} + \text{(the higher term } y'^i\text{'s and } x' y'^i\text{'s)}.$$

Final Step. Equisingular locus

For the equation

$$z^2 + x^2 y + xy^n + xy^{n-r} z + \sum_{i=0}^{n+r-1} a_i y^i + \sum_{i=0}^{n-r-1} (b_i xy^i + c_i y^i z + d_i xy^i z),$$

we translate $x \mapsto x + \alpha, y \mapsto y + \beta, y \mapsto z + \gamma$, which gives

$$(z+\gamma)^2 + (x+\alpha)^2(y+\beta) + (x+\alpha)(y+\beta)^n + (x+\alpha)(y+\beta)^{n-r}(z+\gamma)$$

$$+ \sum_{i=0}^{n+r-1} a_i(y+\beta)^i$$

$$+ \sum_{i=0}^{n-r-1} \{b_i(x+\alpha)(y+\beta)^i + c_i(y+\beta)^i(z+\gamma) + d_i(x+\alpha)(y+\beta)^i(z+\gamma)\}$$

$$= z^2 + \gamma^2 + x^2 y + \beta x^2 + \alpha^2 y + \alpha^2 \beta + x(y+\beta)^n + \alpha(y+\beta)^n$$
$$+ x(y+\beta)^{n-r} z + \gamma x(y+\beta)^{n-r} + \alpha(y+\beta)^{n-r} z + \alpha\gamma(y+\beta)^{n-r}$$
$$+ \sum_{i=0}^{n+r-1} a_i (y+\beta)^i + \sum_{i=0}^{n-r-1} b_i x(y+\beta)^i + \sum_{i=0}^{n-r-1} b_i \alpha(y+\beta)^i$$
$$+ \sum_{i=0}^{n-r-1} c_i (y+\beta)^i z + \sum_{i=0}^{n-r-1} c_i \gamma(y+\beta)^i + \sum_{i=0}^{n-r-1} d_i x(y+\beta)^i z$$
$$+ \sum_{i=0}^{n-r-1} d_i \gamma x(y+\beta)^i + \sum_{i=0}^{n-r-1} d_i \alpha(y+\beta)^i z + \sum_{i=0}^{n-r-1} d_i \alpha\gamma(y+\beta)^i.$$

First, we see that $\alpha = 0$, since the coefficient of $y^{n-r} z$ is α. Moreover, from computations in the previous steps, we see all $c_i = 0$. Hence we have

$$F = z^2 + \gamma^2 + x^2 y + \beta x^2 + x(y+\beta)^n + x(y+\beta)^{n-r} z + \gamma x(y+\beta)^{n-r}$$
$$+ \sum_{i=0}^{n+r-1} a_i (y+\beta)^i + \sum_{i=0}^{n-r-1} b_i x(y+\beta)^i$$
$$+ \sum_{i=0}^{n-r-1} d_i x(y+\beta)^i z + \sum_{i=0}^{n-r-1} d_i \gamma x(y+\beta)^i.$$

Now observing the coefficient of $xy^i z$ $(i = 0, 1, \ldots, n-r-1)$, we have

$$d_i = \begin{cases} \beta^{n-r-i} & \text{if } \dbinom{n-r}{i} \text{ is odd,} \\ 0 & \text{if } \dbinom{n-r}{i} \text{ is even.} \end{cases}$$

This means we can write $xy^{n-r} z + \displaystyle\sum_{i=0}^{n-r-1} d_i xy^i z = x(y+\beta)^{n-r} z$. We can also deduce that

$$a_{2i} = \begin{cases} \beta^{2(r-i)} & \text{if } \dbinom{n}{r-i} \text{ is odd,} \\ 0 & \text{if } \dbinom{n}{r-i} \text{ is even,} \end{cases} \qquad a_{2i+1} = 0 \qquad (1 \leq i \leq r-1, \, r \geq 2).$$

a_0 and a_1 can be written as the functions of $a_{2r}, \ldots, a_{n+r-1}$ and β. First, we have

$$a_1 = \begin{cases} a_{n+r-1}\beta^{n+r-2} + a_{n+r-3}\beta^{n+r-4} + \cdots + a_{2r+1}\beta^{2r} & \text{(if } n+r \text{ is even)}, \\ a_{n+r-2}\beta^{n+r-3} + a_{n+r-4}\beta^{n+r-5} + \cdots + a_{2r+1}\beta^{2r} & \text{(if } n+r \text{ is odd)}. \end{cases}$$

For a_0 and b_i $(i = 0, 1, \ldots, n-r-1)$, we divide into two subcases according to parity of $\binom{n}{r}$.

Case (a). $\binom{n}{r}$ is even.

Observing the coefficient of xy^{n-r}, we have $\gamma = 0$. Moreover, we have

$$
b_i = \begin{cases}
\beta^{n-i} & \text{if } \displaystyle\sum_{\substack{k=i+1 \\ b_k \neq 0}}^{n-r-1} \binom{k}{i} + \binom{n}{i} \text{ is odd,} \\
0 & \text{if } \displaystyle\sum_{\substack{k=i+1 \\ b_k \neq 0}}^{n-r-1} \binom{k}{i} + \binom{n}{i} \text{ is even.}
\end{cases}
$$

Note that each b_i is determined from $i = n-r-1$ to $i = 0$ inductively. We also have

$$
a_0 = \begin{cases}
a_{n+r-2}\beta^{n+r-2} + a_{n+r-4}\beta^{n+r-4} + \cdots + a_{2r}\beta^{2r} & \text{(if } n+r \text{ is even),} \\
a_{n+r-1}\beta^{n+r-1} + a_{n+r-3}\beta^{n+r-3} + \cdots + a_{2r}\beta^{2r} & \text{(if } n+r \text{ is odd).}
\end{cases}
$$

Case (b). $\binom{n}{r}$ is odd.

Observing the coefficient of xy^{n-r}, we have $\beta^r = \gamma$. Moreover, we have

$$
b_i = \begin{cases}
\beta^{n-i} & \text{if } \displaystyle\sum_{\substack{k=i+1 \\ b_k \neq 0}}^{n-r-1} \binom{k}{i} + \sum_{\substack{k=i+1 \\ d_k \neq 0}}^{n-r-1} \binom{k}{i} + \binom{n}{i} + \binom{n-r}{i} \text{ is odd,} \\
0 & \text{if } \displaystyle\sum_{\substack{k=i+1 \\ b_k \neq 0}}^{n-r-1} \binom{k}{i} + \sum_{\substack{k=i+1 \\ d_k \neq 0}}^{n-r-1} \binom{k}{i} + \binom{n}{i} + \binom{n-r}{i} \text{ is even.}
\end{cases}
$$

Note that each b_i is determined from $i = n-r-1$ to $i = 0$ inductively. We also have

$$
a_0 = \begin{cases}
a_{n+r-2}\beta^{n+r-2} + a_{n+r-4}\beta^{n+r-4} + \cdots + a_{2r}\beta^{2r} + \beta^{2r} & \text{(if } n+r \text{ is even),} \\
a_{n+r-1}\beta^{n+r-1} + a_{n+r-3}\beta^{n+r-3} + \cdots + a_{2r}\beta^{2r} & \text{(if } n+r \text{ is odd).}
\end{cases}
$$

By the result above, we have

$$\dim \mathrm{ES}(D_{2n}^r) = n - r + 1.$$

2.2.7   Odd dimension in characteristic 2

Theorem. *For one-dimensional simple singularities in characteristic 2, thus odd dimensional simple singularities in characteristic 2, the equisingular loci are as follows.*

$A_{2m-1}(m \geq 1)$ $x^2 + x(y + \beta)^m$ $(m = 2l \ odd)$

   Trivial $(m = 2l + 1 even)$

$A_{2m}^0(m \geq 1)$ $x^2 + (y + \beta)^{2m+1} + \beta y^{2m}$

   $+ A_{2m-2}y^{2m-2} + A_{2m-4}y^{2m-4} + \cdots + A_2 y^2 + A_0$

$A_{2m}^r(m \geq 1)$ $m - r$ *parameters, too complicated to express (r odd)*

   $m - r + 1$ *parameters, too complicated to express (r even)*

$D_{2m}(m \geq 2)$ 0

$D_{2m+1}^0(m \geq 2)$ $x^2 y + y^{2m} + a_{2m-1}y^{2m-1} + a_{2m-3}y^{2m-3} + \cdots + a_1 y$

$D_{2m+1}^r(m \geq 2)$ *same as above with some r terms being 0*

E_6^0 $x^3 + y^4 + a_8$

E_6^1 *Trivial*

E_7 *Trivial*

E_8 *Trivial*

inside the versal deformation spaces

$A_{2m-1}(m \geq 1)$ $x^2 + xy^m + a_{2l-1}y^{2l-1} + a_{2l-2}y^{2l-2} + \cdots + a_0$

   $+ (b_{2l-1}y^{2l-1} + b_{2l-2}y^{2l-2} + \cdots + b_0)x$ $(m = 2l \ even)$

   $x^2 + xy^m + a_{2l}y^{2l} + a_{2l-1}y^{2l-1} + \cdots + a_0$

   $+ (b_{2l-1}y^{2l-1} + b_{2l-2}y^{2l-2} + \cdots + b_0)x$ $(m = 2l + 1 \ odd)$

$A_{2m}^0(m \geq 1)$ $x^2 + y^{2m+1} + a_{2m-1}y^{2m-1} + a_{2m-2}y^{2m-2} + \cdots + a_0$

   $+ (b_{2m-1}y^{2m-1} + b_{2m-1}y^{2m-1} + \cdots + b_0)x$

$A_{2m}^r(m \geq 1)$ $x^2 + y^{2m+1} + xy^{2m-r} + a_{2m-r-1}y^{2m-r-1}$

$(1 \leq r \leq m - 1)$ $+ a_{2m-r-2}y^{2m-r-2} + \cdots + a_0$

   $+ (b_{2m-r-2}y^{2m-r-2} + b_{2m-r-3}y^{2m-r-3} + \cdots + b_0)x$

   $(r \ odd)$

   $x^2 + y^{2m+1} + xy^{2m-r} + a_{2m-r-1}y^{2m-r-1}$

   $+ a_{2m-r-2}y^{2m-r-2} + \cdots + a_0$

   $+ (b_{2m-r-1}y^{2m-r-1} + b_{2m-r-2}y^{2m-r-2} + \cdots + b_0)x$

   $(r \ even)$

$$D_{2m}(m \geq 2) \quad x^2y + xy^m + a_{m-1}y^{m-1} + a_{m-2}y^{m-2} + \cdots + a_0$$
$$+ (b_{m-1}y^{m-1} + b_{m-2}y^{m-2} + \cdots + b_0)x$$
$$D_{2m+1}^0(m \geq 2) \quad x^2y + y^{2m} + a_{2m-1}y^{2m-1} + a_{2m-2}y^{2m-2} + \cdots + a_0$$
$$+ (b_{2m-1}y^{2m-1} + b_{2m-2}y^{2m-2} + \cdots + b_0)x$$
$$D_{2m+1}^r(m \geq 2) \quad x^2y + y^{2m} + xy^{2m-r} + a_{2m-r-1}y^{2m-r-1}$$
$$(1 \leq r \leq m-1) \quad + a_{2m-r}y^{2m-r-2} + \cdots + a_0$$
$$+ (b_{2m-r-1}y^{2m-r-1} + b_{2m-r-2}y^{2m-r-2} + \cdots + b_0)x$$

$$E_6^0 \quad x^3 + y^4 + (a_1y^3 + a_2y^2 + a_3y + a_4)x + a_5y^3 + a_6y^2 + a_7y + a_8$$
$$E_6^1 \quad x^3 + y^4 + xy^3 + (a_1y + a_2)x + a_3y^3 + a_4y^2 + a_5y + a_6$$
$$E_7 \quad x^3 + xy^3 + (a_1y + a_2)x + a_3y^4 + a_4y^3 + a_5y^2 + a_6y + a_7$$
$$E_8 \quad x^3 + y^5 + (a_1y^3 + a_2y^2 + a_3y + a_4)x + a_5y^3 + a_6y^2 + a_7y + a_8.$$

The higher odd dimensional case can be given as

$$X_2X_3 + \cdots + X_{n-1}X_n + Equation.$$

2.2.8 *Applications*

As applications of the classification of equisingular loci of simple singularities, we give two typical examples.

2.2.8.1 *3-dimensional singularities*

Example. *For E_6^0 singularity in characteristic 3, its equisingular locus is given by*

$$z^2 + x^3 + y^4 + a_8y + a_9$$

inside the versal deformation space (cf. 2.2.4).

   Putting $a_8 = w^3, a_9 = 0$ and taking the coordinate change by $Y := y + w$ gives 3-dimensional canonical singularity

$$z^2 + x^3 + Y^4 + Y^3w \in k[[x, Y, z, w]].$$

This singularity has the remarkable properties that it is not a product of a rational double point and $\mathbb{A}^1$, but it will be a product after Frobenius base *change by* $w \to w^3$. *Furthermore, it has* E_6^0 *singularity when one cut it by any general hyperplane. (For the definition and more precise statements, see Hirokado-Ito-Saito [30] and [28].)*

Example. *For* E_6^0 *singularity in characteristic 3, its equisingular locus is given by*

$$z^2 + x^3 + y^4 + x^2 y^2 + a_4 y^3$$

inside the versal deformation space (cf. 2.2.4).

Putting $a_8 = w$ *gives 3-dimensional canonical singularity*

$$z^2 + x^3 + y^4 + x^2 y^2 + y^3 w \ \in \ k[[x, y, z, w]].$$

This singularity has the same properties as the previous one. (See also [30] and [28].)

For many interesting examples related these loci in characteristic 2, we refer Hirokado-Ito-Saito [30] and [29].

2.2.8.2   *Quasi-fibration*

When we cut a equisingular locus by a curve, it gives a so-called quasi-fibration which is a fibered variety whose general fiber has a singular point, provided that the total space of the family of singularities is smooth.

Example. *The equisingular loci of the 1-dimensional singularity of type* A_2^0 *in characteristic 2 gives a quasi-elliptic fibration over* $k(\beta, A_0)$

$$x^2 + y^3 + \beta^2 y + (\beta^3 + A_0).$$

Note that all examples in the paper by S. Schröer [86] can be given by the equisingular loci of rational double points in 2.2.1, 2.2.2, 2.2.3, 2.2.4 and 2.2.5. One can get many other quasi-fibrations from the list, details are left to the reader.

Chapter 3

Open problems on rational double points in positive characteristics

3.1 Tables

Even though rational double points are basic and fundamental objects in the theory of algebraic surfaces and singularity theory, we are not fully aware of them in positive characteristic. Hence natural questions arise of them from observations and treatments which we made so far in the previous chapters.

In closing Part III on rational double points in positive characteristic, we present some of these questions to motivate further study. Before stating various problems, let us summarize some properties into the form of Tables, which are based on and originated from Artin's classification [5]. Each table has some comments about rational double points being realized as quotient singularities by finite group schemes, most of which are known to or circulated as folklores among specialists in the field. The readers are advised to consult with references or consider proofs by themselves as problems to start their research with.

Below, by the notations $\mathcal{C}_{n+1}, \mathcal{D}, \mathcal{T}, \mathcal{O}, \mathcal{I}$, we denote respectively a cyclic group of order $n+1$, the binary dihedral group of order $2(2n-4)$, the binary tetrahedral group of order 24, the binary octahedral group of order 48, and the binary icosahedral group of order 120. Further, the notations μ, τ, e are for Milnor number, Tjurina number and dimension of the equisingular locus. For a finite group G, the prime-to-p part of G is denoted by $(G)'$.

3.1.1 Complex rational double points

Type	Normal form	π_1	$\mu = \tau$	e
$A_n (n \geq 1)$	$z^2 + x^2 + y^{n+1}$	C_{n+1}	n	0
$D_n (n \geq 4)$	$z^2 + x^2 y + y^{n-1}$	$\mathcal{D}_{n-2}$	n	0
E_6	$z^2 + x^3 + y^4$	$\mathcal{T}$	6	0
E_7	$z^2 + x^3 + xy^3$	$\mathcal{O}$	7	0
E_8	$z^2 + x^3 + y^5$	$\mathcal{I}$	8	0

The notation π_1 is the local fundamental group at the singular point. Note also that all of them are realized as the quotient singularities $\mathbb{A}^2/G$ by finite groups $G = \pi_1$. (See [13], for example.)

3.1.2 Rational double points in characteristic $p > 5$

Type	Normal form	π_1	τ	e	$\tau - e$
$A_n (p \nmid (n+1))$	$z^2 + x^2 + y^{n+1}$	C_{n+1}	n	0	n
$A_n (p \mid (n+1))$		$(C_{n+1})'$	$n+1$	1	n
$D_n (n \geq 4)$	$z^2 + x^2 y + y^{n-1}$	$(\mathcal{D}_{n-2})'$	n	0	n
E_6	$z^2 + x^3 + y^4$	$\mathcal{T}$	6	0	6
E_7	$z^2 + x^3 + xy^3$	$\mathcal{O}$	7	0	7
E_8	$z^2 + x^3 + y^5$	$\mathcal{I}$	8	0	8

For the case of type A_n when p does not divide $n + 1$, it is a (tame) finite group quotient singularity $\mathbb{A}^2/C_{n+1}$. On the other hand, for the case of type A_n with p dividing $n + 1$, it is a group scheme quotient singularity. Note that this is the case for any positive characteristic.

More precisely, if $n + 1 = p$, then it is a μ_p quotient of $\mathbb{A}^2 = \operatorname{Spec} k[x, y]$ which is given by a p-closed derivation of multiplicative type

$$\delta = x\frac{\partial}{\partial x} + (p-1)y\frac{\partial}{\partial y},$$

and this is generalized to the case for $n + 1 = p^e$ in terms of a higher derivation defined by the group scheme μ_{p^e} of multiplicative type. (See [26], [33], and also 2.9 for μ_p-quotients, and Chapter 6 of Part II for higher derivations.) Furthermore, if $n + 1 = p^e m$ with $(m, p) = 1$ then it is a quotient by finite group scheme which is a direct product of μ_{p^e} and C_m.

3.1.3   *Rational double points in characteristic $p = 5$*

Type	Normal form	π_1	τ	e	$\tau - e$
A_n $(5 \nmid (n+1))$	$z^2 + x^2 + y^{n+1}$	$\mathcal{C}_{n+1}$	n	0	n
A_n $(5 \mid (n+1))$		$(\mathcal{C}_{n+1})'$	$n+1$	1	n
D_n $(n \geq 4)$	$z^2 + x^2y + y^{n-1}$	$(\mathcal{D}_{n-2})'$	n	0	n
E_6	$z^2 + x^3 + y^4$	$\mathcal{T}$	6	0	6
E_7	$z^2 + x^3 + xy^3$	$\mathcal{O}$	7	0	7
E_8^0	$z^2 + x^3 + y^5$	0	10	1	9
E_8^1	$z^2 + x^3 + y^5 + xy^4$	$\mathcal{C}_5$	8	0	8

Except for the type A_n with $5 \mid (n+1)$ and the type E_8^0, all rational double points in characteristic 5 are finite group quotient singularities as in the case of characteristic 0. Meanwhile, the type E_8^1 is a wild quotient singularity divided by $\mathbb{Z}/5\mathbb{Z}$, and an explicit description of the action is left to the readers. The type E_8^0 is given as a quotient of $k[[x, z]]$ by a 5-closed derivation of additive type, e.g.,

$$\delta = 2z\frac{\partial}{\partial x} - 3x^2\frac{\partial}{\partial z}.$$

We can say that this is a quotient by a finite group scheme α_5 of additive type.

3.1.4   *Rational double points in characteristic $p = 3$*

Type	Normal form	π_1	τ	e	$\tau - e$
A_n $(3 \nmid (n+1))$	$z^2 + x^2 + y^{n+1}$	$\mathcal{C}_{n+1}$	n	0	n
A_n $(3 \mid (n+1))$		$(\mathcal{C}_{n+1})'$	$n+1$	1	n
$D_n(n \geq 4)$	$z^2 + x^2y + y^{n-1}$	$(\mathcal{D}_{n-2})'$	n	0	n
E_6^0	$z^2 + x^3 + y^4$	0	9	2	7
E_6^1	$z^2 + x^3 + y^4 + x^2y^2$	$\mathcal{C}_3$	7	1	6
E_7^0	$z^2 + x^3 + xy^3$	$\mathcal{C}_2$	9	1	8
E_7^1	$z^2 + x^3 + xy^3 + x^2y^2$	$\mathcal{C}_6$	7	0	7
E_8^0	$z^2 + x^3 + y^5$	0	12	2	10
E_8^1	$z^2 + x^3 + y^5 + x^2y^3$	0	10	1	9
E_8^2	$z^2 + x^3 + y^5 + x^2y^2$	$\mathcal{T}$	8	0	8

Except for the type A_n and the type D_n, the types E_6^0, E_7^0, E_8^0 are all quotient singularities by derivations of additive type. Thus these are quotient singularities by the finite group scheme α_3 of additive type. Given these derivations, singularities are easily worked out, and computations are left to the readers. The type A_n is the same as in characteristic $p > 3$ above.

3.1.5 *Rational double points in characteristic $p = 2$*

The list of rational double points in characteristic 2 is in the next page. As we remarked before Theorem 1.1.7, this list is an exhaustive list of rational double points in characteristic 2 by the algorithm in 1.6.

Since it is convenient to use the alternative form of equation for the type of E_7^1 as explained in 2.2.5, we give two equations for the singularity of type E_7^1.

Remark. Dr. Yuya Matsumoto proposed to allow half-integers r as the exponents in the case of type D_{2n+1}^r. This enables us to treat both cases D_{2n}^r and D_{2r+1}^r uniformly. So, careless of N being even or odd, we define the type D_N^r by the equation

$$D_N^r : \; z^2 + x^2 y + y^{2r} + xy^{\frac{N-2r}{2}} z = 0, \quad N \ge 4,$$

where

$$r = \frac{N-2}{2}, \frac{N-2}{2} - 1, \ldots, \begin{cases} 0 \text{ if } N \equiv 0 \pmod 2 \\ \frac{1}{2} \text{ if } N \equiv 1 \pmod 2 \end{cases}.$$

3.1.5.1 D_{4m}^0 *and* D_{4m}^m

We give here some examples to show how these singularities can be obtained as *quotient* singularities in characteristic 2.

First of all, we treat the case D_{4m} with $m = 1$, that is, the case of type D_4. Let δ be a p-closed derivation of additive type on $A = k[[X, Y]]$ defined as below.

$$\delta := X^2 \frac{\partial}{\partial X} + Y^2 \frac{\partial}{\partial Y}.$$

Then $\operatorname{Ker} \delta$ is a subring $k[[X^2, Y^2, X^2 Y + XY^2]]$, and the quotient ring is therefore isomorphic to $k[[x, y, z]]/(z^2 + x^2 y + xy^2)$, which is of type D_4^0.

On the other hand, let σ be an action of $\mathbb{Z}/2\mathbb{Z}$ on $A = k[[X, Y]]$ defined by

$$\begin{cases} \sigma(X) := \frac{X}{1+X} = X + X^2 + X^3 + \cdots = X + \frac{X^2}{1+X} \\ \sigma(Y) := \frac{Y}{1+Y} = Y + Y^2 + Y^3 + \cdots = Y + \frac{Y^2}{1+Y}. \end{cases}$$

Then the ring of invariants is given by

$$A^{\langle \sigma \rangle} = k[[\frac{X^2}{1+X}, \frac{Y^2}{1+Y}, \frac{X^2 Y + XY^2}{(1+X)(1+Y)}]]$$

$$\cong k[[x, y, z]]/(z^2 + xyz + x^2 y + xy^2).$$

This has a singularity of type D_4^1.

type	normal form	π_1	τ	e	$\tau - e$
$A_n(n:\text{even})$	$z^{n+1}+xy$	$(C_n)'$	n	0	n
$A_n(n:\text{odd})$	$z^{n+1}+xy$	C_n	$n+1$	1	n
$D^0_{2n}\ (n\geq 2)$	$z^2+x^2y+xy^n$	0	$4n$	$n+1$	$3n-1$
$D^r_{2n}\ (n\geq 2)$ $(1\leq r\leq n-1)$	$z^2+x^2y+xy^n+xy^{n-r}z$	$*$	$4n-2r$	$n+1-r$	$3n-1-r$
$D^0_{2n+1}(n\geq 2)$	$z^2+x^2y+y^nz$	0	$4n$	n	$3n$
$D^r_{2n+1}(n\geq 2)$ $(1\leq r\leq n-1)$	$z^2+x^2y+y^nz+xy^{n-r}z$	$†$	$4n-2r$	$n-r$	$3n-r$
E^0_6	$z^2+x^3+y^2z$	C_3	8	1	7
E^1_6	$z^2+x^3+y^2z+xyz$	C_6	6	0	6
E^0_7	$z^2+x^3+xy^3$	0	14	4	10
E^1_7	$z^2+x^3+xy^3+y^4z$	0	12	3	9
	$\overset{c}{\sim}\ z^2+x^3+xy^3+x^2yz$				
E^2_7	$z^2+x^3+xy^3+y^3z$	0	10	2	8
E^3_7	$z^2+x^3+xy^3+xyz$	C_4	8	1	7
E^0_8	$z^2+x^3+y^5$	0	16	4	12
E^1_8	$z^2+x^3+y^5+xy^3z$	0	14	3	11
E^2_8	$z^2+x^3+y^5+xy^2z$	C_2	12	2	10
E^3_8	$z^2+x^3+y^5+y^3z$	0	10	1	9
E^4_8	$z^2+x^3+y^5+xyz$	$\mathbb{Z}/3\mathbb{Z}\times\mathbb{Z}/4\mathbb{Z}$	8	0	8

* The group is trivial for r with $0<2r<n$, the dihedral group $D_{(2r-n)'}$ of order $2(2r-n)'$ for r with $n<2r<2n$ and $\mathbb{Z}/2\mathbb{Z}$ for $n=2r$, where $(2r-n)'$ stands for the greatest divisor of $2r-n$ prime to 2.

† The group is trivial for r with $0<2r<n$, the dihedral group $D_{4r-2n+1}$ of order $2(4r-2n+1)$ for r with $n<2r<2n$.

By II.7.3.2, we can unify these quotients. Namely, by solving simultaneous recursive equations

$$\begin{cases} \delta_\lambda(X) = f(X^2 + \lambda\delta_\lambda(X)X, Y^2 + \lambda\delta_\lambda(Y)Y) \\ \delta_\lambda(Y) = g(X^2 + \lambda\delta_\lambda(X)X, Y^2 + \lambda\delta_\lambda(Y)Y), \end{cases}$$

for $f(X,Y) = X, g(X,Y) = Y$, we obtain by Lemma II.7.3.2 a pseudo-derivation δ_λ such that

$$\delta_\lambda(X) = \frac{X^2}{1 + \lambda X}, \quad \delta_\lambda(Y) = \frac{Y^2}{1 + \lambda Y}.$$

The invariant subring $\operatorname{Ker} \delta_\lambda$ in $A = k[[\lambda, X, Y]]$ is computed as [1]

$$k[[\frac{X^2}{1 + \lambda X}, \frac{Y^2}{1 + \lambda Y}, \delta_\lambda(X)Y + X\delta_\lambda(Y)]]$$

$$= k[[\frac{X^2}{1 + \lambda X}, \frac{Y^2}{1 + \lambda Y}, \frac{XY^2 + X^2Y}{(1 + \lambda X)(1 + \lambda Y)}]].$$

Setting $\lambda = 0$ and $\lambda = 1$, we have the α_2-invariant subring and the $\mathbb{Z}/2\mathbb{Z}$-invariant subring. Hence one can say that the singularity D_4^1 has the singularity D_4^0 as a reduction (or specialization).

Now let us consider a more general case, i.e., a singularity of type D_{4m}. We will use arguments parallel to the previous case. Consider a p-closed derivation of additive type on $A = k[[X, Y]]$

$$\delta := X^2 \frac{\partial}{\partial X} + Y^{2m} \frac{\partial}{\partial Y}.$$

The invariant subring $\operatorname{Ker} \delta$ is

$$k[[X^2, Y^2, X^2Y + XY^{2m}]] \cong k[[x, y, z]]/(z^2 + x^2y + xy^{2m}).$$

This ring gives a singularity of type D_{4m}^0. Then solving same simultaneous recursive equations as above with $f(X,Y) = X, g(X,Y) = Y^m$, then we get the quotient ring as below by pseudo-derivation which gives a singularity of type D_{4m}^m when $\lambda \neq 0$ and D_{4m}^0 when $\lambda = 0$.

$$k[[\frac{X^2}{1 + \lambda X}, \dots, \delta_\lambda(X)Y + X\delta_\lambda(Y)]] = k[[x, y, z]]/(z^2 + \lambda xy^{2m}z + x^2y + xy^{2m})$$

Remark. Other than the singularities of type D_4^0 and D_4^1, the singularities of type E_8^0 and E_8^2 can be treated unified way same as above by using pseudo-derivations. The ones of type D_N^0 and E_7^0 can be also obtained as α_2 quotient, that is, quotient by 2-closed derivations of additive type.

---

[1] Let δ be a pseudo-derivation of $R := k[[\lambda, x, y]]$ in characteristic 2. Then the following information will help computations. For $a, b \in R$, we have $\delta(ab) = a\delta(b) + b\delta(a) + \lambda\delta(a)\delta(b)$. In particular, $\delta(a^2) = \lambda\delta(a)^2$. Further, $\delta(1) = 0$ because $1 + \lambda\delta(1) \neq 0$ and $\delta(\frac{1}{a}) = \frac{\delta(a)}{a(a + \lambda\delta(a))}$.

3.1.5.2 E_6^0 and E_6^1

Consider a derivation given by

$$\delta = Y^2 \frac{\partial}{\partial X} + X^2 \frac{\partial}{\partial Y}$$

which is different from the one in 3.1.5.1. This derivation also gives a rational double point of type D_4^0 as a α_2-quotient because the kernel of this derivation is $k[[X^2, Y^2, X^3 + Y^3]]$ which is isomorphic to $k[[X^2, Y^2, X^2Y + XY^2]]$. This derivation gives a unified pseudo-derivation

$$\begin{cases} \delta_\lambda(X) = \frac{Y^2 + \lambda X^2 Y}{1 + \lambda X Y} \\ \delta_\lambda(Y) = \frac{X^2 + \lambda X Y^2}{1 + \lambda X Y}. \end{cases}$$

On the other hand, a singularity of type A_2 is given as a quotient of $k[[X, Y]]$ by the multiplicative group μ_3 action

$$\begin{cases} \tau(X) = \omega X \\ \tau(Y) = \omega^2 Y, \end{cases}$$

where ω is a primitive cubic root of unity. This μ_3 action on $k[[X, Y]]$ is compatible with the pseudo-derivation δ_λ. Therefore we can take quotients of type D_4^r by μ_3 and of type A_2 by the pseudo-derivation, and obtain a singularities of type E_6^r ($r = 0, 1$) as shown below.

$$\begin{array}{ccccc} \mathbb{A}^2 & \xrightarrow{G_\lambda = \langle \delta_\lambda \rangle} & D_4^r & \longrightarrow & \mathbb{A}^2 \\ {\scriptstyle \mu_3 = \langle \tau \rangle} \downarrow & & {\scriptstyle \mu_3} \downarrow & & \downarrow \\ A_2 & \xrightarrow[G_\lambda]{} & E_6^r & \longrightarrow & A_2 \end{array}$$

Precise calculations are left to the readers. Therefore singularities of type E_6^r ($r = 0, 1$) can be understood as the group scheme quotients.

3.2   Open problems

3.2.1   *Rational double points as quotients*

Problem. Can all rational double points be obtained as quotient singularities even in positive characteristic?

The rational double points in characteristic zero are quotient singularities by finite groups which are subgroups of $\mathrm{SL}_2(\mathbb{C})$. But this is not the case clearly in positive characteristic. There are some obstructions to treat quotients as in characteristic zero such as wild actions, purely inseparable coverings and so on.

By examples given in 3.1.5.1 and 3.1.5.2 and by discussions in Part II, we know that the category of finite groups must be enlarged so that it includes finite group schemes or p-closed derivations. By this interpretation of finite group, we can observe that many more rational double points are obtained as their quotients. Hence the problem can be restated as follows.

Problem.   Define the notion of *quotient* in more adequate way so that all rational double points in arbitrary characteristic can be obtained as quotient singularities. Further, describe an explicit construction of quotient singularity.

There are more dreamlike problems. In considering finite group scheme quotients, describe the relationship between finite subgroup schemes of $SL_2(k)$ and the McKay correspondence using these group schemes.

3.2.2   *Miscellaneous questions*

(1) Find a systematic treatment of quotients in the curve case and in characteristic 2. (See 1.6 and [45] for the list.)

(2) Complete a possibility list of Adjacency in characteristic 2 and in dimension 2 which will relate the moduli of rational double points. (See [21] 4.5.)

(3) Understand the invariant $\tau - e$ in the tables 3.1, and relate it to the τ constant locus inside the versal deformation space. (See [30], [27].)

(4) Apply these local theories such as the study of equisingular loci to the global study of singularities in higher dimensional case and the geometry of pathological varieties specially exist in positive characteristic.

(5) Apply a theory to be obtained as a solution of problems in 3.2 to other singularities.

Bibliography

[1] E. Artin, Algebraic numbers and algebraic functions, Gordon and Breach, 1967, New York.

[2] M. Artin, Some numerical criteria for contractability of curves on algebraic surfaces, Amer. J. Math. **84**, No. 3, 485–496.

[3] M. Artin, On isolated rational singularities of surfaces, Amer. J. Math, **88** (1966), 129–136.

[4] M. Artin, Wildly ramified $\mathbb{Z}/2$ actions in dimension two, Proc. AMS, **52** (1975), 60–64.

[5] M. Artin, Coverings of rational double points in characteristic p, In Complex Analysis and Algebraic Geometry, Cambridge Univ. Press, Cambridge (1977), 11-22.

[6] M. Artin, Supersingular K3 surfaces, Ann. Sci. École Norm. Sup. (4) **7** (1974), 543–567.

[7] P. Blass and J. Lang, Zariski surfaces and differential equations in characteristic $p > 0$, Monographs and Textbooks in Pure and Applied Mathematics, **106**, 1987, New York: Marcel Dekker Inc.

[8] E. Bombieri and D. Husemoller, Classification and embeddings of surfaces, Algebraic geometry (Proc. Sympos. Pure Math., Vol. **29**, Humboldt State Univ., Arcata, Calif., 1974), pp. 329–420. Amer. Math. Soc., Providence, R.I., 1975.

[9] E. Bombieri and D. Mumford, Enriques' classification of surfaces in char. p . Part I, by D. Mumford in *Global Analysis*, Princeton University press, 1969; Part II, in *Complex Analysis and Algebraic Geometry*, pp. 23–42, Cambridge University Press, 1977; Part III, Invent. Math. **35** (1976), 197–232.

[10] Y. Boubakri, G. Greuel and T. Markwig, Invariants of hypersurfaces singularities in positive characteristic, Rev. Mat. Complut. **25**, No. 1, 61–85 (2012),

[11] M. Demazure, Lectures on p-divisible groups, Lecture Notes in Mathematics, Vol. **302**, Springer-Verlag, Berlin-New York, 1972. v+98 pp.

[12] M. Demazure and P. Gabriel, Groupes algébriques. Tome I: Géométrie algébrique, généralités, groupes commutatifs. Avec un appendice Corps de

classes local par Michiel Hazewinkel, Masson & Cie, Editeur, Paris; North-Holland Publishing Co., Amsterdam, 1970. xxvi+700 pp.

[13] A.H. Durfee, Fifteen characterizations of rational double points and simple critical points. Enseign. Math. (2) **25** (1979), no. 1-2, 131–163.

[14] T. Ekedahl, Canonical models of surfaces of general type in positive characteristic. Inst. Hautes Etudes Sci. Publ. Math. No. **67** (1988), 97–144.

[15] X. Faber, Finite p-irregular subgroups of $\mathrm{PGL}_2(k)$, arXiv: 1112.1999v2, 2012.

[16] J. Fogarty, On the depth of local rings of invariants of cyclic groups, Proc. Amer. Math. Soc. **83** (1981), no. 3, 448–452.

[17] G. Freudenburg, Algebraic theory of locally nilpotent derivations (Second edition), Encyclopaedia of Math. Sci. **136**, Invariant theory and algebraic transformation groups VII, Springer Verlag, 2017.

[18] A. Fujiki, On resolutions of cyclic quotient singularities, Publ. Res. Inst. Math. Sci. **10** (1974/75), no. 1, 293–328.

[19] R. Ganong, Plane Frobenius sandwiches. Proc. Amer. Math. Soc. **84** (1982), no. 4, 474–478.

[20] R. Ganong, The pencil of translates of a line in the plane, Affine algebraic geometry, 57–71, CRM Proc. Lecture Notes **54**, Amer. Math. Soc., Providence, RI, 2011.

[21] G.-M. Greuel, H. Kröning, Simple singularities in positive characteristic, Mathematische Zeitschrift, **203** (1990), 339–354.

[22] G.-M. Greuel, C. Lossen and E. Shustin, Introduction to Singularities and Deformations, Springer-Verlag, 2007.

[23] A. Grothendieck, Sur quelques points d'algèbre homologique, Tôhoku Math. J. **9** (1957), 119–221.

[FGA] A. Grothendieck, Fondements de la géométrie algébrique. [Extraits du Séminaire Bourbaki, 1957–1962.] Secretariat mathématique, Paris 1962 ii+205 pp.

[SGA1] A. Grothendieck, Revêtements étales et groupe fondamental, Séminaire de Géométrie Algébrique, 1960/61. Institut des Hautes Études Scientifiques, Paris 1963.

[SGA2] A. Grothendieck, Cohomologie locale des faisceaux coherents et theoremes de Lefschetz locaux et globaux. Fasc. I: Exposes 1–8; Fasc. II: Exposes 9–13. Séminaire de Géométrie Algébrique 1962. Institut des Hautes Études Scientifiques, Paris 1965.

[EGA] A. Grothendieck, Éléments de géométrie algébrique. **IV**. Étude locale des schemas et des morphismes de schemas. I. Inst. Hautes Études Sci. Publ. Math. No. **20** (1964), 259 pp.

[24] R. Hartshorne, Algebraic geometry. Graduate Texts in Mathematics, No. **52**. Springer-Verlag, New York-Heidelberg, 1977. xvi+496 pp.

[25] R. Hartshorne, Stable reflexive sheaves, Math. Ann. **254** (1980), 121–176.

[26] M. Hirokado, Singularities of multiplicative p-closed vector fields and global 1-forms of Zariski surfaces, J. Math. Kyoto Univ. **39**, (1999) 455–468.

[27] M. Hirokado, Further evaluation of Wahl vanishing theorems for surface singularities in characteristic p, Michigan J. **68**, (2019) 621–636.

[28] M. Hirokado, H. Ito and N. Saito, Calabi-Yau threefolds arising from the fiber products of rational quasi-elliptic surfaces, I, Ark. Mat. **45**(2), 279-296 (2007),

[29] M. Hirokado, H. Ito and N. Saito, Calabi-Yau threefolds arising from the fiber products of rational quasi-elliptic surfaces, II, Manuscripta Math. **125**, 325-343 (2008),

[30] M. Hirokado, H. Ito and N. Saito, Three dimensional canonical singularities in codimension two in positive characteristic, J. of Algebra **373**, 207-222 (2013),

[31] F. Hirzebruch, Über vierdimensionale Riemannsche Flächen mehrdeutiger analytischer Funktionen von zwei komplexen Veranderlichen, Math. Ann. **126** (1953). 1–22.

[32] G. Horrocks, Vector bundles on the punctured spectrum of a local ring, Proc. London Math. Soc. (3), **14** (1964), 689–713.

[33] Y. Ishibashi and N. Onoda, On the ring of constants of a diagonalizable higher derivation, J. of Pure and Applied Algebra, **81** (1992) 39–47.

[34] H. Ito, The Mordell-Weil groups of unirational quasi-elliptic surfaces in characteristic 3, Math. Z. **211** (1992), no. 1, 1–39.

[35] H. Ito, The Mordell-Weil groups of unirational quasi-elliptic surfaces in characteristic 2, Tohoku Math. J. (2) **46** (1994), no. 2, 221–251.

[36] H. Ito and N. Saito, The equisingular loci of simple singularities in positive characteristic, preprint.

[37] H. Ito and S. Schröer, Wild quotient surface singularities whose dual graphs are not star-shaped, Asian J. Math. **19** (2015), no.5, 951–986.

[38] N. Jacobson, Restricted Lie algebras of characteristic p, Trans. Amer. Math. Soc. **50** (1941), 15–25.

[39] N. Jacobson, Lectures in abstract algebra, Vol. III, Theory of fields and Galois theory, Van Nostrand, Princeton, N. J., 1964. 323 pp.

[40] T. Kambayashi, On Fujita's strong cancellation theorem for the affine plane, J. Fac. Sci. Univ. Tokyo Sect. IA Math. **27**(1980), no. 3, 535–548.

[41] T. Kambayashi and M. Miyanishi, On forms of the affine line over a field, Lectures in Mathematics, vol. **10**, Department of Mathematics, Kyoto University, Kinokuniya Publ. Co., 1977.

[42] T. Kambayashi and V. Srinivas, On étale coverings of the affine space, Algebraic geometry (Ann Arbor, Mich., 1981), 75–82, Lecture Notes in Math., **1008**, Springer, Berlin-New York, 1983.

[43] T. Kambayashi, M. Miyanishi and M. Takeuchi, Unipotent algebraic groups, Lecture Notes in Mathematics, Vol. **414**, Springer-Verla, Berlin-Heidelberg-New York, 1974, 165 pp.

[44] T. Kimura and H. Niitsuma, On Kunz's conjecture, J. Math. Soc. Japan **34** (1982), no. 2, 371–378.

[45] K. Kiyek and G. Steinke, Einfache Kurvensingularitäten in beliebiger Charakteristik, Arch. Math. **45** (1985), 565–573.

[46] K. Kodaira, On compact analytic surfaces, II, Ann. of Math. (2) **77** (1963), 563–626.

[47] J. Kollàr and S. Mori, Birational geometry of algebraic varieties, Cambridge

Univ. Press **134**, 1988.

[48] W. E. Lang, Quasi-elliptic surfaces in characteristic three. Ann. Sci. Ecole Norm. Sup. (4) **12** (1979), no. 4, 473–500.

[49] J. Lipman, Rational singularities, with applications to algebraic surfaces and unique factorization, Inst. Hautes Etudes Sci. Publ. Math. No. **36** (1969), 195–279.

[50] Yu. I. Manin, Cubic forms. Algebra, geometry, arithmetic. Translated from the Russian by M. Hazewinkel. Second edition. North-Holland Mathematical Library, **4**. North-Holland Publishing Co., Amsterdam, 1986. x+326 pp.

[51] J.S. Milne, Etale cohomology, Princeton Mathematical Series, **33**, Princeton University Press, Princeton, N.J., 1980. xiii+323 pp.

[52] M. Miyanishi, G_a-action of the affine plane, Nagoya Math. J. **41** (1971), 97–100.

[53] M. Miyanishi, On the vanishing of the Demazure cohomologies and the existence of quotient preschemes, J. Math. Kyoto Univ. **11** (1971), 399–414.

[54] M. Miyanishi, Curves on rational and unirational surfaces, Tata Institute of Fundamental Research Lectures on Mathematics and Physics, **60**, Tata Institute of Fundamental research, Bombay; by the Narosa Publishing House, New Delhi, 1978. ii+302 pp.

[55] M. Miyanishi, Unirational quasi-elliptic surfaces in characteristic 3, Osaka J. Math. **13** (1976), no. 3, 513–522.

[56] M. Miyanishi, Unirational quasi-elliptic surfaces, Japan. J. Math. (N.S.) **3** (1977), no. 2, 395–416.

[57] M. Miyanishi, Open algebraic surfaces, CRM monograph series **12**, Amer. math. Soc. 2001.

[58] M. Miyanishi, Frobenius Sandwiches of affine algebraic surfaces, Affine algebraic geometry, 243–260, CRM Proc. Lecture Notes **54**, Amer. Math. Soc., Providence, RI, 2011.

[59] M. Miyanishi, Wild $\mathbb{Z}/p\mathbb{Z}$-actions on algebraic surfaces, J. Algebra **417** (2017), 360–389.

[60] M. Miyanishi and P. Russell, Purely inseparable coverings of exponent one of the affine plane. J. Pure Appl. Algebra **28** (1983), no. 3, 279–317.

[61] T.-T. Moh, On the classification problem of embedded lines in characteristic p, Algebraic geometry and commutative algebra, Vol. I, 267–279, Kinokuniya, Tokyo, 1988.

[GIT] D. Mumford, Geometric invariant theory, Ergebnisse der Mathematik und ihrer Grenzgebiete, Neue Folge, Band **34**, Springer-Verlag, Berlin-New York 1965 vi+145 pp.

[62] D. Mumford, Enriques' classification of surfaces in char. p : I, Global Analysis, Papers in honor of K. Kodaira, University of Tokyo Press-Princeton University Press, 1969.

[63] D. Mumford, The topology of normal singularities of an algebraic criterion for simplicity, Publ. Math. I.H.R.S., **9** (1961), 229–246.

[64] D. Mumford and K. Suominen, Introduction to the theory of moduli, Proc.

of the fifth Nordic Summer School in Mathematics, F. Oort, editor, Wolters-Noordorff Publ., Groningen, 1970.

[65] J.P. Murre, On contravariant functors from the category of pre-schemes over a field into the category of abelian groups (with an application to the Picard functor), Inst. Hautes Etudes Sci. Publ. Math. No. **23** (1964), 5–43.

[66] M. Nagata, Local rings, Interscience Tracts in Pure and Applied Mathematics, No. 13 Interscience Publishers a division of John Wiley & Sons, New York-London 1962 xiii+234 pp.

[67] M. Nagata, Lectures on the fourteenth problem of Hilbert, Tata Institute of Fundamental Research, Bombay (1965), ii+78+iii pp.

[68] M. Nagata, Field theory, Pure and Applied Mathematics, No. **40**, Marcel Dekker, Inc., New York-Basel, 1977. vii+268 pp.

[69] M. Nagata, Flatness of an extension of a commutative ring. J. Math. Kyoto Univ. **9** (1969), 439–448.

[70] Y. Namikawa, K. Ueno, The complete classification of fibers in pencils of curves of genus two, Manusctripta Math. **9** (1973), 143–186.

[71] A. P. Ogg, On pencils of curves of genus two, Topology, **5** (1966), 355–362.

[72] B. Peskin, On rings of invariants with rational singularities, Proc. Amer. Math. Soc. **87** (1983), no. 4, 621–626.

[73] B. Peskin, Quotient-singularities and wild p-cyclic actions, J. Algebra **81** (1983), 72–99.

[74] M. Reid, Canonical 3-folds, Journées de géometrie algébrique d'Angers, ed. A. Beauville, Sijthoff and Noordhoff, Alphen (1980), 273-310

[75] M. Rosenlicht, Generalized Jacobian varieties, Ann. Math. **59** (1954), 505–530.

[76] M. Rosenlicht, Automorphisms of function fields, Trans. Amer. Math. Soc. **79** (1955), 1–11.

[77] A.N. Rudakov and I.R. Shafarevich, Inseparable morphisms of algebraic surfaces, Izv. Akad. Nauk SSSR Ser. Mat. **40** (1976), no. 6, 1269–1307 (Russian); English transl. Math. USSR-Izv. **40** (1976), no. 6, 1205–1237.

[78] P. Russell, Forms of the affine line and its additive group, Pacific J. Math. **32** (1970), 527–539.

[79] P. Samuel, Méthodes d'algèbre abstraite en géométrie algébrique, Seconde edition, corrigee. Ergebnisse der Mathematik und ihrer Grenzgebiete, Band **4**, Springer-Verlag, Berlin-New York 1967 xii+133 pp.

[80] P. Samuel, Lectures on old and new results on algebraic curves, Tata Lecture Notes, 1966.

[81] S. Schröer, On genus change in algebraic curves over imperfect fields, Proc. Amer. Math. Soc. **137** (2009), 1239–1243.

[82] J.-P. Serre, Sur la topologie des variétés algébriques en caractéristique p, 1958, Internat. symposium on algebraic topology pp. 24–53.

[83] J.-P. Serre, Groupes algébriques et corps de classes, Hermann 1959, Paris.

[84] J.-P. Serre, Corps locaux, Publications de l'Institut de Mathematique de l'Universite de Nancago, VIII Actualites Sci. Indust., No. **1296**, Hermann, Paris 1962, 243 pp.

[85] I.R. Šafarevič, et al., Algebraic surfaces. (Russian) Trudy Mat. Inst.

Steklov. **75** (1965), 1–215; English Transl., Proc. Steklov Inst. Math. **75** (1965).

[86] S. Schröer, Singularities appearing on generic fibers of morphisms between smooth schemes, Michigan Math. J. **56** (2008), 55–76.

[87] M. Schütt and T. Shioda, Elliptic surfaces, Algebraic geometry in East Asia–Seoul 2008, 51–160, Adv. Stud. Pure Math., **60**, Math. Soc. Japan, Tokyo, 2010.

[88] T. Shioda, Mordell-Weil lattices and Galois representation. I, Proc. Japan Acad. Ser. A Math. Sci. **65** (1989), no. 7, 268–271.

[89] T. Shioda, Mordell-Weil lattices and Galois representation. II, III, Proc. Japan Acad. Ser. A Math. Sci. **65** (1989), no. 8, 296–303.

[90] T. Shioda, On the Mordell-Weil lattices, Comment. Math. Univ. St. Paul. **39** (1990), no. 2, 211–240.

[91] Y. Takeda, Artin-Schreier coverings of algebraic surfaces, J. Math. Soc. Japan **41** (1989), no. 3, 415–435.

[92] J. Tate, Genus change in inseparable extensions of function fields, Proc. Amer. Math. Soc. **3** (1952), 400–406.

[93] J. Tate, The arithmetic of elliptic curves, Invent. math. **23** (1974), 179–206.

[94] H. Yanagisawa, On determinacy of simple singularities, Master Thesis at Tokyo University of Science (in Japanese), March 2018.

[95] O. Zariski and P. Samuel, Commutative algebra, Volume I. With the co-operation of I. S. Cohen. The University Series in Higher Mathematics, D. Van Nostrand Company, Inc., Princeton, New Jersey, 1958. xi+329 pp.

[96] F. Cossec, I. Dolgachev, S. Kondo and C. Liedtke, Enriques surfaces I, II (new edition), forthcoming.

Index

Printed in the United States
By Bookmasters